Infrared Technology

WILEY SERIES IN MICROWAVE AND OPTICAL ENGINEERING

KAI CHANG, Editor
Texas A&M University

A complete list of the titles in this series appears at the end of this volume.

Infrared Technology

Applications to Electrooptics, Photonic Devices, and Sensors

A. R. JHA

A WILEY-INTERSCIENCE PUBLICATION
JOHN WILEY & SONS, INC.
NEW YORK / CHICHESTER / WEINHEIM / BRISBANE / SINGAPORE / TORONTO

This book is printed on acid-free paper. ♾

Published simultaneously in Canada.

For ordering and customer service, call 1-800-CALL-WILEY.

Library of Congress Cataloging-in-Publication Data is available.

ISBN 0-471-35033-8

Printed in the United States of America

10 9 8 7 6 5 4 3 2 1

Contents

Foreword

Infrared science has played a revolutionary role in the development of the modern technology age. The infrared range of the electromagnetic spectrum spans from approximately 0.75 μm to 1000 μm. The discovery of infrared radiation is credited to the British Royal Astronomer, Sir William Herschel who demonstrated this radiation in 1800 by using sunlight dispersed through a prism and detecting it with a sensitive thermometer. This work laid the foundation of the field of infrared spectroscopy. For more than thirty years, the enormous potential of this new form of radiation was not realized until the intervention of more sensitive detectors such as the radiation thermocouple and the diffraction grating spectrometer. Since then, there has been phenomenal progress in both basic and applied research in infrared science and technology. Infrared spectroscopy has played a leading role in the achievement of this progress

Throughout the nineteenth century and the early part of the twentieth century, infrared studies were primarily concerned with basic research. The experimental results of these studies led to the formulation of almost all of the fundamental theories and laws of thermal radiation. For example, the studies of spectral and temperature dependence of thermal radiation led to the validation of Planck's quantum theory, the Stefan–Boltzmann distribution law, and Wien's displacement law. Heinrich Herz's experimental studies of the propagation of thermal radiation through empty space provided the verification of Maxwell's classical theory of electromagnetic radiation. Infrared spectroscopic studies of molecular and atomic systems in the gaseous, liquid, or solid phase provide insight into their structure and establish their electronic, vibrational, and rotational energy level structure. Infrared spectroscopy has made possible an important tool for chemists and biochemists for material identification. The study of infrared properties such as absorption, emission, reflectivity, refractive indices, electrooptic and nonlinear optical coefficients of materials is important for exploring the potential of these materials for use in devices such as lasers, detectors, optical amplifiers, optical parametric oscillators, electrooptic modulators, and many others.

With regard to applications other than basic research, the list is indeed a long one. One of the earliest applications of infrared technology was in the development of infrared imaging systems for defense purposes during World War II. Even today, infrared technology plays a crucial role in the area of defense applications

This book provides engineers, scientists, and graduate students with a comprehensive guide and state-of-the-art technology to the analysis and development of infrared, photonic, and electrooptic devices and subsystems for commercial, industrial, military, and space applications. However, the biggest boost to the applications of infrared technology was given by the invention of the laser. It is of interest to note that the three most widely used lasers, namely, YAG, neodymium, carbon dioxide, and III-V semiconductor based lasers, all operate in the infrared. The use of the laser has made possible the observation of many new phenomena such as multiphoton processes, second and third harmonic generation, stimulated Raman scattering, and many more. The laser has revolutionized optical communication, which in turn has opened up exciting opportunities for the growth of information technology and the Internet. If past performance is any rule, our future will continue to be profoundly influenced by newer applications of this versatile technology.

DR. SHOBHA SINGH

Preface

The recent maturity of infrared, photonic, and electrooptic technologies has opened the door for such potential applications as space surveillance sensors, airborne reconnaissance sensors, drone electronics, premises security systems, covert communications systems, telecommunications systems, data transmission systems, IR countermeasures equipment, missile warning systems, IR lasers for medical treatment, remote space sensors, pollution monitoring sensors, high resolution imaging sensors, multispectral airborne sensors and a host of other systems used in commercial, industrial, and military applications. Deployment of the above sophisticated sensors in space, medicine, and battlefields is possible owing to the rapid development and availability of state-of-the-art photonic, electrooptic, and optoelectronic devices and components. Integration of emerging technologies, such as superconductors, MMIC, and acousto-optics in some sensors has been identified.

This book summarizes performance capabilities of infrared, photonic, and electrooptic devices and systems backed by mathematical analysis wherever necessary. The book has been designed to have a balanced mix of theory and practical applications. It is well organized and covers numerous topics and a wide range of applications including commercial, industrial, space, and military applications involving cutting-edge infrared and photonic technologies. Relevant mathematical expressions and derivations are provided for the benefits of students who wish to expand their knowledge in the IR and photonic technology areas. The book is written for easy comprehension by both undergraduate and graduate students and contains many numerical examples demonstrating the unique performance capabilities of IR and photonic sensors. The book has been prepared especially for graduate students, engineers, scientists, researchers, and product development managers who wish to, or are actively engaged in, the design and development of state-of-the-art IR and photonic devices and sensors for specific applications. It will be most useful as a compact reference for physicists, research scientists, project managers, educators, and clinical researchers. In brief, this book will be most beneficial to those who wish to broaden their knowledge in the application of infrared and photonic technologies to various devices and sensors. The author has made every attempt to provide well organized material using conventional nomenclature, a constant set of symbols, and identical units for rapid comprehension. State-of-the-art performance parameters for some IR and photonic devices and sensors are provided from various reference sources with due credit given to the authors or organizations involved. The bibliographies include significant contributing sources. It is important to mention that this book includes the latest data on research, design, and development activities in the field of infrared and photonic devices and sensors.

The book is comprised of eleven chapters. Numerical examples are included at the end of each chapter to provide the analytical aspects of important mathematical expressions. Chapter One presents the infrared (IR) theory in the simplest format for better comprehension by students and readers not familiar with IR theory. Quantities, functions, symbols, and units commonly used for describing the performance of photometric, radiometric, photonic, and IR devices are provided. Computed values of radiance exitance, relative radiance exitance, relative photon density, spectral radiant exitance, and spectral band radiance contrast as a function of temperature and emission wavelength are provided for clear understanding of the performance capabilities and limitations of IR sources and systems. Derivation of important functions and quantities commonly used in IR radiation theory are provided wherever necessary.

Chapter Two summarizes the transmission characteristics of optical signals through the atmosphere as a function of altitude, operating wavelength, and climatic conditions. Scattering, absorption, and diffraction coefficients as a function of atmospheric parameters and emission wavelength are provided. Impact of atmospheric turbulence on high-power laser beams is described under various turbulent intensities. Performance degradation of IR missiles and airborne surveillance systems owing to sever effects from absorption, scattering, diffraction, thermal blooming, gas breakdown, and turbulence-induced beam-spreading is discussed. Various models capable of computing the atmospheric transmission characteristics are identified. Impact of the target-to-background contrast on weapon delivery performance in the presence of atmospheric back scattering is described with emphasis on range capability.

Chapter Three describes the performance characteristics and capabilities of various IR sources including man-made sources, natural sources, laboratory sources, commercial sources, and industrial sources. Performance capabilities of both coherent and incoherent sources are discussed with emphasis on cost and complexity. Capabilities of state-of-the-art semiconductor lasers, diode-pumped solid state lasers, fiber lasers, optical parametric oscillators, erbium-doped amplifiers, and high-power chemical and gas lasers including CO_2 and COIL lasers are described with particular emphasis on reliability, safety, and cooling requirements.

Chapter Four focuses on performance capabilities of IR detectors and focal planar arrays (FPAs) using both CCD and CMOS technologies. Performance parameters of cooled and cryogenically cooled detectors and FPAs are summarized with emphasis on sensitivity and spectral bandwidth. Performance specification requirements of ADP detectors and photomultiplier tubes (PMTs) for various applications are described with emphasis on gain and bandwidth-efficiency product. Radiation hardness levels for IR detectors for space applications are specified. Critical performance parameters for one-dimensional and two-dimensional FPAs are described with emphasis on built-in read-out devices. IR detectors with higher sensitivity have been identified.

Chapter Five describes the state-of-the-art performance capabilities of passive infrared and electrooptic devices including optical fibers, nonlinear optical crystals, optical resonators and cavities, microlenses, A/D converters using optical technology, optical filters such as bandpass and tunable filters, fiber optic links, optical delay lines, optical isolators and circulators, optical switches, optical displays, and high-speed IR digital cameras. Potential applications of passive IR devices and sensors have been identified for commercial, industrial, space, and military applications. Performance degradation of these devices under severe operating environments is identified.

Chapter Six describes the performance capabilities and operational limitations of active IR devices and components. Critical performance parameters and design aspects of electrooptic (EO) modulators are described with emphasis on cost and complexity. Performance requirements for optical correlators, photonic-based time-stretch (PBTS) A/D converters (ADC) and optical surveillance receivers are identified. Applications of PBTS-ADC devices in high-resolution airborne radar, space-based side looking radar, and electronic warfare (EW) systems operating under severe electromagnetic environments are described. Performance capabilities of optically controlled phased array antennas including multibeam formation, fast response time, squint-free antenna patterns, deep null steering, true time delay, and wide bandwidth are summarized.

Chapter Seven describes potential applications of infrared and photonic techniques in commercial, industrial, and military systems. Integration of these technologies in high-resolution TV, DPSS lasers, optical projectors, commercial printers, laser marking systems, optical spectrum analyzers, high-quality imaging sensors, high-resolution IR cameras, detection sensors for biological and chemical weapons, smoke/fire detection sensors, battlefield sensors, and unmanned vehicle systems are discussed in detail with emphasis on reliability, performance, and cost-effectiveness. Critical performance parameters of IR-based process monitoring sensors, laser-based airport security systems, laser-based alignment systems for automobile and aircraft industries, and optic links for phased array antennas, communications, and telecommunications systems are described with emphasis on reliability and cost-effectiveness.

Chapter Eight describes the potential applications of infrared and photonic technologies in medicine, telecommunications, and space surveillance. Performance capabilities and critical parameters of IR-based and photonic-based sensors such as IR digital cameras for biotechnology image analysis, photodynamic therapy (PDT), noninvasive revascularization (TMR), lumpectomy, stomatology, and ophthalmology, hyperspectral-imaging spectrometers, IR sensors for environmental research, space surveillance systems, and optical communications and telecommunications equipment involving WDM and dense-WDM techniques are described. Application of photonic technology for DNA analysis, battlefield use, and environmental research is discussed in greater detail.

Chapter Nine deals with the infrared and photonic sensors most attractive for deep space research and military applications. The devices and sensors described in this chapter are best suited for three-dimensional ocean surveillance, unmanned aerial vehicles (UAVs), IR countermeasures, battlefield reconnaissance, target acquisition, IR search and track, and target recognition and identification under severe clutter environments. IR and photonic sensors for military and space applications include space-based antimissile systems, IR line scanners, laser rangefinders, high-resolution imaging sensors, and multispectral sensors. Performance parameters of selected IR and photonic sensors widely used for space and military applications are summarized.

Chapter Ten focuses on potential methods for predicting the IR signature of man-made sources and computer analysis for estimating IR radiation levels from complex aircraft surfaces as a function of surface emissivity, temperature, and surface conditions. Spectral radiance, radiation intensity, and radiant emittance as a function of exhaust temperature and exit nozzle area are calculated for commercial and military jet engines using a MathCad program. Skin temperature and radiation intensity from aircraft and missile surfaces are computed as a function of speed, surface area, and emissivity. A computer simulation method for predicting the IR signature of a jet engine as a function of exhaust temperature, thrust, aspect angle, and exit nozzle area is described. Computed detection ranges for short-range, medium-range, and intercontinental ballistic missiles are provided as a function of detector IFOV, optic size, source intensity, detector sensitivity, S/N ratio for a given probability of detection and false alarm rate, and target radar cross-section. Passive and active IR countermeasure methods are discussed with emphasis on IR background clutter rejection and false alarm reduction techniques.

The last chapter describes future applications of IR and photonic technologies and discusses performance requirements for auxiliary circuits and equipment. Critical requirements for the auxiliary circuits and components such as electronic control circuits, power supplies, thermoelectric coolers, temperature controllers, cryocoolers, and monitoring devices are summarized. The capabilities of selected optical software programs for modeling and analysis of IR systems, photonic devices, and optoelectronic components are described. Future applications of IR and photonic technologies for free-space intrasatellite communication, IR data transfer, industrial process control, underwater mine detection, and medical diagnosis and treatment are identified. In summary, this book provides cutting-edge IR technology aspects most useful in the design and development of infrared and photonic devices and sensors for commercial, industrial, military, and space applications. The book contains enough background and advanced material for graduate students and even for entry-level optical engineers.

I wish to thank Cassie Craig, Editorial Assistant, and Andrew Prince, Managing Editor, at John Wiley & Sons, and Paul Schwartz of Ampersand Graphics, who have been very patient in accommodating my last-minute additions and changes to the text. Last, but not least, I thank my wife Urmila Jha and daughter Sarita Jha, who inspired me to complete this book on time under a tight time schedule. Finally, I wish to express my sincere thanks to my wife, who has been very patient and supportive throughout the preparation of this book.

CHAPTER ONE

Infrared Radiation Theory

1.0 INTRODUCTION

The coming of age of Infrared (IR) technology has opened the door to a variety of applications to optoelectronic devices, photonic components, electrooptic devices, space surveillance and reconnaissance sensors, premises security systems, covert observation systems, communication systems, data transmission systems, missile warning systems, medical diagnostic equipment, pollution monitoring sensors, high-resolution image sensors, multispectral airborne/space sensors, and host of other commercial and military systems. Understanding of IR radiation theory is of paramount importance prior to application of IR technology to various devices and systems. Familiarity with symbols, dimensions, units, and definitions of important photometric quantities is absolutely necessary. This chapter will provide derivations of various relevant functions and expressions that have significant impact on critical performance parameters of the electrooptic and photonic devices.

This book identifies applications of cutting-edge IR technology in commercial and military fields. Various functions and expressions are identified that are relevant in summarizing the performance capabilities of IR and electrooptic devices. Symbols and units of various radiometric quantities are identified. Relevant performance parameters such as flux, source intensity, irradiance, radiant emittance, and energy for photometric, radiometric and photonic systems or devices are summarized.

1.1 FUNCTIONS, DEFINITIONS, AND UNITS

1.1.1 Radiometry

Radiometric quantities and their symbols and units commonly used in the IR device or system design analysis are summarized in Tables 1.1–1.3.

TABLE 1.1 Flux-related quantities, symbols, and definitions [1]

Name	Symbol	Definition
Flux	Φ*	Time rate ($d\Phi/dt$)
Flux Density	Φ/A	Time rate per unit normal to flux flow
Excitance or radiant excitance	M	Flux density emitted from source
Incidence	E	Flux density received
Radiation Intensity	I	Flux per unit solid angle

*Φ indicates flux quantity.

TABLE 1.2 Compatible systems of IR Radiation units

	IR Systems system dealing with		
Parameters	Radiometric	Photometric	Photonic
Flux	Watt	Lumen	Photon/sec
Source Intensity	Watt/sr	Lumen/sr	Photon/sec/sr
Irradiance (or illumination)	Watt/m^2	Lumen/m^2	Photon/sec/m^2
Emittance	Watt/m^2	Lumen/m^2	Photon/sec/m^2
Radiance (or luminance)	Watt/sr/m^2	Lumen/sr/m^2	Photon/sec/m^2
Energy	Watt·sec (joule)	Lumen·sec	Photon

TABLE 1.3 Radiometric quantities, symbols, and units

Quantity	Symbol	Unit
Energy	Q_e	J
Energy density	W_e	J/m^3
Flux power	Φ_e	W
Flux density	$\Phi_{e/A}$	W/m^2
Radiant excitance	M_e	W/m^2
Irradiance	E_e	W/m^2
Radiance	L_e	W/m^2

1.2 VARIOUS FUNCTIONS, DEFINITIONS, AND UNITS

Blackbody functions are of paramount importance in dealing with design analysis and performance prediction of IR sensors and sources. A blackbody is defined as a perfect radiation source, i.e., one that radiates the maximum number of photons per unit time from a unit area in a specified spectral interval into a hemispherical region that any body can radiate at the same temperature and under thermodynamic equilibrium. The blackbody function is represented by the Plank function, which is defined as the number of photons per unit k-interval per unit volume. This can be expressed as

$$N_{k^-} = \left(\frac{k}{\pi}\right)^2 \left(\frac{1}{e^x - 1}\right) \tag{1.1}$$

where, $k = (2\pi/\lambda)$, λ is the wavelength (cm), and $x = (1.4388/\lambda T)$, a dimensionless quantity when temperature T is expressed in Kelvin and wavelength λ im cm.

The number of photons per unit area per unit time radiated into a hemisphere by a blackbody can be written as

$$M_q = \frac{cn}{4} \tag{1.2}$$

where, c is the velocity of light ($c = 3 \times 10^{10}$ cm/sec) and n is the photon volume density per cm^3. If parameter c is expressed in m/sec, then n must also be expressed in meters.

The number of photons per unit area per unit time per unit solid angle radiated from a blackbody can be expressed as

$$L_q = \frac{cn}{4\pi} \tag{1.3}$$

Note: flux-related radiometric parameters and symbols are shown in Table 1.1 [1]. Radiometric, photometric and photonic parameters and their units are summarized in Table 1.2. Specific radiometric quantities and their symbols and units are shown in Table 1.3. Various physical constants, optical parameters, symbols, and units frequently used in photometric and radiometric system analysis and design are summarized in Tables 1.4–1.8 [2]. Metric unit prefixes and symbols are shown in Table

TABLE 1.4 Physical constants commonly used for electrooptic analysis [2] and their values and units

Physical constant	Symbol	Value	Unit
Avogadro's number	N	6.022169×10^{23}	mol^{-1}
Boltzmann's constant	k	1.380622×10^{-23}	$J\ K^{-1}$
Electron charge	e	$1.6021917 \times 10^{-19}$	C
Electron charge to mass ratio	e/m	1.7588028×10^{11}	$C\ kg^{-1}$
Energy of 1 electron volt	eV	$1.6021917 \times 10^{-19}$	J
Voltage–wavelength conversion factor	hc/e	1.2398541×10^{-6}	V m
First radiation constant ($8\pi hc$)	C_1	4.992579×10^{-24}	J m
kT value at room temperature	—	0.0259	eV
Luminous efficacy at 555 nm	$K(\lambda)_{max}$	673	$lm\ W^{-1}$
Mass of electron in free space	m	9.109558×10^{-31}	kg
Permittivity of free space	ε_0	8.86×10^{-12}	$F\ m^{-1}$
Planck's constant	h	6.626196×10^{-34}	J s
Second radiation constant (hc/k)	C_2	0.01438833	m K
Speed of light in vacuum	c	2.9979250×10^{8}	$m\ s^{-1}$
Stefan–Boltzmann constant	σ	5.66961×10^{-8}	$W\ m^{-2}\ K^{-4}$

TABLE 1.5 Units commonly used in radiometric and photometric applications [2]

Unit	Symbol	Notes
bit per second	b/s	
calorie (International Table calorie)	cal_{IT}	1 cal_{IT} = 4.1868J. The 9th Conférence Générale des Poids et Mesures adopted the joule as the unit of heat. Use of the joule is preferred.
calorie (thermochemical calorie)	cal	1 cal = 4.1840 J (See note for International Table calorie above.)
candela	cd	SI unit of luminous intensity
candela per square meter	cd/m^2	SI unit of luminance. The name nit is sometimes used for this unit.
candle	cd	The unit of luminous intensity has been given the name candela; use of the name candle for this unit is deprecated.
circular mil	cmil	1 cmil = $(\pi/4)\ 10^{-6}\ in^2$
coulomb	C	SI unit of electronic charge
curie	Ci	1 Ci = 3.7×10^{10} disintegrations per second. Unit of activity in the field of radiation dosimetry.
cycle per second	Hz	The name hertz is internationally accepted for this unit; the symbol Hz is preferred to c/s.
decibel	dB	One tenth of a bel
degree (temperature)		
degree Celsius	°C	The use of the word centigrade for the Celsius temperature scale was abandoned by the Conférence Générale des Poids et Mesures in 1948. Note there is no space between the symbol ° and the letter.
degree Fahrenheit	°F	
degree Kelvin	K	
degree Rankine	°R	
dyne	dyn	The CGS unit of force
electronvolt	eV	The energy received by an electron in falling through a potential difference of one volt

1.9. It is important to mention that the symbols for photometric quantities are the same as those for corresponding radiometric quantities, as evident from the various quantities shown in various tables.

1.2.1 Generalized Plank Function

Several functions, such as the Plank distribution function, blackbody function, Wien distribution function, zeta function, and total isothermal Plank distribution function

TABLE 1.6 Units commonly used in photometric and radiometric applications [2]

Unit	Symbol	Notes
kilogram-force	kg_f	In some countries the name kilopond (kp) has been adopted for this unit.
knot	kn	kn = nmi hr^{-1}
lambert	L	1L(l/π) cd cm^{-2}. A CGS unit of luminance. Use of the SI unit of luminance, the candela per square meter, is preferred.
liter	l	1l = 10^{-3} m^3
lumen	lm	SI unit of luminous flux
lumen per square foot	lm/ft^2	A unit of illuminance and also a unit of luminous excitance. Use of the SI unit, lumen per square meter, is preferred.
lumen per square meter	lm/m^2	SI unit of luminous excitance
lumen per watt	lm/W	SI unit of luminous efficacy
lumen second	lm s	SI unit of quantity of light, also known as the talbot
lux	lx	lx = lm m^{-2}. SI unit of illuminance
maxwell	Mx	The maxwell is the electromagnetic CGS unit of magnetic flux. Use of the SI unit, the weber, is preferred.
meter	m	SI unit of length
mho	mho	CIPM (Commission Internationale Photometric Measurements) has accepted the name siemens (S) for this unit and will submit it to the 14th Conférence Générale des Poids et Mesures (CGPM) for approval.
micrometer	μm	
micron	μm	See micrometer. The name micron was abrogated by the CGPM in 1967.

are used in solution of IR-related problems. However, the Plank distribution function plays a key role in computation of radiant exitance from a source as a function of temperature. Radiant excitance is expressed as

$$M(x) = 2\pi hc^2 \left(\frac{k_B T}{ch} \right)^4 \left(\frac{x^3}{e^x - 1} \right) \tag{1.4}$$

Inserting the values of various parameters and constants given in various tables, the above equation is reduced to

$$M(x) = (0.872 \times 10^{-12})(T^4) \left(\frac{x^3}{e^x - 1} \right) \tag{1.5}$$

TABLE 1.7 Units commonly used in photometric and radiometric applications [2]

Unit	Symbol	Notes
ounce (avoirdupois)	oz	
pascal	Pa	Pa = N m^{-2}. SI unit of pressure or stress. This name accepted by the CIPM in 1969 for submission to the 14th CGPM.
phot	ph	ph = lm cm^{-2}. CGS unit of illuminance. Use of the SI unit, the lux (lumen per square meter), is preferred.
poise	P	P = dyn s cm^{-2}. Unit of coefficient of viscosity.
rad	rd	Unit of absorbed dose in the field of radiation dosimetry
radian	rad	SI unit of plane angle
rem	rem	Unit of dose equivalent in the field of radiation dosimetry
roentgen	R	Unit of exposure in the field of radiation dosimetry
second (time)	s	SI unit of time
siemens	S	S = Ω^{-1}. SI unit of conductance. This name and symbol were accepted by the CIPM in 1969 for submission to the 14th CGPM. The name mho is also used for this unit in the United States.
steradian	sr	SI unit of solid angle
stilb	sb	sb = cd cm^{-2}. A CGS unit of luminance. Use of the SI unit, the candela per square meter, is preferred.
stokes	St	Unit of viscosity

where the dimensionless variable x has a value of $(1.4388/\lambda T)$ when the wavelength is given in cm and the temperature in K. Note that the general radiometric function (R) is a function of temperature (T) and a spectral variable that is proportional to either optical frequency or emission wavelength.

Spectral radiance excitance is the most important parameter in designing an IR system to meet performance requirements. This parameter as a function of wavelength and temperature can be expressed as

$$M(\lambda) = \frac{c_1(\lambda)^{-5}}{e^x - 1}, \qquad \text{watt/cm}^2\text{/cm} \tag{1.6}$$

where $c_1 = 3.74 \times 10^{-12}$ W·cm^2 and x is a dimensionless parameter defined earlier.

1.3 ILLUMINANCE, LUMINOUS EXITANCE, AND LUMINANCE

These quantities can be best explained by the schematic representation depicting the relationships of common photometric units as shown in Figure 1.1 The symbol "X"

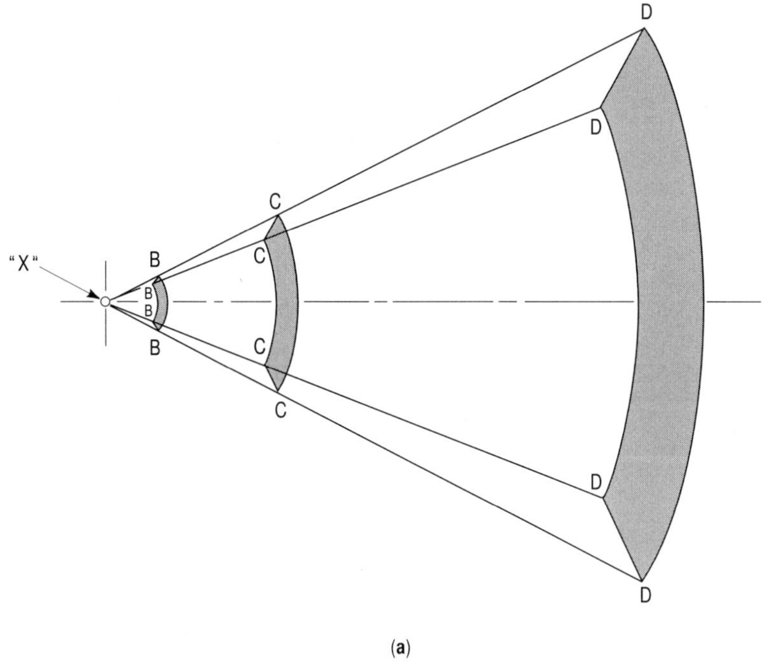

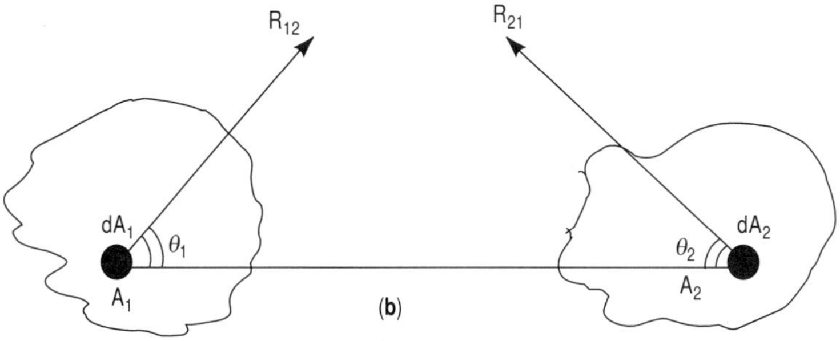

FIGURE 1.1 (A) Schematic representation of units. "X" represents point source having a linear intensity of one candela. Solid angle shown represents one steradian. Point "X" to and point "B" is 1 cm; the surface represented by "BBBB" is 1 cm^2. Point "X" to any point "C" is 1 ft; the surface represented by "CCCC" is 1 ft ^{2}. Point "X" to any point "D" is 1 m; the surface represented by "DDDD" is 1 m^2. (B) Surface radiance levels from two IR sources is separated by a distance r_{12}.

represents a point source with a linear intensity of one candela over a solid angle of one steradian (sr). The Illuminance parameter (E) is defined as the flux received per unit area from a source, and is also known as illumination. A luminous exitance is defined as the flux emitted by a unit area and can be expressed in either lumen per square meter or lumen per square centimeter. The lux is the (SI unit) unit of illuminance shown in Table 1.6 resulting from the flux of 1 lumen falling on the surface

TABLE 1.8 Units and symbols for various quantities used in radiometry and photometry [2]

Unit	Symbol	Notes
tesla	T	$T = N\,A^{-1}\,m^{-1} = Wb\,m^{-2}$. SI unit of magnetic flux density (magnetic induction)
tonne	t	1 t = 1000 kg
var	var	IEC name and symbol for the SI unit of reactive power
volt	V	SI unit of voltage
volt per meter	V/m	SI unit of electric field strength
voltampere	VA	IEC name and symbol for the SI unit of apparent power
watt	W	SI unit of power
watt per meter kelvin	W/(m·K)	SI unit of thermal conductivity
watt per steradian	W/sr	SI unit of radiant intensity
watt per steradian and square meter	$W/(sr{\cdot}m^{-2})$	SI unit of radiance
watthour	Wh	
weber	Wb	Wb = Vs. SI unit of magnetic flux

"DDDD" as shown in Figure 1.1. Parameter luminance (L) is known as brightness, which is defined as the luminous intensity (I) per projected unit area normal to the line of observation (Figure 1.1). This means one can write the expression for the luminance as

$$L = \frac{dI}{dA} \tag{1.7}$$

where, dI is the incremental luminous intensity on the projected area dA.

1.4 TRANSPARENCY, OPACITY, AND OPTICAL DENSITY PARAMETERS

These parameters are of critical importance, particularly, when an IR radiated signal passes through a layer of a material. Transparency of a layer in the material is defined as the ratio of the intensity of the transmitted light to that of the incident light. Opacity is the reciprocal of the transparency. Optical density is the common logarithm of the opacity. This means that

$$\text{Transparency} = \frac{I_t}{I_I} \tag{1.8a}$$

TABLE 1.9 Metric unit prefixes and symbols commonly used in photometric and radiometric applications

Prefix (multiple)	Symbol
tera (10^{12})	T
giga (10^{9})	G
mega (10^{6})	M
kilo (10^{3})	k
hecto (10^{2})	h
deka (10)	da
deci (10^{-1})	d
centi (10^{-2})	c
milli (10^{-3})	m
micro (10^{-6})	μ
nano (10^{-9})	n
pico (10^{-12})	p
femto (10^{-15})	f
atto (10^{-18})	a

$$\text{Opacity} = \frac{I_I}{I_t} \tag{1.8b}$$

$$\text{Optical density} = \log\left(\frac{I_I}{I_t}\right) \tag{1.8c}$$

where subscripts t and I stand for transmission and incident, respectively.

1.4.1 Conversion of One Parameter into Another

Converted values of the parameters summarized in Table 1.10 will be found most useful in the design and development of optical systems.

TABLE 1.10 Conversion between transparency, opacity, and optical density

Transparency	Opacity	Optical density
0.1	10	1.000
0.2	5	0.699
0.3	3.333	0.523
0.4	2.500	0.398
0.5	2.000	0.301
0.6	1.667	0.222
0.7	1.429	0.155
0.8	1.250	0.0.097
0.9	1.111	0.046
1.0	1.000	0.000

1.5 RADIATION GEOMETRY

The geometrical transfer function of radiation is independent of spectral characteristics and is strictly based on the transfer equation for a medium free from absorption and scattering properties [1]. Consider a radiation interaction between two radiation signals R_{12} and R_{21} shown in Figure 1.1. Body surface A_1 with radiance of R_{12} and A_2 with radiance R_{21} are joined by a ray of length r_{12} . The net radiative change between the two sources can be given as

$$\Delta\Phi = (\Phi_{21} - \Phi_{12})$$
$$= \frac{\iint \Delta R \cos(\theta_1) \cos(\theta_2)\, dA_1\, dA_2}{r_{12}^2} \tag{1.9}$$

where $\Delta R = (R_{12} - R_{21})$
θ_1 = angle for radiance level R_{12}
θ_2 = angle for radiance level R_{21}
dA_1 = smallest surface area of A_1
dA_2 = smallest surface area of A_2

Note that the integral in this equation is most useful for many radiative interaction problems and has been known by number of different names, such as radiation interaction factor, solid angle projection, and radiation configuration factor.

1.5.1 Flux Density and Radiance Distribution from a Lambertian Surface

A surface whose radiance is independent of angle is called a Lambertian surface. For a Lambertian surface with a differential area, the flux density or the flux per unit area radiated into a hemisphere can be expressed as

$$M = \pi R \tag{1.10}$$

where M is the flux density (W/cm^2) and R is radiance (W/cm^2/sr). The flux density radiated into a solid angle designated by a cone with half-angle of θ is given by

$$M(\theta) = \pi \sin^2(\theta) \tag{1.11}$$

1.5.2 Radiating Surfaces with Cosine Radiance Distribution Functions [1]

Radiating surfaces with cosine distribution or variable angular distributions of emission or reflection can be characterized as the power of cos θ. Assuming a surface with radiance of $R(\theta) = R(\theta) \cos^n(\theta)$, one gets the expression for the flux density

$$M(\theta) = \left(\frac{2\pi}{n+2}\right) R(\theta)\cos^2(\theta) \tag{1.12}$$

1.5.3 Irradiance of Lambertian Disc

The surface irradiance (E) of a Lambertian disc of radius (r) at given distance (d) away from a radiating surface can be easily computed using appropriate expression. Note that a Lambertian surface is perfectly diffused surface that has a constant radiance level (R) independent of the viewing angle. The surface irradiance follows approximately the inverse square law and that applies only when the distance (d) is less than the radius (r). This means that the distance away from the source must be much greater than the dimension of the IR radiating source.

1.5.4 Optical Temperature and Radiation Temperature

Several methods are available [3] to determine the thermodynamic temperature of a body radiometricallly. These methods include precise measurements of the total radiation level, the radiation at a specific wavelength, the wavelength distribution of the radiation, and the apparent color of the radiation. The temperature measured at a particular wavelength is known as the optical temperature.

The radiation temperature (T) is defined as the temperature of a blackbody that yields the same total radiance level (R) or radiant exitance (M). The radiant exitance for a blackbody is given as

$$M = \sigma(T_B^4) \tag{1.13}$$

$$T_B = \left(\frac{M}{\sigma}\right)^{1/4} \tag{1.13a}$$

$$T_G = \varepsilon^{1/4} \cdot T_B \tag{1.13b}$$

$$T_c = \left(\frac{\varepsilon}{\sigma}\right)^{1/4} \left(\frac{\int C_1 \lambda^{-5} dx}{e^x - 1}\right)^{1/4} \tag{1.13c}$$

where subscript B stands for blackbody, C stands for colored body, G stands for gray body, symbol ε indicates emissivity of the surface, dimensionless variable x has been defined previously ($1.4388/\lambda T$) and parameter σ has a value of (5.67×10^{-12}) W/cm^2/K^4. Note that symbol T is generally used for blackbody temperature instead of T_B. Note the above equations (1.13a, 1.13b, and 1.13c) give the color temperature value for a blackbody, gray body, and colored body, respectively.

Relative error (RE) in radiation temperature calculations can be determined from the following expressions:

For a blackbody, $(RE)_{BB} = 1 - T_B/T_B = 0$
For a gray body, $(RE)_{GB} = 1 - (\varepsilon)^{1/4}$
For a colored body, $(RE)_{CB} = [1 - c_2(\lambda T \log_e \varepsilon + c_2 T)^{-1}]$

Note that when ε is equal to one, it is no longer a gray body but a blackbody. The radiation temperature for black and gray bodies can be easily calculated using equations (1.13a) and (1.13b). However, radiation temperature calculations for a colored body involve more variables and new parameters, including the spectral radiant exitance $M(\lambda)$. Calculated values of colored body radiation temperatures as a function of various parameters are summarized in Table 1.11. Spectral radiant exitance is given as

$$M(\lambda) = \int\left(\frac{C_1\lambda^{-5}dx}{e^x - 1}\right) \tag{1.14}$$

Where $C_1 = 3.74 \times 10^{-12}$ (W·cm²), $x = 1.4388/\lambda T$, when wavelength is given in cm and temperature in K.

1.6 EMISSIVITY AS A FUNCTION OF TEMPERATURE AND WAVELENGTH FOR VARIOUS MATERIALS

Emissivity (ε) is a function of wavelength, temperature, body material, and surface conditions of the radiating body. Estimated values of the emissivity parameter for various materials as a function of temperature and wavelength are summarized in Tables 1.12–1.13. It is important to mention that the reliable emissivity data on most materials as a function of surface conditions are not readily available.

Spectral emissivity of a material is a function of both the temperature and the operating wavelength and is generally expressed by a complex function $\varepsilon(\lambda, T)$. The spectral emissivity parameter truly indicates the emissivity of the material, excluding the effects due to surface conditions. Estimated values of spectral emissivity for tungsten as a function of temperature and wavelength [4] are summarized in Table 1.14.

These spectral emissivity values indicate that spectral emission decreases with the increase in both temperature and wavelength. Spectral emissivity $\varepsilon(\lambda, T)$ and total emissivity ε_T of tungsten are given in Table 1.15.

TABLE 1.11 Radiation temperature for colored body (T_C) as a function of wavelength, emissivity, and surface temperature

T(K)	Band (μm)	$M(\lambda)$(W/cm²)	ε	$[M(\lambda)]^{1/4}$(W/cm²)	T_C(K)
2000	1.5–1.0	18.74	0.260	2.081	963
1000	3.0–2.0	1.17	0.114	1.012	379
500	6.0–4.0	0.073	0.053	0.520	162

TABLE 1.12 Total emissivity of various materials at 0.65 μm [4]

Material	Melting point (K)	Emissivity (ε)
Beryllium	2151	0.61
Carbon	3895	0.80/0.93
Chromium	2163	0.34
Copper	1356	0.10
Iron	1808	0.35
Nickel	1726	0.36
Platinum	2042	0.30
Titanium	1948	0.63
Tungsten	3683	0.43
Zirconium	2125	0.32

TABLE 1.13 Low-temperature emissivity of various materials at 100 °C [4]

Material	Emissivity (ε)
Carbon	0.81
Chromium (unoxidized)	0.08
Copper (unoxidized)	0.02
Iron (unoxidized)	0.05
Iron (oxidized)	0.74
Nickel (oxidized)	0.32
Steel (unoxidized)	0.08
Tungsten (unoxidized)	0.032 at 100 °C (373 K)
	0.071 at 500 °C (773 K)
	0.150 at 1000 °C (1273 K)
	0.230 at 1500 °C (1773 K)
	0.281 at 2000 °C (2273)

TABLE 1.14 Spectral emissivity of tungsten as a function of temperature and wavelength

	Wavelength (μm)				
Temperature (K)	0.6	0.8	1.0	1.5	2.0
1000	0.458	0.450	0.403	0.267	0.186
1500	0.448	0.434	0.392	0.278	0.206
2000	0.438	0.419	0.381	0.288	0.227
2500	0.428	0.406	0.371	0.297	0.248
3000	0.418	0.396	0.365	0.304	0.263

TABLE 1.15 Spectral emissivity and total emissivity of tungsten [1]

Temperature (K)	Spectral emissivity at 0.65 μm	Total emissivity
500	0.463	0.053
1000	0.458	0.114
1500	0.448	0.191
2000	0.438	0.260

These data indicate that the spectral emissivity decreases with the increase in surface temperature, while the total emissivity increases with the increase in temperature. Note that for colored bodies, the total emissivity value must be used to achieve reliable results.

1.7 BRIGHTNESS TEMPERATURE

Brightness temperature is defined as the temperature of a blackbody that gives the same radiance level ($R = M/\pi$) in the same narrow spectral band. This means that

$$e^{(A/T_B)} - 1 = \frac{e^{A/T} - 1}{\varepsilon} \tag{1.15}$$

where $A = 1.4388/\lambda$ (cm·K), T_B is the brightness temperature (K), T is the body temperature (K), and ε is the emissivity of the material. The equation for the brightness temperature can be rewritten after rearranging the terms and taking the natural logarithm of both sides:

$$T_B = \left(\frac{1.4388}{\lambda}\right)\left[\log_e\left(\frac{e^{(1.4388/\lambda T} - 1}{\varepsilon}\right) + 1\right]^{-1} \tag{1.16}$$

Computed values of brightness temperature as a function of emissivity, surface temperature, and wavelength are shown in Table 1.16. These calculated values assume the total emissivity and the emission wavelength at the specified blackbody temperature (T). The brightness temperature varies between 65 to 75% of the blackbody temperature under the assumed values of emission wavelength and total emissivity.

1.8 DISTRIBUTION TEMPERATURE [1]

The distribution temperature (T_D) is defined as the temperature of the blackbody source that best matches the spectral distribution (M_λ) of the body under consideration. A minimum of two narrow spectral bands at wavelengths λ_1 and λ_2 is required to determine the magnitude of this temperature. Under two wavelength assumption

TABLE 1.16 Brightness temperature as a function of wavelength, emissivity, and surface temperature

	Assumed parameters		
Temperature (K)	λ (μm)	ε	T_B (K)
1000	3.0	0.114	689
1500	2.0	0.191	1116
2000	1.5	0.260	1563

and employing the equations previously used for the spectral distributions under radiation temperature expression derivation, the equation for the distribution temperature now can be written as

$$T_D = \frac{M(\lambda_1)}{M(\lambda_2)} = \left(\frac{\varepsilon_1}{\varepsilon_2(\lambda_2/\lambda_1)^5}\right)\left(\frac{e_2^A T - 1}{e_1^A T - 1}\right) \tag{1.17}$$

where $A_2 = 1.4388/\lambda_2 T$, $A_1 = 1.4388/\lambda_1 T$, and the wavelength and temperature are specified in cm and Kelvin units.

Assuming $T = 2000$ K, $\lambda_1 = 1.5$ μm, $\lambda_2 = 1.0$ μm, $\varepsilon_1 = 0.288$, $\varepsilon_2 = 0.381$, and constant 1.4388 (cm·K), the computed value of distribution temperature comes to 1.104. However, its value comes to 1.138 at a blackbody temperature of 1000 K, at wavelengths of 3 and 2 μm with corresponding emissivities of 0.114 and 0.145, respectively.

1.9 COLOR TEMPERATURE [1]

The color temperature (T_C) is defined as the temperature of a blackbody that has the same normalized chromaticity coordinates (x, y) as the body under consideration. The relative radiance levels or spectral distributions can be expressed in the normalized chromaticity coordinates x and y, which can be written as

$$\frac{M(\lambda_1)}{M(\lambda_2)} = \left(\frac{\lambda_2}{\lambda_1}\right)^5\left(\frac{e_2^A T_c - 1}{e_1^A T_c - 1}\right) \tag{1.18}$$

where coefficients A_1 and A_2 are same as defined in equation (1.17).

Using the equation (1.14) for relative radiance after integrated over two wavelength bands and corresponding emissivities as a function of two wavelengths, the color temperature expression can be rewritten as,

$$\frac{e_2^A T_c - 1}{e_1^A T_c - 1} = \left(\frac{\varepsilon_1}{\varepsilon_2}\right)\left(\frac{e_2^A T - 1}{e_1^A T - 1}\right) \tag{1.19}$$

where the coefficients A_1 and A_2 are defined in previous equations.

Example to Compute a Color Temperature for Given Parameters

Assumed parameters are: $T = 2000$ K, $C_2 = 1.4388$ (cm·K), $\lambda_1 = 1.5$ μm, $\lambda_2 = 1.0$ μm, $\varepsilon_1 = 0.288$, $\varepsilon_2 = 0.381$. The color temperature equation is written as

$$\frac{e^{14388/T_C} - 1}{e^{9593/T_C} - 1} = \left(\frac{0.288}{0.381}\right)\left(\frac{e^{7.195} - 1}{e^{4.796} - 1}\right) = 8.38$$

Solving this equation yields a color temperature value of 2270 K based on the above assumed values of various parameters and constants.

1.10 NORMALIZED CHROMATICITY COORDINATES AS A FUNCTION BLACKBODY TEMPERATURE

Estimated values of normalized chromaticity [1] coordinates as a function of blackbody temperature are summarized in Table 1.17.

1.11 COMPUTATIONS OF VARIOUS IR QUANTITIES

This section deals with derivation of expressions for various IR quantities and functions that play key roles in the design and development of photonic circuits, electrooptic devices, and IR sensors. Calculated values and plots of various IR quantities and parameters such as photon flux density, radiant exitance, spectral band radiance contrast, photons per unit interval per unit volume, spectral radiant exitance, and radiance contrast for specific IR bands are provided, allowing readers to visualize the effects of operating temperature and emission wavelength on their magnitudes.

1.11.1 Radiant Exitance Based on the Stefan–Boltzmann Law

The radiant exitance (M) is the IR radiation level expected from a source operating at a specified temperature. The radiant exitance expression is given as

TABLE 1.17 Normalized chromaticity coordinates values [1, Table 1.30]

Temperature (K)	x	y
1000	0.652	0.344
1500	0.585	0.393
2000	0.526	0.413
2500	0.476	0.412
3000	0.436	0.404
3500	0.405	0.390
4000	0.380	0.376

$$M = \sigma T^4, \qquad \text{W/cm}^2 \tag{1.20}$$

where $\sigma = 5.67 \times 10^{-12}$ W/cm^2/K^4 and T is the absolute temperature of the source (K).

Calculated values of radiant exitance as a function of source temperature are plotted in Figure 1.2. One can estimate the radiant exitance from this curve for various sources operating at specified temperature. One can expect a radiant exitance of about 300 W/cm^2 for a military jet engine operating under afterburner condition with an assumed temperature of 2500 K.

1.11.2 Relative Radiant Exitance

Relative radiant exitance $M(\lambda)$ is a most important parameter that indicates the relative radiant exitance level as a function of temperature and emission wavelength. Its expression can be written as

$$M(\lambda) = \int^{\lambda} \left(\frac{M_\lambda d\lambda}{\sigma T^4} \right) \tag{1.21}$$

where $M_\lambda = [C_1 \lambda^{-5}/(e^x - 1)]$, $C_1 = 3.742 \times 10^{-12}$ W/cm^2, $x = 1.4388/\lambda T$(1/cm K), and the constant σ has been defined previously. Calculated values of relative radiant exitance as a function of temperature and wavelength are shown in Figure 1.3.

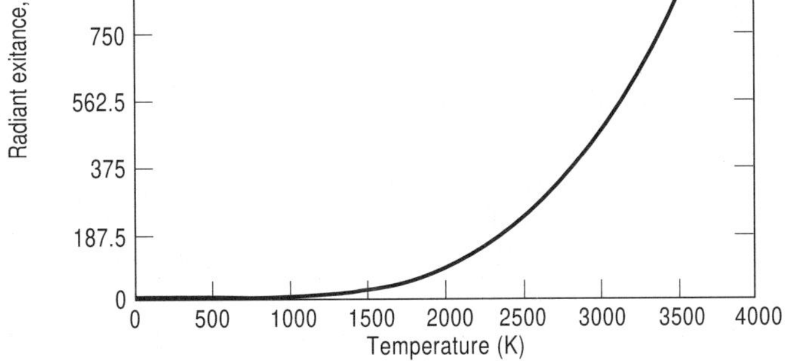

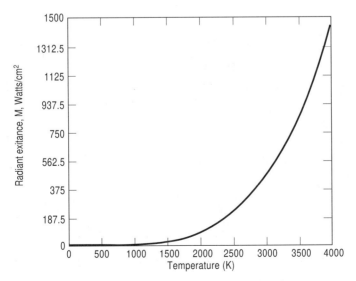

FIGURE 1.2 Computed values of radiant exitance as a function of source temperature.

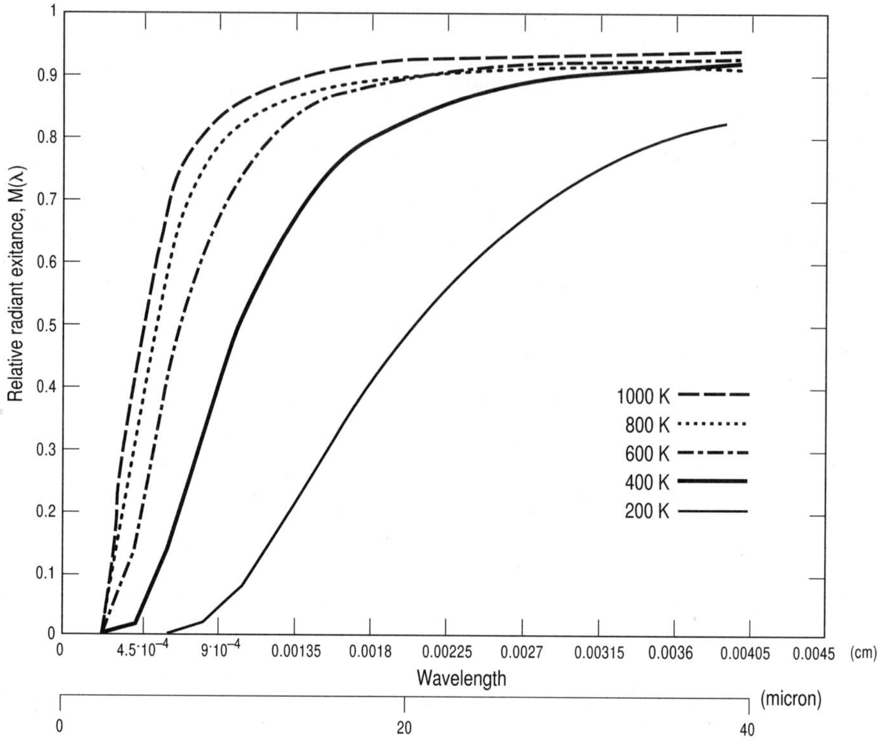

FIGURE 1.3 Relative radiant exitance as a function of wavelength and temperature.

1.11.3 Spectral Radiant Exitance [2]

Spectral radiant exitance indicates the radiant exitance from a blackbody as a function of its temperature and emission wavelength. The expression for the spectral radiant exitance can be written as

$$M_\lambda = \frac{(3.472 \times 10^{-12})\,\lambda^{-5}}{e^x - 1}, \qquad \text{W/cm}^2/\mu\text{m} \tag{1.22}$$

where λ is the emission wavelength and constant $x = 1.4388/\lambda T$.

Computed values of spectral radiant exitance as a function of temperature and wavelength are plotted in Figure 1.4. It is important to mention that the wavelength parameter up to the fourth power is expressed in centimeters, whereas the fifth power is expressed in microns to meet the units requirement as specified after equation (1.22).

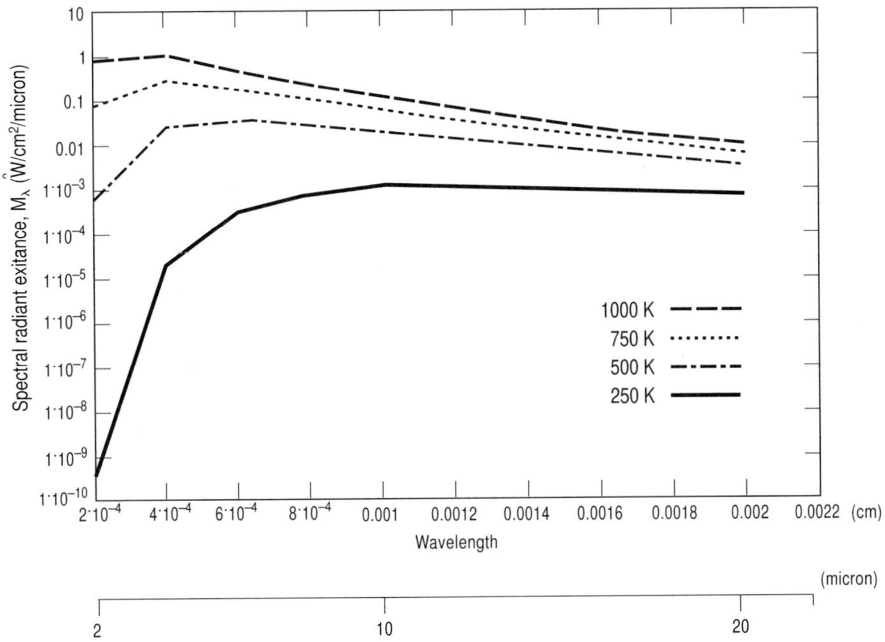

FIGURE 1.4 Spectral radiant exitance for a blackbody as a function of wavelength and temperature.

1.11.4 Number of Photons per Unit Time per Unit Area per Steradian per Micron [1]

In certain applications such as photon detectors, photon count per unit time per unit area is of critical importance. Computer-based calculations have been performed as a function of temperature and emission wavelength to determine the number of photons per second per unit area per steradian per micron. Calculated values of number of photons as a function of temperature and wavelength using equation (1.23) are shown in Figure 1.1.

$$N_p(\lambda) = \frac{2c}{\lambda^4(e^x - 1)} \tag{1.23}$$

where, c is the velocity of light (3×10^8m/sec) and parameters λ and x and their units are the same as specified in previous examples. Note that a rapid drop in number of photons per unit time per unit area per steradian per micron occurs at wavelengths ranging from 2–10 μm, when the temperature varies from 1000–4000 K (Figure 1.5). However, a rapid drop in photon count occurs over the spectral region

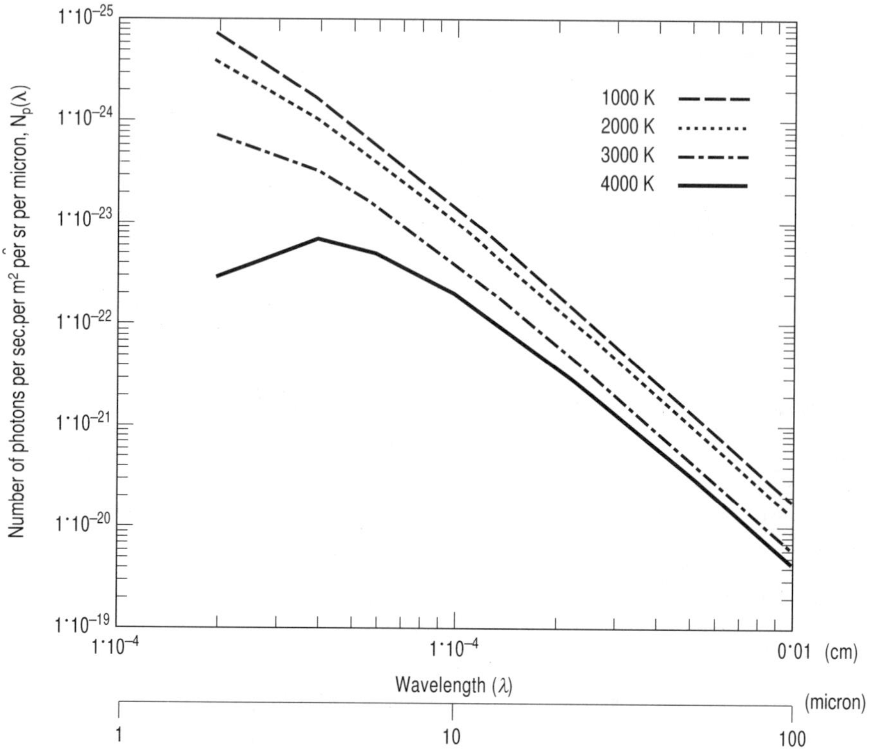

FIGURE 1.5 Number of photons as a function of wavelength and temperature.

from 1–5 μm, when the temperature varies from 250–1000 K. One can see a flat photon response at wavelengths exceeding 10 μm at lower temperatures, as evident from the curves shown in Figure 1.6. An extremely fast drop in the number of photons per unit time per unit area per steradian per micron at 1000 K can be seen in Figure 1.7 over spectral region from 1–3 μm.

1.11.5 Relative Photon Flux Density

The performance of some IR sensors depends on relative photon flux density, which can be calculated from the modified expression shown below.

$$M_p(\lambda) = \int^{\lambda}\left(\frac{(1.24)(\lambda)^{-4}d\lambda}{T^3(e^x - 1)}\right) \tag{1.24}$$

where λ, T, and x parameters defined previously remain the same. Plots showing the

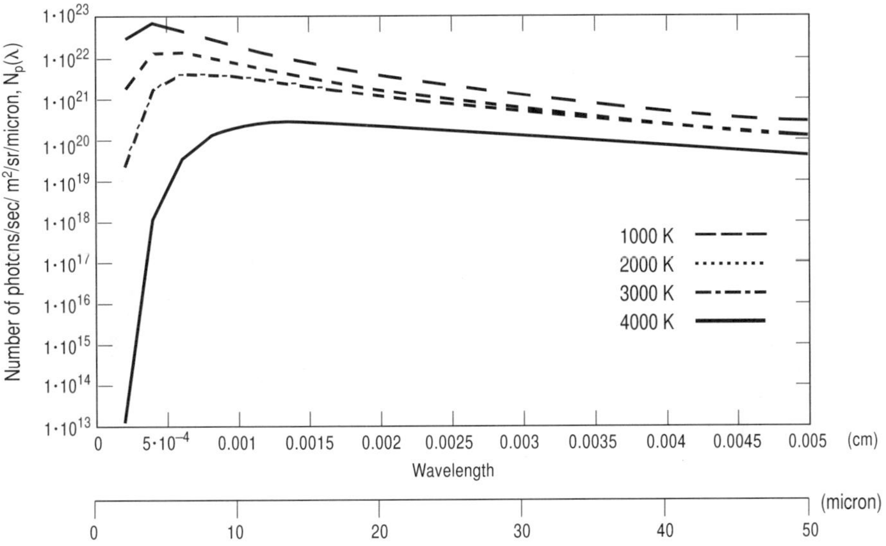

FIGURE 1.6 Number of photons as a function of wavelength and temperature.

computed values of relative photon flux density as a function of temperature and emission wavelength are shown in Figure 1.8.

1.11.6 Photons per Unit *k*-Interval per Unit Volume

In some applications, IR sensor performance is measured in terms of photons per unit interval per unit volume. The magnitude of this parameter $N_k(\lambda)$, based on Plank function can be computed using the following modified equation:

$$N_k(\lambda) = \frac{4/\lambda^2}{e^x - 1} \tag{1.25}$$

where, $x = 1.4388/\lambda T$, when the wavelength is expressed in centimeters and the temperature in Kelvin (K). Calculated values of photons per unit *k*-interval per unit volume as a function of temperature and wavelength are shown in Figure 1.9. Sharp increase in the magnitude of this parameters occurs between 1–4 μm at a temperature of 250 K. Its response remains practically flat beyond 5 μm regardless of the temperature exceeding 500 K.

1.11.7 Percentage of Total Radiance Emittance over a Specified Band at Various Temperatures

A certain amount of IR energy over a specified spectral band is required to meet the performance requirements of IR surveillance sensor or IR countermeasure equip-

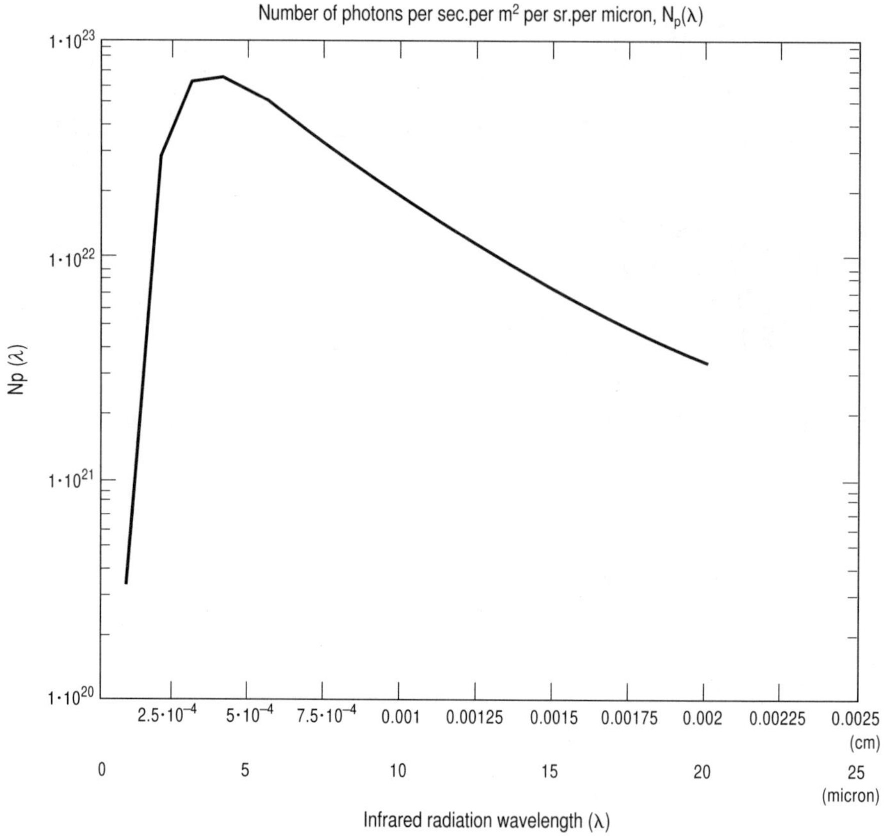

FIGURE 1.7 Number of photons as a function of IR wavelength at 1000 K temperature.

ment. This parameter is expressed in terms of percentage of total radiant exitance given by the Stefan–Boltzmann equation. In brief, this parameter can be characterized as (R_p) and is mathematically defined as

$$R_p = \frac{M(\lambda)}{\sigma T^4} \tag{1.26}$$

where, $M(\lambda) = \int [(0.6608)\, \lambda^{-5}/T^4] d\lambda$, with desired values of λ_2 and λ_1. Note that parameters σ and T have been defined previously. Inserting the relevant constants and parameters in equation (1.26), several values of this parameter as a function of temperature and specified spectral bandwidth are calculated; they are summarized in Table 1.18.

1.11.8 Spectral Band Radiance Contrast

Spectral band radiance contrast [1] is of paramount importance to IR imaging sensors and has potential applications where high resolution images with high contrast

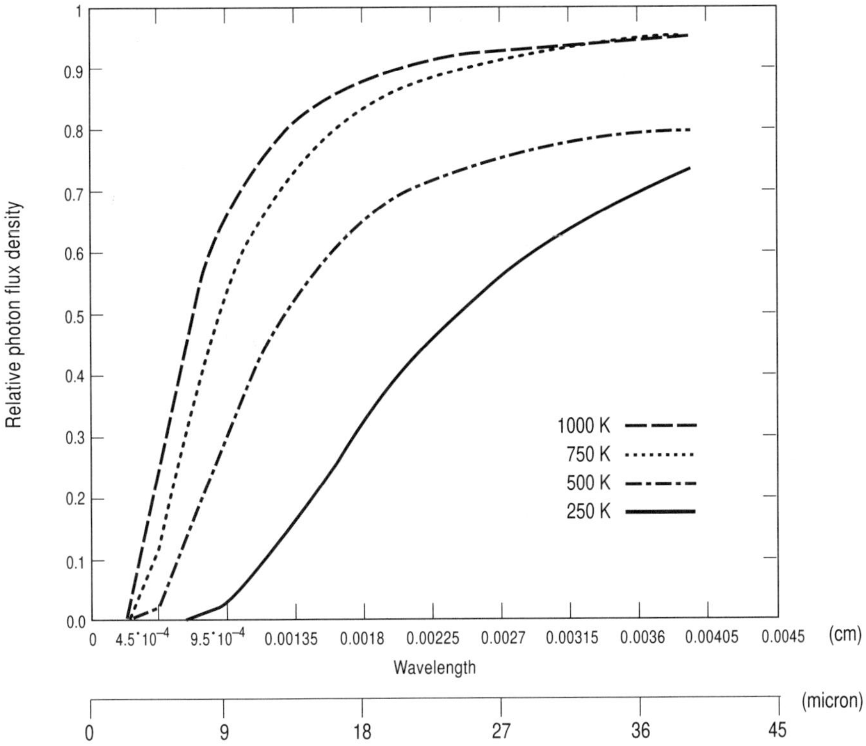

FIGURE 1.8 Relative photon flux density as a function of wavelength and temperature.

are required. Spectral band radiance contrast is defined by a complex function $L(A, B)$ as described below, where each parameter A and B are functions of three independent variables as specified below. The expression for the spectral band radiance contrast is given as

$$L(A, B) = A(T, m, u) + B(T, m, u) \tag{1.27}$$

where, m = 1, 2, 3; T = temperature (K); $u(x) = m{\cdot}x(T)$; $x = 1.4388/\lambda T$; λ = wavelength (cm); $x_1 = 1.4388/\lambda_1 T$; $x_2 = 1.4388/\lambda_2 T$; and so on. For the spectral band $(\lambda_2 - \lambda_1)$, parameters $A(T, m, u)$ and $B(T, m, u)$ are defined by

$$A(T, m, u) = (2.78 \times 10^{-13}\, T^3) \sum_{m=1}^{3} (e^{-u(x_2)}\, m^{-4}) \times [u(x_2)^4 + 4u(x_2)^3 + 12u(x_2)^2 + 24u(x_2) + 24] \tag{1.28}$$

$$B(T, m, u) = (2.78 \times 10^{-13}\, T^3) \sum_{m=1}^{3} (e^{-u(x_1)}\, m^{-4}) \times [u(x_1)^4 + 4u(x_1)^3 + 12u(x_1)^2 + 24u(x_1) + 24] \tag{1.29}$$

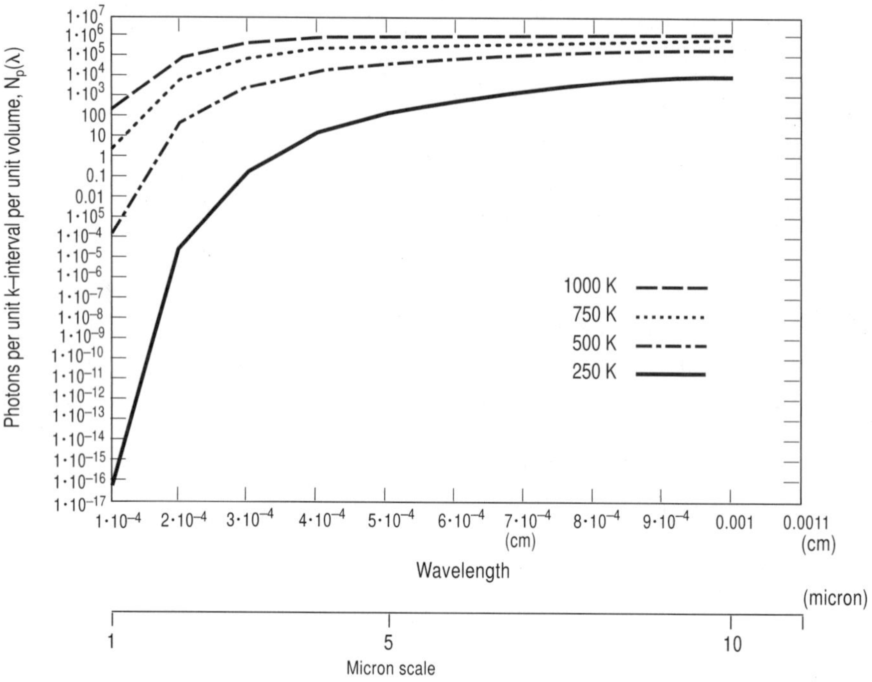

FIGURE 1.9 Photons per unit interval per unit volume as a function of temperature and wavelength.

TABLE 1.18 Percentage of total radiant excitance over a specified spectral bandwidth at various source temperatures

Temperature (K)/(λ_{max}), micron	Bandwidth ($\lambda_2 - \lambda_1$), micron	$R_{c(\%)}$
3000/(0.966)	1.00–0.88	8.13
	1.00–0.75	0.16
	1.00–0.50	0.26
2000/(1.45)	1.50–1.20	1.32
	1.50–1.25	11.24
	1.50–1.00	20.63
1000/(2.90)	3.00–2.25	16.36
	3.00–2.50	11.24
	3.00–2.00	20.63
500/(5.80)	6.00–4.00	20.63
	6.00–5.00	11.36
	6.00–5.50	5.64

Note: the product of $(\lambda_{max})(T)$ = 2898, micron·K, where λ_{max} represents the wavelength of maximum radiation in microns at temperature T (K).

1.12 SUMMARY

Infrared theory has been presented in the simplest format for better comprehension by students and readers with minimum exposure to IR technology. Quantities, functions, symbols and units commonly used for describing the performance of photometric, radiometric, and photonic devices and IR sources were provided. Blackbody functions along with their symbols and units commonly used to specify the capabilities of IR sensors were defined. Plots of computed values of radiance exitance, relative radiance exitance, relative photon density, spectral radiant exitance from blackbodies, photons per unit per unit area per steradian per micron, and spectral band radiance contrast as a function of temperature and emission wavelength were provided for clear understanding the performance capabilities and limitations of various IR sources and systems. Derivation of important functions commonly used in IR radiation theory are provided. A citation of numerical examples using Mathcad software is provided for readers with limited knowledge of IR theory and its applications in electrooptic and photonic devices. Computed values of radiance contrast parameters at various temperatures ranging from 200–1000 K over spectral bands of 8–10 μm, 8–11 μm, 8–12 μm, 8–13 μm, and 8–14 μm are shown in Figures 1.10 and 1.11. Smooth curves are made possible by taking higher values of variable m up to about 5 or 6, which will require higher computational times. With higher values of variable m, one can obtain smooth contrast curves as shown in Figure 1.11 [4].

Numerical Example to Compute the Function *L*(*A, B*)

Compute spectral band radiance contrast $(w/\text{cm}^2\cdot\text{sr}\cdot\text{K})$ 1000 over 8–12 μm band assuming the following parameters and using equations (1.27) through (1.29):

$T = 200, 400,$
$\lambda_1 = 0.0008$
$\lambda_2 = 0.0012$
$x_1(T) = 1.439/(\lambda_1 \cdot T)$
$x_2(T) = 1.439/(\lambda_2 \cdot T)$
$m = 1, 2, 3$
$u(x_1) = m \cdot x_1(T)$
$u(x_2) = m \cdot x_2(T)$
$L(A, B) = A(T, m, u) - B(T, m, u)$

$x_1(T)$	$x_2(T)$	$u(x_1)$	$u(x_2)$
8.994	5.996	8.994	5.996
4.497	2.998	4.497	2.998
2.998	1.999	2.998	1.999
2.248	1.499	2.248	1.499
1.799	1.199	1.799	1.199
		17.988	11.992
		8.994	5.996

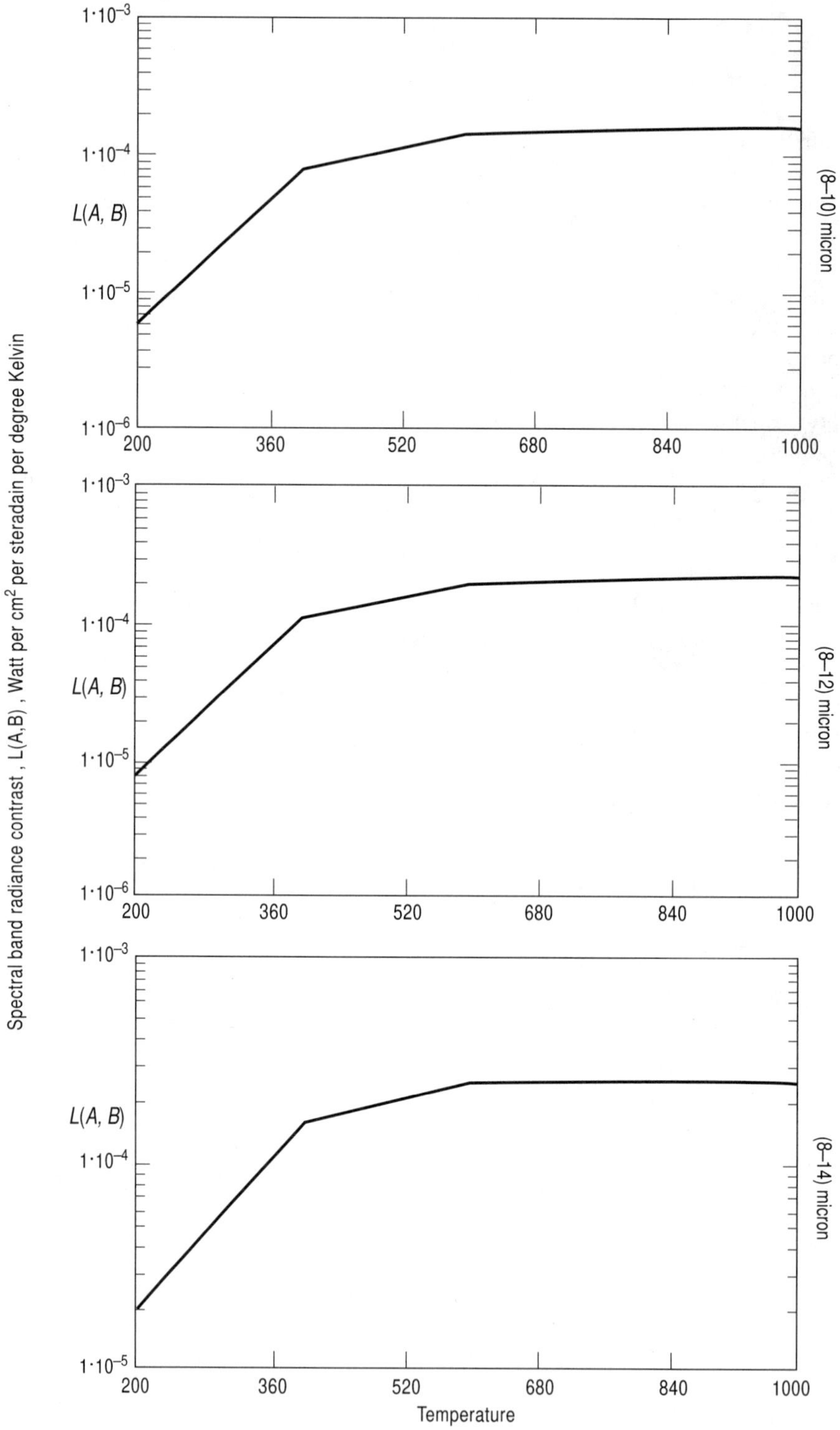

FIGURE 1.10 Radiance contrast as a function of wavelength band and temperature (K).

$$A(T, m, u) = (2.78 \times 10^{-13}\ T^3) \sum_{m=1}^{3} (e^{-u(x_2)}\ \mathrm{m}^{-4}) \times [u(x_2)^4 + 4u(x_2)^3 + 12u(x_2)^2 + 24u(x_2) + 24]$$

$$B(T, m, u) = (2.78 \times 10^{-13}\ T^3) \sum_{m=1}^{3} (e^{-u(x_1)}\ \mathrm{m}^{-4}) \times [u(x_1)^4 + 4u(x_1)^3 + 12u(x_1)^2 + 24u(x_1) + 24]$$

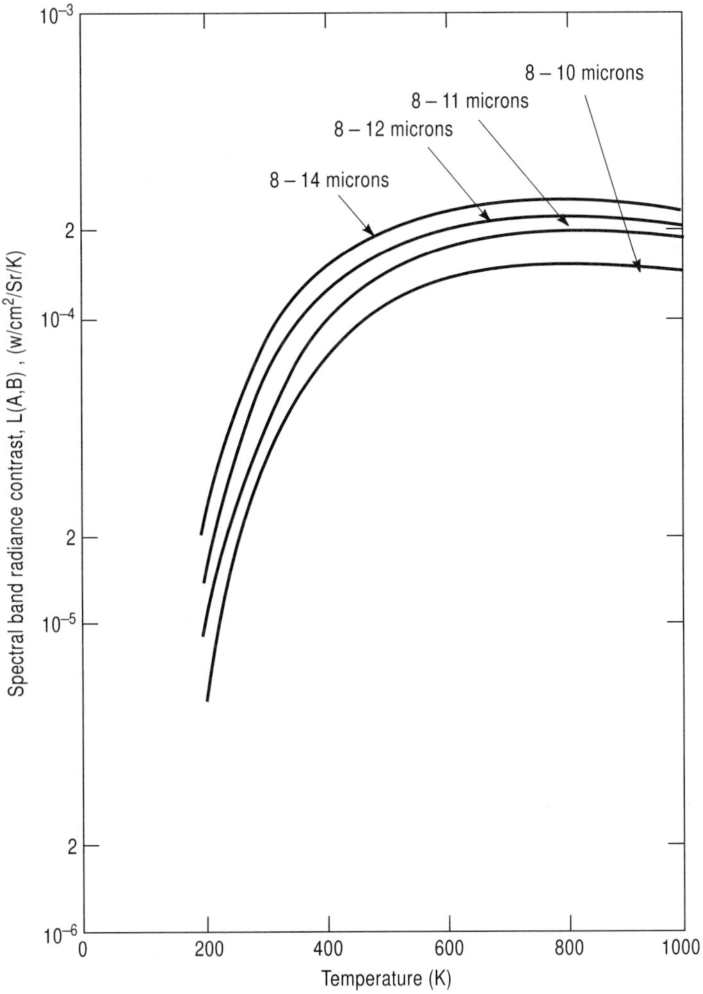

FIGURE 1.11 Spectral band radiance contrast at various temperatures [4].

$A(T, m, u)$	$B(T, m, u)$	$L(A, B)$	T	$L(A, B)$
1.639×10^{-5}	3.165×10^{-6}	1.322×10^{-5}	200	1.322×10^{-5}
4.386×10^{-7}	4.877×10^{-9}	4.337×10^{-7}	400	1.298×10^{-4}
4.877×10^{-9}	2.829×10^{-12}	4874×10^{-9}	600	2.403×10^{-4}
3.743×10^{-4}	2.445×10^{-4}	1.298×10^{-4}	800	2.178×10^{-4}
1.311×10^{-4}	2.532×10^{-5}	1.058×10^{-4}	1000	2.050×10^{-4}
2.532×10^{-5}	1.203×10^{-6}	2.412×10^{-5}		
1.468×10^{-3}	1.263×10^{-3}	2.043×10^{-4}		
9.749×10^{-4}	4.424×10^{-4}	5.325×10^{-4}		
4.424×10^{-4}	8.547×10^{-5}	3.57×10^{-4}		
3.604×10^{-3}	3.386×10^{-3}	2.178×10^{-4}		
2.995×10^{-3}	1.956×10^{-3}	2.178×10^{-4}		
1.956×10^{-3}	7.252×10^{-4}	1.231×10^{-3}		
7.116×10^{-3}	6.911×10^{-3}	2.05×10^{-4}		
6.485×10^{-3}	5.07×10^{-3}	1.416×10^{-3}		
5.07×10^{-3}	2.681×10^{-3}	2.388×10^{-3}		

Plot of radiant contrast function $L(A, B)$ as a function of temperature over a 8–12 micron band is shown in Figure 1.12.

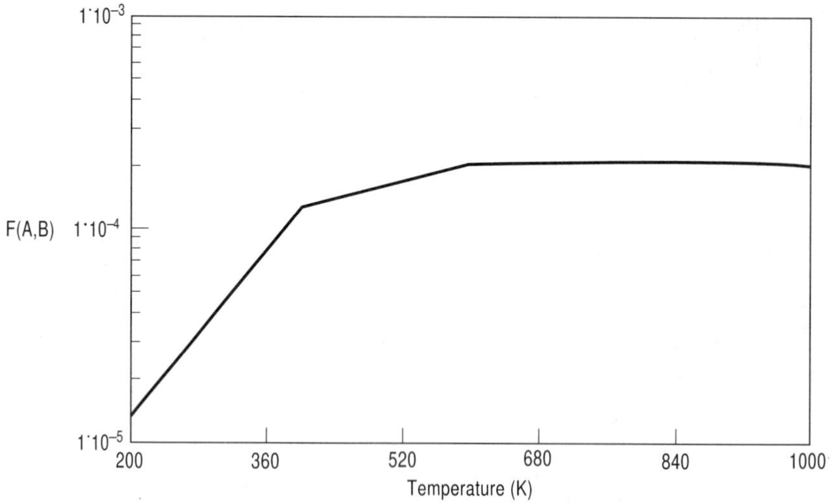

FIGURE 1.12 Radiance contrast as a function of temperature over a 8–12 μm band.

REFERENCES

1. W. L. Wolfe and G. J. Zissis, *The Infrared Handbook*, Environmental Research Institute of Michigan, Ann Arbor, Michigan, 1978.
2. *Electro-optics Handbook*, (EOH-11), RCA, Commercial Engineering Division, Harrison, NJ, 1974.
3. M. A. Bramson, *Infrared: A Handbook for Applications*, Plenum Press, New York, 1966.
4. R. C. Weast (Ed.), *Handbook of Physics and Chemistry*, 51st ed., CRC Press, Boca Raton, Florida, 1970.

CHAPTER TWO

Transmission Characteristics of IR Signals in the Atmosphere

2.1 INTRODUCTION

All electrooptical signals experience absorption, scattering, reflection, and diffusion while travelling through the atmosphere. However, high-power laser beams are seriously affected during the transmission in the various regions of the atmosphere. High-power lasers for space applications include gaseous lasers and chemical lasers, such as CO_2, DF, and HF lasers. Transmission of IR radiation in the atmosphere is very complex due to the dependence of scattering and absorption effects on a number of physical properties of the atmosphere. In a clear troposphere, IR radiation is attenuated due to absorption and scattering by atmosphere gases and aerosol particles. Atmospheric absorption is caused predominantly by the presence of molecule absorption bonds and is strictly a function of radiation wavelength. The amount of attenuation introduced by atmospheric absorption and scattering is an important factor in the design of IR sensors.

Atmospheric attenuation reduces system performance, and in the case of tracking and acquisition molecules could seriously degrade the system's performance level, leading to mission failure. Even in clear air with no suspended particles such as dust, smoke, fog, or rain, atmospheric transmission is a complex function that is highly dependent on IR laser wavelength, propagation path, altitude, seasonal conditions, and variations in atmospheric temperature and density. High-power IR laser beam propagation is subjected to more prominent effects, including absorption, scattering, diffraction, turbulence-induced beam spreading and wandering, thermal blooming, and gas breakdown. Simplified models are currently available to indicate their impact on the transmission characteristics as a function of atmospheric parameters, laser wavelength, and temporal mode of operation (i.e., CW or pulsed modes)

The most important effects, which must be considered for the wavelength and polar levels of intersect, are:

1. The linear absorption and scattering due to molecule and aerosol contacts of the atmosphere.
2. The random spreading, wandering, and distortion of the beam due to atmospheric turbulence. Note that these three effects are frequently observed in cases of high-power airborne and space lasers.
3. The nonlinear thermal blooming phenomenon that results from the absorption of small amounts of the laser beam's power.
4. The severe attenuation effects of the plasma resulting from gas breakdown are observed at high optical intensities. Note that these effects are dependent on the prevailing atmospheric conditions, laser wavelength, laser power level, and the temporal mode of operations (CW or pulsed).

2.2 ATMOSPHERIC MODELS USED FOR CALCULATION OF PARAMETERS

Severe models such as the Goody model (1964) and Elterman model (1968, 1970) are available to compute the absorption and scattering coefficient due to molecular and aerosol effects in the atmosphere [1]. These models take into account a number of different physical properties of the atmosphere over various geographical locations. The most general expression for the monochromatic transmittance of radiation along a path in the atmosphere can be written as follows:

$$\tau = \exp(-\gamma \Delta L) \tag{2.1}$$

where $\gamma = k + \sigma$

γ = attenuation coefficient per km

k = absorption coefficient per km

σ = scattering coefficient per km

$k = k_m + k_a$ and $\sigma = \sigma_m + \sigma_a$, where m stands for molecule and a for aerosol.

Note all these coefficients are defined in $(\text{km})^{-1}$. The scattering and absorption coefficients as defined above depend on both the molecules and aerosols present in the atmospheric path and on the geographic location, such as tropical, midsummer (MS), midwinter (MW), and subarctic summer (SS). The molecular scattering coefficient (σ_m) depends on the density of the molecule in the radiation path, whereas the molecular absorption coefficient (k_m) is dependent on atmospheric parameters such as pressure, temperature, and gaseous content as a function of altitude. Computed atmospheric optical properties, namely, pressure, temperature, density, water vapor, and ozone, are shown in Tables 2.1 and 2.2 for tropical and midaltitude summer (MS) climates, respectively. It is evident from the data shown in Tables 2.1 and

TABLE 2.1 Computation of atmospheric optical properties using the standard atmospheric model for various altitudes (tropical)

Height (km)	Pressure (mb)	Temperature (K)	Density (g/m^3)	Water vapor (g/m^3)	Ozone (g/m^3)
0	1.013E+03	300.0	1.167E+03	1.9E+01	5.6E–05
1	9.040E+02	294.0	1.064E+03	1.3E+01	5.6E–05
2	8.050E+02	288.0	9.689E+02	9.3E+00	5.4E–05
3	7.150E+02	284.0	8.756E+02	4.7E+00	5.1E–05
4	6.330E+02	277.0	7.951E+02	2.2E+00	4.7E–05
5	5.590E+02	270.0	7.199E+02	1.5E+00	4.5E–05
6	4.920E+02	264.0	6.501E+02	8.5E–01	4.3E–05
7	4.320E+02	257.0	5.855E+02	4.7E–01	4.1E–05
8	3.780E+02	250.0	5.258E+02	2.5E–01	3.9E–05
9	3.290E+02	244.0	4.708E+02	1.2E–01	3.9E–05
10	2.860E+02	237.0	4.202E+02	5.0E–02	3.9E–05
11	2.470E+02	230.0	3.740E+02	1.7E–02	4.1E–05
12	2.130E+02	224.0	3.316E+02	6.0E–03	4.3E–05
13	1.820E+02	217.0	2.929E+02	1.8E–03	4.5E–05
14	1.560E+02	210.0	2.578E+02	1.0E–03	4.5E–05
15	1.320E+02	204.0	2.260E+02	7.6E–04	4.7E–05
16	1.110E+02	197.0	1.972E+02	6.4E–04	4.7E–05
17	9.370E+01	195.0	1.676E+02	5.6E–04	6.9E–05
18	7.890E+01	199.0	1.382E+02	5.0E–04	9.0E–05
19	6.660E+01	203.0	1.145E+02	4.9E–04	1.4E–04
20	5.650E+01	207.0	9.515E+01	4.5E–04	1.9E–04
21	4.800E+01	211.0	7.938E+01	5.IE–04	2.4E–04
22	4.090E+01	215.0	6.645E+01	5.IE–04	2.8E–04
23	3.500E+01	217.0	5.618E+01	5.4E–04	3.2E–04
24	3.000E+01	219.0	4.763E+01	6.0E–04	3.4E–04
25	2.570E+01	221.0	4.045E+01	6.7E–04	3.4E–04
30	1.220E+01	232.0	1.831E+01	3.6E–04	2.4E–04
35	6.000E+00	243.0	8.600E+00	1.1E–04	9.2E–05
40	3.050E+00	254.0	4.181E+00	4.3E–05	4.1E–05
45	1.590E+00	265.0	2.097E+00	1.9E–05	1.3E–05
50	8.540E–01	270.0	1.101E+00	6.3E–06	4.3E–06
70	5.790E–02	219.0	9.2 10E–02	1.4E–07	8.6E–08
100	3.000E–04	210.0	5.000E–04	1.0E–09	4.3E–11

2.2 that both water vapor and ozone content are well below 10^{-9} g/m^{3}at an altitude of 100 km.

Atmospheric pressure is the most critical parameter and is strictly dependent on altitude and climatic conditions. Atmospheric pressure plots for tropical, midaltitude summer (MS), midaltitude winter (MW), and subarctic winter (SW) as a function of altitude are depicted in Figure 2.1. It is clear from Figure 2.1 that the atmospheric pressure values for MS, MW, and SW climatic conditions drop rapidly as the altitude exceeds 100 km. As far as temperature is concerned, its value drops from 300 K at sea level to 210 K at 100 km. However, data shown later indicate that the temperature suddenly increases to 1404 K at 200 km, to 1576 K at 500 km, and to 1812 K at 700 km. These data suggest that IR devices and sensors used by the low

TABLE 2.2 Computation of the atmospheric optical properties using the standard model (midlatitude summer)

Height (km)	Pressure (mb)	Temperature (K)	Density (g/m³)	Water vapor (g/m³)	Ozone (g/m³)
0	1.013E+03	294.0	1.191E+03	1.4E+01	6.0E–05
1	9.020E+02	290.0	1.080E+03	9.3E+00	6.0E–05
2	8.020E+02	285.0	9.757E+02	5.9E+00	6.0E–05
3	7.100E+02	279.0	8.846E+02	3.3E+00	6.2E–05
4	6.280E+02	273.0	7.998E+02	1.9E+00	6.4E–05
5	5.540E+02	267.0	7.211E+02	1.0E+00	6.6E–05
6	4.870E+02	261.0	6.487E+02	6.1E–01	6.9E–05
7	4.260E+02	255.0	5.830E+02	3.7E–01	7.5E–05
8	3.720E+02	248.0	5.225E+02	2.1E–01	7.9E–05
9	3.240E+02	242.0	4.669E+02	1.2E–01	8.6E–05
10	2.810E+02	235.0	4.159E+02	6.4E–02	9.0E–05
11	2.430E+02	229.0	3.693E+02	2.2E–02	1.1E–04
12	2.090E+02	222.0	3.269E+02	6.0E–03	1.2E–04
13	1.790E+02	216.0	2.882E+02	1.8E–03	1.5E–04
14	1.530E+02	216.0	2.464E+02	1.0E–03	1.8E–04
15	1.300E+02	216.0	2.104E+02	7.6E–04	1.9E–04
16	1.110E+02	216.0	1.797E+02	6.4E–04	2.1E–04
17	9.500E+01	216.0	1.535E+02	5.6E–04	2.4E–04
18	8.120E+01	216.0	1.305E+02	5.0E–04	2.8E–04
19	6.950E+01	217.0	1.110E+02	4.9E–04	3.2E–04
20	5.950E+01	218.0	9.453E+01	4.5E–04	3.4E–04
21	5.100E+01	219.0	8.056E+01	5.1E–04	3.6E–04
22	4.370E+01	220.0	6.872E+01	5.1E–04	3.6E–04
23	3.760E+01	222.0	5.867E+01	5.4E–04	3.4E–04
24	3.220E+01	223.0	5.014E+01	6.0E–04	3.2E–04
25	2.770E+01	224.0	4.288E+01	6.7E–04	3.0E–04
30	1.320E+01	234.0	1.322E+01	3.6E–04	2.0E–04
35	6.520E+00	245.0	6.519E+00	1.1E–04	9.2E–05
40	3.330E+00	258.0	3.330E+00	4.3E–05	4.1E–05
45	1.760E+00	270.0	1.757E+00	1.9E–05	1.3E–05
50	9.510E–01	276.0	9.512E–01	6.3E–06	4.3E–06
70	6.710E–02	218.0	6.706E–02	1.4E–07	8.6E–08
100	3.000E–04	210.0	5.000E–04	1.0E–09	4.3E–11

earth orbit (LEO) and medium earth orbit (MEO) satellites would require special cooling schemes to ensure reliable and efficient performance levels under these thermal conditions. These sensors will not be adversely affected by water vapor and ozone because their lowest levels occur at altitudes of 100 km and above. Optical properties of the atmosphere at higher altitudes are summarized in Table 2.3. It is evident from Table 2.3 that the pressure magnitude drops to 1.6×10^{-4} mb at an altitude of 100 km to 1.53×10^{-9} mb at 700 km. One can notice a radical increase in atmospheric temperature from 200 K at an altitude of 100 km to 1812 K at 700 km [1]. In general, pressure, density, and molecular weight all decrease at moderate rates with increasing operating altitude.

Both the pressure and density decrease rapidly as the altitude increases. Howev-

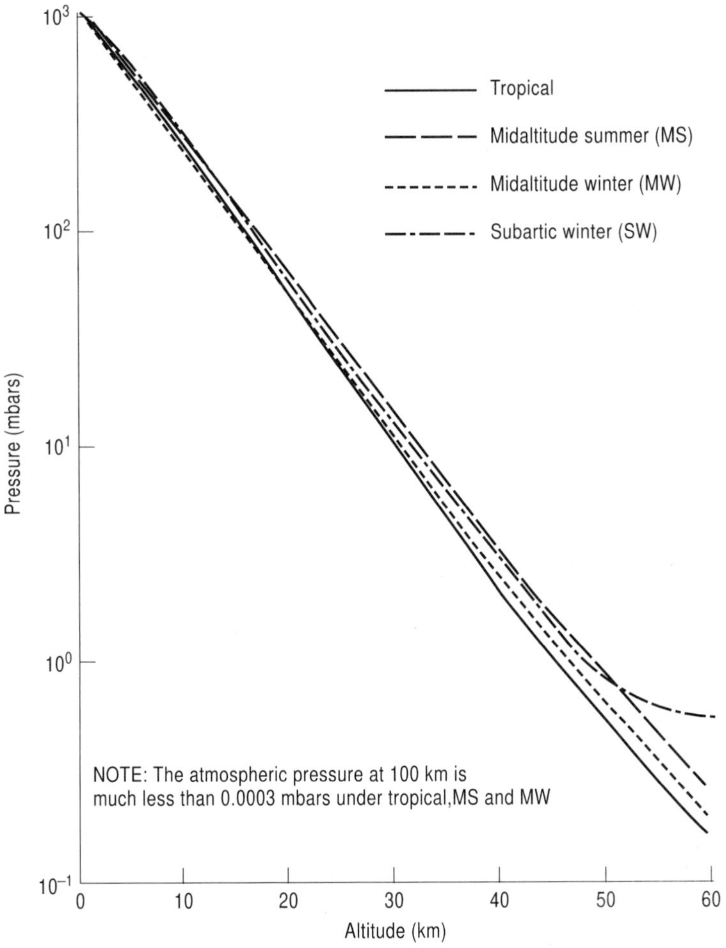

FIGURE 2.1 Atmospheric pressure as a function of altitude under various climatic conditions.

TABLE 2.3 Optical properties of the atmosphere based on AR DC model at higher altitudes [1].

Altitude (km)	Temperature (K)	Pressure (mb)	Density (g/m^3)	Molecule (M)
100	200	1.6×10^{-4}	3.73×10^{-4}	28.90
200	1404	1.22×10^{-6}	3.67×10^{-7}	26.32
300	1423	1.92×10^{-7}	4.75×10^{-8}	21.95
400	1480	4.25×10^{-8}	9.00×10^{-9}	19.56
500	1576	1.19×10^{-8}	2.20×10^{-9}	18.28
600	1691	3.98×10^{-9}	6.62×10^{-13}	17.52
700	1812	1.53×10^{-9}	2.30×10^{-13}	17.03

TABLE 2.4 Attenuation coefficients (absorption and scattering) for molecules and aerosols at 1.06 μm based on atmospheric model as a function of operating altitude ($\lambda = 1.06$ μm)*

	Tropical		Midlatitude summer		Midlatitude winter		Subarctic summer		Subarctic winter		Aerosol			
											Clear		Hazy	
Height	k_m(km^{-1})	σ_m(km^{-1})	k_m(km^{-1})	σ_m(km^{-1})	k_m(km^{-1})	σ_m(km^{-1})	k_m(km^{-1})	σ_m(km^{-1})	k_m(km^{-1})	σ_m(km^{-1})	k_a(km^{-1})	σ_a(km^{-1})	k_a(km^{-1})	σ_a(km^{-1})
0	<E–06	8.04E–04	<E–06	8.20E–04	<E–06	8.91E–04	<E–06	8.38E–04	<E–06	939E–04	1.98E–02	6.79E–02	9.63E–02	3.31E–01
0–1		7.6;E–04		7.81E–04		8.43E–04		7.98E–04		8.77E–04	1.31E–02	4.50E–02	5.82E–02	2.00E–01
1–2		6.99E–04		7.06E–04		7.52E–04		7.21E–04		7.70E–04	5.71E–03	1.96E–02	2.13E–02	7.31E–02
2–3		6.33E–04		6.38E–04		6.70E–04		6.50E–04		6.82E–04	2.43E–03	8.36E–03	7.78E–03	2.67E–02
3–4		5.72E–04		577E–04		5.99E–04		5.84E–04		6.06E–04	1.15E–03	3.94E–03	2.84E–03	9.76E–03
4–5		5.19E–04		5.21E–04		5.37E–04		5.24E–04		5.40E–04	7.23E–04	2.49E–03	1.04E–03	3.56E–03
5–6		4.69E–04		4.69E–04		4.80E–04		4.71E–04		4.82E–04	5.27E–04	1.81E–03	5.27E–04	1.81E–03
6–7		4.22E–04		4.21E–04		4.27E–04		4.23E–04		4.29E–04	4.27E–04	1.47E–03	4.27E–04	1.47E–03
7–8		3.80E–04		3.78E–04		3.80E–04		3.79E–04		3.81E–04	4.18E–04	1.44E–03	4.18E–04	1.44E–03
8–9		3.41E–04		3.38E–04		3.36E–04		3.38E–04		3.34E–04	4.15E–04	1.43E–03	4.15E–04	1.43E–03
9–10		3.04E–04		3.02E–04		2.97E–04		3.01E–04		2.88E–04	4.01E–04	1.38E–03	4.01E–04	1.38E–03
10–11		2.72E–04		2.69E–04		2.59E–04		2.64E–04		2.46E–04	3.84E–04	1.32E–03	3.84E–04	1.32E–03
11–12		2.41E–04		2.39E–04		2.22E–04		2.26E–04		2.10E–04	3.81E–04	1.31E–03	3.81E–04	1.31E–03
12–13		2.13E–04		2.11E–04		1.90E–04		1.95E–04		1.80E–04	3,75E–04	1.29E–03	3.75E–04	1.29E–03
13–14		1.88E–04		1.83E–04		1.63E–04		1.67E–04		1.54E–04	3.56E–04	1.22E–03	3.56E–04	1.22E–03
14–15		1.66E–04		1.56E–04		1.40E–04		1.44E–04		1.31E–05	3.42E–04	1.18E–03	3.42E–04	1.18E–03
15–16		1.44E–04		1.33E–04		1.20E–04		1.23E–04		1.12E–05	3.23E–04	1.11E–03	3.23E–04	1.11E–03
16–17		1.24E–04		1.14E–04		1.03E–04		1.06E–04		9.63E–05	3.13E–04	1.08E–03	3.13E–04	1.08E–03
17–18		1.05E–04		9.72E–05		8.80E–05		9.14E–05		8.25E–05	3.06E–04	1.05E–03	3.06E–04	1.05E–03
18–19		8.63E–05		8.29E–05		7.53E–05		7.86E–05		7.06E–05	2.77E–04	9.51E–04	2.77E–04	9.51E–04
19–20		7.16E–05		7.07E–05		6.45E–05		6.75E–05		6.05E–05	2.18E–04	7.48E–04	2.18E–04	7.48E–04
20–21		5.96E–05		6.02E–05		5.51E–05		5.80E–05		5.17E–05	1.59E–04	5.45E–04	1.59E–04	5.45E–04
21–22		4.98E–05		5.14E–05		4.70E–05		4,99E–05		4.42E–05	1.17E–04	4.03E–04	1.17E–04	4.03E–04
22–23		4.19E–05		4.38E–05		4.01E–05		4.29E–05		3.78E–05	8.89E–G4	3.06E–04	8.89E–04	3.06E–04
23–24		3.55E–05		3.74E–05		3.43E–05		3.69E–05		3.23E–05	6.93E–05	2.38E–04	6.93E–05	2.38E–04
24–25		3.02E–05		3.19E–05		2.93E–05		3.15E–05		2.76E–05	5.66E–05	1.94E–04	5.66E–05	1.94E–04
25–30		2.01E–05		2.15E–05		1.95E–05		2.13E–05		1.84E–05	2.85E–05	9.79E–05	2.85E–05	9.79E–05
30–35		9.20E–06		9.89E–06		8.79E–06		9.98E–06		8.14E–06	8.02E–06	2.76E–05	8.02E–06	2.76E–05
35–40		4.37E–06		4.71E–06		3.95E–06		4.73E–06		3.65E–05	2.11E–06	7.26E–06	2.11E–06	7.26E–06
40–45		2.15E–05		2.31E–05		1.83E–06		2.33E–05		1.68E–06	<E–06	1.91E–06	<E–06	1.91E–06
45–50		1.09E–06		1.19E–06		<E–06		1.21E–06		<E–06		<E–06		<E–06
50–70		<E–06		<E–06				<E–06						
70–100														

*Symbols: k = absorption coefficient, σ = scattering coefficient, m = molecular component, a = aerosol component.

TABLE 2.5 Attenuation coefficients (absorption and scattering)for molecules and aerosols at 3.3925 μm based on atmospheric model as a function of operating altitude (3.39225 μm)*

	Tropical		Midlatitude summer		Midlatitude winter		Subarctic summer		Subarctic winter		Aerosol: Clear		Aerosol: Hazy	
Height	k_m(km^{-1})	σ_m(km^{-1})	k_m(km^{-1})	σ_m(km^{-1})	k_m(km^{-1})	σ_m(km^{-1})	k_m(km^{-1})	σ_m(km^{-1})	k_m(km^{-1})	σ_m(km^{-1})	k_a(km^{-1})	σ_a(km^{-1})	k_a(km^{-1})	σ_a(km^{-1})
0	1.77E+00	7.79E-06	1.83E+00	7.95E-06	1.98E.00	8.53E-06	1.85E+00	8.12E-06	2.03E+00	8.89E-06	1.25E-02	1.65E-02	7.71E-02	8.05E-02
0-1	1.75E+00	7.44E-06	179E+00	7.56E06	1.92E+00	8.17E-06	1.84E+00	7.72E-06	1.97E+00	8.50E-06	1.05E-02	1.09E-02	4.66E-02	4.86E-02
1-2	1.68E+00	6.77E-06	1.70E+00	6.83E-06	1.79E+06	7.29E-06	1.75E+00	6.98E-05	1.84E+00	7.46E-06	4.57E-03	4.77E-03	1.70E-02	1.78E-02
2-3	1.63E+00	6.13E-06	1.65E+00	6.18E-06	1.75E+00	6.49E-06	1.65E+00	6.29E-06	1.75E+00	6.69E-06	1.95E-03	2.03E-03	6.24E-03	6.58E-03
3-4	1.69E+00	5.54E-06	1.60E+00	5.59E-06	1.55E+00	5.80E-06	1.66E+00	5.65E-06	1.67E+00	5.86E·35	9.18E-04	9.57E-04	2.27E-03	2.37E-03
4-5	1.54E+00	5.02E-06	1.54E+00	5.05E-06	1.50E+00	5.20E-07	1.54E+00	5.07E-05	1.62E+00	5.23E-05	5.79E-04	6.04E-04	8.31E-04	8.56E-04
5-6	1.52E+00	4.54E-06	1.54E+00	4.56E-06	1.54E+00	4.64E-06	1.61E+00	4.56E-06	1.57E+00	4.67E-06	4.22E-04	4.40E-04	4.22E-04	4.40E-04
6-7	1.47E+00	4.09E-06	1.46E+00	4.08E-06	1.54E.00	4.14E-06	1.47E+00	4.09E-06	1.54E+00	4.16E-05	3.42E-04	3.57E-04	3.42E-04	3.57E-04
7-8	1.45E+00	3.69E-06	1.52E+00	3.56E-06	1.46E+00	3.68E-06	1.45E+00	3.67E-06	1.50E+00	3.68E-05	3.35E-04	3.49E-04	3.35E-04	3.49E-04
8-9	1.46E+00	3.30E-06	1.41E+00	3.27E-36	1.44E+00	3.26E-06	1.51E+00	3.28E-06	1.47E+00	3.23E-05	3.33E-04	3.47E-04	3.33E-04	3.47E-04
9-10	1.41E+00	2.95E-06	1.43E+00	2.92E-06	1.49E+00	2.88E-06	1.40E+00	2.92E-06	1.44E+00	2.79E-06	3.22E-04	3.35E-04	322E-04	3.35E-04
10-11	1.42E+00	2.63E-06	1.40E+00	2.60E-06	1.41E-00	2.51E-06	1.42E+00	2.55E-06	1.41E+00	2.38E-04	3.07E-04	3.21E-04	3.07E-04	3.21E-04
11-12	1.37E+00	2.34E-06	1.39E+00	2.31E-06	1.39E+00	2.15E-06	1.35E+00	2.19E-06	1.38E+00	2.04E-06	2.05E-04	3.18E-04	3.05E-04	3.18E-04
12-13	140E+00	2.06E-06	1.37E+00	2.04E-0F	1.36E-00	1.81E-06	1.38E+00	1.89E-05	1.34E+t00	1.74E-05	3.00E-04	3.13E-04	3.00E-04	3.13E-04
13-14	1.31E+00	1.82E-05	1.34E+00	1.77E-06	1.30E+00	1.58E-06	1.31E+00	1.62E-06	1.31E+00	1.49E-06	2.86E-04	2.98E-04	2.86E-04	2.98E-04
14-15	1.35E+00	1.60E-06	1.36E+00	1.51E-06	1.29E+00	1.36E-06	1.32E+00	1.39E-06	1.27E+00	1.27E-06	2.74E-04	2.86E-04	2.74E-04	2.86E-04
15-16	1.32E+00	1.40E-06	1.28E+00	1.29E-06	1.24E+00	1.16E-06	1.21E+00	1.19E-06	1.21E+00	1.09E-06	259E-04	2.70E-04	2.59E-04	2.70E-04
16-17	121E+00	1.20E-06	1.20E+00	1.10E-06	1.19E+00	<E-06	1.19E+00	1.03E-06	1.16E-00	<E-06	2.51E-04	2.62E-04	2.51E-04	2.62E-04
17-18	1.17E+00	1.01E-06	1.14E+00	<E-05	1.13E+00		1.13E+00	<E-06	1.09E+00		2.45E-04	2.56E-04	2.45E-04	2.56E-04
18-19	1.10E+00	<E-06	1.09E+00		1.03E+00		1.07E+00		1.01E+00		2.27E-04	2.31E-04	2.22E-04	2.31E-04
19-20	9.95E-01		1.00E-00		9.47E-01		9.97E-01		9.10E-01		1.74E-04	1.82E-04	1.74E-04	1.82E-04
20-21	8.96E-01		9.09E-01		8.64E-01		8.92E-01		8.07E-01		1.27E-04	1.33E-04	1.27E-04	1.33E-04
21-22	7.82E-01		8.14E-01		7.19E-01		7.99E-01		6.99E-01		9.39E-05	9.79E-05	9.39E-05	9.79E-05
22-23	6.55E-01		6.62E-01		6.52E.01		6.99E-01		5.86E-01		7.10E-05	7.43E-05	7.12E-05	7.43E-05
23-24	6.48E-01		6.23E-01		6.06E-01		5.92E-01		4.84E-01		5.55E-05	5.79E-05	5.55E-05	5.79E-05
24-25	4.51E-01		4.76E-01		4.41E-01		4.68E-01		3.89E-01		4.53E-05	4.73E-05	4.53E-05	4.73E-05
25-30	2.33E-01		2.63E-01		2.18E-01		2.61E-01		1.93E-01		2.28E-05	2.38E-05	2.28E-05	2.38E-05
30-35	1.93E-01		2.07E-01		1.84E-01		7.02E-02		1.71E-01		5.43E-05	5.70E-06	6.43E-06	6.70E-06
35-40	9.62E-02		1.03E-01		8.70E-02		1.04E-01		8.04E-02		1.69E-04	1.76E-06	1.69E-06	1.75E-06
40-45	4.86E-02		5.30E-02		4.14E-02		5.22E-02		3.79E-02		<E-05	<E-05	<E-06	<E-06
45-50	2.46E-02		2.67E-02		2.05E-02		2.73E-02		1.83E-02					
50-70	6.75E-03		7.50E-03		4.63E-03		7.77E-03		4.51E-03					
70-100	3.21E-04		3.72E-04		2.68E-04		3.91E-04		2.25E-04					

TABLE 2.6 Absorption and scattering coefficients for a CO_2 laser (10.59 μm) under various geographical and weather conditions*

											Aerosol			
	Tropical		Midlatitude summer		Midlatitude winter		Subarctic summer		Subarctic winter		Clear		Hazy	
Height	k_m(km^{-1})	σ_m(km^{-1})	k_m(km^{-1})	σ_m(km^{-1})	k_m(km^{-1})	σ_m(km^{-1})	k_m(km^{-1})	σ_m(km^{-1})	k_m(km^{-1})	σ_m(km^{-1})	k_a(km^{-1})	σ_a(km^{-1})	k_a(km^{-1})	σ_a(km^{-1})
0	6.094E–01	0.00	3.8525–01	0.00	9.575E–02	0.00	2.238E–01	0.00	5.214E–02	0.00	5.406E–03	4.587E–03	2.634E–02	2.235E–02
0–1	4.586E–01	0.00	2.977E–01	0.00	8.576E–02	0.00	1.802E–01	0.00	5.315E–02	0.00	3.690E–03	3.130E–03	1.661E–02	1.409E–02
1–2	2.766E–0l	0.00	1.841E–01	0.00	7.137E–02	0.00	1.247E–01	0.00	5.083E–02	0.00	1.610E–03	1.366E–03	5.348E–03	4.533E–03
2–3	1.640E–01	0.00	1.218E–01	0.00	6.096E–02	0.00	9.268E–02	0.00	4.492E–02	0.00	6.870E–04	5.828E–03	1.856E–03	1.575E–03
3–4	1.045E–01	0.00	8.901E–02	0.00	5.093E–02	0.00	7.322E–02	0.00	3.917E–02	0.00	3.191E–04	2.707E–04.	8.107E–04	6.879E–04
4–5	7.809E–02	0.00	6.849E–02	0.00	4.179E–02	0.00	5.808E–02	0.00	3.178E–02	0.00	1.987E–04	1.685E–04	2.951E–04	2.512E–04
5–6	6.346E–02	0.00	5.745E–02	0.00	3.416E–02	0.00	4.774E–02	0.00	2.527E–02	0.00	1.451E–04	1.231E–04	1.451E–04	1.231E–04
6–7	5.143E–02	0.00	4.879E–02	0.00	2.810E–02	0.00	3.711E–02	0.00	1.984E–02	0.00	1.169E–04	9.918E–05	1.169E–04	9.918E–05
7–8	4.174E–02	0.00	3.948E–02	0.00	2.273E–02	0.00	2.97E–02	0.00	1.540E–02	0.00	1.344E–04	9.704E–05	1.144E–04	9.704E–05
8–9	3.454E–02	0.00	3.123E–02	0.00	1.853E–02	0.00	2.348E–02	0.00	1.266E–02	0.00	1.137E–04	9.645E–05	1.137E–04	9.645E–05
9–10	2.729E–02	0.00	2.568E–02	0.00	1.428E–02	0.00	1.812E–02	0.00	1.179E–02	0.00	1.019E–04	9.326E–05	1.099E–04	9.326E–05
10–11	2.177E–02	0.00	2.073E–02	0.00	1.295E–02	0.00	1.577E–02	0.00	1.178E–02	0.00	1.051E–04	8.917E–05	1.051E–04	8.917E–05
11–12	1.698E–02	0.00	1.637E–02	0.00	1.252E–02	0.00	1.623E–02	0.00	1.176E–02	0.00	1.043E–04	8.845E–05	1.043E–04	8.845E–05
12–13	1.366E–02	0.00	1.259E–02	0.00	1.235E–02	0.00	1.559E–02	0.00	1.153E–02	0.00	1.027E–04	8.712E–05	1.027E–04	8.712E–05
13–14	9.747E–03	0.00	1.101E–02	0.00	1.233E–02	0.00	1.623E–02	0.00	1.203E–02	0.00	9.760E–05	8.280E–05	9.750E–05	8.280E–05
14–15	7.725E–03	0.00	1.149E–02	0.00	1.189E–02	0.00	1.613E–02	0.00	1.174E–02	0.00	9.363E–05	7.944E–05	9.368E–05	7.944E–05
15–16	5.717E–03	0.00	1.121E–02	0.00	1.155E–02	0.00	1.540E–02	0.00	1.158E–02	0.00	8.850E–05	7.509E–05	8.850E–05	7.509E–05
16–17	4.379E–03	0.00	1.104E–02	0.00	1.132E–02	0.00	1.606E–02	0.00	1.130E–02	0.00	8.579E–05	7.278E–05	8.579E–05	7.278E–05
17–18	4.695E–03	0.00	1.118E–02	0.00	1.129E–02	0.00	1.589E–02	0.00	1.099E–02	0.00	8.390E–05	7.118E–05	9.390E–05	7.118E–05
18–19	5.743E–03	0.00	1.130E–02	0.00	1.089E–02	0.00	1.583E–02	0.00	1.083E–02	0.00	7.584E–05	6.434E–05	7.534E–05	6.434E–05
19–20	6.857E–03	0.00	1.178E–02	0.00	1.057E–02	0.00	1.605E–02	0.00	1.040E–02	0.00	5.979E–05	5.073E–05	5.979E–05	5.073E–05
20–21	8.279E–03	0.00	1.212E–02	0.00	1.080E–02	0.00	1.565E–02	0.00	1.027E–02	0.00	4.362E–05	3.701E–05	4.362E–05	3.710E–05
21–22	9.857E–03	0.00	1.282E–02	0.00	1.081E–02	0.00	1.594E–02	0.00	9.925E–03	0.00	3.219E–05	2.731E–05	3.219E–05	2.731E–05
22–23	1.102E–02	0.00	1.333E–02	0.00	1.077E–02	0.00	1.593E–02	0.00	9.511E–03	0.00	2.442E–05	2.072E–05	2.442E–05	2.072E–05
23–24	1.193E–02	0.00	1.469E–02	0.00	1.069E–02	0.00	1.581E–02	0.00	9.668E–03	0.00	1.900E–05	1.612E–05	1.900E–05	1.612E–05
24–25	1.307E–02	0.00	1.466E–02	0.00	1.105E–02	0.00	1.682E–02	0.00	9.019E–03	0.00	1.551E–05	1.316E–05	1.5516–05	1.316E–05
25–30	1.587E–02	0.00	1.750E–02	0.00	1.067E–02	0.00	1.916E–02	0.00	9.551E–03	0.00	8.2776–05	7.022E–05	8.2776–05	7.022E–06
30–35	1.366E–02	0.00	1.523E–02	0.00	7.821E–03	0.00	2.648E–02	0.00	6.468E–03	0.00	2.363E–06	2.005E–05	2.368E–06	2.005E–06
35–40	1.192E–02	0.00	1.381E–02	0.00	7.221E–03	0.00	1.518E–02	0.00	5.460E–03	0.00	0.	0.	0.	0.
40–45	9.253E–03	0.00	1.113E–02	0.00	6.251E–03	0.00	1.245E–02	0.00	4.243E–03	0.00	0.	0.	0.	0.
45–50	6.178E–03	0.00	7.711E–03	0.00	4.490E–03	0.00	8.396E–03	0.00	3.135E–03	0.00	0.	0.	0.	0.
50–70	9.095E–04	0.00	1.088E–03	0.00	2.765E–04	0.00	1.109E–03	0.00	7.810E–04	0.00	0.	0.	0.	0.
70–100	1.535E–05	0.00	1.743E–05	0.00	1.580E–04	0.00	1.762E–05	0.00	1.785E–05	0.00	0.	0.	0.	0.

*Symbols: k = absorption coefficient, σ = scattering coefficient, m = molecular component, a = aerosol.

TABLE 2.7 Attenuation coefficients (both absorption and scattering) for molecules and aerosols at 337 μm based on atmospheric model as a function of operating altitude (λ = 3.39225 μm)*

											Aerosol			
	Tropical		Midlatitude summer		Midlatitude winter		Subarctic summer		Subarctic winter		Clear		Hazy	
Height	k_m(km^{-1})	σ_m(km^{-1})	k_m(km^{-1})	σ_m(km^{-1})	k_m(km^{-1})	σ_m(km^{-1})	k_m(km^{-1})	σ_m(km^{-1})	k_m(km^{-1})	σ_m(km^{-1})	k_a(km^{-1})	σ_a(km^{-1})	k_a(km^{-1})	σ_a(km^{-1})
0	2.67E+01	<E–06	2.03E+01	<E–06	5.39E+00	<E–06	1.37E+01	<E–06	1.71E+00	<E–06	<E–06	<E–06	<E–06	<E–06
0–1	2.12E+01		1.60E+03		4.35E+00		1.07E+03		1.60E–00					
1–2	1.33E+01		9.66E+00		2.84E+00		6.39E+00		1.39E+00					
2–3	8.72E+00		5.58E+00		1.82E+00		4.09E+00		9.79E–01					
3–4	4.00E+00		2.87E+00		1.10E+00		2.39E+00		6.33E–03					
4–5	1.75E+00		1.50E+00		5.46E–01		1.36E+00		3.45E–01					
5–6	1.06E+00		7.17E–02		2.83E–01		7.24E–01		1.52E–02					
6–7	5.52E–01		3.97E–01		1.42E–01		3.56E–01		6.68E–02					
7–8	2.78E–03		2.17E–02		5.13E–02		1.73E–01		3.29E–02					
8–9	1.34E–02		1.12E–01		1.87E–02		6.98E–02		5.96E–03					
9–10	5.78E–02		5.72E–02		1.73E–03		2.02E–02		3.91E–03					
10–11	2.16E–02		2.75E–01		3.10E–03		6.39E–03		2.20E–03					
11–12	6.64E–03		8.43E–03		2.49E–03		3.43E–03		1.30E–03					
12–13	2.07E–03		2.05E–03		1.85E–03		1.88E–03		7.57E–04					
13–14	5.59E–04		5.41E–04		4.81E–04		4.88E–04		4.53E–04					
14–15	2.66E–04		2.47E–04		2.21E–04		2.24E–04		2.08E–04					
15–16	1.79E–04		1.61E–04		1.46E–04		1.48E–04		1.36E–04					
16–17	1.35E–04		1.20E–04		1.09E–04		1.11E–04		1.02E–04					
17–18	9.63E–05		8.73E–05		7.91E–05		8.58E–05		7.41E–05					
18–19	6.98E–05		7.00E–05		6.38E–05		6.85E–05		5.98E–05					
19–20	6.14E–05		5.59E–05		5.12E–05		5.28E–05		4.80E–05					
20–21	4.13E–05		4.44E–05		4.08E–05		4.24E–05		3.84E–05					
21–22	4.22E–05		4.33E–05		3.98E–05		4.17E–05		3.52E–05					
22–23	3.54E–05		3.45E–05		3.19E–05		3.59E–05		3.21E–05					
23–24	3.18E–05		3.33E–05		3.09E–05		3.27E–05		2.92E–05					
24–25	2.85E–05		3.17E–05		2.94E–05		2.95E–05		2.64E–05					
25–30	1.64E–05		1.74E–05		3.58E–05		1.74E–05		1.50E–05					
30–35	4.02E–06		4.32E–06		3.83E–06		4.36E–06		3.55E–06					
35–40	<E–06		<E–06		<E–06		<E–06		<E–06					
40–45														
45–50														
50–70														
70–100														

*Symbols: k = absorption coefficient, σ = scattering coefficient, m = molecular component, a = aerosol component.

er, the temperature increases at a moderate rate as the altitude increases from 200 to 700 km. Familiarization with these atmospheric properties at various altitudes is necessary to understand the impact of these parameters on the performance levels of IR sensors. Various atmospheric models were examined and evaluated to obtain accurate and reliable values of absorption and scattering coefficients for molecular and aerosol distributions as a function of wavelength and altitude. Absorption and scattering coefficients at 1.06 μm, 3.392 μm, and 10.59 μm (CO_2 laser wavelength) wavelengths as a function of altitude, climatic conditions, and environments are summarized in Tables 2.4, 2.5, and 2.6, respectively. Examination of these coefficients indicates that the lowest values of aerosol coefficients are possible under clear atmospheric conditions. Molecular and aerosol absorption and scattering coefficients for long-wave IR (LWIR) operation at 335 μm as a function of altitude and climatic conditions are shown in Table 2.7. These tabulated data indicate that the IR sensors operating at 100 km and above will experience relatively much lower attenuation due to scattering and absorption.

2.2.1 Combined Effects of Scattering and Absorption on Atmospheric Attenuation

Combined effects of scattering and absorption due to aerosol and molecular distributions will be briefly discussed as a function of altitude and wavelength for tropical climatic conditions. The tropical climate has been selected because IR sensors operating in up to 10 km or lower altitudes experience serious performance degradation due to higher values of scattering and absorption coefficients at lower altitudes, regardless of the operating wavelength. Close examination of the coefficients shown in Tables 2.4–2.7 reveals that the IR sensors operating at and under 10 km do suffer from higher degradation under tropical environments irrespective of IR spectral regions.

2.2.2 Scattering and Absorption Coefficients Due to Aerosol Distributions

Combined effects of scattering and absorption due to aerosols as a function of altitude and radiation path altitude are shown in Figure 2.2 for the tropical climate. It is evident from Figure 2.2 that atmospheric attenuation undergoes radical change over 1–6 km of radiation path altitude, regardless of operating wavelength. It is further evident that the overall attenuation due to aerosols remains fairly constant over the 8–12 km range at wavelengths of 1.06, 3.85, 5.05, and 10.58 μm. Note that IR sensors operating at 1.06 μm will experience maximum attenuation, whereas sensors operating at 10.59 μm will experience minimum attenuation, regardless of radiation path range. It is important to mention that high-power airborne or space-based lasers must operate either at 3.85 μm (DF laser) or 10.59 μm (CO_2 laser) to avoid higher attenuation at other wavelengths.

2.2.3 Scattering and Absorption Effects Due to Molecular Distributions

Combined effects of scattering and absorption due to molecular distributions for a tropical climate are shown in Figure 2.3. It is evident from the curves shown in Fig-

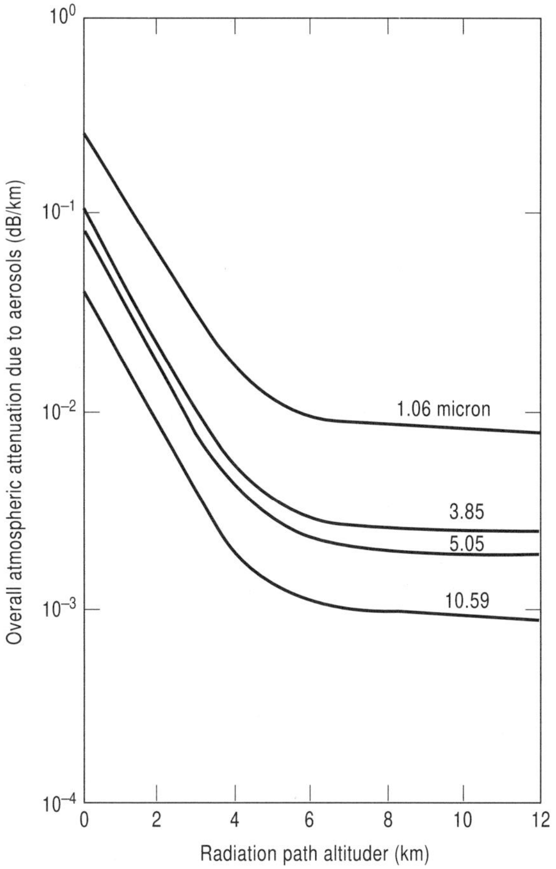

FIGURE 2.2 Combined effect due to scattering and absorption due to atmospheric aerosols as a function of wavelength (μm) and altitude (km) for a tropical climate.

ure 2.3 that the overall attenuation due to molecular distributions is maximum for the 1.06 μm wavelength and minimum for the 10.59 μm wavelength, even up to a projected path altitude of 20 km. These curves further indicate that a high-power CO_2 laser must be operated well above 20 km if optimum performance is the principal requirement. The curves further indicate that a DF chemical laser operating at 3.85 μm is best suited for high-power airborne or spaced-based laser applications because of minimum attenuation at higher altitudes. Performance under various atmospheric conditions, system complexity, weight, size, and power consumption must be seriously considered prior to selection of a specific wavelength for high-power system applications.

2.2.4 Total Attenuation from Both Aerosol and Molecular Distributions

The overall attenuation from both aerosol and molecular distributions as a function of wavelength and radiation altitude is evident from the curves shown in Figure 2.4.

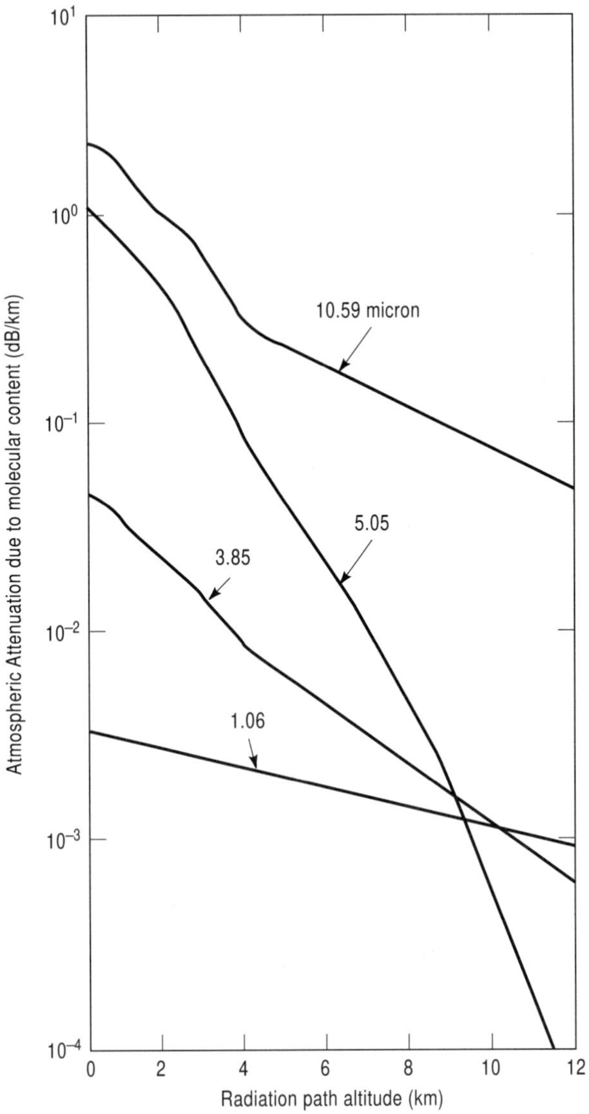

FIGURE 2.3 Composite scattering and absorption attenuation effects caused by the molecular content as a function of altitude for a tropical climatic environment.

These curves indicate that the overall attenuation from both sources is maximum at 10.59 μm and minimum at 5.05 and 3.85 μm. This means that a laser operating at 10.59 μm will experience maximum performance degradation even at altitudes approaching 20 km. However, an acceptable performance from a gaseous laser is possible at radiation path altitudes exceeding 100 km. One can see that a DF chemical laser operating at 3.85 μm is best suited for operations beyond 100 km path alti-

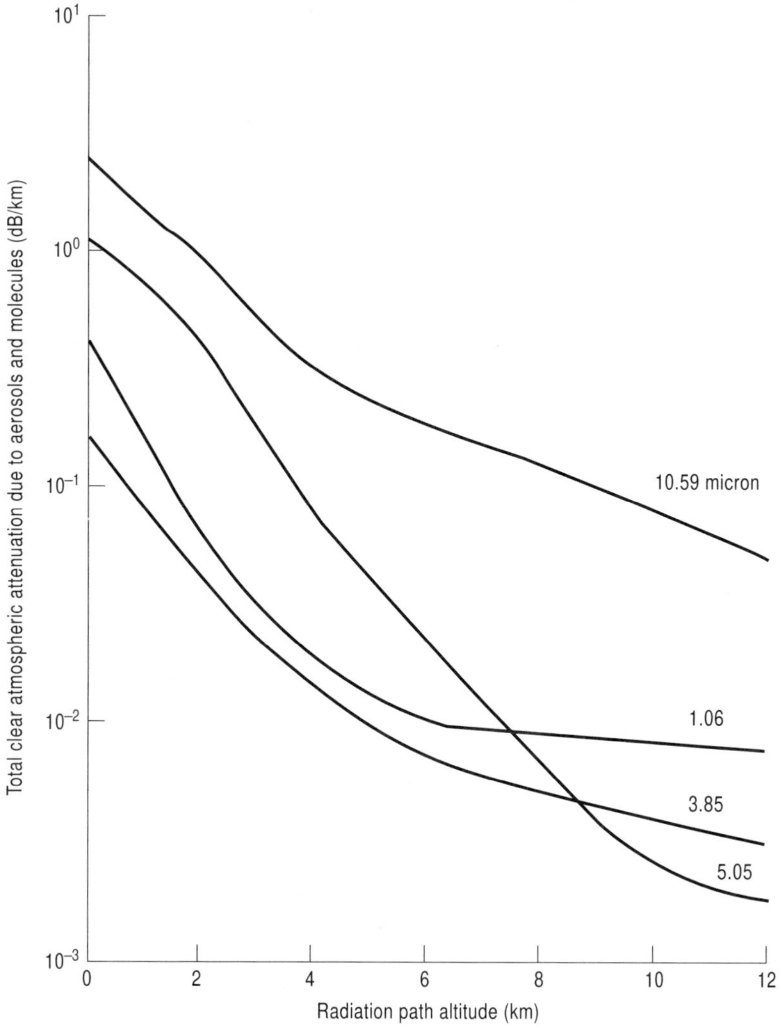

FIGURE 2.4 Total clear atmospheric attenuation effects due to both aerosols and molecules as a function of altitude and wavelength for a tropical climate.

tudes because of minimum overall attenuation from both aerosol and molecular distributions at 3.85 μm. Note that the curves shown in Figure 2.4 are valid only for a tropical climate and that a DF laser at 3.85 μm will have much better range performance compared to that of a CO_2 laser operating at 10.59 μm under other climatic conditions. However, its deployment in a spacecraft or satellite depends on other factors such as weight, size, power consumption, overall efficiency, cooling requirements, and reliability.

2.2.5 Impact of Various Atmospheric Properties on Airborne and Space IR Sensors

Low values of clear weather atmospheric attenuation coefficients from both the molecular and aerosol distributions as a function of radiation path length and most common wavelengths are necessary to achieve reliable and acceptable performance levels for airborne and space sensors. As stated earlier, both the absorption and scattering coefficients are strictly dependent on climatic environment, operational wavelength, molecular distribution, and atmospheric layer height from the earth's surface. IR sensors operating below 10 miles or 50,000 ft will experience levels of absorption and scattering that could lead to serious performance degradation of the electrooptical sensors. However, IR sensors operating at an altitude of 100 km will experience minimum performance degradation because of lower absorption. This means that space-based communication systems operating at optical frequencies (1–10 μm) may experience slight reductions in range because of absorption and scattering coefficients associated with various layers of the atmosphere. Note that the tropospheric scatter coefficient will be a dominant component of the overall attenuation coefficient. In the case of a low-orbiting earth (LOE) satellite operating at 125–250 miles above the earth, absorption and scattering coefficients will not adversely affect the IR sensor performance. However, ionospheric scattering from the lower D region of the ionosphere, extending from 600–1200 miles, could be a factor in maintaining an acceptable performance level of the IR sensors operating abroad a medium-orbiting earth (MOE) satellite. It is important to note that the aerosol absorption coefficient depends on aerosol distribution at a given point in the atmosphere. However, the molecular atmospheric absorption coefficient (k_m) is a function not only of the gas pressure, but also the local temperature and pressure. Furthermore, the parameter k_m is a highly oscillatory function of wavelength, due to the presence of numerous molecular absorption bands. The molecular attenuation or extinction coefficient (γ_m) is strictly a function of optical path altitude. The atmospheric model assumed for a tropical climate represents the worst-case climatic condition and contains high water vapor concentration. Overall attenuation levels for four useful wavelengths—1.06 μm, 3.85 μm, 5.05 μm, and 10.59 μm—under tropical climate condition are shown in Figure 2.2. The dominant factor in the weakly attenuated 1.06 μm line is the molecular scattering, absorption being negligible. However, for the other three wavelength lines, absorption is pronounced and scattering is negligible.

2.2.5.1 Impact of Laser Line Position on Absorption and Scattering Coefficients

To illustrate the importance of the wavelength dependence of the absorption and scattering coefficients, it must be noted that a even slight shift in the CO_2 laser line will lead to an increase in attenuation as high as 20 db/km at sea level. Thus, it is extremely important to carefully select the appropriate laser line in order to match the atmospheric window for minimum attenuation. It is important to mention that absorption characteristics of multimode lasers are much more complex because of the reasons sited above. As mentioned earlier, the attenuation levels at various wavelengths due to the presence of aerosols in a clear atmosphere can limit the sensor performance level, as shown in Figure 2.2. Furthermore, in a clear atmosphere, aerosol scattering and absorption effects are roughly of equal magnitude. These coefficients are less sensitive to variations in a single operating wavelength mode. The combined effects of molecular and aerosol attenuation as a function of wavelength and altitude are of critical importance in case of multiple IR transmission. The curves shown in Figure 2.2 can be used to estimate composite attenuation effects in a straight path direction. If attenuation over a slanted path is desired, a total path may be incrementally subdivided into the average attenuation determined graphically for each slant path increment. Attenuation due to each increment is summed up to determine a total path attenuation [2]. The additive effects of haze, fog, clouds, and rain must be incorporated into the clear weather attenuation to predict realistic performance from a sensor whether in single-line or multimode sensor operation [3].

2.3 IMPACT OF ATMOSPHERIC TRANSMISSION CHARACTERISTICS ON HIGH-POWER LASER SYSTEMS

High-power IR laser reconnaissance and surveillance sensors operating aboard LOE and MOE satellites will experience severe performance degradation in lower atmospheric regions. High-power laser beams experience more pronounced effects, including absorption, scattering, diffraction, beam spreading, beam wandering, thermal blooming, and gas breakdown. Experimental data available to date indicate that absorption causes serious degradation in high-power laser performance, particularly when the satellites are operating in the lower atmosphere. As stated earlier, the absorption coefficient is dependent on radiation wavelength, line position, operating altitude, and climatic conditions. Absorption coefficients for a CO_2 laser [3] operating at sea level and at 12 km altitude as a function of line positions and climatic conditions are shown in Table 2.8. Calculated values of molecular absorption and scattering coefficients for a high-power CO_2 laser as a function of altitude under various climatic conditions, including midaltitude summer (MS), midaltitude winter (MW), subarctic summer (SS), and subarctic winter (SW) are summarized in Table 2.9. It is evident from Table 2.9 that a CO_2 laser beam will experience maximum absorption at sea level under tropical conditions and minimum absorption under SW conditions. Data shown in Table 2.8 reveal that higher line positions, namely, P36, P34, P32, R36, R34, and R32, experience minimum absorption compared to lower line positions, namely, P16, P14, P12, R16, R14, and R12. Absorption and scattering coefficients due to aerosol distributions as a function of altitude are shown in Table 2.9 for clear, hazy, and other atmospheric conditions. It is important to mention that for a CO_2 laser, both the absorption and scattering coefficients due to aerosols remain unchanged above 5 km operating altitude, whether in clear or hazy atmospheric environments. However, no such evidence is available for HF and DF laser beams under similar operating conditions.

TABLE 2.8 Attenuation coefficients for various CO_2 laser wavelengths at different line positions*

CO_2 Laser parameters		Atmospheric absorption coefficient (km^{-1})*			
		Height = 0 km (sea level)			Height = 12 km
Line position	ν (cm^{-1})	k_{trop}	k_{mw}	k_{sw}	k_{mw}
P36	929.013	.744	0.0584	.0190	.00211
P34	930.997	.538	0.0536	.0227	.00311
P32	932.956	.557	0.0650	.0302	.00520
P30	934.890	.572	0.0737	.0360	.00677
P28	936.800	.588	0.0852	.0440	.00887
P26	938.684	.583	0.0853	.0447	.00955
P24	940.544	.603	0.0955	.0517	.0118
P22	942.380	.606	0.1021	.0569	.0136
P20	944.190	.609	0.0958	.0521	.0125
P18	945.976	.635	0.1223	.0717	.0186
P16	047.738	.572	0.0747	.0378	.00897
P14	949.476	.607	0.1101	.0642	.0173
P12	951.189	.591	0.1058	.0619	.0171
P10	952.877	.596	0.1008	.0580	.0161
P8	954.541	.553	0.0817	.0452	.0123
P6	956.181	.513	0.0615	.0314	.00810
P4	957.797	.484	0.0498	.0236	.00573
P2	959.388	.978	0.0753	.0282	.00609
R0	961.729	.456	0.0347	.0130	.00234
R2	963.260	.461	0.0401	.0170	.00367
R4	964.765	.478	0.0502	.0241	.00590
R6	966.247	.519	0.0614	.0308	.00783
R8	967.704	.505	0.0663	.0352	.00931
R10	969.136	.510	0.0714	.0389	.0104
R12	970.544	.578	0.0788	.0418	.0109
R14	971.927	.556	0.0796	.0427	.0110
R16	973.285	.554	0.0799	.0425	.0106
R18	974.618	.522	0.0755	.0405	.0101
R20	975.927	.194	0.2140	.0740	.0109
R22	977.210	.674	0.0871	.0398	.00803
R24	978.468	.503	0.0641	.0318	.00690
R26	979.701	.484	0.0579	.0280	.00585
R28	980.909	.474	0.0529	.0245	.00471
R30	982.091	.552	0.0587	.0240	.00378
R32	983.248	.454	0.0436	.0183	.00324
R34	984.379	.455	0.0439	.0158	.00229
R36	985.484	.436	0.0357	.0133	.00176

*Symbols: *trop* = tropical, *mw* = midaltitude winter, *sw* = subarctic winter.

2.3.1 Transmission Characteristics for a Deuterium Fluoride (DF) Laser

High-power chemical DF lasers operating over several bands in the 4.499–5.047 μm spectral region will be subjected to absorption, scattering, and turbulence while transmission beams through the lower atmospheric regions. Atmospheric absorption

TABLE 2.9 Calculated values of attenuation coefficients (both absorption and scattering) for a CO_2 laser operating at 10.6 μm as a function of altitude and weather

											Aerosol			
	Tropical		Midlatitude summer		Midlatitude winter		Subarctic summer		Subarctic winter		Clear		Hazy	
Height	k_m(km^{-1})	σ_m(km^{-1})	k_m(km^{-1})	σ_m(km^{-1})	k_m(km^{-1})	σ_m(km^{-1})	k_m(km^{-1})	σ_m(km^{-1})	k_m(km^{-1})	σ_m(km^{-1})	k_a(km^{-1})	σ_a(km^{-1})	k_a(km^{-1})	σ_a(km^{-1})
0	5.788E–01	<1.0E–06	3.582E–01	<1.0E–06	7.937E–02	<1.0E–06	2.006E–01	<1.0E–00	4.118E–02	<1.0E–06	5.48E–03	4.65E–03	2.67E–02	2.27E–02
0–1	5.172E–01		3.256E–01		7.312E–02		1.818E–01		4.147E–02		3.64E–03	3.05E–03	1.61E–02	1.37E–02
1–2	2.845E–01		1.877E–01		5.895E–02		1.137E–01		4.002E–02		1.55E–03	1.34E–03	5.90E–03	5.01E–03
2–3	1.807E–01		1.452E–01		4911E–02		8.152E–02		3.516E–02		6.75E–04	5.73E–04	2.16E–03	1.83E–03
3–4	9.616E–02		7.582E–02		4.043E–02		6.090E–02		3.048E–02		3.18E–04	2.70E–04	7.88E–04	6.68E–04
4–5	6.290E–02		5.544E–02		3240E–02		4.663E–02		2.453E–02		2.0lE–04	1.70E–04	2.88E–04	2.44E–04
5–6	5.019E–02		4.468E–02		2.622E–02		3.737E–02		1.932E–02		1.46E–04	1.24E–04	1.46E–04	1.24E–04
6–7	3.998E–02		3.752E–02		2.147E–02		2.861E–02		1.509E–02		1.18E–04	1.00E–04	1.18E–04	1.00E–04
7–8	3.200E–02		3.018E–02		1.728E–02		2.277E–02		1.171E–02		1.16E–04	9.83E–05	1.16E–04	9.83E–05
8–9	2.634E–02		2.378E–02		1.405E–02		1.788E–02		9.593E–03		1.15E–04	9.77E–05	1.15E–04	9.77E–05
9–10	2.074E–02		1.952E–02		1.083E–02		1.375E–02		8.932E–03		1.11E–04	9.45E–05	1.11E–04	9.45E–05
10–11	1.651E–02		1.574E–02		9.813E–03		1.195E–02		8.921E–03		1.06E–04	9.04E–05	1.06E–04	9.04E–05
11–12	1.287E–02		1.241E–02		9.484E–03		1.229E–02		8.908E–03		1.06E–04	8.96E–05	1.06E–04	8.96E–05
12–13	1.035E–02		9.534E–03		9.358E–03		1.181E–02		8.733E–03		1.04E–04	8.83E–05	1.04E–04	8.83E–05
13–14	7.382E–03		8.337E–03		9.340E–03		1.229E–02		9.109E–03		9.89E–05	8.39E–05	9.89E–05	8.39E–05
14–15	5.850E–03		8.700E–03		9.001E–03		1.222E–02		8.891E–03		9.49E–05	8.06E–05	9.49E–05	8.05E–05
15–16	4.330E–03		8.491E–03		8.749E–03		1.166E–02		8.773E–03		8.97E–05	7.61E–05	8.97E–05	7.61E–05
16–17	3.316E–03		8.364E–03		8.573E–03		1.217E–02		8.559E–03		8.69E–05	7.38E–05	8.69E–05	7.38E–05
17–18	3.556E–03		8.467E–03		8.556E–03		1.203E–03		8.324E–03		8.50E–05	7.21E–05	8.50E–05	7.21E–05
18–19	4.350E–03		8.560E–03		8.249E–03		1.199E–02		8.209E–03		7.68E–05	6.51E–05	7.68E–05	6.51E–05
19–20	5.194E–03		8.929E–03		8.011E–03		1.217E–02		7.884E–03		6.04E–05	5.12E–05	6.04E–05	5.12E–05
20–21	6.273E–03		9.186E–03		8.186E–03		1.186E–02		7.784E–03		4.40E–05	3.74E–05	4.40E–05	3.74E–05
21–22	7.471E–03		9.729E–03		8.194E–03		1.208E–02		7.523E–03		3.25E–05	2.76E–05	3.25E–05	2.76E–05
22–23	8.351E–03		1.010E–02		8.161E–03		1.208E–02		7.209E–03		2.47E–05	2.09E–05	2.47E–05	2.09E–05
23–24	9.041E–03		1.114E–02		8.107E–03		1.199E–02		7.329E–03		1.92E–05	1.63E–05	1.92E–05	1.33E–05
24–25	9.909E–03		1.112E–02		8.378E–03		1.275E–02		6.837E–03		1.57E–05	1.33E–05	1.57E–05	1.33E–05
25–30	1.203E–02		1.327E–02		8.087E–03		1.453E–02		7.238E–03		750E–06	6.71E–06	7.90E–06	6.71E–06
30–35	1.190E–02		1.319E–02		6.848E03		2.007E–02		5.785E03		2.23E–05	1.89E–06	2.23E–06	1.89E–06
35–40	1.101E–02		1.269E–02		6714E–03		1.395E–02		5.099E–03		<1.0E–06	<1.0E–06	<1.0E–6	
<1.0E–06														
40–45	8.865E–03		1.063E–02		6.023E–03		1.189E–02		4.100E–03					
45–50	6.038E–03		7.522E–03		4.405E–03		8.186E–03		3.082E–03					
50–70	9.007E–04		1.077E–03		2.744E–04		1.097E–03		7.761E–04					
70–100	1.535E–05		1.743E–05		1.580E–04		1.762E–05		1.785E–05					

*Symbols: k = absorption coefficient, σ = scattering coefficient, m = molecular component, a = aerosol component.

coefficients at sea level and 12 km altitude for various bands under various climatic conditions are shown in Table 2.10 [4]. It is important to note that line position P8 at sea level will experience the highest absorption in bands 8–7 and 9–8 regardless of the weather conditions. However, at 12 km and above, one will find minimum absorption under midaltitude winter conditions irrespective of emission band or line position. DF laser single-line performance level is shown in Table 2.11. These data indicate which band offers the maximum power performance at a specific wavelength. Band P2 offers optimum performance, whereas band P3 delivers marginal performance. Furthermore, line position 10 (3.8757 μm) in band P2 yields the highest power performance, whereas line position 5 (3.5806 μm) in band P1 offers minimum power performance. It is important to point out that at altitudes greater than 100 km, a DF laser beam will experience much lower absorption compared to a CO_2 laser beam.

2.3.2 Transmission Characteristics for a Hexafluoride (HF) Laser

HF is a chemical laser generally used for a high-power applications. Calculated values of absorption coefficients for a high-power laser at sea level and 12 km altitude under various climatic conditions are summarized in Table 2.12. This particular laser beam transmission consists of six bands, namely 1-0, 2-1, 3-2, 4-3, 5-4, and 6-5, and operates in the 2.9103–3.5229 μm range. Note that band 3-2 has only one laser line position, P6, operating at 2.9643 μm. This single line in band 3-2 offers lower absorption at sea level under moderate weather conditions. It is important to mention that HF lasers experience minimum absorption at altitudes of 12 km and up, as supported by the data summarized in Table 2.12.

Absorption coefficients shown in Table 2.12 indicate that line positions P6, P7, and P8 offer the lowest possible values of the absorption coefficient under MW climatic conditions. Furthermore, HF lasers operating at 3.4714 μm at line position P7 should be preferred because of minimum loss under most climatic conditions. HF laser single-line performance at various positions is described in Table 2.13. Line positions P1 and P2 are of significance because of lower loss. These data indicate that line position 7 in band P1 offers the highest power performance at 2.7441 μm and line 2 in the same band yields the lowest power at 2.5786 μm. It is important to mention that line position 6 in band P2 offers the highest power performance compared to four lines in band P1. Thus band P2 must be preferred if maximum power performance is the principal requirement.

2.4 ABSORPTION AND SCATTERING EFFECTS FROM ATMOSPHERIC CONSTITUENTS

Carbon dioxide is the most dominant constituent among the fixed and uniformly mixed gases found in the atmosphere, as shown in Table 2.14 [5].

TABLE 2.10 Attenuation coefficients for a DF laser as a function of geographical location, weather, wavenumber, and laser line position*

DF Laser parameter			Atmospheric absorption coefficients			
			Height = 0 km (Sea level)			Height = 12 km
Band	Line Position	$\nu(\text{cm}^{-1})$	k_{trop}	k_{mw}	k_{sw}	k_{mw}
1–0	P1	2884.934	.414	.123	.0772	.00316
	P2	2862.652	.0540	.0115	.00485	.00316
	P3	2839.779	.0386	.00725	.00266	.000038
	P4	2816.362	.0837	.0190	.0104	.00108
	P5	2792.437	.0471	.0106	.00496	.000157
	P6	2767.914	.0719	.0184	.00952	.000672
	P7	2743.028	.0352	.00801	.00352	.000043
	P8	2717.536	.114	.0204	.00718	.000034
	P9	2691.409	.0248	.00485	.00252	.000053
	Pl0	2665.20	.0237	.00752	.00489	.000307
	P11	2638.396	.337	.0664	.0247	.000187
	P12	2611.125	.0133	.00394	.00302	.000090
	P13	2584.91	.0145	.0102	.00981	.00390
	P14	2557.09	.0176	.0180	.0185	.00335
	P15	2527.06	.0145	.0155	.0161	.000565
	Pl6	.2498.02	.0261	.0282	.0295	.00103
2–1	P3	2750.05	.0401	.00898	.00403	.000074
	P4	2727.38	.0378	.00653	.00272	.000033
	P5	2703.98	.00528	.00171	.00118	.0000307
	P6	2680.28	.0600	.0139	.00611	.000069
	P7	2655.97	.0535	.0134	.00667	.000733
	P8	2631.09	.00950	.00348	.00293	.000761
	P9	2605.87	.0311	.00776	.00455	.000110
	P10	2580.16	.0282	.0295	.0311	.00180
	P11	2553.97	.0144	.0163	.0177	.000883
	P12	2527.47	.0140	.0152	.0158	.000554
	P13	2500.32	.0240	.0265	.0278	.000972
	P16	2417.27	.0811	.0901	.0943	.00330
3–2	P3	2662.17	.0354	.00790	.00361	.000047
	P4	2640.04	.0437	.00914	.00424	.000075
	P5	2617.41	.00490	.00276	.00253	.000090
	P6	2594.23	.0118	.00557	.00480	.000152
	P7	2570.51	.0507	.0560	.0613	.00557
	P8	2546.37	.0322	.0356	.0379	.00228
	P9	2521.81	.0150	.0164	.0171	.00599
	Pl0	2496.61	.0319	.0298	.0307	.00107
	P11	2471.34	.0509	.0491	.0508	.00184
	P12	2445.29	.0659	.0728	.0756	.00266
	P13	2419.02	.0797	.0885	.0927	.00325
	P14	2392.46	.141	.119	.115	.00369
4–3	P5	2532.50	.0134	.0143	.0148	.000528
	P6	2509.86	.0199	.0218	.0228	.000795
	P7	2486.83	.0318	.0349	.0356	.00129
	P8	2463.25	.0681	.0563	.0571	.00198
	P9	2439.29	.0686	.0758	.0794	.00279
	Pl0	2414.89	.0829	.0921	.0964	.00338

*Symbols: *trop* = tropical, *mw* = midaltitude winter, *sw* = subarctic winter.

TABLE 2.11 DF laser single-line performance level

Line position	Wavelength (μm)	Power level (W)
P1 (5)	3.5806	0.01
(6)	3.6128	0.025
(7)	3.6456	0.04
(8)	3.6798	0.10
(9)	3.7155	0.15
(10)	3.7520	0.20
(11)	3.7902	0.15
(12)	3.8298	0.10
(13)	3.8708	0.05
P2 (3)	3.6363	0.025
(4)	3.6665	0.06
(5)	3.6983	0.10
(6)	3.7310	0.15
(7)	3.7651	0.20
(8)	3.8007	0.25
(9)	3.8375	0.30
(10)	3.8757	0.32
(11)	3.9155	0.27
(12)	3.9565	0.20
P3 (7)	3.8903	0.01
(8)	3.9272	0.02
(9)	3.9654	0.05
(10)	4.0054	0.14
(11)	4.0464	0.07

TABLE 2.12 Attenuation coefficients for HF lasers as a function of wavelength and weather conditions [5]*

			Atmospheric absorption coefficient $(km)^{-1}$			
			Sea level			12 km
Band	Laser line position	Wavelength (μm)	k_{trop}	k_{ms}	k_{sw}	k_{mw}
1–0	P11	2.9103	2.210	0.221	0.054	0.000029
	P12	2.9573	0.496	0.075	0.023	0.000022
2–1	P8	2.9111	2.010	0.209	0.051	0.000027
3–2	P6	2.9643	0.364	0.054	0.017	0.000029
4–3	P8	3.2427	1.120	0.211	0.081	0.000295
	P9	3.1948	0.801	0.148	0.055	0.000806
5–4	P4	3.1739	0.498	0.126	0.074	0.002290
6–5	P6	3.4226	0.586	0.045	0.010	0.000077
	P7	3.4714	0.043	0.004	0.001	0.000006
	P8	3.5229	0.369	0.065	0.022	0.000049

*Symbols: *trop* = tropical, *mw* = midaltitude winter, *sw* = subarctic winter.

TABLE 2.13 HF laser single-line performance level at various lines

Line position	Wavelength (μm)	Power level (W)
P1 (2)	2.5788	0.12
(3)	2.6085	0.45
(4)	2.6398	1.50
(5)	2.6727	0.15
(6)	2.7075	1.50
(7)	2.7441	2.40
(8)	2.7826	1.40
(9)	2.8Z31	1.00
(10)	2.8657	0.30
P2 (3)	2.7275	0.20
(4)	2.7604	0.50
(5)	2.7853	1.30
(6)	2.8318	2.20
(7)	2.8706	2.20
(8)	2.9111	1.60
(9)	2.9539	0.90
(10)	2.9989	0.50
(11)	3.0461	0.30

2.4.1 Carbon Dioxide

This is the most dominant gas component among the fixed gases. Its fractional value is 325 ppm in the region at the top of the stratosphere and is uniformly mixed up to the stratosphere, where it can be destroyed by photochemical decomposition phenomena. The undisturbed value of the parameter is 330 ppm, as shown in Table 2.14.

2.4.2 Nitric Oxide

This has been given a fractional volume abundance value of 0.28 ppm at ground level, confirmed by both the aggregate and LOWTRAN methods. However, the mean value is 0.14 ± 0.04 ppm for altitudes ranging from 5–13 km [5].

2.4.3 Carbon Monoxide (CO)

This is considered to be a uniformly mixed gas vertically up to the troposphere (the region at the top of the atmosphere), above which it is oxidized by CO_2. The gas concentration increases in the stratosphere (upper portion of the atmosphere extending form 7–31 miles) and in the mesosphere. There is no strong evidence of significant long-term increases in CO, because short-term temporal and spatial variations tend to cancel long-term increases, if any. Concentration varies from 0.01–0.20 ppm in remote locations and 0.5–2.2 ppm in urban areas. The aggregate method uses a concentration value of 0.12 ppm, whereas the LOWTRAN method uses an approximate value of 0.075 ppm .

TABLE 2.14 Concentration of uniformly mixed gases in the atmosphere [5]

Constituent	Molecular weight	ppm by volume	Horizontal path $(cm\ atm)_{STP}/km$ at sea level	Vertical path $(cm\ atm)_{STP}/km$ at sea level	Absorbent amount $(g/cm^2/mb)$
Methane (CH_4)	16	1.6	0.16	1.28	9.01×10^{-7}
Carbon Monoxide (CO)	28	0.075	0.0075	0.06	7.39×10^{-8}
Air (N_2)	28.97	10^6	10^5	8×10^5	1.02
Oxygen (O_2)	32	2.095×10^5	2.095×10^4	1.680×10^5	0.236
Carbon dioxide (CO_2)	44	330	33	264	5.11×10^{-4}
Nitrous oxide (N_2O)	44	0.280	0.028	0.220	4.37×10^{-7}

2.4.4 Methane (CH_4)

This gas is found in the troposphere region of the atmosphere, which is below the stratosphere and extends outward 7–10 miles from Earth's surface. The volume concentration ratio of methane is 1.6 ppm, which remains constant with altitude up to the troposphere region (7–10 miles from earth surface). However, its value decreases rapidly in the lower stratosphere. The aggregate method uses a value of 1.1 ppm, whereas the LOWTRAN method use a value of 1.6 ppm.

2.4.5 Nitric Acid (HNO_3)

The mixing ratio of nitric acid to air in percentage by volume varies from 10^{-11} at 50 km altitude to 10^{-7} at 25 km.

2.4.6 Water Vapor (H_2O)

Water vapor is the most gaseous constituent as far as atmospheric transmittance is concerned. Profiles of water vapor in conjunction with the seasonal–temporal profiles of pressure, temperature, and air density under various climatic conditions are summarized in Table 2.1. Data shown in Table 2.1 are for tropical climate; Table 2.2 contains data for MS, SW, and MW climatic conditions. Variations of atmospheric pressure as a function of altitude and climatic environments are evident from the curves shown in Figure 2.1. One can clearly see from the curve that pressure under SW climatic environment decreases linearly above 50 km altitude.

2.4.7 Ozone (O_3)

Ozone is another highly variable atmospheric constituent, depending on both the time of the year and geographical location. The concentration peaks around 30 km, where the mixing ratio of ozone to air is better than 10 ppm. Adverse affects due to ozone on IR signals are not fully known.

2.5 CRITICAL ATMOSPHERIC PARAMETERS FOR MODERATE AND HIGH ALTITUDE OPERATIONS

2.5.1 Parameters at Moderate Altitudes

As stated earlier, molecular and aerosol absorption and scattering coefficients are strictly dependent on the atmospheric properties at a given altitude. Airborne sensors operating at low altitude will experience severe performance degradation regardless of climatic conditions. Critical parameters such as pressure (P), temperature (T), and water vapor density (d_w) as a function of altitude and climatic conditions are summarized in Table 2.15. These parameters are valid for moderate altitudes ranging from 50–100 km. Atmospheric characteristics at lower altitudes have been summarized in Table 2.1. These data indicate that sea level water vapor density is about 19 g/m^3 under tropical environments, 14 g/m^3 under MW climate, 1.2 g/m^3 under SW climate, and 9.2 g/m^3 under SM climate. The data further indicate that the water vapor density is maximum under tropical, MS, and SM climatic conditions. This means that an IR sensor operating at sea level or low altitude will experience severe performance degradation due to high attenuation levels.

Note that the pressure and temperature remain practically constant at 0.0003 mb and 210 K, respectively, at 100 km altitude, regardless of climatic conditions. However, moderate variations or changes in the magnitude of the above parameters have been observed at sea level under various climatic environments. Familiarization with atmospheric parameters such as pressure and temperature as a function of altitude and climatic conditions is absolutely necessary in predicting the exact performance degradation of IR sensors. In addition, one must understand the impact on attenuation or extinction coefficients of particle distribution from atmospheric aerosols and molecule distributions.

2.5.2 Atmospheric Parameters at High Altitude

Radiation transport in the infrared coupled with turbulent convection cools the atmosphere near the Earth's surface, resulting in a decrease in temperature with altitude through the troposphere at a rate of approximately 7 K/km. At the tropopause (approximately 12 km altitude), the temperature gradient reverses and the tempera-

TABLE 2.15 Critical atmospheric parameters at moderate altitudes under various climatic conditions [6]*

	Tropical		MS		MW		SW		SS	
Altitude	P	T	P	T	P	T	P	T	P	T
50	0.854	270	0.951	276	0.682	265.7	0.572	259	0.99	277
70	0.579	219	0.067	218	0.0467	230.7	0.040	246	0.071	216
100	0.0003	210	0.0003	210	0.0003	210.2	0.0003	210	0.0003	210

*Symbols: P = pressure (mb), MS = midsummer, SW = subarctic winter, T = Temperature (K), MW = midwinter, SS = subarctic summer ture continues to rise through the stratosphere, where solar ultraviolet (UV) radiation is absorbed by ozone. At 50 km (the stratopause), the temperature starts to decrease, until a minimum of 130–190 K is reached. Below 100 km, turbulent mixing in the atmosphere maintains a uniform relative composition of the major constituents, namely, molecular nitrogen (N_2), oxygen (O_2), argon (Ar), and carbon dioxide (CO_2). Helium eventually takes over at approximately 600 km. The ionosphere is the charged component of the atmosphere between 70 and 1000 km. The electron density profile forms the basis for dividing the ionosphere into various layers. Solar radiation is responsible for the formation of the ionospheric F-region (150–1000 km) and part of the E-region (90–150 km). It is important to mention that the atmospheric parameters, namely, pressure and temperature, which can impact the performance level of IR sensors, undergo radical variations at altitudes exceeding 100 km. Satellite-based IR reconnaissance and surveillance sensors operating in LEO and MEO regions generally experience higher temperature, lower pressure, and lower density. Variations in temperature, pressure, air density, and molecular weight as a function of altitude are shown in Table 2.16.

It is evident from the data shown in Table 2.16 that radical change in temperature occurs from 200 K at 100 km to 1404 K at 200 km altitude. Temperature parameters continue to undergo radical change from 1576 K at 500 km to 1812 K at 700 km altitude. In contrast, both the air density and pressure decrease rapidly over the 100–700 km altitude range. Note that molecular weight decreases with the increase in altitude, but at a moderate rate. IR sensors operating aboard LEO satellites at 125–200 km altitude most likely will not experience temperatures above 250 K. However, MEO satellite-based IR sensors operating in the 350–700 km altitude range will be subjected to much higher temperatures. IR sensors installed on an outer surface of the satellite will experience much higher temperatures, which can impact on their reliability and longevity. IR sensors aboard geostationary or synchronous satellites will experience solar clutter problems in addition to adverse effects due to atmospheric characteristics.

2.6 AEROSOL PROPERTIES AND CLASSIFICATIONS

Aerosols are considered normal constituents of the atmosphere, with variable mass–mixing ratios depending on whether the air is very clear or polluted. Aerosols

TABLE 2.16 Critical atmosphere parameters based on ARDC model

Altitude (km)	Pressure (mb)	Temperature (K)	Density (g/m^3)	Molecular weight
100	2.1×10^{-4}	200	0.00037	28.90
200	1.61×10^{-6}	1404	3.67×10^{-7}	26.32
300	2.52×10^{-7}	1423	4.75×10^{-8}	21.95
400	5.59×10^{-8}	1480	9.00×10^{-9}	19.56
500	1.56×10^{-8}	1576	2.20×10^{-9}	18.28
600	5.23×10^{-9}	1691	6.62×10^{-10}	17.52
700	2.01×10^{-9}	1812	2.30×10^{-10}	17.03

are introduced into the various layers of atmosphere by wind-raised dust and sea salt, combustion products such as ash and soot, and products formed by the chemical reactions within the atmosphere involving gaseous elements such as nitrates, sulfates, ammonia, etc. These products are removed from the atmosphere by gravitational fallout and falling precipitation. Aerosol residence times vary from minutes to weeks in the troposphere and up to years in the stratosphere region. Studies performed on various layers of the atmosphere indicate that sulfuric acid is an important component of aerosols in the stratosphere region of the atmosphere above 30 km. Properties of the aerosols and their molecular constituents must be known to compute radiative transfer properties or optical properties [7]. In various layers of the atmosphere, such properties include particle size and height distribution, chemical composition, and complex index of refraction. It is important to mention that there can be wide spatial and temporal variations in these properties because of a variety of source and sink mechanisms.

2.6.1 Atmospheric Aerosol Models

Aerosol modeling requires specific parameters to describe the interaction of atmospheric aerosols with electromagnetic and optical radiation [8]. However, aerosol particle size is of critical importance. Certain theoretical models have been developed to fit different environmental conditions and seasons and are divided in to following distinct latitude regions.

(1) *Junge Model:* This model is defined as

$$\frac{dN(r)}{dr} = Cr^{-(\beta+1)} \tag{2.1}$$

where $dN(r)$ = number of particles per unit volume of the radius between r and $(r + dr)$, C = normalized constant, and β = shaping constant (its value varies from 2 to 4) for aerosols with radius lines between 0.1 and 10 μm.

This model shows that the aerosol numbers drop off rapidly below same radius less than 0.1 μm. This model offers good results for aerosols having radius (r) less than 1 μm, which includes sea salt spray, jet exhaust, and Saharan dust particles.

(2) *Deirmendjian Model:* This model is defined as

$$\frac{dN(r)}{dr} = (ar^{\propto})(e^{-br^{\beta}}) \tag{2.2}$$

where a = total N unit density and α, β, and b = shaping constants.

This model offers excellent results as the particle radius (r) approaches zero. This model is best suited for stratosphere, continental, and marine hazes. The Deirmendjian model [1] parameters are shown in Table 2.17.

By letting $\propto$ = 2.0 and β = 1.0, the continental and marine distributions are not greatly changed as far as their effects on atmospheric scattering are concerned.

(3) *Zold Model:* The stratospheric aerosol can be best described by the Zold model, which is based on a log-normal distribution function, written as

TABLE 2.17 Deirmendjian model parameters

Environment	a	B	$\propto$	β
Stratosphere	4.0×10^5	20.0	2	1.0
Continental	5.0×10^6	15.1	2	0.5
Marine	5.3×10^4	8.9	1	0.5

$$\frac{dN(r)}{dr} = \left(\frac{C}{\sigma r \sqrt{(2\pi)}}\right)\left(\frac{e^{-(\log \frac{r}{r_m})^2}}{2\sigma^2}\right) \tag{2.3}$$

where r = particle radius, r_m = mean radius, σ = standard deviation, $\log_e$ = natural logarithm, C = normalized constant.

This model [8] offers the best fit and provides excellent results for values of r equal to 0.3 μm and σ equal to 1.3. Authors Shellbe et al. [9] have selected a modified gamma function to represent the stratosphere aerosol with different values of the parameters than those used by the Deirmendjian model.

2.6.2 Aerosol Models Best Suited for Various Altitude Regimes

Shellbe et al. have constructed a series of models for different environmental conditions and seasons best suited for specific altitude regimes, as described below.

1. The atmosphere layer below 2 km includes 10 models describing several surface visibilities in rural, urban, and marine environments.
2. Troposphere regimes above 2 km spring–summer and fall–winter models are fully described .
3. Background conditions as well as volcanic conditions for spring–summer and fall–winter models are described for the stratosphere regions up to 30 km (generally known as lower atmosphere).
4. Two models are described, one for the background conditions and one for high aerosol concentrations for altitudes above 30 km.

Note that the "rural model' is intended to represent clean conditions containing a layer composed of about 70% water soluble atmospheric contents such as ammonium and calcium sulfate and 30% dust-like aerosols. Combustion and industrial aerosols are the major constituents of a rural model in the lower 2 km atmosphere layer. The added aerosols generally have the same size distribution as the rural model and are mixed in the proportion of 35% urban to 65% rural aerosol. A mixture of evaporation products of sea spray and continental-produced aerosols is assumed, with humidity as high as 80%. The rural aerosol model must be used above the boundaries in the troposphere without large-particle distribution. In the case of the stratosphere aerosol model, the background aerosol products can be perturbed by a factor of 0.100 or more by volcanic eruption, and the main constituent is about 75%

sulfuric acid solution. Size distribution parameters for stratosphere aerosols under volcanic conditions are summarized in Table 2.18 [1]. The Elterman model is best suited for moderate volcanic conditions in the 10–30 km region. The aerosols in the upper atmospheric layer, which are 30 km or more above the Earth's surface, represent only a very small fraction of the total aerosol contents. These aerosols are primarily meteoric dust with particles size not exceeding 0.03 μm.

IR sensors operating over urban areas will experience high levels of aerosol attenuation because of various impurities present in the lower atmosphere regions. Performance degradation of IR sensors is expected due to active gases present even in clear atmosphere with visibility of 23 km. Chapter 4 of *The Infrared Handbook* by Wolfe and Zissis [1] is the best reference for comprehensible information on aerosol extinction, absorption, and scattering coefficients as a function of wavelength and altitude based on urban, rural, marine, and troposphere aerosol models.

2.7 MODELS FOR COMPUTATION OF OPTICAL PROPERTIES OF THE ATMOSPHERE

This section deals with various mathematical expressions and models to compute both the aerosol and molecular scattering coefficients needed to estimate spectral transmittance and radiance with high accuracy. The theory of molecular band atmosphere offers the best tool to compute spectral transmittance in narrow regions of the spectrum using well-developed models and empirical techniques. Comprehensive review of relevant publications indicate that only the aggregate and LOWTRAN methods are considered most reliable for computation of optical properties of the atmosphere over the 1–30 μm range. Potential aerosol and scattering models will be briefly described to compute the optical properties of the atmosphere.

2.7.1 Scatter Phenomena

Scattering can be a single scattering or a multiple scattering phenomenon. Radiated energy transfers through optically scattering media can be approximated accurately with a single scattering transfer equation. Single scattering can be calculated using the aggregate and LOWTRAN models with 5 and 23 km visibility. When scattering occurs simultaneously due to both aerosols and molecules in the atmosphere, it is called multiple scattering. Molecular scattering is negligible compared to aerosol

TABLE 2.18 Size distribution for stratospheric aerosols under volcanic conditions

Conditions	a	b	α	β
Stratospheric background	324.00	18	1	1
Fresh volcanic	341.32	8	1	0.5
Aged volcanic	5461.33	16	1	1.5

scattering outside the visible region of the spectrum. Scattering by molecules can be confined mainly to the visible region. In the case of solar emission, molecular scattering is the most dominant, because the peak of the atmospheric radiation, emitted solely by molecular distribution, occurs beyond 10 μm. On the other hand, scattering by aerosols, which depends on their sizes, is less dependent on wavelength. However, under heavy hazy environments, scattering can be effective in both the visible and infrared regions. Note that under clear sky conditions, aerosol scattering in the long wavelength infrared (LWIR) regions can be neglected. In the real world the effects of scattering by atmospheric gases and scattering and absorption by aerosols must be given serious consideration in the IR spectrum region of interest.

2.7.2 Scattering from the Molecular Atmosphere

Molecular scattering of visible (Table 2.19) radiation can be defined by a volume scattering coefficient and can be written as

$$K_s(\lambda, h) = \frac{\{(8\pi^3)[n^2(\lambda, h) - 1]^2\}}{3\lambda^4 N^2{}_{\mathrm{STP}}} \cdot N(h)\left(\frac{6 + 3\delta}{6 - 7\delta}\right) \tag{2.4}$$

where δ = polarization factor (0.035), $n(\lambda, h)$ = index of refraction function of emission wavelength (λ) and altitude (h), N_{STP} = molecule number density under standard temperature and pressure (STP), and $N(h)$ = molecular number density at height (h).

TABLE 2.19 Volumetric scattering coefficient for a purely molecular atmosphere as a function of wavelength (λ) and altitude (*h*)*

Value of K_s (λ, *h*),volumetric scattering coefficient			
	Wavelength (μm)		
Height (*h*) (km)	0.8	0.9	1.06
0	0.0025	0.0016	0.0008
2	0.0021	0.0013	0.0007
5	0.0015	0.0010	0.0005
10	0.0009	0.0005	0.0003
20	0.0002	0.0001	0.0001
Values of *q*(*s*) (λ, *h*), molecular optical depth			
0	0.021	0.013	0.007
2	0.017	0.010	0.005
5	0.011	0.005	0.004
10	0.006	0.004	0.002
20	0.001	0.001	0.000

**Note:* both the $K_s(\lambda, h)$ mass scattering coefficient and the $q(s)(\lambda, h)$ molecular optical depth parameters are functions of wavelength and altitude.

2.8 NORMALIZED PARTICLE DISTRIBUTION FOR AEROSOL MODELS

The LOWTRAN model is capable of handling two aerosol models describing clear and hazy atmospheric conditions with down-level visibility of 23 and 5 km, respectively (Table 2.20). In brief, the aerosol model for clear atmosphere offers a visibility of 23 km, whereas the aerosol model for hazy atmosphere offers visibility of 5 km above sea level. This model assumes that the size distribution is the same for all altitudes under continental haze environments. The normalized particle size distribution for aerosol models is given as

$$N_r(r) = (8.83 \times 10^{-4})(r^{-4}) \tag{2.5}$$

where r is the particle size ranging from 0.1–10 μm.

Vertical distribution of particle density is a complex parameter, which is responsible for establishing the visibility range as a function of altitude and is expressed in number of particles per unit volume. The particle density is a function of the complex refractive index of the aerosols [11]. The imaginary part of this parameter is calculated to be zero below $\lambda = 0.6$ μm and to increase linearly from approximately 0.6–2 μm. The Elterman hazy model is the same as the clear model above 5 km, but it increases exponentially below 5 km to yield an efficient visible range at sea level to 5 km.

2.9 CONCENTRATION OF UNIFORMLY MIXED ATMOSPHERIC GASES

Table 2.14 provides the reader the data needed to determine the concentration of all uniformly mixed atmosphere gases in any horizontal or vertical path. The fourth column lists the number of $(\text{cm atm})_{STP}$ of gas in a 1 km path at sea level. The fifth column gives the total of gas in a vertical path through the entire atmosphere. The number of $(\text{cm atm})_{STP}$ is a horizontal path at any other altitude and can be given by the following equation:

TABLE 2.20 Vertical Distributions of Particle Density for Clean and Hazy Atmospheric Conditions [10]

	Particle density (d), particle/cm^3	
Altitude (km)	Clear visibility (23 km)	Hazy visibility (5 km)
0	2,828	13,780
10	56.75	56.75
20	26.67	26.67
30	2.238	2.238
40	0.155	0.155
50	0.01078	0.01078
100	1.969×10^{-8}	1.969×10^{-8}

$$(\text{cm atm})_{\text{STP}} = L\left(\frac{\text{ppm}}{10}\right)\left(\frac{PT_0}{P_0 T}\right) \tag{2.6}$$

where L is the horizontal path length (km), P_0 and T_0 are pressure and temperature at STP (1 atm and 273 K, respectively), and P and T are the atmospheric pressure and temperature at operating altitude (km).

Spectral transmittance and radiance level calculations require that relevant atmospheric parameters and their values be inserted in the model equations. Generally, all calculations assume a horizontally uniform flat earth with standard atmospheric conditions (STP), which include a pressure of 760 mm of mercury and a temperature of 273 K. The slant path or range is called the equivalent absorption path and uses the same expressions for uniform conditions. A slant range is obtained by dividing the calculated vertical path or range by the cosine of the zenith angle (θ_{Ze}).

Several LOWTRAN methods generated by the USAF Geophysics Laboratory are empirically derived and are easier to use than the aggregate method, but a LOWTRAN method offers slightly less accurate results according to *The IR Handbook* [1]. LOWTRAN graphs are used to determine the equivalent horizontal or slant path absorber amount and to calculate the spectral transmittance due to absorption by atmospheric gases. As stated earlier, the slant path can be obtained by dividing the vertical path by the cosine of the zenith angle. As of 1976, the LOWTRAN program had evolved thoroughly for major stages. The latest LOWTRAN program (after 1980) will provide reliable data with minimum errors .

2.10 ABSORPTION BY A SINGLE LINE

Aerosol-related absorption and scatter coefficients as a function of wandering for a moderately clear atmosphere with a visibility of 23 km are shown in Figure 2.2.

2.10.1 Band Model For Atmospheric Absorption

The most reliable method for computing atmosphere absorption is to use a well-developed mathematical model of the band structure. The most popular band models are described below.

1. The Elasser model assumes spectral lines of equal intensity, equal spacing, and identical half width or identical full width at half maximum (FWHM) .
2. The statistical or random model, which was originally developed for water vapor, assumes that the positions (or locations) and strengths (intensities) of the lines are defined by a probability function. The statistical model was verified independently by Goody and, thus, this model is also known as Goody model.
3. The random-Elasser model is a generalization of the Elasser and random

models. It assumes a random superposition of any number of Elasser bands of different intensities, spacings, and half widths.

4. The quasi-random model is the most accurate model, provided that the average interval is made sufficiently small. This model requires the maximum computation effort compared to all other models.

All of these models are based on the Elasser model; therefore, emphasis will be placed on mathematical expressions for this model. This model of an absorption band is formed by repeating a single Lorentz line periodically throughout the frequency interval. This generates a series of lines with equal spacing, constant intensity, and FWHM throughout the interval. This arrangement of spectral lines was first discovered by Elasser in 1938 [10]. The absorption coefficient given by an Elasser band can be written as

$$A = (\sinh \beta) \int^{y} (I_{0(y)})(e^{-y \cosh \beta)}) \, dy \tag{2.7}$$

where $\beta = (2\pi\alpha)/d$, A = absorption by a single line, $y = \beta\psi/\sinh \beta$, $\Psi = sw/2\pi\alpha$ = band model parameter, w = absorber amount (g/cm^2), s = line strength, d = spacing between spectral lines, α = Lorentz line half width, and I_0 = Bessel function of imaginary argument.

Weak Line Approximation

The weak line approximation is independent of the position of the spectral lines within the band. Absorption curves as a function of β and $\beta\Psi$ are shown in Figure 2.5. The expression for atmospheric paths at low altitude, which offers fairly good results, can be written as $A = 1 - e^{\beta\Psi}$ when $\Psi \geq 1$. This expression yields good approximation results wherever the absorption is small at the line centers, regardless of the parameter β.

2.10.2 Strong Line Approximation

Long atmospheric paths at high altitudes will have large values of absorption w and small values of parameter β. Under these conditions, the absorption at the line center is essentially complete at large values of Ψ. The half widths are narrow and the spectral lines do not overlap strongly (e.g., when β is small). Thus, for large values of Ψ and small small values of β, the equation can be expressed by an error function,

$$A = \text{erf} \sqrt{0.5\beta^2\Psi} \tag{2.8}$$

$$\text{erf}(a) = \left[\left(\frac{2}{\pi} \right) \int_0^a e^{-t^2} \, dt \right] \tag{2.9}$$

where $a = \sqrt{0.5\beta^2\Psi}$ and $\Psi = \text{SW}/2\pi\alpha = \text{SW}/\beta d$.

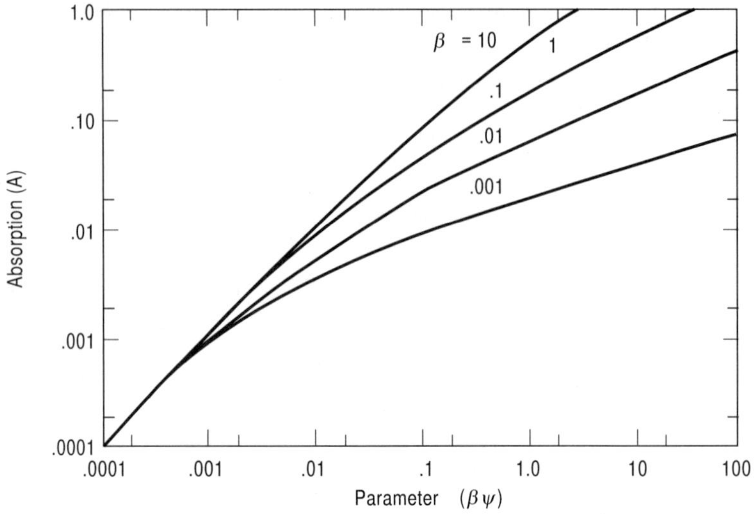

FIGURE 2.5 Absorption function for low altitude paths based on the Elsasser model (the weak-line approximation is indicated by the uppermost curve).

Equation (2.8) is referred to as the strong line approximation to the Elasser band model. Equations (2.7) and (2.8) yield nearly equal results when $\beta \leq 0.01$ and $\Psi\beta > 0.003$. When parameter $\beta \leq 0.1$, Equation (2.8) is valid whenever $0.1 \leq A \leq 1.0$ which includes most values of absorption (A) that are of great interest. Absorption curves as a function of band $\beta^2\Psi$ for constant pressure based on the Elsasser model are shown in Figure 2.6. The stray line absorption approximation is shown in the up-

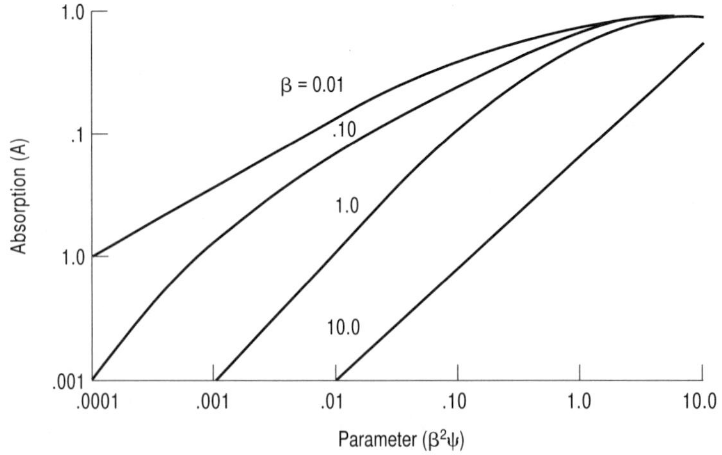

FIGURE 2.6 Absorption function for constant pressure (β = constant) based on the Elsasser model (the strong-line absorption approximation is indicated by the uppermost curve).

permost curves for $\beta = 0.01$. The general expression for absorption by an Elasser band given by Equations (2.7) and (2.8) are useful to compute absorption created by CO_2 because the bands consist of fairly regularly spaced lines. Note the bands of H_2O and O_3 have a highly irregular fine structures and, thus cannot be well described by Equation (2.7). Some reliable statistical methods must be used to develop analytical expressions for the transmittance functions for water (H_2O) and ozone (O_3).

2.10.2.1 Statistical Band Model

Case A. When lines have equal intensity and each line has weak absorption:

$$A_{\text{weak}} = 1 - e^{-\beta\psi} \tag{2.10A}$$

Case B. When each line absorbs strongly:

$$A_{\text{strong}} = 1 - e^{-2\sqrt{S_0 W/d}} \tag{2.10B}$$

Case C. When lines are of different or unequal strengths with exponential probability distribution functions:

$$A_{\text{unequal}} = 1 - e^{(-\beta\Psi_0/\sqrt{1+2\psi_0})} \tag{2.10C}$$

where $\psi_0 = (S_0 \mathrm{W}/2\pi\alpha)$, S_0 = mean line strength, W = absorption weight (g/cm^2), α = Lorentz half width, β = constant, and Ψ_0 = band model parameter.

The formula given by Equation (2.10C) was developed by Goody and, thus, is known as the Goody band model. The weak line approximation to the Goody model is obtained when $\Psi_0 \ll 1$ and Equation (2.10 C) is reduced to

$$A = 1 - e^{-\beta\Psi_0} \tag{2.11}$$

The strong-line approximation to the model with line strength defined by an exponential distribution can now be rewritten as

$$A = [1 - e\sqrt{-0.5\beta^2\Psi_0}] \tag{2.12}$$

The nonoverlapping approximation is obtained when the exponent in Equation (2.10C) is small and, therefore, can be given by the first two terms of the equation. Thus,

$$A = \frac{\beta\Psi_0}{\sqrt{1 + 2\Psi_0}} \tag{2.13}$$

For other models, the readers are advised to refer to Chapter 5 of *The Infrared Handbook* [1].

TABLE 2.21 Performance summary of band models [1, 12]

Gas	Model	Spectral range (μm)	Resolution (micron)
H_2O	Strong line Goody	1–2	0.10
	Goody	2–4.3	0.05
	Strong line Goody	4.3–15	0.50
	Goody	15–30	1.0
CO_2	Empirical	1.37–2.64	0.20
	Classical Elasser	2.64–2.88	0.01
	Classical Elasser	4.18–4.45	0.02
	Temperature-dependent Elasser	11.67–19.92	0.10
O_3	Modified classical Elasser	9.398–10.19	0.10
	Goody	11.7–15.4	0.50
N_2	Strong line Elasser	4.23–4.73	0.50
	Classical Elasser	7.53–8.91	0.50
	Goody	15.4–19.3	0.50
CH_4	Classical Elasser	5.91–9.1	0.10

2.11 PRACTICAL METHOD TO COMPUTE SPECTRAL TRANSMITTANCE

Comprehensive knowledge of atmospheric transmittance is absolutely necessary because it has a significant impact on the performance level of the airborne IR imaging sensors. The aggregate and LOWTRAN methods are two practical methods for calculating the spectral transmittance. The aggregate method is so-called because it uses a selective collection of the conventional two-parameter models described earlier. This method is more adaptive to a wide variety of different atmospheric conditions than the LOWTRAN method. LOWTRAN method is strictly based on an empirical method and depends essentially on a single adjustable or variable parameter. For transmittance values between 10 and 90%, the results obtained by either method are fairly identical. However, the LOWTRAN method offers simplicity and is more attractive for routine computations. The aggregate method must be used for different spectral regions with various atmospheric models. Performance summaries of various band models and their assumed parameters are shown in Table 2.22. Both the aggregate and LOWTRAN methods are best suited to calculate spectral transmittance. Two transmission media will be considered, namely, vacuum and gaseous.

Vacuum. When a source emits radiation with a source intensity (I) over a path through a vacuum, the irradiance (E) at some distance R from the source can be calculated using the following expression:

$$E = \frac{(T_a)(I)}{R^2} \tag{2.14}$$

TABLE 2.22 Impact of atmospheric transmittance and transference on IR sensors

Parameter	Definition / Equivalent	Application
Transmittance (T_a)	Transference of S/N ratio when the limiting noise is independent of path radiance	(1) Passive IR sensors (2) Gated viewing EO sensors
	Transference of the contrast of object seen against the horizontal sky ($K = 1$)	Airborne EO sensors
Transference (τ_c)	Transference of the contrast function $[(L_0 - L_b)/L_b]$	Night vision sensors at daylight levels
	Determinants of performance for all contrast limited sensors, $\tau_c = 1/[1 + K(1 - T_a)/T_a)]$	Photographic EO systems

where I = intensity of source (Watt/sr), R = distance from the source (m), T_a = atmospheric transmittance, and E = irradiance (Watt/sr/m^2).

Gaseous Atmosphere. If the radiation path is through a gaseous atmosphere, some of the radiation is lost by scattering and some by absorption. Therefore, the above equation has to be multiplied by a modified parameter T_a to represent the new atmosphere transmittance over a designated, which can be significantly less than unity (between 0.5 and 0.7).

The atmospheric transmittance parameter (T_a) over a path is a function of several variables, namely, wavelength, path length, pressure, temperature, humidity, zenith angle, and chemical composition of the atmosphere. The parameter T_a causes a reduction in radiant intensity due to absorption, scattering, diffraction, and reflection characteristics of the atmospheric gases. The atmospheric transmittance is defined as

$$T_a = e^{-\sigma_0 R} \tag{2.15}$$

where σ_0 = overall spectral attenuation coefficient in km^{-1} and R = range in km.

$$\sigma_0 = \sigma_a + \sigma_R + \sigma_s + \sigma_d \tag{2.16}$$

where a stands for absorption, r for reflection, s for scattering, and d for diffraction. Assuming negligible contributions from reflection and diffraction to the overall attenuation coefficient, the above equation is reduced to

$$\sigma_0 = \sigma_a + \sigma_s \tag{2.17}$$

It is important to mention that the coefficients σ_a and σ_s include the absorption and scattering contributions both from the gaseous media and aerosol distributions. Absorption due to water vapor, carbon dioxide, and ozone is most serious. One will notice later on that the absorption due to other atmospheric constituents is negligible (<5%). It is important to point out that ozone absorption is most pronounced in the ultraviolet region of the spectrum and is negligible at longer radiation wavelengths exceeding 8 μm. The Rayleigh scattering coefficient is proportional to $1/\lambda^4$. The aerosol coefficient is a complex function of particle shape and size, wavelength, refraction index of the medium, and scattering angle.

2.11.1 Transmission through the Entire Atmosphere

Plots of spectral transmittance (%) through the entire atmosphere [12, 13] from sea level to outer space along travel paths as a function of zenith angles of 0°, 60°, and 70.5° are shown in Figure 2.7. The plots are a function of various zenith angles and air mass ratios ranging from 1 to 3. Those curves indicate the net loss from all scattering mechanisms in a clear atmosphere. Scattering due to air molecules is called Rayleigh scattering and scattering from large aerosol particles is known as Mie scattering. Regardless of the scattering type, the scattering coefficient is directly proportional to the size of the particle (or radius) and inversely proportional to the transmission wavelength.

2.11.2 Horizontal Path Transmittance

The atmospheric spectral transmittance over a 1000 ft horizontal distance at sea level in the 0.5–25 μm spectral region is shown in Figure 2.8. The pronounced absorption effects due to water, carbon dioxide, and N_2O as a function of wavelength are clearly visible in Figure 2.8 [13]. As stated earlier, atmospheric transmittance over a path length at a specified wavelength is only valid for very narrow wavelength bands such as laser transmission and IR transmission on a horizontal path through the atmosphere with uniform composition. The attenuation coefficients for each of the several atmospheric constituents such as ozone, aerosol, and oxygen must be calculated separately and combined to obtain the total or overall effect of the transmittance. Figure 2.9 shows the sea level attenuation coefficient for a horizontal path in a clear standard atmosphere with a sea level visibility of 23.5 km. Absorption effects due to water vapor and carbon dioxide are included in the model shown in Figure 2.9. These effects are generally negligible for narrow band radiation at specific wavelengths shown in Figure 2.9. The total sea level attenuation coefficient is indicated by the curve with small circles. This coefficient is the sum of the ozone absorption coefficient, the Rayleigh scattering coefficient, and the aerosol scattering coefficient. It is important to remember that the effects of ozone absorption are most serious in the UV region of the spectrum and become quite negligible at longer radiation wavelengths exceeding 10 μm or so.

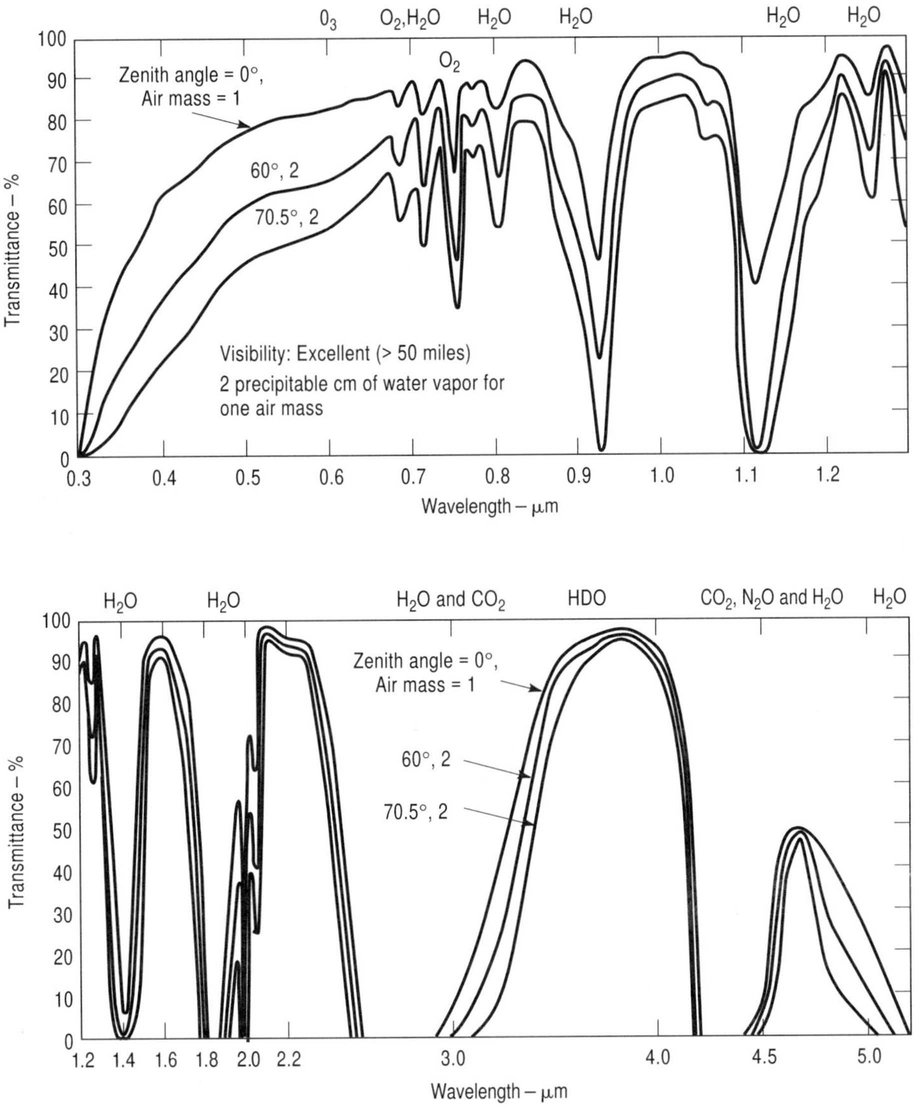

FIGURE 2.7 Spectral transmittance of the atmosphere as a function of wavelength and zenith angle assuming air mass value of 1.0.

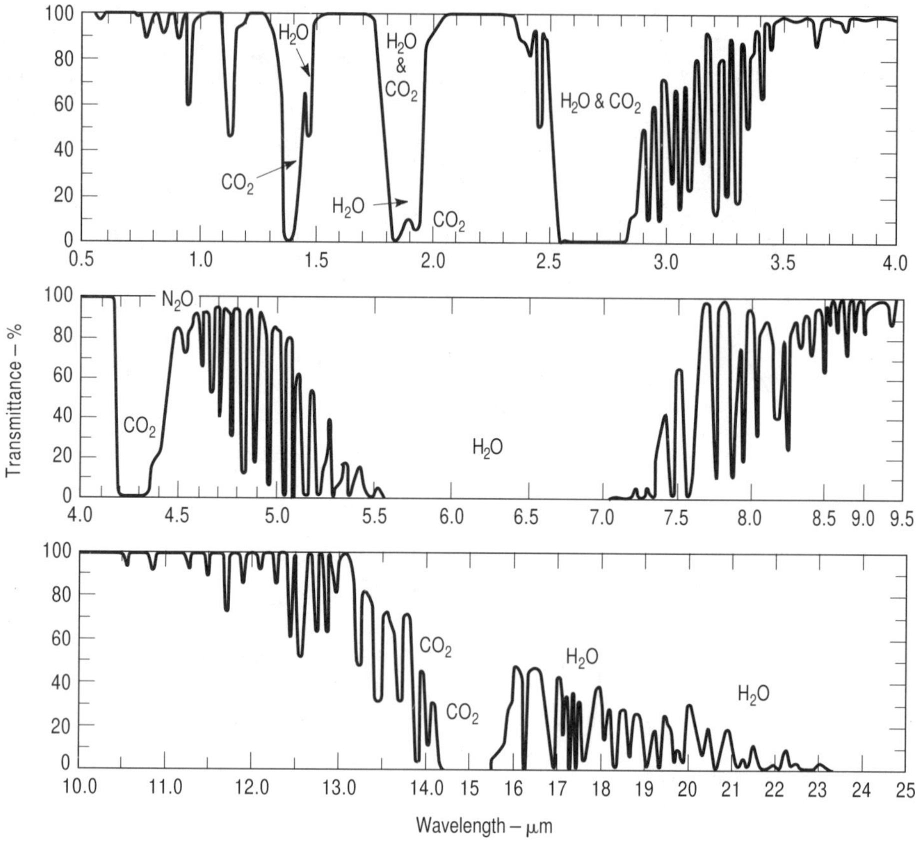

FIGURE 2.8 Transmittance of a 1000 ft horizontal air path at sea level as a function of operating wavelength and assuming a water temperature of 79° F.

2.11.3 Transference of Contrast Function [14]

$$\tau_c = \left[\frac{1}{\left\{ 1 + K\left(\frac{1 - T_a}{T_a} \right) \right\}} \right] \tag{2.18}$$

where τ_c = transference, K = sky/background ratio, and $T_a = \sigma_0 R$, which has been defined in the previous section. From Equation (2.18), $\tau_c = T_a$ when $K = 1$. The following values of parameter K are used for different environments:

For fresh snow: $K = 0.2$ (clear)
$K = 0.9$ (overcast)

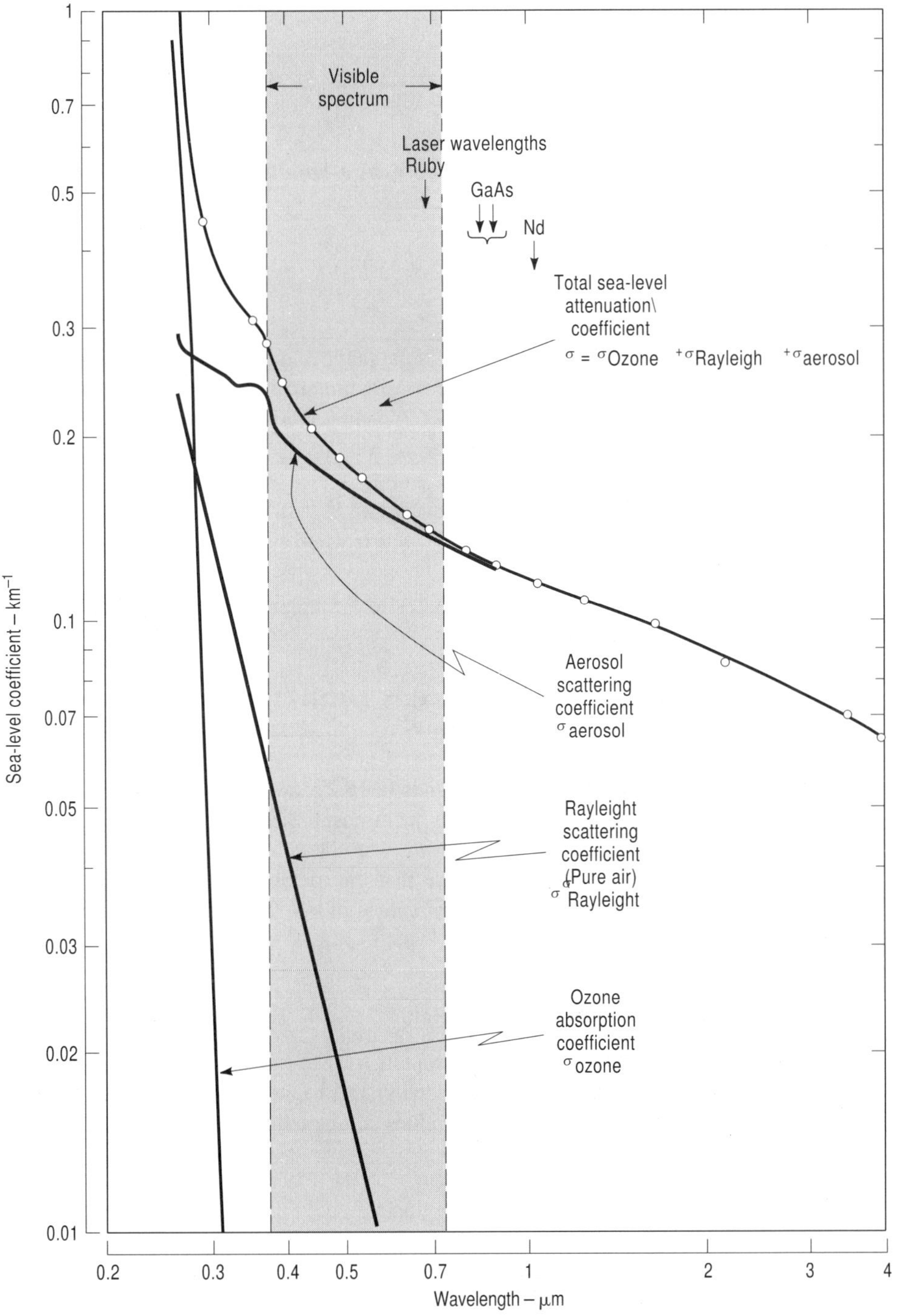

FIGURE 2.9 Atmospheric attenuation coefficient for horizontal transmission at sea level under clear atmospheric conditions.

For desert: $K = 1.4$ (clear)
$K = 7$ (overcast)

For forest: $K = 5$ (clear)
$K = 25$ (overcast)

2.11.3.1 Computations of Transference τ_c as a Function of Sky/Background Radiance Ratio

For fresh snow: $\tau_c = [0.978]$, when $K = 0.2$ and $T_a = 0.9$
For overcast: $\tau_c = 0.909$, when $K = 0.9$ and $T_a = 0.9$

In the equation for $T_a = e^{-\sigma_0 R}$ (where R is the range and σ_0 is the transmission coefficient) a value of 0.125/km is assumed for the attenuation coefficient parameter (σ_0). Using this value and a value of 0.9 for T_a, one gets a maximum range of 0.833 km. Note that when parameter K is unity, τ_c is equal to T_a.

For desert: $K = 1.45$ (clear) and 7.0 (overcast); $\tau_c = 0.417$ (clear with $K = 1.4$), 0.125 (overcast with $K = 7.0$), and 0.5 (clear when $K = 1$).

For the 0.5 value, the maximum range comes to 5.5 km, assuming the same value equal to 0.125 /km for the parameter σ_0.

For forest: $K = 5$ (clear) and 25 (overcast); $\tau_c = 0.0217$ (clear with $K = 5$), 0.0044 (overcast with $K = 25$), and 0.1 when $K = 1$

For the 0.1 value, the maximum range comes to 18.27 km. These calculations indicate that the maximum range occurs when the atmospheric transmittance approaches zero.

The above calculations further indicate that the maximum range is 0.833 km, 5.500 km, and 18.270 km for T_a parameter values of 0.9, 0.5, and 0.1, respectively, with a value of 1.0 for parameter K.

2.11.3.2 Scattering from Water Droplets

The scattering coefficient for the water droplets (due to rain, heavy fog, and snow) is independent of operating wavelength in the visible to far-infrared (FIR) region of the spectrum. However, its approximate values can be calculated using the following equation:

$$\sigma_{\text{rain}} = 0.248 r^{0.67} \tag{2.19}$$

where σ_{rain} = the scattering coefficient from rain (km^{-1}) and r = the rainfall rate (mm/hr).

2.12 HORIZONTAL VISIBILITY FOR WEAPON DELIVERY

Horizontal visibility is the most critical requirement in case of IR sensors capable of providing high-quality imagery needed for precision weapon delivery. "Visibility range," "horizontal visibility," and "meteorology range" are used to define the horizontal distance represented by a symbol R_v, for which the contrast transference is 2%. The apparent contrast decays exponentially with range according to the $e^{-\sigma_v R}$ law, where σ_v is the average attenuation coefficient for the visible spectrum range from 0.38–0.72 μm. The effective value of σ_v as a function of the visible range R_v is expressed empirically by the Golly equation:

$$\sigma_v = \frac{3.912}{R_v} \tag{2.20}$$

Atmospheric attenuation coefficients as a function of daylight visibility range are shown in Figure 2.10 for various atmospheric environments. This plot is valid for standard clear atmosphere with a visibility of 23.5 km and for radiation wavelength <1 μm. The coefficient has been corrected for different visibilities by a factor shown in Figure 2.10. The line structure of atmospheric absorption due to water vapor, carbon dioxide, and other absorbers is not included in the empirical method

2.12.1 Correction Factor for σ_v Parameters as a Function of Altitude

Correction factors as a function of altitude for two cases are shown in Figure 2.11. The lower curve gives the correlation factors for horizontal paths from sea level to specific altitudes, and the upper curve is for slant paths from sea level to specific altitudes. The horizontal path curve assumes that nonclear atmospheres have similar profiles. The slant path curve is based on integration of an exponential approximation of the standard atmosphere over the appropriate paths. This (upper) curve is less accurate than the horizontal path curve (lower curve). Accurate values of slant paths can be computed if values of "extinction optical thickness" are used to compute the atmosphere transmittance from a point at sea level to points at various altitudes and horizontal ranges from that point. Contours of constant transmittance at wavelengths 1.06 μm and 2.12 μm in a clear standard atmosphere with a visibility of 23.5 *K* are shown in Figure 2.12. These curves offer most accurate values of constant atmospheric transmittance values (T_a) as a function of altitude (h) and horizontal range at two different wavelengths.

2.13 ATMOSPHERIC TRANSMITTANCE OVER VARIOUS PATH LENGTHS

This section provides simplified method to compute the atmospheric transmittance (T_a) over various path lengths as a function of altitude for both horizontal and slant paths. Values of the attenuation coefficient at sea level as a function of wavelength

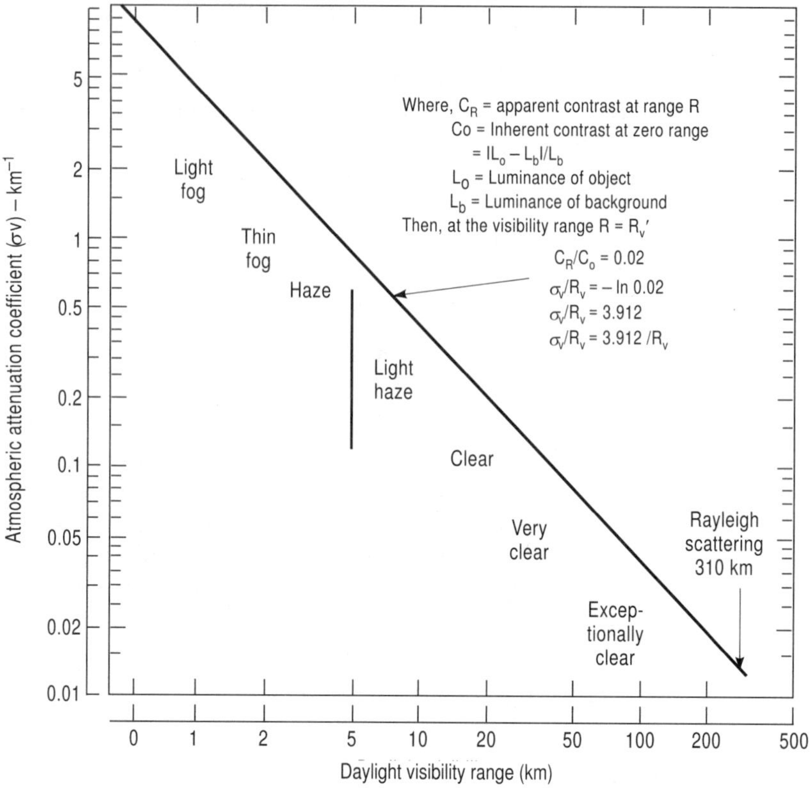

FIGURE 2.10 Atmospheric attenuation coefficient as a function of daylight visibility range and climatic environments.

and atmospheric conditions are shown in Figure 2.13. The atmospheric conditions are identified both by the visibility range and by a general description of atmospheric conditions such as clear, haze, very clear, etc. The curves shown in Figure 2.13 neglect absorption from water vapor and carbon dioxide. Correction factors for horizontal paths and for slant paths from sea level as a function of altitude are shown in Figure 2.11. As stated earlier, the horizontal path curve is based on a standard clear atmosphere model that assumes that nonclear atmosphere have similar profiles; the slant path curve is based on the integration of the exponential approximation of the standard atmosphere over appropriate paths and offers less accurate values than the horizontal path curve.

Water vapor concentration, which is a function of temperature and relative humidity, has a great impact on the detection range capability of an IR sensor. Values of water vapor concentration as a function of temperature and relative humidity are shown in Figure 2.14. Exponential approximations for the ratios of water vapor concentration (p) to sea level concentration (p_0) for slant paths and horizontal paths are

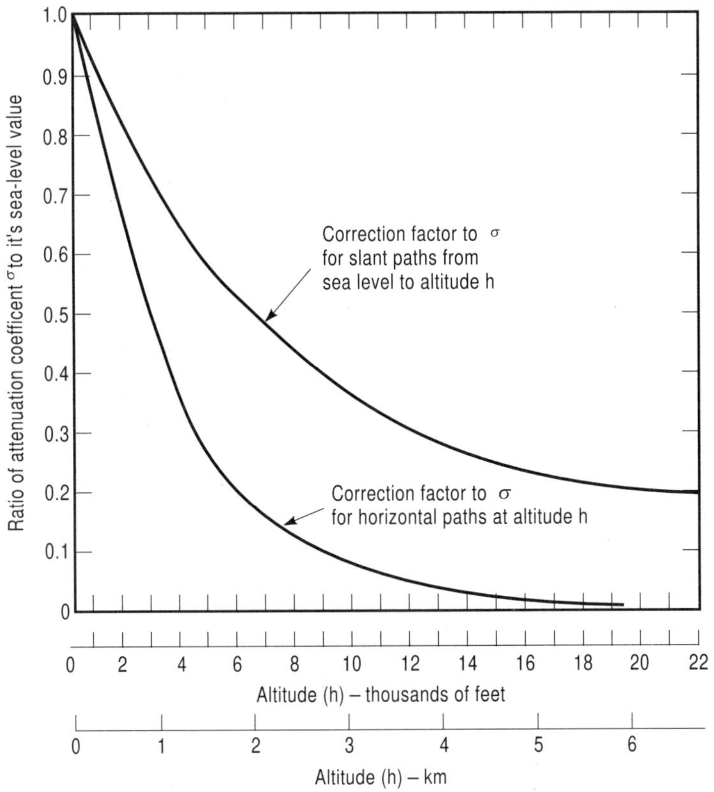

FIGURE 2.11 Correction factors to attenuation coefficient for slant and horizontal paths as a function of altitude.

shown in Figure 2.15 for tropical, MS, MW, and SS climatic conditions. The correlation factors for horizontal paths and for slant paths from sea level to altitude are given in Figure 2.15 as a function of altitude.

2.14 EFFECTS OF ATMOSPHERIC SCATTERING AND ABSORPTION ON IR SENSOR PERFORMANCE LEVEL

This section deals with the mathematical equations capable of predicting the effects of atmospheric scattering and absorption on the performance level of electrooptical sensors and devices, including the photoelectron devices, photonic devices, optoelectronic components, imaging sensors, and passive IR sensors. The apparent radiance L_0' of an object as seen by an observer at a horizontal range R [2] is given as

$$L_0' = L_0 T_a + L_{hs}(1 - T_a) \tag{2.21}$$

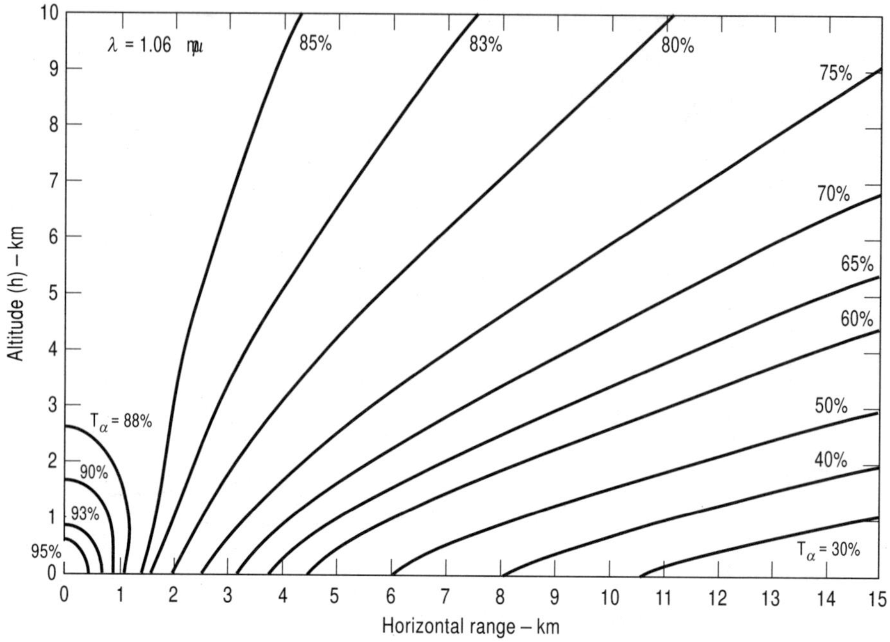

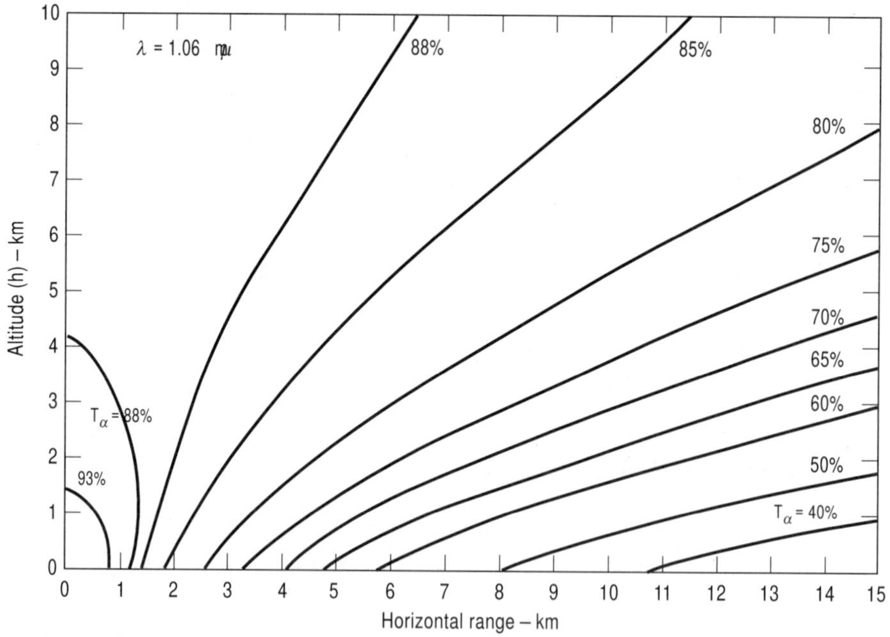

FIGURE 2.12 Contours of constant atmospheric transmittance for radiation at 1060 nm (top) and 2170 nm (bottom) in clear atmosphere with 23.5 km visibility.

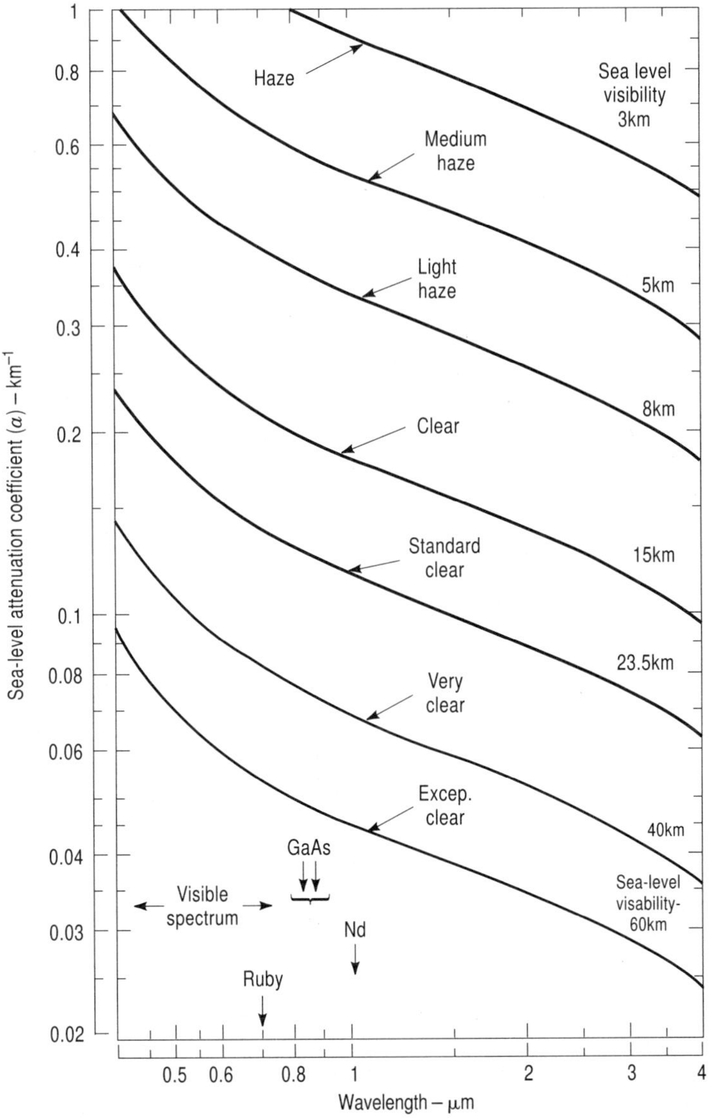

FIGURE 2.13 Sea-level attenuation coefficient variation as a function of wavelength and atmospheric conditions.

where L_0 = inherent radiance at zero range
T_a = atmospheric transmittance of the path R
L_{hs} = radiance of the horizon sky

The term $L_{hs}(1 - T_a)$ indicates the path radiance of the atmosphere between the source and the IR sensor. Note that equation (2.21) is valid for cases with uniform

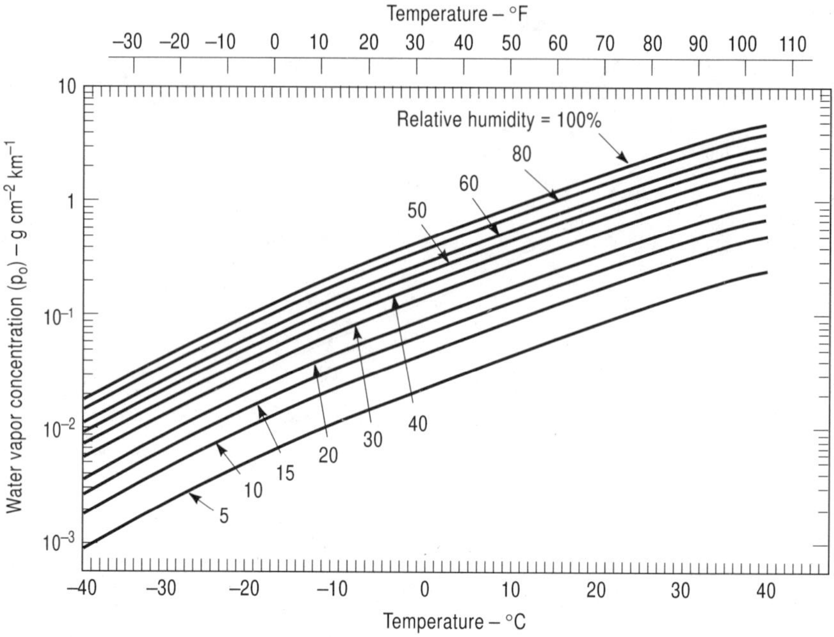

FIGURE 2.14 Water vapor concentration per km path length as a function of temperature (F and C) and relative humidity (%).

irradiance of the path and for a spatially homogeneous spectrum involving particle sizes and types with negligible absorption in the spectral band of interest. The apparent radiance of an object is different from its inherent radiance because of atmosphere transmittance, sky radiance effects, apparent contrast of the object (or target), and visual perception. The transference contrast function (τ_c) represents the ratio of an apparent contrast observed at the same range through the atmosphere to the same contrast observed at zero range. In certain cases the transference ratio is simply equal to the transmittance of the atmosphere ($\tau_c = T_a$). Note that the transference of an optical property is a complex function of the atmospheric transmittance, radiance of the object (L_o), and radiance of the background (L_b). This means that the ratio of horizon sky radiance (L_{hs}) to background radiance (L_b) can be written as

$$K = \frac{L_{hs}}{L_b} \tag{2.22}$$

where L_{hs} = radiance of the horizon sky and L_b = radiance of inherent background. Based on equation (2.22) one can now show that

$$\left[\frac{|L'_o - L'_b|}{L'_b}\right] = \tau_c\left[\frac{|L_o - L_b}{L_b|}\right] \tag{2.23}$$

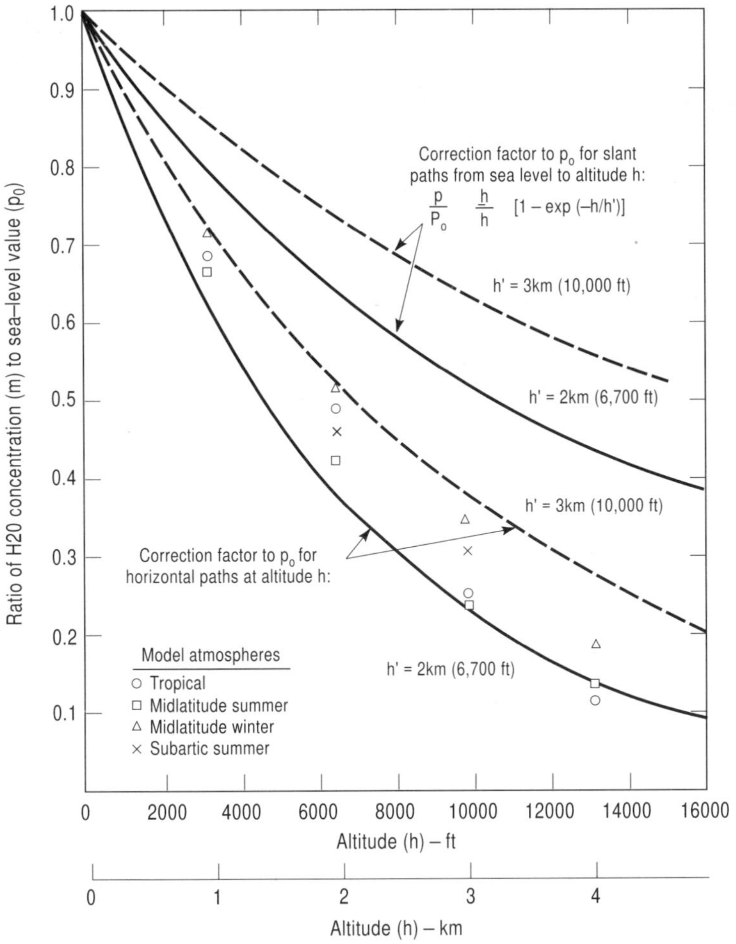

FIGURE 2.15 Ratios of water vapor concentration (p) to sea level value for slant and horizontal ranges based on various atmospheric models listed.

where L'_o = observed radiance of the object at horizontal range, R

L'_b = observed radiance of the background at the horizontal range, R

τ_c = transference of the contract function (also know as transference)

$$\tau_c = \frac{1}{[1 + (1 - T_a)/T_a]} \tag{2.24}$$

where T_a = atmosphere transmittance of the radiation

The effect of the sky radiance ratio (K) on the transference and the transmittance is evident from the curves shown in Figure 2.16. The values of parameter K for three types of terrain—snow, desert, and forest—under a variety of atmospheric conditions are also shown. The transference parameter τ_c as a function of T_a for three distinct values of K, namely, $K = 0.2$, 1.0, and 25, are shown in Figure 2.17.

2.14.1 Prediction of Back Scattering Effects on IR Sensors

This section outlines a reliable method to predict the effects of atmospheric back scattering on sensors' image quality when an artificial source near the image sensor is deployed to illuminate a distant scene or target. Assuming the spectrum between the illuminator and imaging system is small compared to the observation range and a constant atmosphere attenuation coefficient (σ), the field of view (FOV) of the sensor can be calculated by summing the back scattering contributions of all illumi-

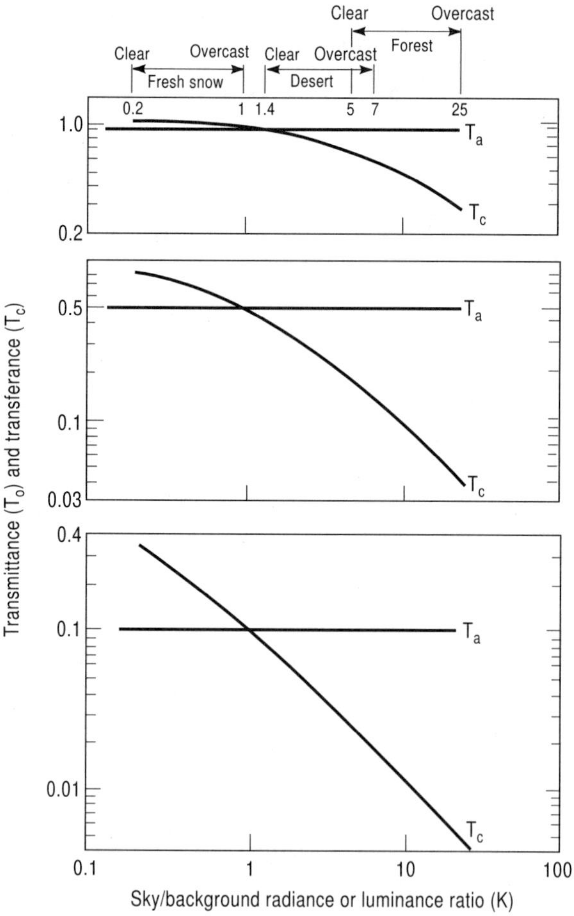

FIGURE 2.16 Transmittance and transference as a function of luminance ratio (*K*).

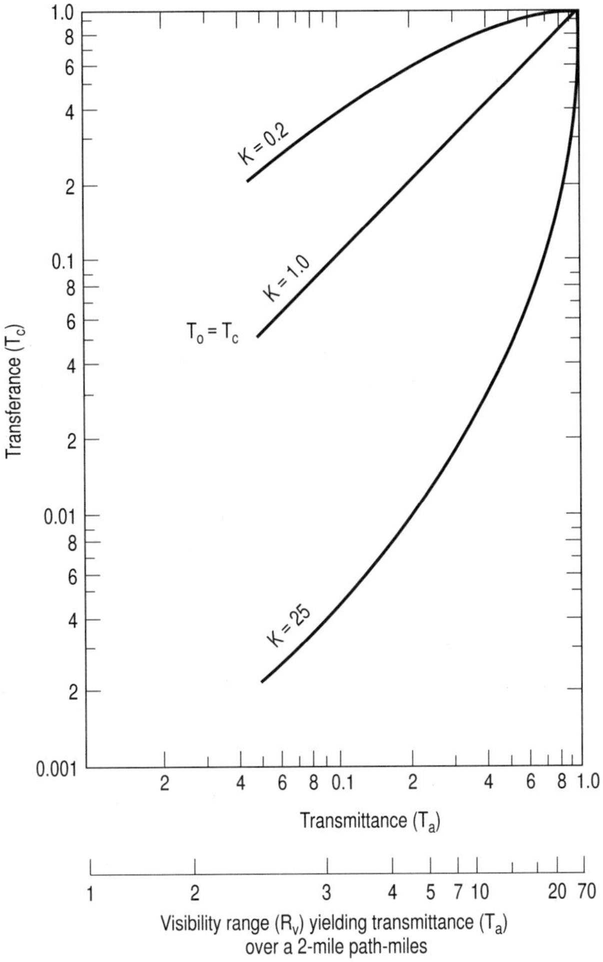

FIGURE 2.17 Transference parameter as a function of transmittance and luminance parameter (K).

nating particles on the path between the sensors and the target. The radiance of the back scattering (L_{bs}) can be expressed as

$$L_{bs} = \left(\frac{G\sigma^2 I}{2\pi}\right)\int_{x_1}^{x_2}\left[\left(\frac{e^{-x}}{x^2}\right)dx\right] \tag{2.25}$$

where L_{bs} = radiance of atmosphere due to back scattering

$X_2 = 2\sigma R_{\max}$

$X_1 = 2\sigma R_{\min}$

G = back scatter gain of atmospheric particles related to isotropic scattering (typical value of 0.4 at $\lambda \leq 1$ μm)

X_2 and X_1 are limits of integration at maximum and minimum ranges
I = radiant intensity of illuminator (W/sr).

Inserting ($\int e^{-x}/x^2\, dx$) = {$1/z[E_2(z)]$} and $t = (x/z)$ in equation (2.25), one gets

$$E_2(z) = \int_{x_1}^{x_2} \frac{e^{-zt}}{t^2}\, dt \tag{2.26}$$

Combining equations (2.25) and (2.26), one gets

$$L_{bs} = \left(\frac{G\sigma I}{4\pi}\right)\left[\frac{E_2(X_1)}{R_{\min}} - \frac{E_2(X_2)}{R_{\max}}\right] \tag{2.27}$$

The plot of exponential integral function $E_2(z)$ is shown in Figure 2.18. This figure provides the plot of the integral $E_2(z)$, which aids in rapid calculation of the atmospheric back scatter effect when artificial illumination is used. This plot may be of significant importance for the design and development of airborne image sensors for military applications requiring images of high quality.

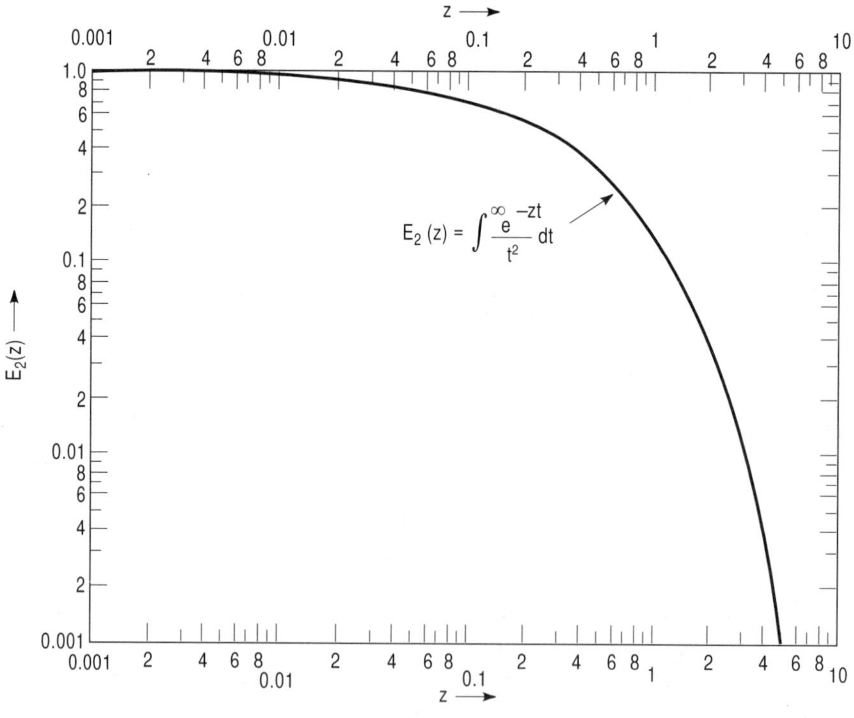

FIGURE 2.18 Plot of exponential integral to compute back scattering effects.

2.14.2 Target-to-Background Contrast

The target-to-background contrast is given by

$$C = \left(\frac{\rho_t}{\rho_b}\right) - 1 \tag{2.28}$$

where ρ_t = target reflectance
ρ_b = background reflectance

Note that the contrast parameter determines the quality of image and plays a significant role in predicting the performance of EO imaging sensors under development. The impact of target and background reflectance on on target-to-background contrast is shown in Figure 2.19 [15].

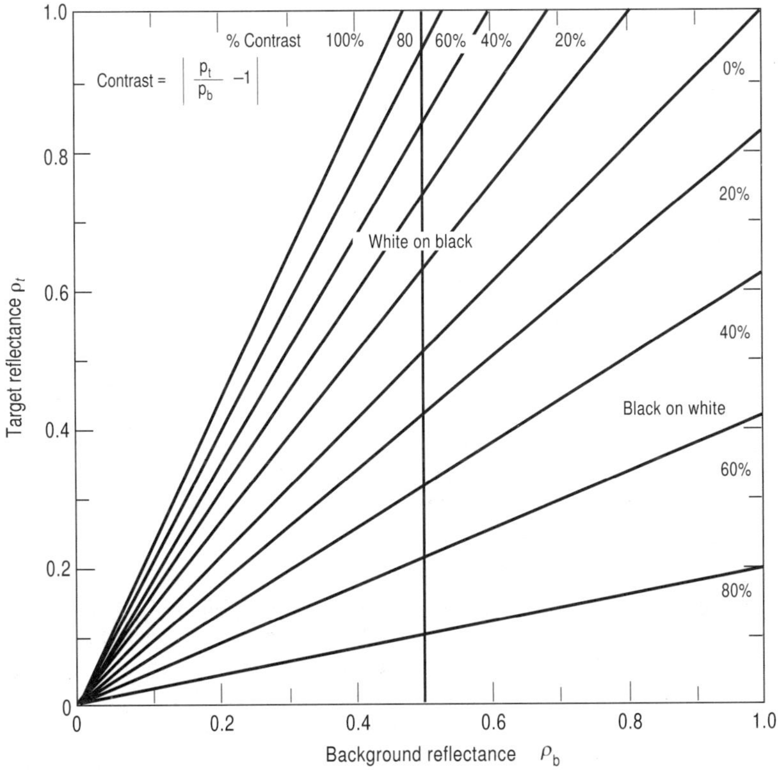

FIGURE 2.19 Target reflectance as a function of background reflectance and contrast ratio at 850 nm wavelength.

Numerical Example

Predict the effects of atmospheric back scattering on image quality assuming the following parameters: wavelength (λ) = 0.85 μm, attenuation coefficient (σ) for horizontal path at sea level = 0.125/km, target reflectance (p_t) = 0.4, background reflectance (p_b) = 0.6, back scattering gain (G) = 0.24, maximum range = 1000 m, minimum range = 15 m.

Solution: Using equation (2.27) and Figure (2.18), the radiance due to atmospheric back scattering can be expressed as

$$L_{bs} = [I/\pi R_{\max}^2][(G\sigma R_{\max})/4][(R_{\max}/R_{\min})E_2(2\sigma R_{\min}) - E_2(2\sigma R_{\max})]$$
$$= [I/\pi R_{\max}^2]\,[(0.24)(0.126)(10^{-3})(10^{-3})/4][(1000/15)E_2(2 \times 0.126 \times 10^{-3} \times 15) -$$
$$E_2(2 \times 0.126 \times 10^{-3} \times 10^3)] = [I/\pi R_{\max}^2][0.00756][(66.66)E_2(0.00375) - E_2(0.252)]$$

Inserting the values of $E_2(0.00378) = 0.98$, and $E_2(0.252) = 0.51$ obtained from the exponential integral plot shown in Figure 2.18, the above expression is reduced to

$$L_{bs} = [1/\pi R_{\max}^2][0.00756][(66.66 \times 0.98) - (0.51)]$$
$$= [I/\pi R_{\max}^2][0.00756][65.33 - 0.51] = [1/\pi R_{\max}^2][0.49]$$
$$= A[0.49]$$

where $A = [I/\pi\, R_{\max}^2]$

Performance comparison with and without atmospheric scattering

Performance parameter	Without back scattering	With back scattering
Apparent target radiance	L'_o	$A[0.49 + 0.4 \times 0.6] = A(0.118)$
Apparent background Radiance	L'_b	$A[0.49 + 0.4 \times 0.6] = A(0.118)$
Apparent target to background contrast (C):	$\lvert(L'_0/L'_b) - 1\rvert$	|SAME|
	$\lvert(p_t/p_b) - 1\rvert$	$\lvert(0.49 + 0.24)/(0.49 + 0.36) - 1\rvert$
	$\lvert(0.4/0.6) - 1\rvert$	$\lvert(0.73/0.85) - 1\rvert$
	[0.333]	[0.141]

These calculations indicate that at a maximum range of 1 km, the apparent target-to-background contrast is reduced to (0.141) from a value of (0.333) in clear atmosphere, which amounts to a reduction by 42.6%. Repeating these calculations for a maximum range of 3000 m shows a reduction of apparent contrast to about 6% of its value at the scene. From this it can be concluded that as the range increases from 1 km to 3 km, there is a drastic reduction in apparent contrast at the scene, which could present a serious problem to a weapons delivery system engineer. This material will be of significant importance to the weapons designers who are seeking weapons delivery capability with high kill probability.

2.15 SUMMARY

This chapter summarizes the transmission characteristics of the electrooptical (EO) signals as a function of altitude, operating wavelength, and climatic environments. EO signals experience absorption, scattering, and diffraction while traveling through various regions of the atmosphere. Particularly, high-power laser beams are seriously degraded due to atmospheric absorption and scattering by the atmospheric gases and aerosol particles. High power laser beams are further affected by the atmospheric turbulence phenomenon. Even in clear air with no suspended particles such as dust, smoke, fog, or rain, the atmospheric transmittance parameter is strictly dependent on sensor wavelength, operating altitude, climatic conditions, and variations in atmospheric temperature, pressure, and density. High power IR missiles or airborne surveillance systems will experience pronounced effects, particularly in lower atmospheric regions, due to severe absorption absorption, scattering, diffraction, thermal blooming, gas breakdown, and turbulence-induced beam spreading. Various models are described to compute atmospheric transmission characteristics as a function of altitude and climatic conditions. Calculated values of absorption and scattering parameters due to aerosols and molecular distributions as a function of altitude and climate are provided to predict IR sensor performance under various operating environments. Absorption and scattering parameters summarized in various tables indicate that IR sensors operating at 100 km or more above the earth are less affected than to those operating in lower atmospheric regions. A numerical example is provided to demonstrate the impact of the target-to-background contrast on weapons delivery range capability in the presence of atmospheric back scattering.

REFERENCES

1. W. L. Wolfe and G. J. Zissis, *The Infrared Handbook,* 1978, Environmental Research Institute of Michigan, Ann Arbor, Michigan.
2. R. A. McClatchey et al., "Atmospheric attenuation of HF and DF laser radiation," Air Force Cambridge Research Lab Bedford, MA (AFCRL), Research paper No. 400, 23 May 1991.
3. R. A. McClatchey et al., "Optical properties of the atmosphere," Air Force Cambridge Research Lab Bedford, MA (AFCRL), Environmental research paper No. 354, 10 May 1971.
4. R. N. Spanbaker et al., *Journal of Molecular Spectroscopy, 16,* 1, 100, 1965.
5. T. F. Deutsch, *Applied Physics Letter, 10,* 234, 1968.
6. *The Infrared Handbook,* Wolfe and Zissis, pp. 105–110, 4.37–4.42, 1978.
7. P. D. Taylor, "Introduction to Ultra-wideband Radar System", pp. 327–329, CRC Press, Boca Raton (Florida).
8. W. L. Wolfe and G. J. Zissis, *The Infrared Handbook,* pp. 4.37–4.42, 1978.
9. E. P. Shethle et al., "Models of atmospheric aerosols and their properties," 22nd Technical Meeting on Optical Properties in the atmosphere, The Technical University of Denmark, Lynby (Denmark), pp. 27–31, October 1975.

10. W. L. Wolfe and G. J. Zissis, *The Infrared Handbook,* pp. 3.12–3.18, 1978.
11. Ibid., pp. 2.5–2.11, 1978.
12. W. M. Elasser, "Mean Absorption and equivalent absorption coefficient of a band spectrum," *Physical Review, 54,* 126–131, 1938.
13. *Electro-Optics Handbook,* pp. 83–84, RCA, Harrison , NJ, 1974.
14. Ibid., pp. 79–89.
15. Ibid., pp. 90–104.

CHAPTER THREE

Potential IR Sources

3.0 INTRODUCTION

This chapter deals with various infrared sources, which fall into two major categories, namely, coherent and noncoherent (incoherent). The incoherent sources are relatively simple and inexpensive. Such sources include incandescent nongaseous sources such as globars and tungsten lamps, discharge lamps, high pressure gaseous sources, and light emitting diodes (LEDs). The coherent sources are complex and costly, but offer superior performance over noncoherent sources. Coherent sources include lasers such as solid state, semiconductor, fiber-coupled diode, chemical, and gas lasers, optical parametric oscillators (OPOs), and erbium-doped fiber amplifiers (EDFAs). Coherent sources are used for applications where high resolution, maximum optical frequency stability, and narrow beam with high quality are the principal requirements. Various commercial sources such as synchrotrons, plasmatrons, sparks, exploding wires, shock tubes, microminiature lamps, and plasma tubes are available for specific applications. High power IR sources include high power CO_2 lasers, COIL lasers, DF lasers and HF lasers, which are used for specific industrial and military applications.

Man-made sources, also known as artificial sources, include surface vehicles such as buses, autos, tanks, and personnel carriers, helicopters, jet and propeller aircraft, and muzzle flash. Both laboratory and commercial sources (Figure 3.1) are based on black body design concepts. Standard sources offer high stability, superior performance, and high-quality beams and are generally used for standardization and calibration of radiation sources.

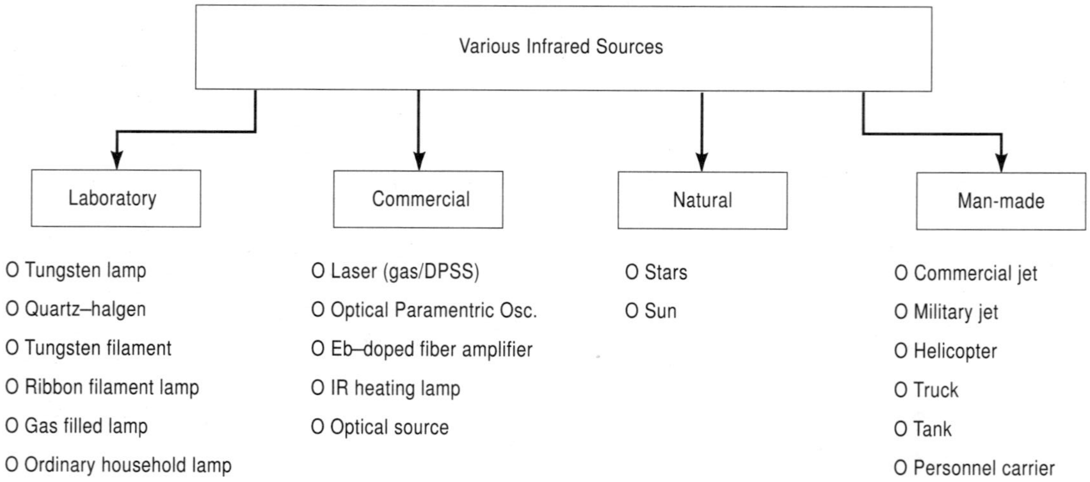

FIGURE 3.1 Summary of various infrared sources.

3.1 CLASSIFICATION OF INFRARED SOURCES

3.1.1 Laboratory Sources

These sources are designed to offer IR radiation of exceptionally high quality that is both reproducible and predictable. These sources generally use cavity configurations capable of producing radiation quality very close to Planckian. Such sources are used for calibration and standardization. Standard sources are widely used by The National Bureau of Standards (NBS). Applications using laboratory and commercial standards sources include photometry, pyrometry, polarimetry, optical radiometry, sensitometry, and spectrometry.

Conical, cylindrical, and spherical cavity configurations are generally used to obtain high-quality radiation level sufficiently close to Planckian. Note that only black body sources are suited to produce radiation that is nearly Planckian. The total emissivity of the cavity forming a black body is the most critical design parameter. This parameter is a function of cavity length-to-radius ratio, surface smoothness factor, and material used. Note that the power emitted by the cavity opening is dependent on spectral emissivity (a function of temperature and wavelength), spectral radiance level, area of emitting source, angle of radiation direction with respect to the normal, and partial reflections from the emitting element. Additionally, opening and temperature variation can have an impact on the IR power emitted from the cavity opening. Emissivity for both cylindrical and spherical cavities vary from 0.965 to 0.999 depending on the physical dimension of the cavity. The emissivity parameter is strictly a function of cavity length-to-exit hole radius (L/r) in the case of a cylindrical cavity and diameter-to-exit hole radius (D/r) in the case of a spherical cavity. Computed values of emissivity for these cavities are summarized in Table 3.1. These values indicate that large dimensional ratios offer the highest emissivities, which is highly desirable for an IR cavity.

3.1.2 Description of Sources Based on Specific Applications for Calibration or Measurement of Specific Parameters

3.1.2.1 Tungsten Strip Lamps

These lamps are manufactured by The General Electric company and are designated as GE 30 A/T24. They are high-temperature (> 2000 K), cylindrical graphite black

TABLE 3.1 Emissivity of cylindrical and spherical cavities as a function of their physical dimensions [1]

(L/r) or (D/r)	Cylindrical cavity	Spherical cavity
5	0.965	0.970
10	0.990	0.992
20	0.997	0.998
30	0.999	0.999

body sources with an emissivity greater than 0.996 over the 0.25–0.75 μm range. They are used as working standard source lamps. Such lamps are also available from other sources.

3.1.2.2 Tungsten–Quartz–Iodine Lamps

These lamps are also available from the General Electric Company and have spectral radiance levels in the 0.25–0.26 μm range. These lamps have optical power ratings varying from 200–1000 W and are widely used as standard sources for calibration.

3.1.2.3 FEL Standard Lamps

These lamps [1] are widely used in irradiance calibration. FEL lamps use either quartz–halogen or tungsten coil filament design configuration and have output ratings around 1000 W. These lamps have a minimum rated life of 500 hours and can be scanned over 0.25–0.80 μm to check the emission lines and absorption bands with bandwidth of 20–30 nm. Using prescribed positioning and alignment techniques, the lamps are slowly brought up to a designated electrical operating point. The lamps must be stabilized at least 15 minutes before making irradiance measurements if an accuracy better than 0.02% is desired.

3.1.2.4 Lamps for Spectral Radiance Calibration

Ribbon filament lamps are calibrated at 33 distinct wavelengths over the 225–2400 nm spectral region at a radiance temperature of about 2675 K for a wavelength calibration of 225 nm, at 2495 K for a wavelength of 650 nm, at 2415 K for a wavelength of 800 nm, and at 1620 K for a wavelength of 2400 nm with a moderate irradiance level. Quartz–halogen, irradiance standard 1000 W DXW, known as modified FEL, provides an irradiance level exceeding 1200 mW/cm^2 at a distance about 50 cm. This particular lamp provides calibrations at 24 wavelengths from 250–1600 nm.

3.1.2.5 Airway Beacon Lamps

Airway beacon lamps (500T-20/13) are widely used for calibration purposes with irradiance levels about 3 mW/cm^2 at a distance of 100 cm (or 3.3 ft).

3.1.2.6 Luminous Intensity Standard (LIS) Lamps

Various types of LIS lamps are available for calibration purposes. Standard lamps with 100 W ratings using tungsten filaments are used for luminous intensity calibration at about 90 candela (cd) and a color temperature of about 2700 K with an uncertainty of 4% of luminous intensity in SI units. Luminous intensity calibration lamps are also available that are capable of producing color temperatures exceeding 2700 K with color temperature uncertainty of 9 K relative to NBS standards. Several luminous intensity standard lamps are available to meet specific requirements of luminous intensity with various color temperatures and color temperature uncertainties.

3.1.2.7 Luminous Flux Standard (LFS) Lamps

Several LFS lamps are used for calibration of luminous flux level. A 25 W vacuum lamp provides calibration for luminous flux of about 270 lumens and a color temperature of about 2500 K with a luminous flux uncertainty of 4.5% relative to SI units. A 60 W gas filled lamp provides calibration for luminous flux of about 870 lumens and a color temperature of 2800 K with luminous uncertainty of 4.5% relative to SI units [1]. A standard 100 W gas filled lamp is used for calibration for luminous flux of greater than 1600 lumens and a color temperature greater than 2900 K with uncertainty of 4.5% relative to SI units. A 200 W gas filled calibration lamp is available with luminous flux capability of 3300 lumens and a color temperature of 3000 K with a luminous uncertainty of 4.5% relative to SI units. Calibration lamps with higher wattage ratings are required for higher luminous flux levels. A 500 W gas filled lamp can used for calibration of luminous flux level of about 10,000 lumens at a color temperature of 3000 K with luminous uncertainty of 4.5% relative to SI units and 1.5% relative to NBS standard. Suppliers for NBS-type standard lamps include Eppley Lab. Inc, Newport, RI, and Optronic Lab. Inc, Silver Spring, MD.

3.2 COMMERCIAL STANDARD SOURCES

Commercial black body standard sources are available that are capable of producing IR radiation with variable intensities using various radiating element shapes including cones, cylinders, spheres, and flat plates. Black body standards and sources with various design configurations including off-axis spherical cavities, reverse conical cavities, and cavities with parallel-groove structures and concentric geometry are described in Table 3.2.

It is important to mention that the critical component of an IR source is the radiating element structure, which can have different geometrical configurations. Separate apertures with air-cooled or water-cooled designs can be used to control the radiation output of the radiating element. Multiple intercepting cavities with conical

TABLE 3.2 Performance capabilities of certain standard sources

	Maximum operating temperature		
Performance parameters	<600 K	>600 K	1900 K
Cavity configuration	recessed cone of 15°	recessed cone of 15°	recessed cone
cavity emissivity	0.99	0.99	0.99
Source housing temperature			
Above ambient (°C)	< 10	< 10	< 10
Ambient temperature range (°C)	–40 to +60	–40 to +60	–4 to +60
Sensitive element	Platinum	Resistance	Thermometer (PRT)
Short-term stability (°C)	0.02	0.25	0.25
Long-term stability (°C)	0.01	0.50	0.05

and spherical shapes are used to achieve high emissivity ($\varepsilon = 0.999$) and high radiation output power. Auxiliary accessories such as multispeed chopper and temperature controlling units are incorporated in the design to yield uniform emissivity with desired controlled temperatures. Typical performance characteristics and physical parameters of these sources are summarized in Table 3.3.

Note that IR sources operating at lower temperatures are best suited for precision radiometry and satellite spectrometers. On the other hand, high-temperature sources are generally used for specific military applications, calibration, monitoring of long-term stability, and in visual optical pyrometers operating at temperatures exceeding 1080 °C.

3.3 INCANDESCENT NONGASEOUS BLACK BODY SOURCES

Nongaseous sources include glower, globar, and gas mantle. The choice of sources depends availability, cost, reliability, and safety requirements. In the long-wavelength IR (LWIR) region, the globar and gas mantle sources have a slight edge over glower sources.

3.3.1 The Nernst Glower [1]

This is rather small, simple, and least expensive source with modest optical output. The glower is made of a cylindrical rod or tube using high-temperature refractory materials such as zirconia, yttria, and beria. The cylindrical elements are available in many sizes. Platinum leads are attached to the ends of the tube which conduct power to the glower from the source. The glower requires application of external heat for the initial starting until it begins to radiate IR energy. The glower with a

TABLE 3.3 Performance and physical parameters of selected black body sources

(A) Refrigerated black body IR sources (Courtesy of Electro-optical Industries, Santa Barbara, CA)

Operating temperature (°C)	Cavity size (in)	FOV (deg)	Time (min), warmup/cooldown	Dimensions (in)	Weight (lb)
–20 to 100	1.0	30	10/10	13 × 12 × 20	50
–50 to 100	1.0	30	10/10	13 × 12 × 20	50
–20 to 600	1.0	30	10/20	13 × 12 × 20	50
–50 to 600	1.0	30	10/20	13 × 12 × 20	50

(B) Nonrefrigerated IR Sources (Courtesy of Infrared Industries, Santa Barbara, CA)

Temperature range (°C)	Cavity diameter (in)	Aperture diameter (in)	FOV (deg)	Warmup (min)	Power output (W)	Weight (lb)
50–600	0.25	0.10	50	10	60	2
50–250	0.50	0.20	12	20	100	5
400–1500	0.50	0.20	12	180	300	15
50–1000	1.00	0.60	20	60	460	17

0.05 in diameter element requires an input power of about 200 W to maintain a color temperature range of 1500–1950 K. A glower made with a long, thin cylindrical element is very useful for illuminating the spectrometer slits. It has an operating range from the visible region to about 30 μm; nevertheless, its effectiveness compared to other sources deteriorates beyond 15 μm. The radiance of a glower is close to that of a gray body source operating at the same temperature with an emissivity greater than 0.75 below 15 μm emission. The operational life of a glower decreases as the operating temperature is increased. The glower is extremely fragile and generally requires a rigid support to remain intact. The glower life is dependent on operating temperature, care in handling, and mechanical environment. Its lifetime varies from 250 to 1200 hours depending on intermittent or constant operation. Its lower cost and simple design with moderate radiation level makes it most attractive for optical laboratory measurements.

3.3.2 Glowbar Lamp

The glowbar is made from a cylindrical rod bonded to a silicon carbide structure capped with metallic caps, which act as electrodes for the conduction of the current through the glowbar from the power source. The supply current causes the glowbar to heat, thereby yielding radiation at temperatures exceeding 1000 °C. Water cooling through the housing containing the rod is required to cool the electrodes. The complex cooling scheme and higher costs are the major shortcomings of this source. A typical glowbar requires an input power exceeding 200 W and provides a color temperature of 1470 K. The spectral emissivity varies from 0.75 to 0.95 over the 2–12 μm range at color temperature range from approximately 1250–1375. Note that the spectral emissivity can change with use.

3.3.3 Gas Mantle

This is a high-intensity gasoline lamp and is widely used where electricity is not readily available such as campgrounds, remote mountainous retreats, etc. The critical element of this lamp is the mantle made from thorium oxide with some additive to enhance its efficiency in the visible region. Its near-IR emission is quite small over the 0.80–5.0 μm spectral range, but increases significantly beyond 10 μm. The gas mantle can be modified to achieve a laboratory experimental source. By using a gas flame on an electrically heated mantle, increased radiation levels can be obtained. This level can be further increased by directing the flame against the mantle at an angle, which will produce an elongated area of intense radiation with color temperature exceeding 1670 K over the 10–40 μm spectral range. This source is also best suited to illuminate the slits of a spectrometer.

3.3.4 Tungsten Filament Lamps

Tungsten filament lamps are widely used in homes, commercial buildings, and other places where cost and longevity are the demanding requirements. These lamps come in various forms and sizes with miniaturized and subminiaturized architectures and are available from several sources. These lamps provide steady sources of numerous types of IR radiation. The filament structure is used to reduce extraneous radiation to ensure quality and stability, which are needed for scientific investigations. Performance capabilities of some tungsten filament lamp sources are summarized in Table 3.4.

Some tungsten filament lamps are specifically designed for scientific applications including photometry, optical radiometry, spectrometry, polarimetry, calorimetry, spectrophotometry, microphotography, and stroboscopy. Photoflood lamps, which belong to this category of lamps, are designed for commercial photometry applications. These lamps come with wattage ranges of 250–1000 W, average life 4–16 hours, and illumination level from as low as 8500 lumens to as high as 75,000 lumens. A 1000 W photoflood lamp will have an illumination level of 31,000 lumens with average life of 10 hours. However, a 650 W lamp with 24,000 lumens illumination has an average life of 100 hours, while the same lamp with 75,000 lumens illumination will have an average life of only 16 hours. This indicates that combination of wattage rating and illumination level determine the average life of a photoflood lamp.

3.3.5 Carbon Arc Lamps

There are three distinct categories of carbon arc lamps

1. Low-intensity arc lamps
2. Flame arc lamps
3. High-intensity arc lamps

The first two types operate from DC power sources, and the last can be operated either from DC or AC power sources. All arc lamps require ballasts and most are considered to be high-intensity sources. In the case of DC arc lamps, brightness at the center of the crater is lower by 20–30% cd/mm^2 depending on the DC current level.

TABLE 3.4 Typical performance characteristics of tungsten filament lamps [4]

Wattage rating (W)	Color temperature (K)	Average life (hours)	Initial luminance (lumens)
10	2850	100	140
20	2900	200	300
30	2900	250	450
60	2850	1000	880
100	2850	200	1660
500	2950	200	11,000
750	3000	200	17,000
1000	3050	200	23,500

Typical applications of DC arc lamps include searchlight projection, motion picture projection, microscope illumination, and motion picture and television background projection. Color temperatures of DC arc lamps vary from 3600 K to 6500 K depending on specific application requirements.

Typical Flame Lamp

Flame-type arc lamps have been widely used in photochemical processes, therapeutic treatment, accelerated plant growth techniques, photocopying, graphic arts, and motion picture and television studio lighting. Spectral radiance levels from these lamps varies from about 29% to 36% of the input power over the 700–1500 nm range. Spectral intensity from these lamps at one meter from the arc axis varies from 1.58 to 3.96 mW/cm^2 over the 700–1125 nm spectral range. Spectral energy distribution of a carbon arc lamp using a soft-carbon core has a sharp and narrow response around 4000 Å compared to an arc lamp with a polymetallic carbon core, as shown in Figure 3.2. It is important to mention that an arc lamp with polymetallic-carbon core yields a spectral energy roughly four times with a 60-AMP arc lamp compared to that of a 30-AMP arc lamp at 4000 Å, as illustrated in Figure 3.2

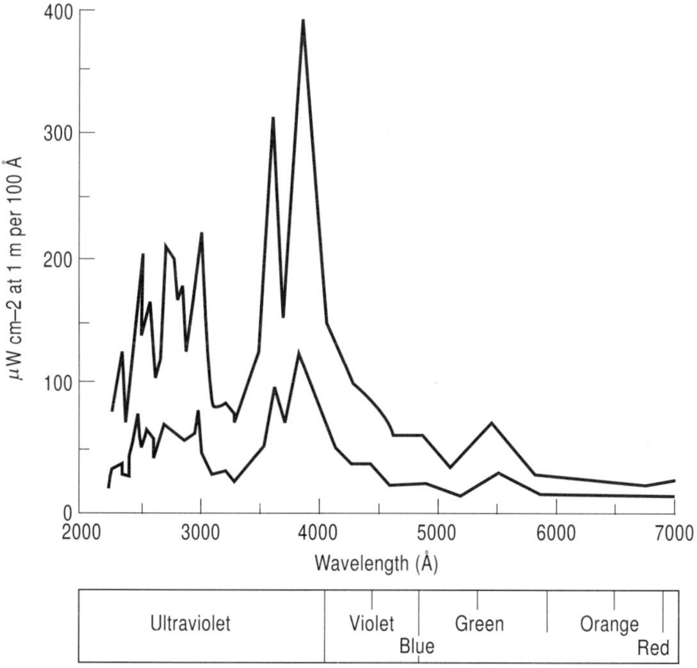

FIGURE 3.2 Spectral energy distribution of carbon arc lamps. (Upper curve: 60 A, 50 V; lower curve: 30 A, 50 V across arc.)

3.3.6 High-Pressure Discharge Lamps

Although arc lamps offer high intensity levels, they suffer from poor efficiency, short life, and produce harmful combustion products. High-pressure discharge (HPD) lamps offer constant radiation level, greater molecular interaction, higher efficiency, and longer life, and do not produce harmful products. Mercury lamps are widely used for domestic and commercial applications. Spectral intensity is dependent on the mercury pressure, which ranges from 30 to 300 atmospheres. High-pressure mercury lamps come in various shapes and sizes [1] as shown in Figure 3.3. Performance characteristics of various lamps are summarized in Table 3.5. The data shown in Table 3.5 indicate that deuterium lamps emit intense white light around 1800 nm, while both the halogen and Cesiwid glow-bar lamps yield radiation at 1 and 2 μm, respectively, as shown in Figure 3.4. Halogen lamps provide an efficient light source with long life, compact size, and constant light intensity.

In multivapor arc lamps, both the argon gas and the mercury vapor provide the needed starting function, while the sodium iodide, thallium iodide, and indium iodide vaporize and dissociate to provide most of the lamp radiation. Ballasts are similar to those used in mercury lamps. Performance data available on GE multivapor lamps indicate an average life of 10,000–15,000 hours and initial lumens ranging from approximately 34,000 to 88,000. Higher brightness and power outputs are possible with capillary mercury arc lamps, which operate at higher mercury pressures and require cooling to avoid catastrophic failures. The bulb walls are made from quartz and an outer quartz jacket is provided for air or water cooling. Water-cooled mercury lamps operate at higher supply voltage and current. Their average brightness varies from 40,000 to 90,000 cd/cm^2, output luminance is in excess of 60,000 lumens, and average life exceeds 80 hours. On the other hand, air-cooled capillary mercury arc lamps operate at relatively low voltages and with reduced output of about 60,000 lumens and lower life of 60 hours.

3.3.7 Compact Arc Lamp Sources

These lamps use mercury, mercury–xenon, or xenon gas in clear, spherical quartz enclosures with electrode terminals filled with inert gas and sealed. Note that higher luminance levels and temperatures require extreme electrical loading and initial gas pressure of several atmospheres. These lamps require initial high-voltage ignition pulses and a ballast to limit the current during the operation. These arc sources provide clear, maintenance-free operation for extended periods. Mercury and mercury–xenon compact arc lamps must be operated near the rated power level using a well-regulated power supply for stable operation. Spectral distribution of radiant intensity from a 5 kW DC xenon lamp varies from 6 to 8 W/Sr/10 nm over the 800–1000 nm spectral region.

3.3.8 Low-Pressure Discharge Sources

Low-pressure mercury (LPM) tubes are specifically designed to provide UV emission around 257 nm. The LPM lamps based on hot cathode design operate at rela-

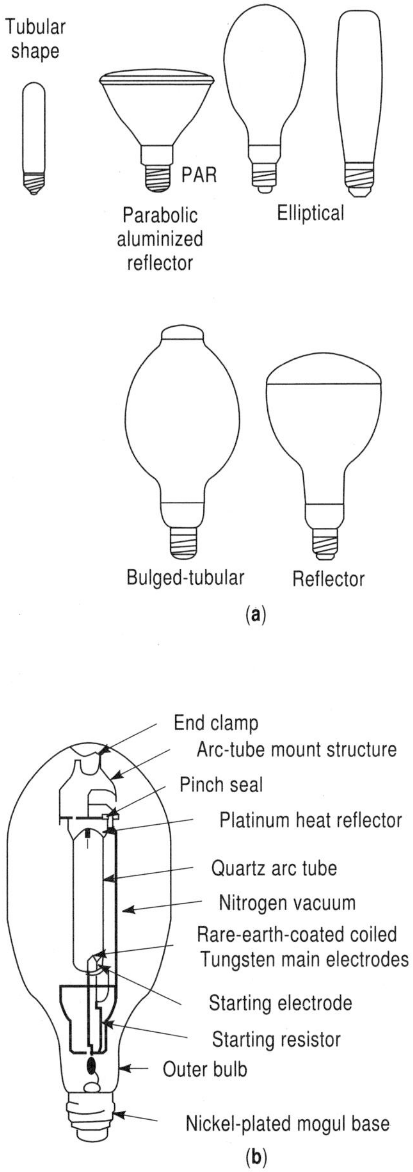

FIGURE 3.3 (A). Various bulb shapes and sizes. (B) High-pressure mercury lamp showing its critical elements.

tively lower voltage, ranging from 105 to 150 V. These lamps differ from ordinary fluorescent lamps, which transmit in the UV range. The walls of the fluorescent lamp are coated with material that absorbs UV radiation and reemits visible light. On the other hand, LPM lamps based on the cold cathode design operate at higher voltage (240–420 V). However, a lamp with a hot cathode design provides much

TABLE 3.5 Typical performance characteristics of arc lamps

Lamp type	Type	Manufacturer	P (W)	I (A)	U (V)	Luminous flux (lm)	Light Intensity (cd)	Luminance average (cd/cm^2)	Arc size W × H (mm^2)
D 902	Deuterium	Heraeus	30	0.2–0.5	75–95				1.0∅
XBO 75 W/2	Xe	Osram	75	5.4	14	1000	100	40,000	0.25 × 0.5
HBO 50 W/3	Hg	Osram	50	2.3	22	1300	150	90,000	0.2 × 0.35
HBO 100 W/2	Hg	Osram	100	5	20	2200	260	170,000	0.25 × 0.25
JQ 50	Halogen	Osram	50	4.17	12	1600			3.2 × 1.5
JQ 100	Halogen	Osram	100	8.35	12	3600			4.2 × 2.3
XBO 150 W/1	Xe	Osram	150	7.5	20	3000	300	15,000	0.5 × 2.2
XM 300-1HS	Xe	ORC	300	18	15–20	5000	700	39,000	0.75 × 1.5
HVN 200-1	Hg–Xe	ORC	200	8–9.5	20–25	4500	600	22,200	0.5 × 1.5
JQ 240	Halogen	Osram	150	6.25	24	6000			5.8 × 2.9
JQ 250	Halogen	Osram	250	10.42	24	10,000			7 × 3.5
Cesiwid 150	Glow-bar	Sign	150	8.2	18				4 × 2.5
XBO 450 W.OFR	Xe	Osram	450	25	18	13,000	1300	35,000	0.9 × 2.7
XBO 450 W/1	Xe	Osram	450	25	18.5	13,000	1300	45,000	0.7 × 2.2
XBO 500 W/H/OFR	Xe	Osram	500	28	18	14,500	1450	40,000	0.9 × 2.5
959 QTZ	Xe	ORC	500	30	14–20	9000	1500	350,000	0.3 × 0.3
L 5269 OFR	Xe	ORC	500	30	14–20	9000	1500	350,000	0.3 × 0.3
XBO 1000 W/HS/OFR	Xe	Osram	1000	50	20	32,000	3000	60,000	1.1 × 2.8
XBO 1000 W/HS	Xe	Osram	1000	50	20	32,000	3000	60,000	1.1 × 2.8
HBO 1002 W/C	Hg	Osram	1000	17 (750 W)	44 (750 W)			80,000	1.1 × 2.5
XBO 1600 W/OFR	Xe	Osram	1600	65	22	60,000	5500	70,000	1.5 × 3.3
XBO 2000 W/HTP/OFR	Xe	Osram	2000	70	29	80,000	7500	75,000	1.3 × 4.8
HBO 200 W/2	Hg	Osram	200	3.1–4.2	65–47	10,000	1000	40,000	0.6 × 2.2
941 B0010	Hg–Xe	ORC	600	26	20–25	23,000	2500	70,000	0.5 × 1
XBO 250 W/OFR	Xe	Osram	250	18	14	4800	530	26,000	0.7 × 1.7
HBO 350W	Hg	Osram	350	4.7–5.9	60–75	19,500	2100	50,000	0.8 × 2.7
HBO 500 W/2	Hg	Osram	500	5.9–7.4	85–67	30,000	2850	30,000	1.1 × 4.1

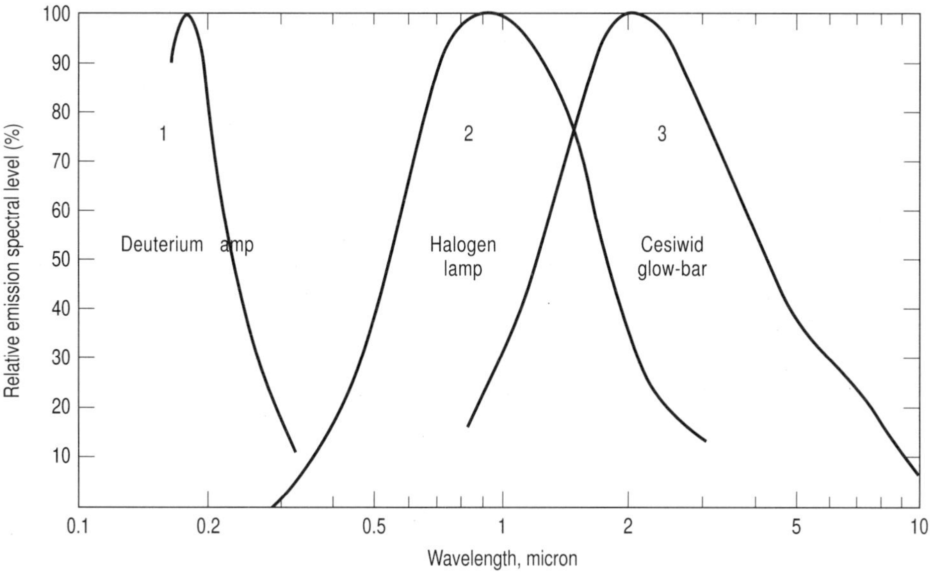

FIGURE 3.4 Relative emission spectral levels from various lamps. (A) deuterium lamp, (B) halogen lamp, and (C) Cesiwid globar lamp.

higher output compared to that of a lamp with a cold cathode design. In addition, the UV irradiation from hot cathode lamps is higher by 3–5 times compared to lamps with cold cathode designs operating at 257 nm wavelength.

3.3.9 Electrodeless Discharge Lamps

Electrodeless discharge (ED) lamps offer a simple and inexpensive radiation source with compact design. These lamps consist of fused-quartz tubes with radiation-producing elements and microwave generators for producing the electrical field within the tube needed for excitation. These tubes are capable of transmitting radiation from UV to near-IR regions. The ED lamp produces a stable radiation with sharp spectral lines, thereby making it most attractive for spectroscopy and interferometry applications. These lamps are available with various variable outputs of different UV spectra. A 10-inch, water-cooled electrodeless discharge lamp can deliver an IR output level of 1000 W with 30% efficiency and lifetime as high as 3000 hours [5].

3.3.10 Spectral Arc Lamp Sources

Spectral arc lamps (SALs) can be filled with different elements and rare gases and are capable of producing optical power throughout most of the UV and visible spectra. The envelope of this lamp is made from quartz or glass, depending on the operating environments and spectral requirements. Discrete radiation sources are possible, operating from 230 to 800 nm. These lamps offer design that are inexpensive, require no regulated power supply, and regulate the spectral intensity in the gaseous media within the tube.

3.3.11 Tungsten Arc Discharge (TAD) Lamps

The essential elements of this discharge lamp include a ring electrode and a pellet electrode both made from tungsten. The arc forms between these two electrodes, thereby causing the pellet to heat incandescently to a temperature as high as 3100 K. This lamp provides an intense source of radiation from a radiating element of very small area. Typical spectral distribution of a GE photomicrographic lamp with 30 A rating is about 160 μW/Sr/0.1 nm over 800–1000 nm.

3.3.12 Hydrogen and Deuterium Arc Sources

These sources are most suitable for applications requiring intense continuous radiation in the UV region (165–200 nm). Both lamps use quartz envelopes to meet stringent operational requirements under intense heat environments. Hydrogen lamps with a few mm of pressure provide useful sources down to 200 nm with minimum cost and power consumption. Note that a deuterium lamp offers a continuous radiation source with higher intensity than the hydrogen arc lamp. These lamp sources are generally used as laboratory standards in lower UV region.

3.3.13 Light Emitting Diodes (LEDs)

These sources are small, simple, and inexpensive. They are mostly used for integration in various systems to identify discrete functions of system operation or a set of specific performance parameters. Spectral ranges for various electroluminescent diode elements are shown in Figure 3.5. The LEDs emit visible and near-IR radiation with high intensity over very narrow spectral bands and are widely used for commercial, industrial, and military display applications. LEDs are classified in various categories based on specific emission bands and brightness levels. Incoherent and spontaneous devices are commercially available that are capable of operating over the 340–940 nm spectral range. LEDs with higher brightness levels [6] in the visible spectral range (370–750 nm) are generally fabricated using GaAsP (red), InGeP (red), GaP (red, yellow, and green) and GaN (blue, green, and yellow) materials. GaAsP and GaP LEDs are mostly used for commercial and military displays. LED sources used for display applications are summarized in Table 3.6. Performance parameters for specific commercially available electroluminescent LED diodes are summarized in Table 3.7. Emission wavelength, quantum efficiency, and power output of most popular LEDs are shown in Table 3.8.

Typical performance characteristics of most LEDs are as follows:

Visual appearance: Medium to wide viewing angles

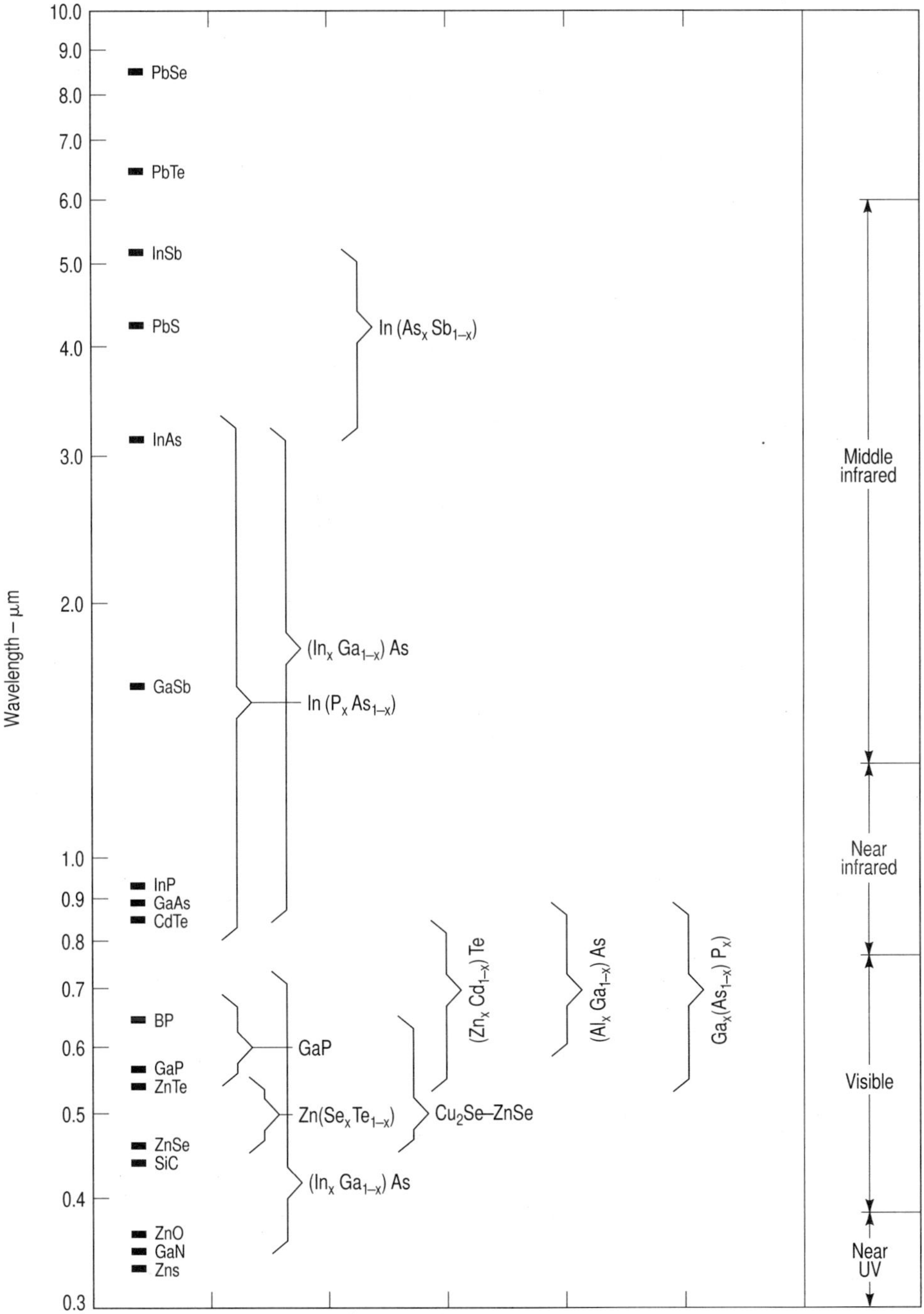

FIGURE 3.5 Typical spectral range for electroluminescent diodes.

TABLE 3.6 LED Sources for Display Applications

LED material	Emission wavelength (nm)
GaAsP	550, 900
CdTe	855
ZnCdTe	530, 830
ZnSeTe	627
ZnTe	620
GaP	565, 680
SiC	456
GaN	340, 700

Visible in dim ambient illumination
Different colors available
Power dissipation: 0.1 to 10 W/cm^2 at 2 V operation
Response time: Varies from 10 to 1000 ns
Operating temperature range: –40 to +100 C
Reliability (MTBF): 50,000 hours (typical)

It is important to mention that LEDs provide highly monochromatic emitting light over a very narrow frequency range. The specific color of light emitted is a function of the LED substrate material, as shown in Table 3.8, and is measured by a peak wavelength expressed in nanometers. The LED luminous intensity is measured in millicandela (mcd) and is directly proportional to the forward current flowing in the LED. The higher the current, the higher the intensity. LEDs are designed to operate at various supply voltages, depending on the internal series resistance. According to various suppliers, LEDs are designed to have lifetimes exceeding 100,000 hours (or 10 years) and are designed with shock and vibration resistance features. LEDs are widely used as replacements for incandescent lamps, front panel indicators, dashboard level displays, and automobile and aircraft front panel displays.

TABLE 3.7 Performance parameters of selected LED diodes [7]

Parameter	GaAs LED	GaAsP LED	GaP LED
Temperature (K)	300	300	300
Wavelength (nm)	940	660, 610	690, 550
Spectral bandwidth (nm)	50	30	90, 30
Typical drive current (mA)	50	15	15
Operating modes (CW or P)	P, CW	P, CW	P, CW
Rise time (ns)	250	10	300
Output power	2 mW (CW)	1700 nits*	600–3500 nits*
Power efficiency (%)	3	< 1	3, 0.1

*One nit = 0.2919 foot-lambert (fL)

TABLE 3.8 Emission wavelength, quantum efficiency, and output of some LEDs

LED material	Color	Peak emission wavelength (nm)	Maximum quantum efficiency (%)	Power efficiency (lumens/W)
GaAs	Red	649	0.005	0.33
GaP: N	Green	570	0.001	0.60
GaAsP: N	Yellow	589	0.008	0.36

3.4 MAN-MADE OR ARTIFICIAL SOURCES

Artificial or man-made IR sources include surface land vehicles such as automobiles, trucks, tractors, tanks, and trains, airborne sources such as missiles, aircraft (commercial and military), helicopters, and rockets, and sea-based sources such as cargo vessels, ships, aircraft carriers, submarines, missile destroyers, hydrofoils, hovercrafts, and high-speed coastal boats. Radiation characteristics from these sources include maximum radiation intensity, peak radiation wavelength, and spectral power density. The characteristics of military man-made sources are generally not available due to security reasons; nevertheless, they are available for the sources used specifically in commercial applications. Specific radiation patterns are of paramount importance to estimate the IR radiation energy released by various man-made sources. The IR target data can be calculated form the prediction models based on certain architectural features, physical parameters, and operating specifications. Most of the complex man-made sources such as aircraft or tanks are assumed to be made up of many facets representing the actual physical shape of the target. These facets are made of radiating materials with known emissivities and operating temperatures based on the conditions under which the vehicle or aircraft has been operating. Since these conditions vary from vehicle to vehicle, aircraft to aircraft, or tank to tank, it is difficult to specify radiative properties except in special cases to meet a specific performance requirement or mission objective.

The IR signature of a vehicle, aircraft, or missile can reveal the emisson wavelength and radiation intensity of the radiating source, depending on the emissivity and temperature of the exhaust pipe or exit nozzle. The spectral emissivity of a source associated with a vehicle, aircraft, or ship, which is a function of temperature and emission wavelength, is generally assumed to be close to 0.9 under normal operating conditions. When a military aircraft is operating under afterburner and extreme maneuvering conditions or is equipped with IR suppression devices to avoid detection by enemy IR sensors, precise prediction or estimation of its IR intensity will be extremely difficult. Furthermore, the radiation intensity of a man-made source is also dependent on weather environments, surface geometry, and surface conditions. Note that different parts of the vehicle or aircraft will be at different temperatures and are made of different materials. The radiation intensity will be maximum from the hottest part of vehicle or aircraft, which is the exhaust pipe or exit nozzle of the man-made source. Furthermore, one can expect higher IR signatures from an aircraft during takeoff or under afterburner, from a missile during firing phase, and from a tank moving at high speed. The radiation intensity (W/Sr) is a function of source temperature, emissivity of the exhaust surface, and area of the exposed surface of the emitting element. The IR signature models can predict the IR radiation level and emission wavelength fairly well, provided the relevant parameters such as surface emissivity, source temperature, rough estimate of source architectural dimensions, and geometrical configuration are accurately known.

3.4.1 Estimation or Prediction of Aircraft Signatures

An aircraft is a complex target and will have specific IR signatures resulting from various surfaces operating under different temperatures, altitudes, and climatic environments. In addition, estimation of IR signature becomes more complicated due to presence of hot gases in large quantities in the vicinity of engine exhaust area. Note that the highest spectral quality of radiation will be found when one observes the tail of the aircraft. The IR emission and absorption from the aircraft depend on the local thermodynamic conditions and molecular properties of the gases exiting from the exhaust pipe. The peak radiation intensity is dependent on engine type and speed, engine thrust, exhaust nozzle area, exhaust temperature, radiation from hot engine parts, radiation from the aircraft surface, and radiation from the hot gas plume. Note that the overall radiation intensity can be impaired due to the effects of the coannular flow, external flow, turbulent kinetic energy, and the emissivities, temperatures and areas of various surfaces contributing to the total radiation level. Radiation intensity, peak radiation wavelength, and other factors contributing to overall IR radiation intensity from commercial jet aircraft are summarized in Table 3.9. Total and net radiation intensities from all four engines have been calculated using equation (3.1) and assuming an emissivity of 0.9, parameter σ of 5.67×10^{-12} W/cm^2/K^4, cruise temperature (T) of 758 K and exhaust nozzle area (A) of 3660 cm^2, as shown in Table 3.9.

TABLE 3.9 Radiation intensity and peak radiation wavelength from jet engines used by commercial aircraft [8]

Parameters	707-320 Engine	707-320 B Engine
Engine type	Four turbojets	Four turbofans
Maximum thrust/engine (lb)	16,800	18,000
Exhaust nozzle area (cm^2)	3660	3502
Exhaust temperature at takeoff (K)	908	828
Radiation wavelength (μm)	3.19	3.50
Energy concentration (%)	16.2	22.3
Cruising temperature (K)	758	718
Cruising wavelength (μm)	3.82	4.03
Radiation intensity (W/Sr) under cruising conditions	1962 (per engine)	1512 (per engine)
Total radiation intensity from four engines (W/Sr)	7848	6048
Assuming IR energy of 29% over 3.2 to 4.2 μm, net radiation intensity from four engines (W/Sr)	2276	1814

$$\text{Radiation intensity, } I = \frac{(\varepsilon)(\sigma)(A)T^4}{\pi} \tag{3.1}$$

$$= \frac{(0.9)(5.62 \times 10^{-12})(3660)\ 758^4}{\pi}$$

= 1962 W /Sr for 707-320 engine and 1512 W/Sr for 707-320 B engine

These calculations indicate the rough estimates of the IR radiation intensity from jet engines as a function of temperature, nozzle area, and percentage of energy concentrated in a specified spectral band.

In case of military turbojet engines such as the Pratt and Whitney JT 40, the maximum thrust at sea level is roughly 15,800 lb per engine, which increases to about 23,500 lb under afterburner operation, leading to a very high exhaust temperature at the engine tail pipe. Preliminary studies performed on gaseous emissions and exhaust temperatures associated with military jets indicate the tail pipe radiation levels are as high as 25 times greater than gaseous radiation levels in the 4.30 to 4.55 μm range. In brief, IR radiation levels in military jets are significantly higher than IR levels produced by commercial aircraft.

It is important to mention that as the speed of the aircraft increases, both the skin temperature due to aerodynamic heating and the IR radiation level increase, which can be detected at large distances by IR sensors . The aircraft skin temperature (T_s) due to aerodynamic heating under laminar flow at 37,000 ft altitude can be computed using the following equation:

$$T_s = T_a(1 + 0.164\ \text{M}^2) \tag{3.2}$$

where T_a = temperature at a given altitude, which is available from various textbooks

M = aircraft speed (in Mach; 1 M = 666 knots)

Note that the temperature at a given altitude is also dependent on the climatic conditions and geographical region. Calculations using the above equation indicate that the highest surface temperatures occur under tropical conditions at a given speed and altitude. For example, the aircraft skin temperature (T_s) is about 602 K at an altitude of 50,000 ft and a speed of 5 M, but increases to 1678 K at an altitude of 100,000 ft and a speed of 10 M under tropical environments. The value of parameter T_a is 243 K at 30,000 ft, 203 K at 50,000 ft, and 232.5 K at 100,000 ft altitude under tropical climatic conditions. Skin temperatures as a function of speed and operating altitude are shown in Figure 3.6.

3.4.2 IR Radiation Levels from Muzzle Flash and Artillery Guns

The hot gases ejected from the muzzle of a gun contain a large amount of combustion products, including carbon monoxide, carbon dioxide, hydrogen, nitrogen, and

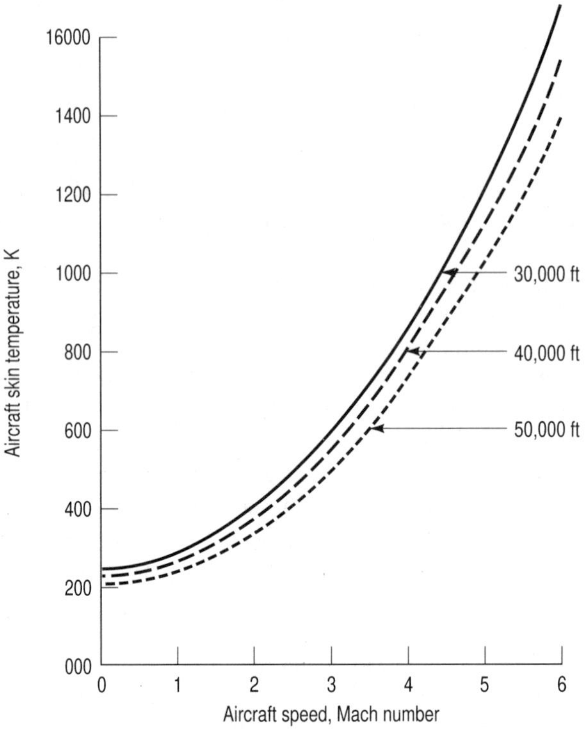

FIGURE 3.6 Aircraft skin temperature as a function of speed and operating altitude.

water vapor, at a sufficiently high temperature capable of producing visible radiation. The radiation from the hot gases near the muzzle exit, is known as primary flash. As the hot gases flow through the normal shock wave zone, they are heated further and radiate in the intermediate flash. The typical muzzle velocity varies from 2100 to 2400 ft/sec, which will produce the normal shock as well as the oblique shock wave zone. On mixing with atmospheric constituents, the hot gases ignite and burn with a large luminous flame known as secondary flash. To avoid detection of the secondary flash, which emits intense IR radiation, the gun is fitted with a device attached to the muzzle to prevent formation of shock waves. Chemical suppression agents can be added to the propellants to prevent ignition of the combustible muzzle gases. The spectral intensity of the secondary flash from a 155 mm artillery gun is inversely proportional to the observation distance from the gun. The spectral intensity is reduced to approximately 50% at a distance of 200 ft, to 25% at 500 ft, and to 5% at 1400 ft from its original IR radiation level at 2.1, 4.2, and 8.4 μm, respectively [9]. However, in case of a 280 mm gun, the radiation intensity will be slightly higher than that of the 155 mm gun level and the IR radiation level will be mostly concentrated over the 1.21–2.87 μm spectral region [9]. Gun size, propellant type, quantity of the charge used, and firing sequence of the gun determine whether a secondary flash will occur or not and determine the IR radiation level over a spectral region.

3.5 VARIOUS IR SOURCES

IR lasers with various power levels and emission wavelengths are available for commercial, industrial, military, and medical applications to meet specific requirements. Semiconductor lasers including VCSEL types with various power levels, CW and pulsed optically pumped solid-state lasers using nonlinear rare earth materials with moderate power levels, gas lasers with high power levels operating either in CW or pulsed mode, tunable dye lasers with moderate power levels, fiber-coupled lasers with moderate power levels, and chemical lasers with high power levels are available for various applications. Regardless of laser type, typical critical performance parameters including power level, beam quality, quantum efficiency, operating temperature, and reliability must be specified to meet specific application requirements. Diode-pumped solid-state (DPSS) lasers have recently received great attention because of their enhanced overall efficiency, high peak power capability, tenfold reduction in cooling requirement, and exceptionally high reliability. Performance characteristic and capabilities of various types of lasers will be briefly discussed with emphasis on cost, reliability, and efficiency.

3.5.1 Semiconductor Diode Lasers

Semiconductor lasers using various semiconductor lasing media fall into various categories, namely, tunable diode lasers, quantum-well (QW) diode lasers, quantum-cascade lasers, injection lasers, and vertical cavity surface emitting diode lasers (VCSELs). Semiconductor diode lasers such as GaAsP, InGaAsP, and AlGaAs offer CW power levels from 50 to 500 mW. CW power levels exceeding 1 W have been reported under cryogenic operating temperatures. Typical performance parameters of semiconductor lasers are summarized in Table 3.10. Injection lasers using semiconductor diodes are widely used in applications requiring lower power levels and high beam quality. In summary, semiconductor lasers are widely used, where low to

TABLE 3.10 Performance parameters of semiconductor pulsed lasers

Performance parameters	SH-GaAs diode laser	GaAs diode laser
Operating temperature (K)	77	300
Emission wavelength (nm)	850	900
Spectral bandwidth (nm)	15	15
Drive current (mA)	4	30
Maximum pulse duration (microsecond)	2	0.2
Duty factor (%)	2	0.1
Peak power output (W)	5	12
Conversion efficiency (%)	35	5

moderate power levels, high reliability, long lifetime, and low cost are the principal requirements.

Note that cryogenic operation offers significant improvement in conversion efficiency. Recent articles published in *Laser Focus World* [18] indicate that there are several suppliers of high power diode lasers in Japan, USA, and Europe. Siemens' high power diode lasers are based on a strained QW in InGaAlAs semiconductor material technology using MOVPE fabrication technology. The company has developed single diode lasers capable of delivering CW power levels exceeding 1 W and pulsed powers up to 10 W at 808 nm, 940 nm, and 980 nm wavelengths. The company can supply pulsed diode lasers with peak power levels exceeding 15 W with pulse width of 100 ns and duty cycle of 0.1% over the temperature range of –20 to + 85 °C. These lasers are best suited for range-finding systems and security sensors. Room temperature power levels from a QW-diode laser as a function of drive current are shown in Figure 3.7. Note that power levels can be significantly increased using lower cryogenic temperatures and higher drive current, as illustrated in Figure 3.8. Further increase in power output is possible using multiple-QW devices operating at cryogenic temperatures. Double-heterojunction (DH) diode lasers with a ternary-InAsSb active layer and AlAsSb cladding layers are capable of delivering high CW power levels at 4 μm wavelength and are best suited for operations in the mid-IR range.

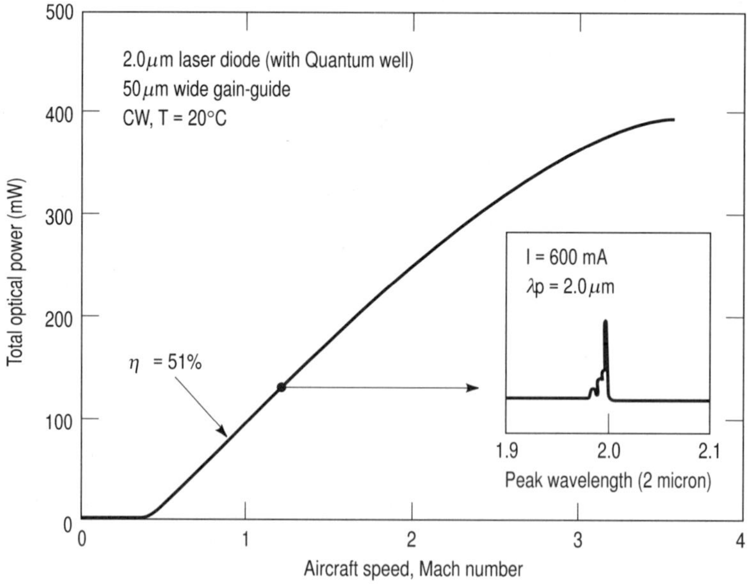

FIGURE 3.7 Opitical power output as a function of drive current at 20 °C for a 2 μm InGaAs/InGaAsP laser diode with 50 μm gain guide (cavity length = 1 mm).

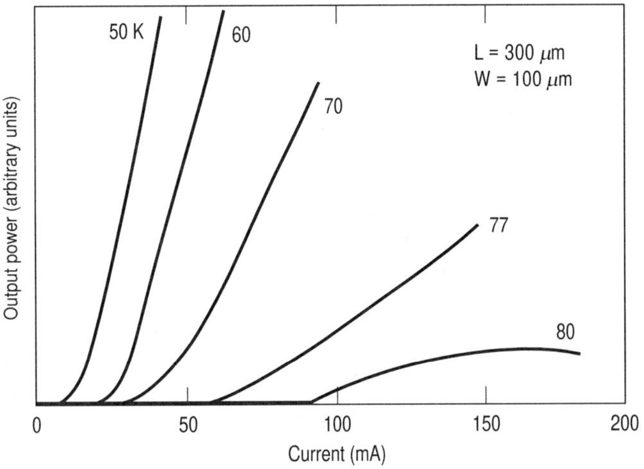

FIGURE 3.8 Optical power output as a function of drive current and cryogenic temperature. Cryogenic cooling offers higher differential efficiency, higher optical power output, and lower threshold current in semiconductor quantum well laser diodes.

3.5.2 Diode-Pumped Solid-State (DPSS) Lasers Using Diode Bars and Arrays

Several diodes can be integrated into bars or arrays to achieve several hundred milliwatts of CW optical power levels even at room temperature. DPSS lasers offer higher powers with improved efficiency, better than flash lamp pumped solid-state (FLSS) lasers, and are best suited for cutting and welding operations currently dominated by FLSS and CO_2 lasers. An article published in *Photonics Spectra* [17] reveals that a DPSS laser using diode arrays has demonstrated a pulsed power of 4 kW or a CW power of 240 W with aperture dimension of 1 × 2 cm and with a beam dimension of 25 × 280 mrad. DPSS lasers using Nd:YAG diode arrays offer high brightness and average power levels with impressive machining and welding capabilities. DPSS lasers generally employ 30 W, 40 W, 60 W, and 100 W diode bars to achieve high average power levels. Comparative performance levels of DPSS and FLSS lasers are summarized in Table 3.11.

High-power solid-state lasers using AlGaAs (810 nm), InGaAs (960 nm), and Nd:YAG (1060 nm) diode bars have potential applications in material processing and medical fields, including urology, gynecology, ENT treatment, photodynamic therapy (PDT), laser disc decompression, liver tumor ablation, and laser angioplasty. In addition, these lasers have potential industrial applications, such as cutting, drilling, and welding. DPSS lasers are best suited for soldering the leads in sensitive electronic components and surface-mounted devices. Brazing is the most critical process, because of the need to eliminate excessive flux from the joint and a DPSS laser with 15 W output will be most suitable for such an application. These lasers are equally good for successful marking applications on low-thermal conductivity materials, namely, plastic, glass, ceramic, and certain composites. DPSS lasers are

TABLE 3.11 Performance comparison between DPSS and FPSS lasers [11]

Performance parameters	DPSS laser	FPSS laser
Electrical-to-light efficiency (%)	50	70
Overall efficiency from input to output (%)	10	12
Size/weight	small/light	large/heavy
Cooling requirements	moderate	heavy
Reliability or MTBF (hours)	>10,000	1,000
Cost per watt	\$400	\$20
Beam quality	excellent	moderate
Intrinsic frequency stability	high	low
Beam pointing accuracy	high	poor
Amplitude stability	high	poor
Potential applications	accurate measurement	cutting, welding

widely used in heat treatment, surface hardening of certain metals, welding of thin titanium, and engraving of surfaces with high reliability and minimum cost.

3.5.3 High-Power, Aluminum-Free Diode (AFD) Lasers

AFD lasers provide stable operation at 730 nm with CW optical power capability exceeding 3 W at room temperature. The AFD laser using InGaAsP active-region devices offers higher brightness and power density from a given aperture size. Figure 3.9 shows the block diagram of a three-level laser system using 15 W or 40 W modified AFD bars, Ho:YLF laser rod, and Tm:Ho:YAG laser rod to achieve reasonable power at 3.914 μm in the mid-IR region. AFD lasers offer increased resistance to dark-line defects, longer lifetimes, ease of manufacturing, and oxidation-free facets. However, AFD lasers suffer from poor confinement at higher operating temperatures, which can degrade the laser efficiency at high power levels if adequate cooling is not provided to the laser cavity. Note that an AFD laser with 2% of the aluminum in the cladding region of the AlGaAs diode bars (Figure 3.9) offers significant improvement in efficiency and optical power output, even at elevated temperatures.

3.5.4 Fiber-Coupled Lasers

Fiber-coupled lasers, also known as fiber lasers, offer extremely high brightness and diffraction-limited output with high optical intensity and a large depth of field. These unique properties along with low-noise CW output make fiber lasers most attractive for laser printers and for legible marking on semiconductor molding compounds, lead frames, wafers, ceramic substrates, and packages. Low-noise CW fiber lasers write more uniformly, thereby making them most attractive for precision marking applications. These lasers are also best suited for micromechanical applications, namely, precision cutting, welding, and bending. Fiber lasers with power ratings of 10 W or 15 W are available, operating at a wavelength of 1.1 μm. Higher-

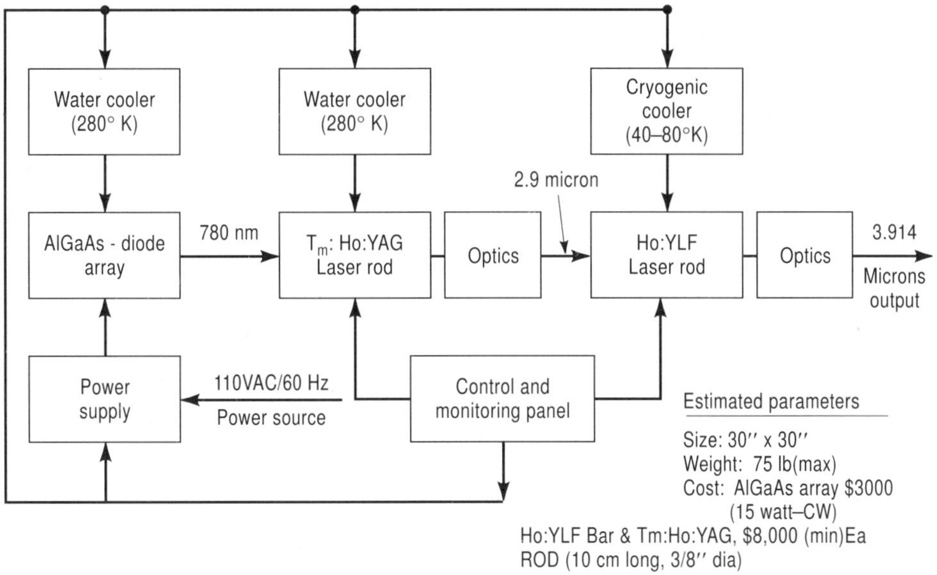

FIGURE 3.9 Block diagram for a three-level system involving laser diode array, T_m:Ho:YAG laser rod, and Ho:YLF laser rod fully equipped with needed accessories.

power levels are obtained using fibers with optimum diameter and numerical aperture.

3.5.4.1 High-Power DPSS Fiber Lasers

Certain applications require a high-power laser source with ultralow noise. As stated earlier, fiber lasers offer high brightness, CW power capability as high as 20 W without cryogenic cooling, tunable narrow-band output using single-mode fibers, high efficiency, and improved reliability. Critical elements of a DPSS-Ho:YLF laser are shown in Figure 3.9. Erbium-doped fluoride single-mode fiber lasers (Er:LiYF$_4$) operating at 3 μm have been used in specific medical applications. Fiber laser efficiency can be improved by concentrating the pump energy from the optical cavity and excited electron population in the optical fiber core. The right combination of pump wavelength and absorption lines is of paramount importance in designing high-power fiber lasers. Erbium (Er) and ytterbium (Yb) are widely used as dopants in the design of fiber lasers. However, the ytterbium-based pump, which absorbs energy at 915 nm and emits it at 1100 nm, is considered most suitable for the fiber laser design. Regardless of the doping material and host optical crystal, an interionic up-conversion process depletes the lower laser level, which is strictly dependent on the host crystal geometry, line-width broadening mechanisms, photon energy levels, dopant concentration, and energy transfer levels. Energy level diagrams for the DPSS-Ho:YLF laser and Er:LiYF$_4$ laser are shown in Figure 3.10 and Figure 3.11, respectively [12] . Studies performed by the author on excitation and loss mechanisms of host geometry and dopant concentration indicate that 970 nm is

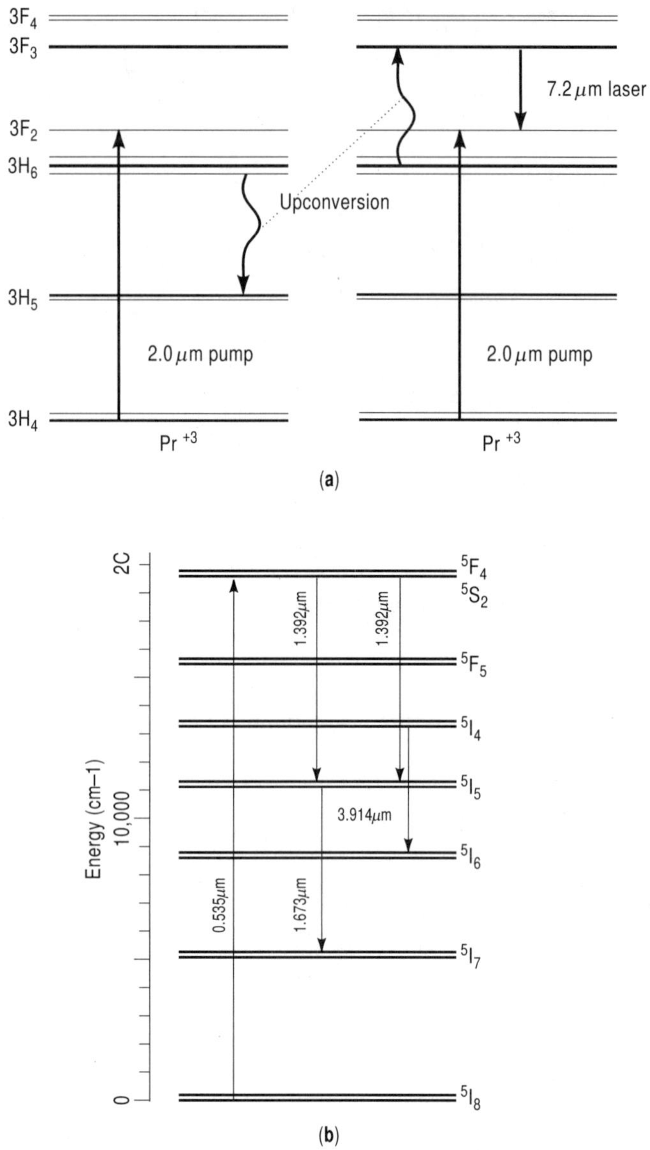

FIGURE 3.10 Energy level diagrams for (A) Pr:$LaCl_3$ and (B) Ho:YLF lasers.

the best pumping wavelength for a Er:$LiYF_4$ laser and 791 nm for the erbium-doped fluorozirconate fiber laser. Laser slope efficiencies of 40% in the Er:$LiYF_4$ laser and 23% in the fluorozirconate fiber laser have been reported in the literature. Scientists at Polaroid research laboratory (near Boston, MA) have demonstrated a CW output of 35 W from a Yb-doped fiber laser using fiber length ranging from a few meters to tens of meters. Potential applications of high-power fiber lasers include

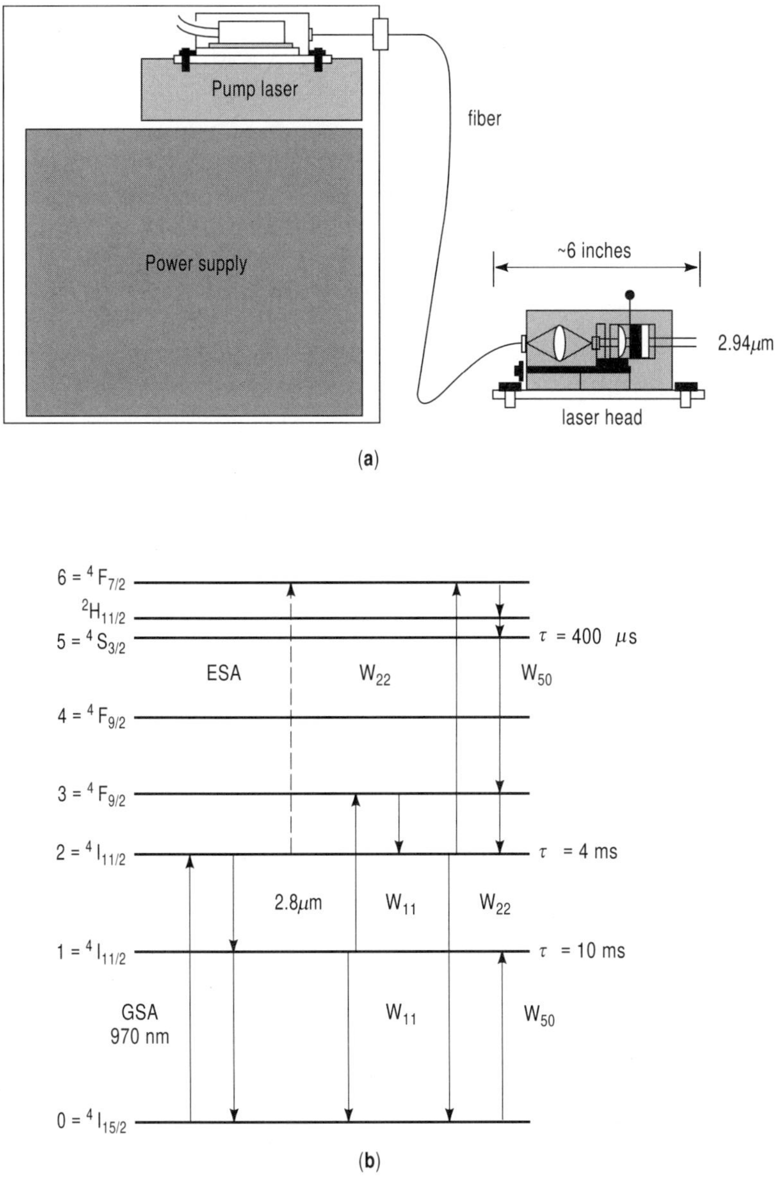

FIGURE 3.11 (A) Block diagram for Er:YLF laser; (B) Energy level diagram for a 2.8 μm Er:$LiYF_4$ laser.

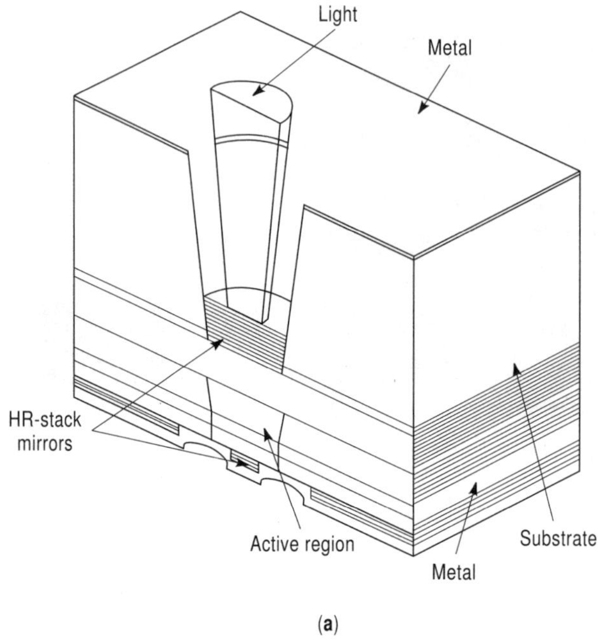

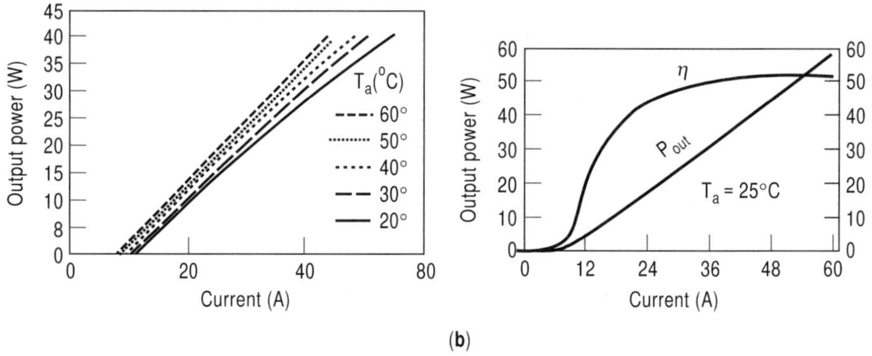

FIGURE 3.12 (A) Block diagram for VCSEL; (B) power output and efficiency for a 40 W, 808 nm laser diode bar at 250 °C.

laser spectroscopy, pollution detection, environmental monitoring, surgery, remote sensing using aircraft or satellite, and wavelength division multiplexing (WDM).

3.5.5 Vertical-Cavity Surface-Emitting Lasers (VCSEL)

A block diagram of a VCSEL laser along with output power and conversion efficiency as a function of temperature is shown in Figure 3.12. As the name implies, the light is emitted from the surface of the laser material rather than from its edge.

VCSELs offer physically small laser assembly, effective fiber coupling, lower fabrication cost, lower threshold current, on-wafer testing capability, longer lifetime, and improved power efficiency. Since the light is emitted perpendicular to the surface, VCSEL mounting is easy and cost-effective. The laser consists of a vertical cavity surrounded by high-reflectivity stack mirrors, as depicted in Figure 3.12. According to published reports, VCSELs operating at 1300 nm and 1500 nm wavelengths provide adequate optical power levels at 2.5 V and 5 mA over the 0 to 70 °C temperature range. The 1300 nm VCSEL is the most cost-effective single-mode laser source for wide-area network (WAN), interoffice communication, and short-distance LAN applications. It is important to mention that the single-mode lasers with coherent detection allow closer channel spacing, leading to higher data transfer rates. VCSELs have potential applications in short-distance optical links and local-area networks (LANs), because of the absorption-free windows at 1300 nm and 1500 nm wavelengths.

3.5.6 Quantum-Cascade Lasers

Quantum-cascade (QC) lasers are fundamentally different from semiconductor diode lasers because the emission wavelength depends on the thickness of the quantum well (QW) and barrier layers of the active region rather than the semiconductor material bandgap. Power levels from these lasers can be high because one electron produces many photons per period and most QC devices have 25 to 30 periods. QC lasers offer optical power with different wavelengths. The wavelength can be varied by altering the thickness of the active region. The outstanding feature of the QC laser is its ability to select the wavelength without being tied to the bandgap of the laser material. This laser incorporates a superlattice architecture using GaInAs/AlInGa (InP) laser materials. QC lasers offer mid-IR emission systems capable of replacing lead–salt diodes, which require costly cryogenic cooling. A typical QC laser provides a CW power of 14 mW and a peak power of 500 mW at 7.6 μm wavelength under room temperature conditions (300 K). A superlattice QC laser developed by the Bell Lab Group [13] demonstrated multiple transitions. This laser provided a peak power of 100 mW simultaneously at 6.6 μm and 8.0 μm wavelengths. The ability of the QC laser to emit simultaneously at multiple wavelengths makes it most attractive for applications such as air pollution monitoring, industrial process control, jet engine combustion monitoring, and cruise missile defense.

3.5.6.1 Ultrafast Lasers [14]

Scientists in the Netherlands have obtained a compressed pulse of 13 fs using an input pulse from a Ti:sapphire laser by first extending its bandwidth from 0.5 to 1.1 μm and then compressing its duration to less than 5 fs [14]. Scientists indicate that 3 fs pulses can be generated using the above-mentioned scheme. Early ultrafast (UF) lasers occupied room-size volumes and required continuous tuning to maintain normal performance. Current UF lasers are compact, reliable, and simple to operate. These lasers can produce short pulses with high peak power levels. A 100 fs pulse of only 1 microjoule energy can produce a peak power of 10 MW. A conventional Q-switch laser would require from 10,000 to 100,000 times more energy to provide the same peak power level. A recent article in *Laser Focus World* [18] stated that UF lasers can now deliver pulses shorter than 5 fs and peak power greater than 1300 TW with a light intensity capability of about 10^{21} W/cm^2 on the target. These lasers are best suited for research in astrophysics, quantum mechanics, accelerators, fusion energy generation, and ophthalmic surgery. The UF laser design got a boost from the chirped pulse amplification (CPA) concept, which produces pulses with short duration and low energy, which are then stretched out, amplified, and recompressed. CPA produces extremely short, high-intensity pulses that are most useful for micromachining and medical applications.

The CPA technique involves six major photonic components, namely, a femtosecond oscillator, oscillator pump, pulse stretcher, amplifier, amplifier pump, and pulse compressing network, in addition to secondary support components or accessories. The combined extremely high intensity magnetic field and temperature generated by the ultrashort pulse laser can ionize nearly all the materials within the focal spot before any significant heat conduction or mass flow takes place.

3.5.7 Tunable lasers

Tunable lasers include organic dye lasers, vibronic solid-state (VSS) lasers, semiconductor diode lasers, excimer lasers, carbon dioxide lasers, free electron lasers (FELs), and distributed feedback (DFB) lasers. The operating wavelength can be changed by varying either the temperature of the laser diode or the drive current. All of the above lasers can be designed to vary the operating wavelengths over a broad or narrow spectral band. Nonlaser devices such as optical parametric oscillators (OPOs) can be used to pump lasers to provide the tuning capability. Tunable lasers operating in near-IR, IR, mid-IR, and far-IR regions will be briefly discussed, with emphasis on their performance levels.

3.5.7.1 Short wavelength Tunable Lasers

Excimer lasers operating in the UV-region use novel gases to achieve tuning capability over a narrow spectral range. A mixture of argon and krypton is allowed to expand through a supersonic nozzle into a vacuum to produce a tunable wavelength output. During this process, the energy is transferred between the two species, leading to generation of continuous radiation covering a spectral range from 120 to 180 nm. Specific wavelengths can be selected by using an optical cavity with adjustable length.

3.5.7.2 Dye Lasers

Organic dye lasers with tuning capability between 300 and 1100 nm have been frontrunners for a long time. These lasers use organic dyes in a liquid solution. The dyes have a wide range of vibrational states and the transitions can occur continuously over a definite spectral region. Different classes of laser dye molecules can be tuned over different wavelength ranges. Merocyanine dye molecules offer wavelength tuning over a narrow spectral region (800–850 nm), whereas cyanine dye molecules can provide a broadband tuning capability over the 700–1100 nm spectral range. Organic dyes are fragile and degrade rapidly after tens or hundreds of hours of use. These dyes are not soluble in water, so methanol or ethanol or other suitable solvents are normally used to limit the dye degradation.

Organic dye lasers (ODLs) are capable of producing extremely broadband spectral output, which can be tuned to various desired wavelengths using an adjustable optical cavity. These cavities can use either a rotating grating configuration or a movable wedge structure of refractive material shown to select a specific tuning frequency, as illustrated in Figure 3.13. Conventional liquid-based dye lasers are mostly used for scientific research, spectroscopy, and ultrashort pulse generation. Because of rapid degradation of the dyes, need for frequent disposal of hazardous materials, and complexity of the devices used, dye lasers are not normally acceptable in commercial applications.

3.5.7.3 Solid-State Dye Lasers (SSDLs)

Research scientists at the Chinese Academy of Science in Shanghai developed a unique SSDL, in 1997 that involves addition of a DCM dye (commercial designation) to the originally modified TiO_2 silicon gel and then drying at a fixed temperature of 50 °C for a week. A slab of this material was incorporated in a laser assembly pumped by a Nd:YAG laser, which in turn was pumped by a laser diode array producing an all SSDL, as shown in Figure 3.13. The use of a prism provides tuning over a 60 nm range from 590 to 650 nm with an efficiency of 18% when using 8 ns pulses. This particular pulsed SSDL demonstrated a lifetime greater than 27,000 shots at 30 Hz pulse repetition frequency (PRF). The relatively high thermal conductivity of the gel material helped to slow down the degradation of the dye.

3.5.8 Solid-State Vibronic Tunable (SSVT) Lasers [16]

SSVT lasers cover much higher ranges of wavelengths than most dye lasers. In erbium, ruby, and neodymium lasers the upper and lower laser states are single states, whereas in vibronic lasers the lower level is a band representing the vibration sublevels of a single electronic energy state. These vibration sublevels are caused by the linkage of the electronic energy level to vibrations of the crystalline lattice. Because of this phenomenon, SSVT lasers can emit radiation over a ± 200% spectral range with respect to the center wavelength. Any laser wavelength within this 40% band can be selected by adjusting the optical cavity length, producing a tunable output. Among the vibronic laser materials, Ce:YLF has the shortest tuning range, from 309 to 325 nm (16 nm), while the Ti:Sapphire has the broadest tuning range, from 660 to 1180 nm (520 nm). However, Tm:YAG offers a tuning range in the mid-IR region, from1870 to 2160 nm (290 nm). These data indicate that the extremely broadband tuning capability of the Ti:Sapphire material makes it most ideal for femtosecond (fs) research activity. Recently, the material Cr:Li $SrAlF_6$ has gotten great attention due to its broad absorption band around 670 nm, which can be

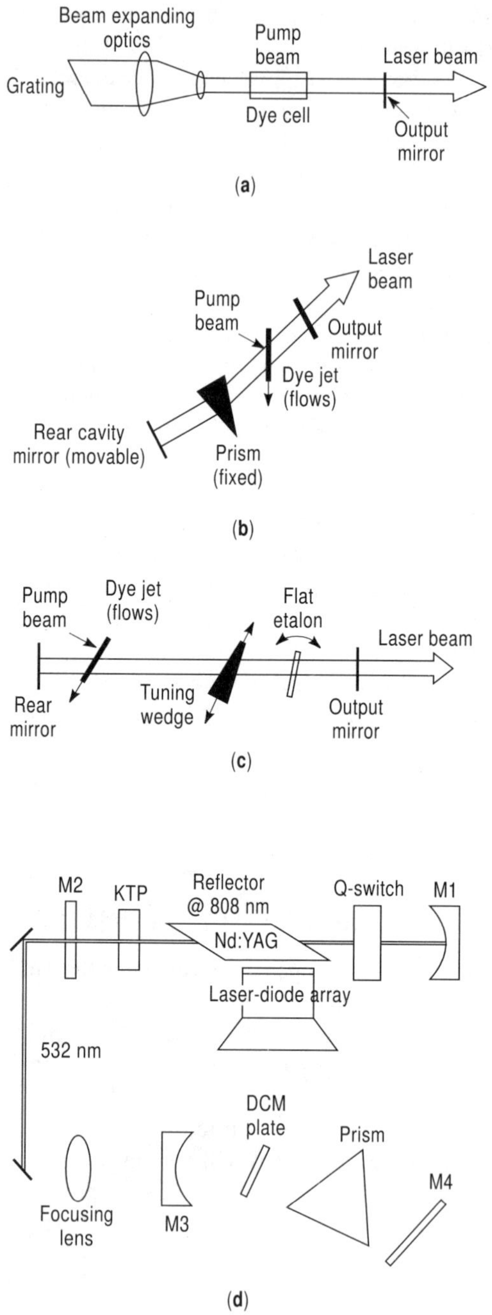

FIGURE 3.13 Various external-cavity tuners (top) and all solid state laser using diode array (bottom).

pumped by diode–laser arrays, thereby leading to a compact, reliable, and efficient tunable laser with a tuning range of 160 nm (760–920 nm), which also makes it a most attractive femtosecond pulse laser source.

3.5.9 Tunable Diode Lasers (TDLs)

Tunable diode lasers operating in the near-to-mid-IR region are most suitable for both communication and environmental monitoring applications. Wavelength division multiplexing (WDM) can be optimized for specific wavelength spacing between the channels, and is only possible with tunable lasers. Tunable lasers operating in the mid-IR region offer higher sensitivity for detection of low levels of atmospheric gases. Different indium based laser systems such as In GaAsP/InP (1–2 μm), InGaAs/AlGaAs/GaAs (0.8–1.1 μm), and InGaAs Sb/AlGaAsSb (1.8–4.5 μm) can operate anywhere from the visible range to all the way out to 4.5 μm in the mid-IR region. There are two practical concepts for tunability. In case of DFB lasers, the optical frequency can be changed by varying either the temperature of the laser diode or the drive current to the diode. However, in case of lasers using external cavities, tuning can be achieved either through a distributed BRAGG reflection device, an acoustic optical filter, or rotating diffraction grating. Tunable diode lasers using external cavities are widely used in WDM and dense WDM systems, telecommunication equipment, meteorology, and spectroscopy.

3.5.10 Free-Electron Lasers (FELs) for Far-IR Regions

FELs are known for their ultrawide wavelength tuning capabilities ranging from 2 μm to all the way out to millimeter wavelengths. The lasing medium in a FEL does not consist of atoms or ions, but a beam of electrons. The electron beam passes through a series of evenly spaced magnets of alternating polarity. Since the energy of the electrons is relatively easy to control or adjust, a FEL can be tuned by varying the electron energy. Higher energy levels provide shorter resonant wavelengths. A FEL requires an electron accelerator to provide energy levels of tens of MeV. However, cost, size, and high computational requirements associated with the electron accelerator have limited the application of FELs to scientific research facilities.

3.5.11 Gas Lasers

Gas lasers offer significant operational advantages such as low power consumption and reliable operation in the visible region of the spectrum. Performance parameters such as wavelength, typical and maximum output power levels, and mode of operation of several gas lasers are summarized in Table 3.12. The performance data indicate that in the IR region, CO and CO_2 lasers offer high efficiencies and optical power levels and, thus, can be directly used in welding or parametric down-conversion to longer wavelengths. Some of the gas lasers are best suited either for pulsed operation or CW operation only. The data further indicate that CO_2, TEA (transverse excitation atmospheric), Ne, and N_2 lasers can be operated in either mode.

TABLE 3.12 Performance capabilities of typical gas lasers

Gas	Principal wavelengths μm	Power Output Typical	Power Output Maximum	Mode of operation
Xe (ionized)	0.3454	15 mW		cw
	0.3781	50 mW		cw
	0.4060	5 mW		cw
	0.4214–0.4272	50 mW		cw
	0.5419–0.6271	10 mW	1 W	cw
Xe (Xe–He)	2.026		1 mW	cw
	3.507		1 mW	cw
	5.575		1 mW	cw
	9.007		1 mW	cw
N2 (ionized)	0.3371		200 kW	p
			100 mW	cw
HeCd (ionized)	0.3250	4 mW	40 mW	cw
	0.4416	20 mW	200 mW	cw
Ne (ionized)	0.3324		10m W	p, cw
Ne (neutral)	0.5401		1 k W	p
Cu (neutral)	0.5106		40k W	p
Ar (ionized)	0.3511	5 mW	0.35 W	cw
	0.3638	5 mW	0.35 W	cw
	0.4579	0.1 W	0.75 W	cw
	0.4658	0.1 W	0.3 W	cw
	0.4727	50 mW	0.4 W	cw
	0.4765	.3 W	1.5 W	cw
	0.4880	1.0 W	5.0 W	cw
	0.4965	.2 W	1.5 W	cw
	0.5107	.1 W	.7 W	cw
	0.5145	1.0 W	6.0 W	cw
He–Se (ionized)	0.4605–0.6490 (24 lines)		100 mW	cw
	0.5228		20 mW	cw
Kr (ionized)	0.3507	20 mW		cw
	0.3564	20 mW		cw
	0.4762	50 mW		cw
	0.4825	30 mW		cw
	0.5208	70 mW		cw
	0.5309	200 mW		cw
	0.5682	150 mW		cw
	0.6471	500 mW		cw
	0.6764	120 mW		cw
	0.7525	100 mW		cw
	0.7931	10 mW		cw
	0.7993	30 mW		cw
He–Ne	0.6328	2 mW	150 mW	cw
	1.1523	2 mW	25 mW	cw
	3.3913	1 mW	10 mW	cw

(continued)

TABLE 3.12 *(continued)*

Gas	Principal wavelengths μm	Power Output Typical	Power Output Maximum	Mode of operation
HF (chemical)	2.6 → 3.5	4500 W		cw
		1 joule		p
CO	4.9 → 5.7		100 W	cw
CN	5.2		30 mW	p
CO_2 (flowing)	10.6	100 W		cw
	9.6	5 kW	200 kW	p
Transverse excitation atmospheric (TEA)	10.6	1000 W		cw
		50 kW		p
HCN	126 → 134	0.5 W	10 W	p
	310	1 W		p
	336	0.6 W		p
	372	0.6 W		p
H_2O	28		5 kW	p

Note that high-power CO_2 lasers operating at 10.6 μm are capable of yielding peak power levels exceeding one gigawatt (GW) with conversion efficiencies greater than 10%. These high-power lasers are best suited for welding, cutting, and brazing applications.

3.5.12 Chemical Lasers [17]

Hydrogen fluoride (HF) and deuterium fluoride (DF) lasers offer high output powers and have potential applications in industrial manufacturing. The CW power level from these lasers was around 1 kW as early as 1972, but this level significantly increased to better than 100 kW in early 1990. Chemical lasers with power levels as low as 1 W or 10 W have been developed for clinical diagnostic use and for the study of atmospheric properties.

A chemical laser consists of five critical regions, namely, discharge, inlet, lasing, heat exchanger, and exhaust regions. The discharge region provides an F-atom flow for sustaining the chemical reaction by means of a high voltage discharge in a water-cooled pyrex tube containing SF_6, He, and O_2 leading to generation of excited HF or DF molecules. The He serves as a thermal diluent, whereas the SF_6 provides the F-atom source. The oxygen is added to the flow to increase the F-atom concentration and to eliminate excessive S deposition through exothermic reaction. Typical efficiency of a 12 W HF chemical laser is about 1%, with a gas pressure of 23.5 torr in the laser tube section. Higher efficiencies have been reported at higher output power levels. Various temperatures and gas mass flow rates are required to meet specific power levels and wavelengths for both HF and DF laser systems. Because of excessive weight, complex cooling schemes, and high operating cost, high-power chemical lasers are only used for research and military applications. Typical operating requirements and laser components parameters for a 12 W HF laser are summarized as follows:

Electrical power requirements:
  Voltage: 15.6 kV
  Current: 350 mA
  Power: 5600 W
Optical cavity parameters:
  Radius of curvature of mirror: 40 in
  Output coupler separation: 16 in
Gas condition in cavity:
  Lasing region pressure: 23.5 torr (1 torr = 1 /50 psi)
  Channel exit pressure: 15.7 torr
  Channel inlet temperature: 1000 K (approximately)
  Lasing region temperature: 700 K (approximately)
Gas mass flow rate (*dm/dt*):
  For oxygen: 0.180 g/sec
  For helium: 0.054 g/sec
  For hydrogen: 0 027 g/sec
  For SF_6: 0.870 g/sec

3.5.13 Ultra-High-Power Laser with Output of Near-Diffraction-Limited Optical Quality

This section will deal with high-power carbon dioxide and chemical oxygen–iodine lasers (COIL) capable of producing outputs of near-diffraction-limited optical (NDLO) quality. These lasers are not limited by the same thermal effects that impact the solid state Nd:YAG lasers.

3.5.13.1 Carbon Dioxide Laser (CO_2) Operating at Mid-IR Wavelengths

Carbon dioxide lasers remain the most practical choice for industrial, medical, and research applications because of their capability to produce high CW power levels exceeding several kilowatts with impressive conversion efficiency. These lasers are also used for nonindustrial applications [17] where mid-IR wavelengths are of critical importance, such as medical diagnostic and surgical applications. These lasers rely on transitions between the vibrational modes of the CO_2 molecules. The upper laser level is the asymmetric stretching mode, which can exist in bending mode with transitions at 10.6 and 9.6 μm. It is important to mention that simultaneous rotational transitions generate several laser lines, so that the laser can either lase in narrow lines or across a specific band. Optimum laser efficiency requires mixing of

CO_2 with nitrogen and helium. The overall efficiency of this laser varies from 5 to 20% depending on the mix of nitrogen and helium. However, the power output depends on the window design, cooling scheme, removal of heat, and optic size. One can get CW power output as high as 500 kW from this laser, provided there is an efficient cooling scheme, proper optic size, and optimum mix of gases.

This laser system consists of a sealed glass tube containing the mix of three specified gases in the right proportion, optical mirrors at each end forming an optical cavity, and the output window. Generally, hydrogen or water vapor is added to react with carbon monoxide and oxygen leading to chemical formation of CO_2. This process increases the laser life by 5,000 hours, depending on the power level. The output power is proportional to the length of the glass tube and typical power output is about 40–50 W/m. Power output of 80–100 W/m can be achieved with slow-flowing gas at low pressure (around 100 torr) along the axis of the lasing tube. Further increase in power level is possible by using multiple parallel tubes connected in series with optical mirrors, producing around 1 kW output for every 4 to 8 tubes.To enhance power output capability, slab technology must be integrated in the design of ultra-high-power CO_2 lasers.

If the sealed glass tube is replaced by a pressurized waveguide, the laser is called a waveguide CO_2 laser. This waveguide laser has small power output capability (not exceeding 50 watts) but has an optical beam of high quality that can be easily tuned to any discrete CO_2 line. Its compact size, high conversion efficiency, and excellent beam quality can compete with HeNe lasers for some applications. Power output can be increased by using the flow gas concept, so that the heat can be removed through convection rather than conduction at much faster rates.

In the case of ultra-high-power CO_2 lasers operating in CW mode, both the gas flow and the electrical discharge must be across the beam axis to allow greater amounts of gas to flow and to allow the greatest amount of heat to be removed. Pulse operation requires pulses ranging from 100 nanoseconds to a few microseconds and high gas pressure near or above one atmospheric pressure equal to 14.7 psi. However, higher gas pressure means more gas per volume, which will increase the energy density available to the laser pulse. Such lasers are known as transversely excited atmospheric (TEA) lasers. In such lasers, the electrical discharge is always perpendicular to the beam axis. High gas pressure broadens the emission lines, but mode locking can be used to generate nanosecond pulses. The TEA laser permits nearly continuous tuning across a broad IR region, which could be most beneficial for many scientific and sensing applications. High-power pulses from a TEA laser can be used for heat treatment and stripping old paint from surfaces. Low-power TEA lasers with 100 W output are widely used in laser printing, remote sensing, air pollution monitoring, medical research on cancerous cells, mineral detection, and spectroscopy. TEA lasers can be used in surgical applications, because the coherent CW beam cauterizes as it cuts, a most desirable advantage.

The most common applications for CO_2 lasers are in manufacturing industry, where high CW output is essential for cutting and welding of thick steel and titanium alloys. Titanium, a tough metal to cut mechanically, absorbs IR energy strongly at 10 μm wavelength, so these lasers are particularly useful in dealing with this metal. CO_2 lasers offer spot sizes as small as 50 μm (0.0050 cm or 0.00197 inch), making it possible to create intricate patterns for prototypes with high accuracy and minimum cost.

3.5.13.2 Chemical Oxygen–Iodine Laser (COIL)

The latest scientific research and development activities on COIL lasers have revealed their outstanding performance capabilities and potential applications. By injecting atomic rather than molecular iodine into a COIL system, Japanese scientists have reported significant increase in laser output power and conversion efficiency. Rocketdyne Propulsion and Power Company in California is developing a system to pipe laser power from a COIL laser through a low-loss optical fiber to aid in dismantlement of decommissioned nuclear facilities. This laser technique offers an efficient process for cutting, welding, and surface decontamination functions required in dismantling of decommissioned nuclear facilities. Based on scientific research studies and available test data, COIL lasers could be better suited for cutting thick metals. Research studies [18] indicate that the cutting capability of a 1 kW COIL system is equal to that of a 1 kW Nd:YAG laser and is about 2.5 times better than that of a 1 kW CO_2 laser in cutting 5 mm thick stainless steel. Scientists at the University of Illinois have successfully cut a 0.5 inch thick stainless steel plate with a 7.36 kW (peak power level) COIL laser that was delivered via 900 μm (core diameter) optical fiber. Scientists have cut several thick metal plates with a 10 kW supersonic COIL system using a Mach-2 single-slit supersonic nozzle [18]. Russian scientists working at the High-Energy Density Research Laboratory have bounced light from a high-power COIL laser off asteroids located as far as 10^7 km away to determine their size and range. In summary, a COIL system is capable of producing both high-power CW and pulsed power outputs of near-diffraction-limited quality and near-perfect circular beam with high efficiency.

3.6 OPTICAL PARAMETRIC OSCILLATORS (OPOs)

Optical parametric oscillators (OPOs) provide tunable coherent optical power over a wide optical frequency range. An OPO uses a nonlinear, birefrigent optical crystal that can be added to a pump laser, often a diode laser. An OPO works by splitting the energy of a photon between two photons, both at longer wavelengths. The division between the photons and the resulting wavelengths are determined by the electric field imposed on the nonlinear crystal. Conservation of momentum law requires that the sum of the products, refractive indices, and optical frequencies of the output photons must be equal to the same products for the pump photons. An OPO essentially acts like a tunable laser over a narrow band in the UV, visible, and IR spectral regions. An OPO is a nonlinear device because it uses a nonlinear crystal and its operating principle is similar to that of a RF parametric amplifier, which uses a varactor diode with nonlinear characteristics. When pumped with a short-wavelength beam at optical frequency ν_p, it generates two outputs, one at the signal frequency ν_s and the other at the idler frequency ν_i. The sum of signal frequency and idler frequency is equal to the pump frequency, as in the case of a RF parametric amplifier. Phase-matching techniques permit tuning of both signal and idler outputs to their respective optical frequencies. Essentially, an OPO shifts the fixed-wavelength output from a near-infrared DPSS laser to idler wavelengths of 3–5 μm.

Stable OPO operation requires singly resonant devices and their relaxed frequency constraints, which demands high pump power levels. A conventional singly resonant OPO design requires the device to be resonant only at the signal frequency, not at the pump or idler frequencies. In an alternate design configuration incorporating the extra cavity concept, the singly resonant OPO is resonant at both the pump and signal frequencies. The extra cavity configuration offers a pump enhanced device analogous to extra cavity, pump enhanced, second harmonic generator, and requires sophisticated electronics to stabilize both the laser frequency and the OPO cavity length.

High-quality nonlinear optical crystals are required for phase matching the output of neodymium (Nd) based lasers to provide idler output in the 3–5 μm range. The latest OPO designs are based on the DPSS lasers; performance parameters of various OPOs are summarized in Table 3.13.

Various frequency conversion schemes using OPO devices and Nd:YAG laser sources capable of generating radiation levels exceeding 1 mJ [19] are illustrated in Figure 3.14. Noncritical phase matching (NCPM), critical phase matching (CPM), and difference frequency generation schemes using $AgGaS_2$, KTP, and KTA nonlinear crystals in the design of OPOs are also shown in Figure 3.14. The block diagrams for several OPOs [19] shown in Figure 3.14 extend solid-state coherent sources deeper into the infrared regions covering from 1.79 to 11 μm.

3.6.1 Nonlinear Optical Crystals for Applications in OPOs

Only a handful of nonlinear crystals [19] that are currently available are described in Table 3.14. Improvement in nonlinear crystal materials will allow OPOs to act as continuously tunable IR lasers using solid-state laser pumping schemes. Efficient

TABLE 3.13 Performance summary of various OPO Devices

	Designer		
Parameters	Stanford University (CA)	Aculight Corp. (CA)	University of St. Andrew (Scotland)
Design concept	Extra cavity (PPLN) (singly resonant)	Extra cavity (PPLN) (singly resonant)	Intracavity (singly resonant)
Crystal type	PPLN	PPLN	KTA
Pump power (W)	2–4	23	—
Efficiency (%)	90	22	—
Power output (W)	1.8–3.6	5.1	1.0
Idler wavelength (μm)	2.3–4.7	3.4	3.5
Pump laser source	DPSS	FLSS (Nd:YAG)	DPSS ($Nd:YV0_4$)

TABLE 3.14 Improved nonlinear optical crystals for application in optical parametric oscillators (OPOs)

Material	$AgGaS_2$	$AgGaSe_2$	$ZnGeP_2$	$CdGeAs_2$	CdSe	GaSe
Nonlinear coefficient, d_{ij} (pm/V)	19	39	75	236	18	54
Transparency range (μm)	0.5–13.2	0.78–18	0.72–12.3	2.6–17.8	0.8–18	0.65–18
Birefringence ($n_e - n_0$)	–0.053	–0.033	+0.039	+0.096	+0.019	–0.373
Thermal conductivity (W/cmK)	0.015	0.011	0.036	0.042	—	0.162, 0.02
Laser damage threshold	0.25 J/cm^2	0.5–3 J/cm^2	2–10 J/cm^2	20–40 mW/cm^2	60 mW/cm^2	3 J/cm^2
Shortest pump wavelength (μm)	0.6 (1.06)	1.27	1.3 (1.7)	2.6 (4.5)	2.37	0.65 (1.3)
Advantages	Phase matches with 1 μm pump	Low loss, Good d_{ij}	High d_{ij}, low thermal lensing	Very large d_{ij}, wide phase matching (PM) range	Low loss, long crystals available	Wide PM range, low loss, large d_{ij}
Disadvantages	Low damage threshold	Absorption peak at 2 μm, creates thermal lensing	Absorption losses <1.7 and >8.8 pm	Long wave pump required	Long wave pump required	Very soft, large walk-off

Note: n_e represents the index of refraction along E-O crystal axix; n_0 represents the index of refraction in the space or region outside the EO crystal.

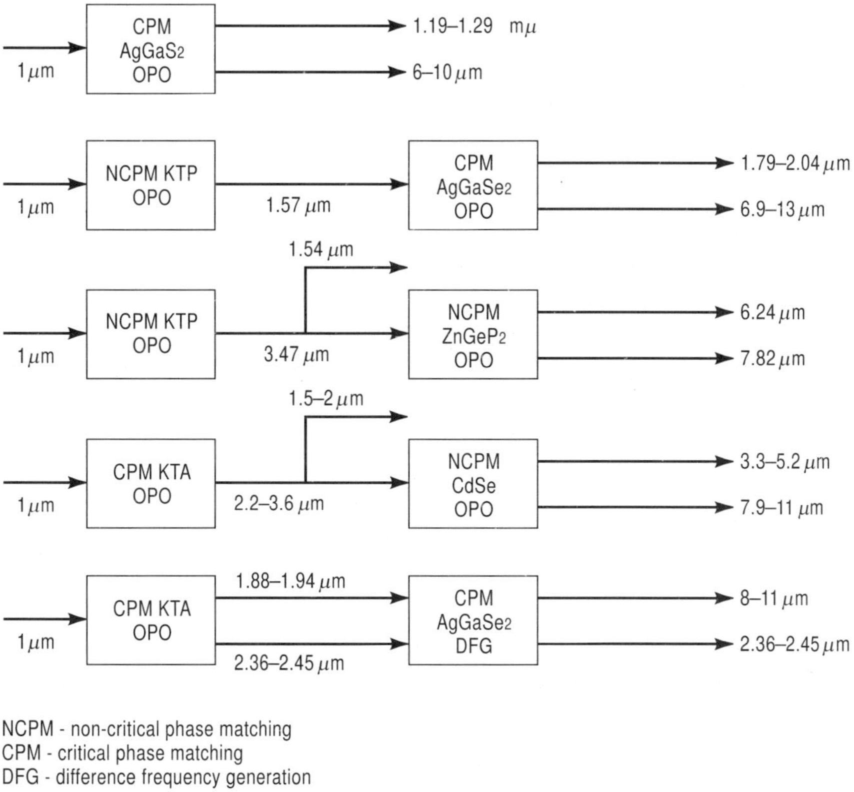

FIGURE 3.14 Various frequency conversion schemes using optical parametric oscillators (OPOs) capable of generating laser energy exceeding 1 mJ over the 8–12 μm spectral region.

frequency schemes are dependent on coherent pump sources such as 1-μm Nd:YAG or 2-μm Ho:Tm lasers and nonlinear optical crystals with high nonlinear coefficients. It is important to mention that pumping a crystal with a 2-μm Ho:Tm laser is the most cost-effective approach for generating radiation over the 8–12 μm range in a single OPO stage because of the lower quantum defect at longer pumping wavelengths. Nonlinear crystals with high nonlinear coefficients and maximum thermal conductivities are best suited for high average power generation. Pumping with a 3 μm erbium laser offers the highest quantum-limited conversion efficiency with minimum cost. However, such laser sources are difficult to Q-switch and can damage optical coatings. Some crystal are most suitable for the second stage of the OPO. For example, CdSe nonlinear crystal is considered most ideal for the second stage of an OPO because inherent pumping at longer wavelengths provides full spectrum coverage from 8 to 12 μm. An OPO using KTA crystal has demonstrated IR energy exceeding 3 mJ per pulse using the $AgGaSe_2$ at 8 μm and pumping at 1.54 μm, with an overall efficiency of 0.7%. Any damage to the OPO output mirror can lower the conversion efficiency. An average output power of 500 mW can be achieved at 8.24 μm from a critically phase matched ZnGe P_2 crystal with a pump power of 9.2 W supplied by a 2.05 μm Tm:Ho:YLF laser. Regardless of frequency schemes, frequency conversion efficiency drives the pump power requirements, which has an impact on the crystal choice and its size and cost.

3.6.2 Low Pulse Energy (LPE) OPOs

Low pulse energy OPOs operating in near-IR region are in great demand for industrial and medical applications. High repetition rate Q-switch DPSS lasers with output energies of several mJ are receiving great attention for applications where use of LPE-OPO is essential. The combination of compact, affordable, doped-pump laser and LPE-OPO can provide a cost-effective, all solid state, wavelength selectable light source in the near-IR region that has potential applications in chemical research, environmental monitoring, and semiconductor processing. The prime design goal must be toward obtaining the required threshold nonlinear drive in the presence of large beam walk-off without exceeding the optical damage threshold.

A LPE-OPO design consists of two BBO optical crystals each 5 mm long, an optical cavity of 5 cm length, a 550 nm pump source with 10 ns pulses to produce a 800 nm signal and an idler at 1590 nm, and an output mirror with 70% reflectivity. The resonant signal at 800 nm demonstrated a beam walk-off angle of about 3.4° in each crystal axis with a pump beam diameter of more than 400 μm to maintain the required threshold pump energy and the optical damage threshold of about 200 MW/cm^2. A 200 μm beam diameter requires a pump intensity of 450 MW/cm^2, which could damage the OPO and its optics. Regardless of whether it is of high or low pulse energy OPO design, the OPO must be pumped to at least twice the threshold pump intensity to obtain reasonable OPO output. The threshold of an OPO is the minimum pump intensity required to exceed the OPO cavity losses and to "turn on" the devices. When pump energy reaches 4 to 6 times above the threshold level, 30 to 50% of the pump photons can be converted into signal and idler photons.

3.6.3 Critical Performance Parameters of OPOs

Critical performance parameters include gain length, which is a function crystal length, walk-off angle, and dephasing effects, and the pump beam intensity. The threshold of the OPO is determined by several factors, namely, nonlinear drive level, walk-off angle, cavity length, optical losses, pump pulse width, pump beam quality, and diffraction effects. Nonlinear drive level as a function of pump beam diameter and pump energy for various LPE-OPO designs using single BBO crystals is shown in Figure 3.15. The following LPE-OPO design issues are of critical importance:

- Selection of the right nonlinear crystals, such as KTP, BBO, and LBO
- Nonlinear drive not to exceed optical damage threshold of the crystal
- Nonlinear drive is proportional to beam diameter area
- Small beam diameter requires both reasonable beam intensity and nonlinear drive

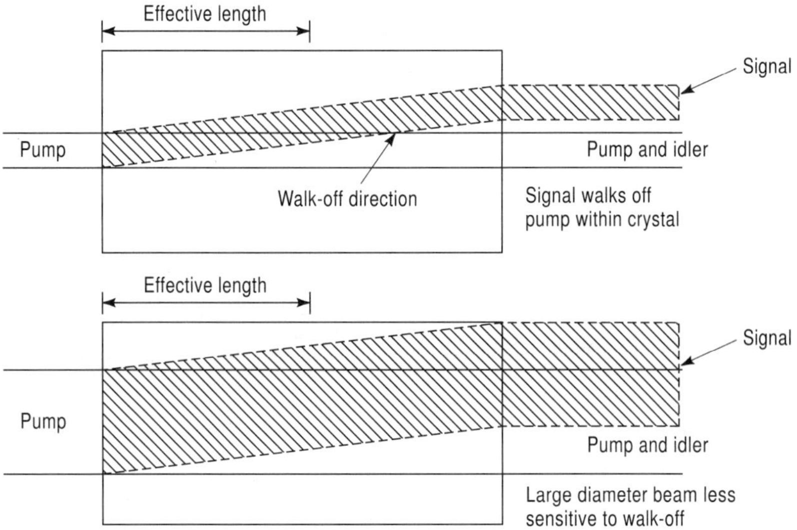

(a) Signal walk-off in critically phasematched nonlinear crystals.

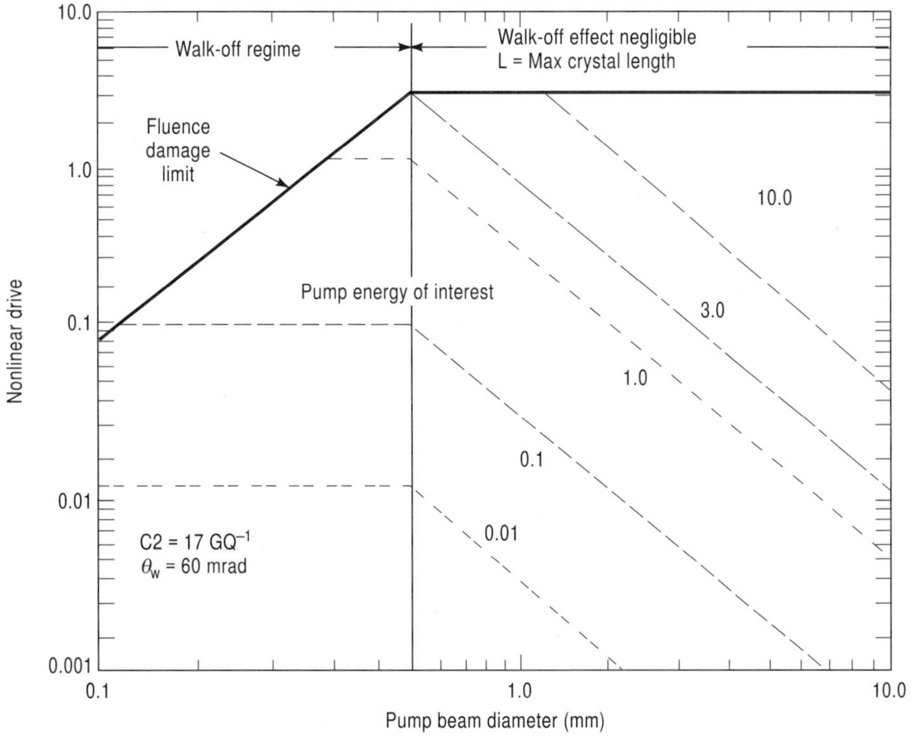

(b) Nonlinear drive level as a function of beam diameter

FIGURE 3.15 (A) Signal walk-off in nonlinear crystals; (B) Nonlinear drive level as a function of pump beam diameter.

- Noncritically phased matched OPOs that use KTP or KTA crystals require pumping from a 1 μm laser with low pulse energy to generate an eye-safe signal over the 1.5–1.6 μm range. Nonlinear drive requirements are different for different crystals
- Damage threshold limit is about 200 MW/cm^2 for a 10 ns pulse
- The parametric gain is proportional to the square of the crystal effective length
- Nd:YAG and Nd:YLF laser pumping schemes offer compact size and cost-effective design with broad tuning capability
- Small walk-off is possible with small pump beam diameter in the case of a single BBO based LPE-OPO

3.7 OPTICAL AMPLIFIERS

The use of optical amplifiers in WDM and dense WDM systems has demonstrated significant increases in bandwidth just when the data-transmission bandwidth (BW) requirements are growing exponentially. The WDM technique was originally implemented in the early 1990s as a "graceful growth" plan to match the 10% annual growth in voice traffic. Since then, with data traffic growing at a 50% annual rate, data bandwidth requirements are expected to increase significantly in the future, which can only be satisfied by advances in the development of silicon-based, erbium-doped fiber amplifiers (EDFAs). The EDFA (Figure 3.16) deploys long-period fiber gratings to flatten the amplifier gain response over the longer-wavelength window from 1565 to 1605 nm, as shown in Figure 3.16. The latest EDFA technology offers extended bandwidth, wider windows with tighter channel spacing, and more than 1000 channels at a 1 Gbit/sec data rate in a WDM system. A praseodymium-doped fiber amplifier (PDFA) has demonstrated even wider bandwidth windows in the 1400 nm spectral range with dramatic reduction in transmission losses. The use of a fiber laser as a pump offers broader transmission windows in a Raman amplifier. Japanese scientists [20] have developed a system combining a fluoride-fiber-based EDFA and Raman amplifiers to obtain broad, flat amplification windows with instantaneous bandwidth of 75 nm in the 1531–1606 nm spectral range. The use of optical filters and fiber gratings with excellent wavelength tolerances in high volume will require minimum acceptable yield and stability over the desired temperature range to keep the manufacturing costs low. The effective use of the full 200–400 nm bandwidth of optical fibers will present challenging problems. Trade-off studies between the transmission speed (moving from the current 10 Gbits/sec to 40 Gbits/sec to higher) and the number of transmission channels must be performed to meet the cost-effective criterion.

3.7.1 Noise Figure (NF) of EDFA Devices

The noise figure of the EDFA is its most critical performance parameter and has an impact on distortion in the amplifier output, as shown in Figure 3.16. With reduc-

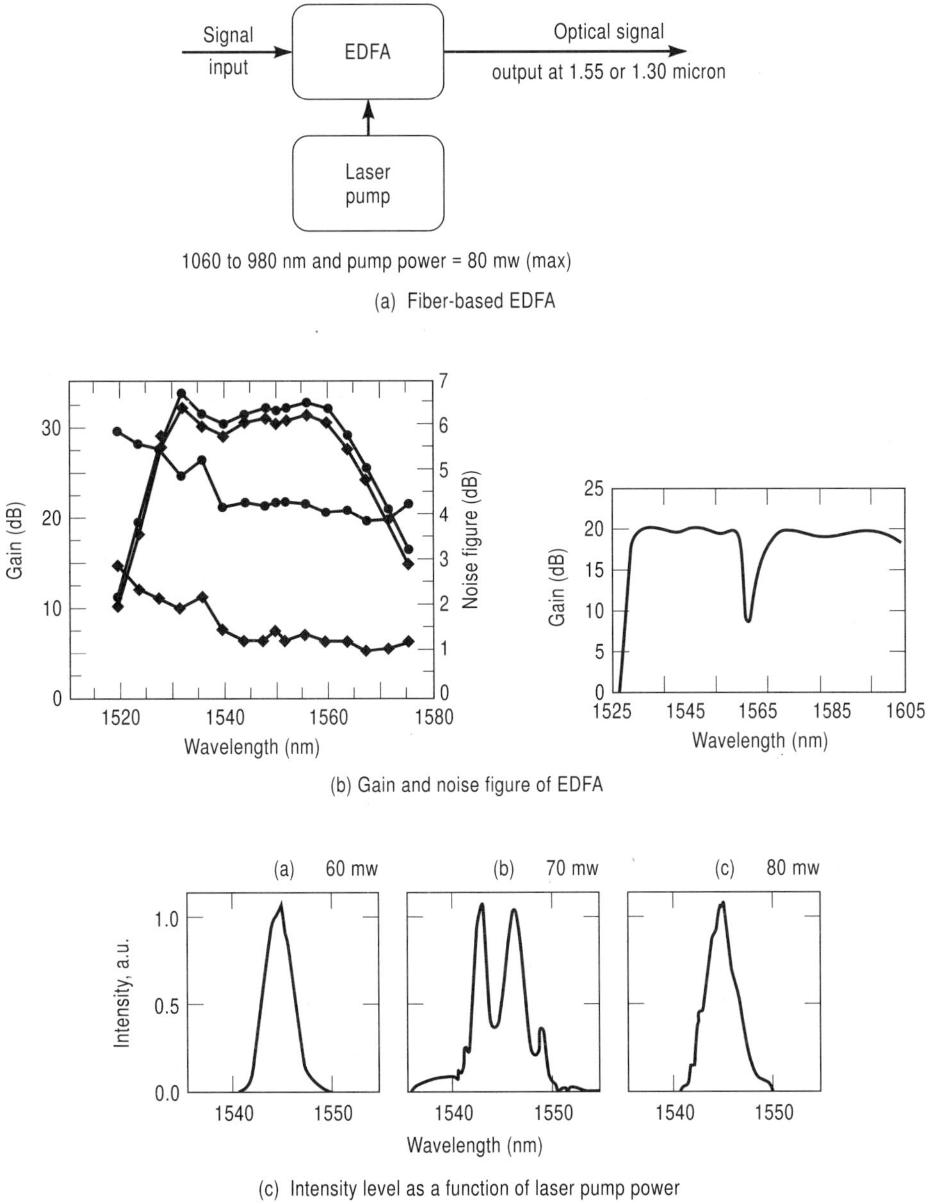

FIGURE 3.16 (A) Fiber-based EDFA; (B) gain/noise figure; (C) intensity level of EDFA.

tion of ASE power by half, the EDFA noise figure is reduced by 3 dB, as evident from the curves shown in Figure 3.16. However, a gain reduction of about 1.4 dB due to beam splitter insertion loss and the residual polarization mismatch loss caused by the error of polarization control cells is also evident from the same figure. The noise figure normally increases with increase in pump power and amplifier gain. The amplifier NF can be reduced using narrow-line width bandpass filters or by incorporating an optical isolator to block the backward propagation. However, the isolator approach will increase the EDFA cost slightly.

3.7.2 Gain and Power Output of EDFA

The EDFA gain is directly proportional to the pump power at optimum wavelength. The latest EDFA designs incorporate two transition devices and can deliver output power as high as 4 W or + 36 dBm at 1558 nm center wavelength using a 1060 nm diode laser pumping the neodymium-doped fiber cladding and co-doped erbium and ytterbium fiber core. No saturation effects are expected in the gain and output of a three-stage, 4 W (CW output), 1558 nm optical amplifier, even up to 12 W of pump power. In the case of an EDFA amplifier using 80 mW of pump power at 1060 nm, a minimum gain of 30 dB is possible over the 1530–1565 nm spectral window, as shown in Figure 3.16. A single stage praseodymium-doped fluoride fiber amplifier (PDFFA) developed by British Telecommunication Corp. in the early 1990s demonstrated a gain of 16 dB at 1.3 μm wavelength using a PDFF with numerical aperture of 0.39. The same amplifier design delivered an output power exceeding + 22 dBm at 1.3 μm using a 1 μm pump. This amplifier is best suited for telecommunication applications under harsh environmental conditions with no maintenance.

3.7.3 Optical Fiber Requirements for EDFAs

Optical fibers must be selected to provide minimum insertion loss, low dispersion, and high mechanical integrity under harsh operating environments. Fibers with high information capacity, superior signal quality (i.e., without pulse dispersion and distortion), minimum weight, capability to operate over wide temperature range and large numerical aperture must be considered for EDFA devices. Studies performed by the author indicate that single-mode fluoride fibers are best suited for EDFA devices operating at 1300 nm wavelength, regardless of geographical location. Silica fibers support optical transmission with minimum loss at 870 nm, 1300 nm, and 1550 nm wavelengths. An erbium-doped silica fiber is easy to make, easy to work with, and can be used in the EDFAs for standard telecommunication applications. Despite these advantages, this fiber suffers from the problem of optical dispersion around 1550 nm, as shown in Figure 3.17. Dispersion-shifted fibers (Figure 3.17) can be used to compensate for the difference, but their higher costs limit their practical usage. In summary, optical fibers for use in EDFA devices must have minimum dispersion over wider bandwidth and high data rate capability within narrow optical frequency bands. Currently, most of the EDFA devices used in fiber optic telecom-

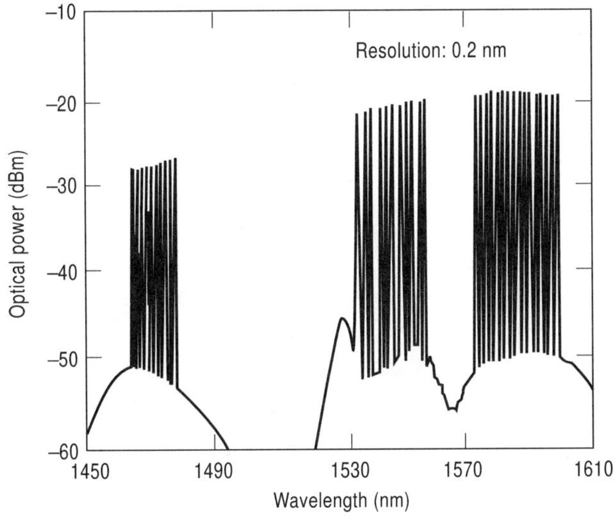

FIGURE 3.17 Three-band wavelength division multiplexer (WDM) spectra in a dispersion-shifted optical fiber.

munication equipment operate at 1300 nm wavelength because of minimum dispersion.

3.8 SUMMARY

This chapter provides detailed descriptions of various IR sources including man-made sources, natural sources, laboratory sources, commercial and industrial sources, and military sources. Performance parameters and operational limitations of both the coherent and incoherent sources are discussed. Incoherent sources include arc lamps, electronic discharge lamps, globars, incandescent lamps, high-pressure gaseous sources, gas mantle lamps, and light emitting diodes (LEDs). Coherent sources include solid-state lasers or DPSS lasers using nonlinear optical crystals, fiber lasers, optical parametric oscillators, EDF amplifiers, PDFF amplifiers, high-power chemical and gas lasers and ultra-high-power COIL lasers. Artificial sources include buses, trucks, tanks, helicopters, commercial and military aircraft, artillery muzzle flashes, and army personnel carriers. In addition, performance capabilities of high-power lasers are identified for space and satellite applications. Performance requirements of some ground-based and space-based IR sources used for the detection and tracking of long range enemy missiles are summarized, with emphasis on range limitations, cost, and complexity. Finally, performance requirements of various optical fibers are identified for application in optical amplifiers and oscillators.

REFERENCES

1. W. L. Wolfe and G. J. Zissis, *The IR Handbook*, p. 2.7, RCA, Commercial Engineering, Harrison, NJ, 1978.
2. Ibid., p. 2.11.
3. Ibid., pp. 2.8, 2.26.
4. Ibid., Table 2.9.
5. Ibid., p. 2.42.
6. Bergh and Dean, "Light emitting diodes," *Proc.. of IEEE*, *6*(2), 1977.
7. *Electro-Optics Handbook* (EOH-11), p. 139, RCA, Commercial Engineering, Harrison, NJ, 1974.
8. W. L. Wolfe and G. J. Zissis, *The IR Handbook*, p. 2.81, 1978.
9. Ibid., pp. 2.84–86.
10. *Electro-Optics Handbook* (EOH-11), p. 143, RCA, Commercial Engineering, Harrison, NJ, 1974.
11. J. L. Lerner (Contributing editor), *Laser Focus World*, p. 97, November, 1988.
12. M. Pollman et al., "Efficiency of erbium-based microcrystal and fiber Lasers," *Journal of Quantum Electronics*, *32*(4), 657–662, 1996.
13. (Contributing editor), *Laser Focus World,* pp. 15–16, January 1999.
14. P. Bado, "Ultrafast lasers escape the lab," *Photonics Spectra*, p. 157, July 1998.
15. A. Baltuska et al., *Optical Letter*, *23*(18), p. 1474, 1998.
16. (Contributing editor), *Laser Focus World*, p. 148, October 1998.
17. (Contributing editor), "High-power Laser operating in mid-IR region," *Photonics Spectra*, p. 5, January 1999.
18. (Contributing editor), "Highlights of a COIL System," *Laser Focus World*, p. 22, December 1998.
19. M. Bowers, "OPOs moving into the Mid-IR," *Lasers and Optronics* pp. 13–19, September 1998.
20. (Contributing editor), "Erbium-doped fiber amplifiers," *Laser Focus World*, p. 30, September 1998.

CHAPTER FOUR

Detectors and Focal Planar Arrays

4.0 INTRODUCTION

This chapter focuses on the performance capabilities and limitations of noncryogenic and cryogenic detectors and focal planar arrays (FPAs). Detectors can be used to measure both the power and energy associated with free space and optical systems. Performance levels of photon detectors, quantum detectors, and other optical detectors are summarized, with emphasis on sensitivity and reliability. Detector design can be optimized for specific wavelength operation to achieve the highest sensitivity. Some applications may require more than one detector, which might involve rapid optical switching devices between the various detectors used. The selection of a detector is strictly dependent on type of application, operating mode (pulsed or CW), and measurement parameters involved. Optical detectors are classified in various categories such as high-speed detectors, low-power detectors, high-power detectors, pulse energy detectors, time domain detectors, frequency domain detectors, and photon detectors. A detector can be used to measure energy with a power detector or vice versa. Some applications may require more than one detector, which will require rapid switching between the low-power and high-power detectors, between the low-energy and high-energy detectors, or between the low-speed and high-speed detectors, depending on the level and type of optical signal. Selection of detector material is dependent on operating wavelength, parameter to be measured, and the operating environments. Semiconductor materials such as Si, PbS, PbSe, InGaAs and Hg:Cd:Te are widely used in the design of infrared detectors because they offer acceptable performance even at room temperature (300 K). Regardless of the type and material, relative responsivity of the detector is the most critical performance parameter. Typical responsivity curves as a function of wavelength for various detectors are shown in Figure 4.1.

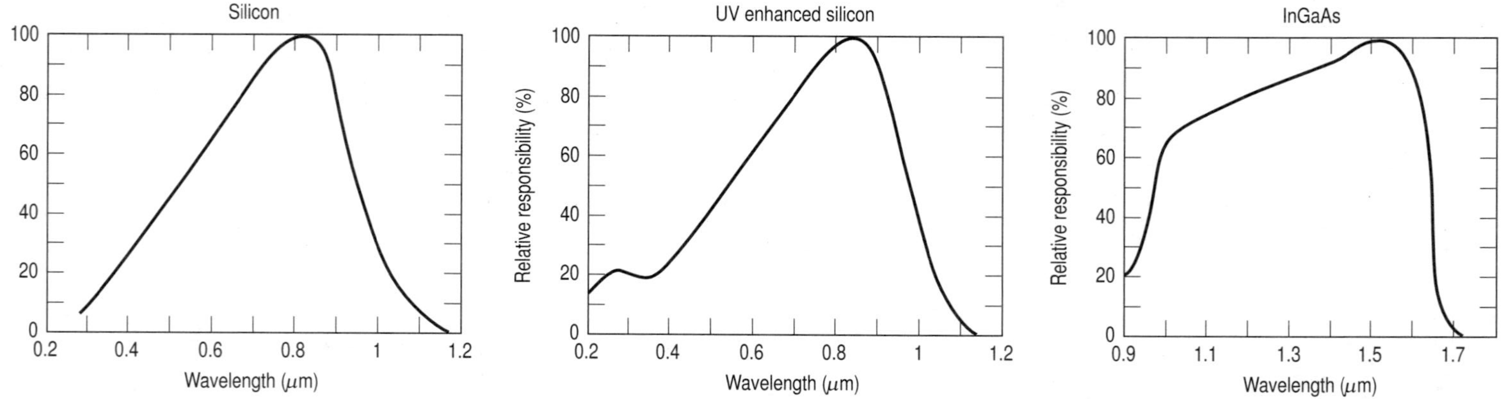

FIGURE 4.1 Relative responsivity of silicon, UV enhanced silicon, and InGaAs optical detectors as a function of operating wavelength. (Courtesy of Newport Corp., Irvine, CA)

4.1 DETECTOR TYPES

Various types of optical detectors are available to measure parameters either in the frequency or time domains. Once the domain is known, the detector material is selected to provide optimum sensitivity.

4.1.1 Time Domain and Frequency Domain Detectors [1]

Time domain detectors provide ring-free pulse output signals at all speed levels. The bit error rate (BER) of an optical system degrades due to ringing of the signal. Time domain detectors are primarily used to characterize temporal waveforms, where no signal distortion can be tolerated. For time domain applications, InGaAs PIN diode detectors are widely used, because of optimum responsivity of 0.6 A/W for microwave signals up to 20 GHz.

Frequency domain detectors require flat frequency response over the required operating bandwidth to maintain uniform responsivity across the entire band. The temporal response is not considered important for this type of detector. Frequency domain detectors are best suited for communication systems and near-IR sensors, where a flat frequency response from DC to 20 GHz is of paramount importance.

4.2 LOW-POWER, HIGH-POWER, AND HIGH-SPEED DETECTORS

4.2.1 Low-Power Detectors

High-quality semiconductor materials are required in the design of low-power detectors. In ordinary detectors, coherent light causes reading errors across the detector surface so the measurements are more sensitive to thermal drift, which leads to error rates between 5 and 8% when making laser power measurements. These errors can be significantly reduced or completely eliminated with stable, uniform detector response, which is achieved through incorporation of a built-in attenuator. The built-in attenuator provides low reflection, high damage threshold, accurate low-power measurements, and spectral flatness, as shown in Figure 4.2 [2]. Spectral responsivity of optical detectors with and without attenuators as a function of wavelength, and the impulse response of a detector using a filter are shown in Figure 4.2.

4.2.2 High-Power Detectors

High-power detectors are designed to provide fast response, low spectral reflection, high damage-threshold surface, flat spectral response over the 1–12 μm range, and reliable power readings of CW or pulsed laser signals. Such detectors using an isothermal disk design offer high accuracy, uniformity, and reliability during the measurements of CW or pulsed laser power levels. Miniaturized pyroelectric high-power detectors come in TO-18 packages with thermally stable black coating, which provides unfiltered detector response over the 1–25 microsecond period. These de-

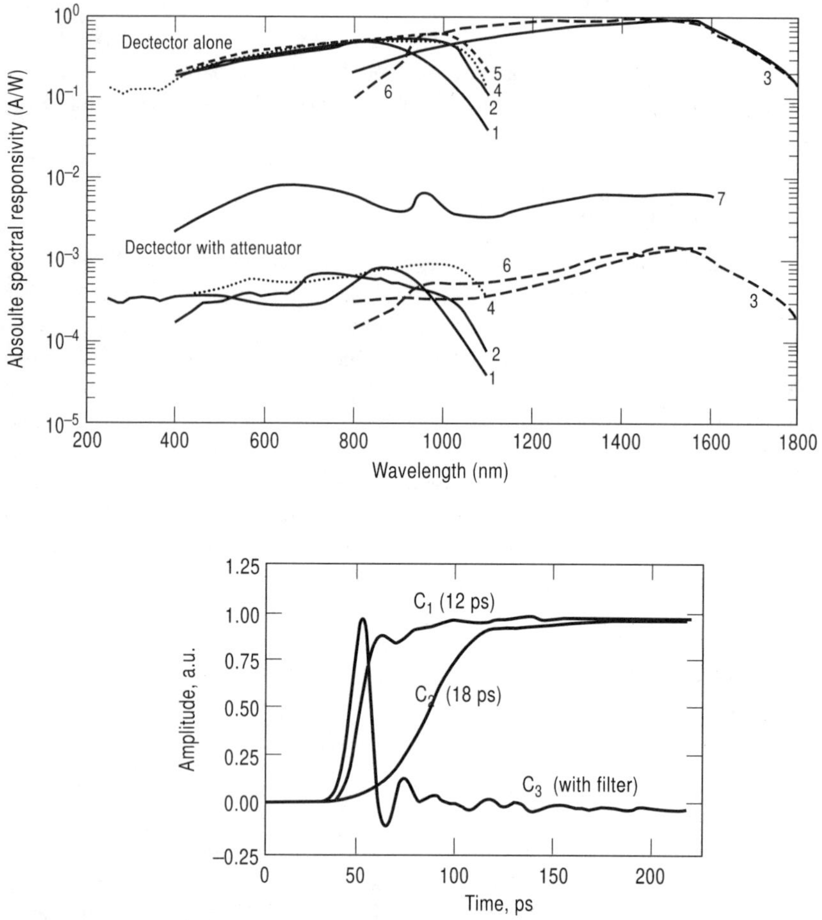

FIGURE 4.2 Spectral responsivity of detectors optimized for specified band and impulse response (top). 1. For optimized UV band; 2. for broadband UV band; 3. for broadband IR band; 4. for optimized over 800–1000 nm; 5. for optimized over 700–1000 nm; 6. for optimized over 900–1600 nm; 7. for optimized over ultrawide spectral bandwidth. Bottom: detectors with FWHM of 12 ps and 18 ps and a detector with filter (C_3). (courtesy of Newport Corp., Irvine, CA)

tectors can be designed for either current-mode operation or voltage-mode operation using postamplifiers and bandpass filters to optimize the detector performance.

High-power, fan-cooled detectors can measure optical power levels of up to 300 W and are designed to meet damage-threshold levels as high as 10 kW/cm^2 and 0.3 J/cm^2 for CW-power and pulsed-power measurements, respectively. Exceeding the rated damage-threshold level can degrade the detector performance, but causes no catastrophic failure. High-power detectors are best suited for power measurements from high-power Nd:YAG, Ti:sapphire, CO_2, and holmium lasers.

4.2.3 High-Energy Detectors (HEDs)

High-energy detectors (HEDs) generally use crystalline pyroelectric elements enclosed in an EMI-shielded cylindrical housing and are capable of providing energy measurements from femtojoule to joule levels with high reliability over a wide spectral region. HED designs must incorporate a black organic coating, which offers maximum resistance to damage. Organic coatings are periodically required to maintain specified damage-threshold capability. Superior low-noise detector performance is possible with battery-powered designs using a preamplifier. HEDs with black organic coatings are most suitable for pulsed flash lamps, diode-pumped Nd:YAG lasers, and tunable dye lasers. HEDs with pyroelectric elements provide broad spectral response up to 20 μm and repetition rates close to 4,000 pulses per second (pps). These detectors, when designed with built-in three-position electronic switches, offer selection of integration time constant and output voltage for greater versatility.

4.2.4 High-Speed (HS) Detectors

High-speed detector design [3] includes large-area fiber coupling, high conversion gain, and broad spectral response, which are possible with InGaAs material using the metal–semiconductor–metal (MSM) design technique. Key performance features of HSDs include high-speed digital and microwave signal detection with sensitivity from 0.2 to 0.8 A/W over the 400–1700 nm spectral range and precision high-speed optical-to-electrical conversion with temporal response from 15 ps at 30 GHz to 30 ps at 15 GHz. These detectors are most suitable for accurate diagnosis of communication systems and time-domain measurements requiring high repeatability. Typical normalized responsivity curves [3] for low-power and high-speed detectors are shown in Figure 4.3 .

As stated earlier, InGaAs PIN diode detectors offer state-of-the art time domain performance and optimum responsivity at each speed level. HSDs with fiber optic coupling, internal regulated voltage, current monitoring circuits, and power-status indicator LEDs provide accurate, ring-free temporal response are best suited for fiber optic communication links and near-IR sensors operating over the 900–1600 nm range. HSDs with DC-coupled transimpedance amplifiers have significantly higher conversion gain, ranging from 200 V/W to 800 V/W with temporal response from 200 ps to 300 ps. Both the detector element and the transimpedance amplifier can be integrated on a common substrate. The amplifier components must be shielded from outside noise sources to achieve improved detector performance.

A completely self-contained high-speed detector [4] with the above features is shown in Figure 4.4. The high-speed detector configuration shown in Figure 4.4 includes two lithium batteries, each with internal voltage regulation capability, a battery-monitoring device that lights an LED indicating that battery replacement is needed and a current-monitoring circuit ensuring the required line voltage output and the transimpedance amplifier gain of 10 mV/μA under normal operating conditions. This detector offers a 12 ps full-width, half-maximum (FWHM) impulse re-

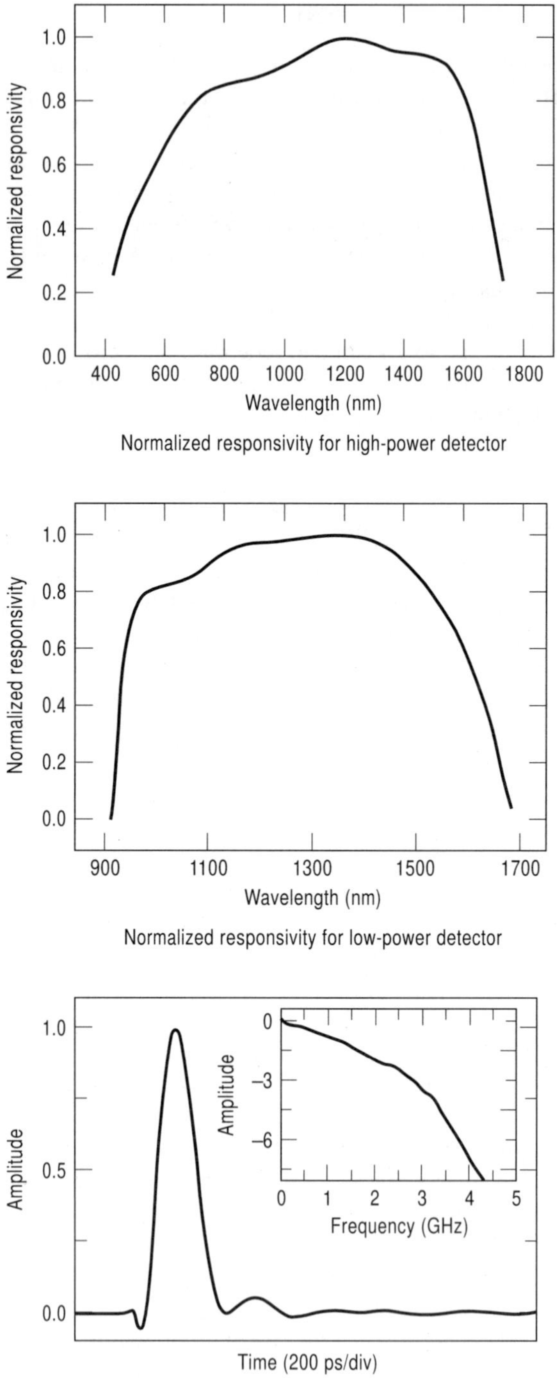

FIGURE 4.3 Normalized responsivity curves and impulse response of detectors.

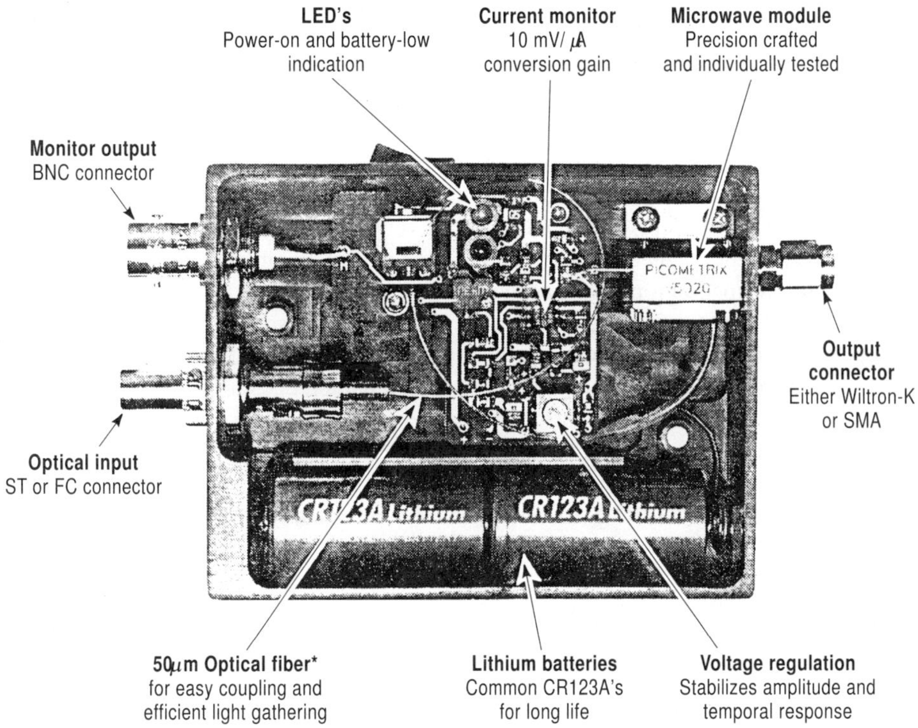

FIGURE 4.4 Fully equipped, high-speed detector system with associated components. (Courtesy of Newport Corp., Irvine, CA)

sponse, clean and fast response with minimum ringing, DC coupling, bandwidth up to 60 GHz in the visible region and 45 GHz in the near-IR region, and high responsivity over the 1310–1550 nm spectral range. These detectors are best suited for digital fiber optic communication; BER testing; measurements of extinction ratios, peak power levels, and short pulse parameters; and characterization of electrooptic modulators and other IR components. These detectors [5] are available with free-space (FS), fiber-coupled (FC), and multimode-fiber-coupled (MFC) inputs and with transimpedance gain exceeding 10^6 V/A.

4.3 SEMICONDUCTOR PHOTOVOLTAIC CELL DETECTORS

Photovoltaic cell (PC) detectors are widely used in satellites and space vehicles to provide the needed electrical power to various sensors aboard. Early solar cells were fabricated using Si and GaAs materials with various geometrical configurations, including the V-groove design capable of yielding conversion efficiencies from 15 to 20% under controlled environments. Solar cells with multilayer and V-groove designs offer higher conversion efficiency (18–22%) and higher power outputs. Solar concentrated arrays with refractive linear element technology and cylindrical Fresnel lenses provide a 7:1 concentration ratio that can allow a single multijunction (SMJ) $GaInP_2$/GaAs/Ge cell to collect the solar energy equivalent to that generated by seven cells. SMJ cells offer conversion efficiency as high as 26% under controlled environments, but offer slightly lower efficiency, around 20% in the field.

Published literature on solar array technology [6] indicate that two solar concentrator arrays of appropriate dimensions can supply more than 2 kW of electrical power, sufficient to meet the electrical requirements of communication satellites orbiting in low, medium, or synchronous orbits. Solar cells are the most critical and expensive elements of a spacecraft or satellite. All the sensors operating aboard are dependent on the electrical power available from the solar panels.

It is important to mention that a solar cell is a photon detector. The photon energy of the sun is collected by the SMJ cell detector and is converted into electrical energy. Silicon solar cells yield conversion efficiencies between 12 to 15%, whereas the GaAs cells with V-groove configurations offer conversion efficiencies between 18 to 22%. Designers of "SMJ"-$GaInP_2$/GaAs/Ge solar cell detectors claim efficiencies between 22 to 26% under control environments with concentration ratios exceeding 7:1.

4.3.1 Solar Concentrator Design Aspects

A solar concentrator using refractive linear element technology allows the semiconductor PV cell design with relatively small area to provide electrical power equivalent to that of a seven-element array. The concentrator is fabricated from 3.3 inch wide [6] Fresnel lenses using an aerospace adhesive. The lenses are produced in rolls of hundreds of feet in length. The prismatic side faces the solar cells and the sunward side is curved into the shape of an arc with a relatively large acceptance angle of about 2°. A 0.003 inch thick ceria-doped glass arch is attached to the curved surface of the 0.010 inch thick Fresnel lenses. The lens length is limited only by practical handling concerns. The solar array [6] consists of 720 Fresnel lenses fabricated on an 8.4 inch long array. The lenses are held 3.62 inches away from the cells. The SMJ cell arrays have demonstrated a conversion efficiency of 22% in concentrated sunlight. The reduced cell area permits use of thicker than normal glass covers, leading to a significant increase in the radiation resistance of the photovoltaic (PV) cells. Such SMJ cell array configurations have potential applications in spacecraft or satellites in high-radiation orbits.

4.4 METAL–SEMICONDUCTOR–METAL (MSM) PHOTON DETECTORS

GaAs-based MSM photo detectors or photon detectors feature extremely fast response time, low dark current, and a cut-off frequency as high as 10 GHz. These detectors have a photosensitive area that is larger than that of the typical photo detector, which provides efficient coupling to an optical system. Its simple structure is best suited for optoelectronic integrated circuits, making them most attractive for optical communications. The MSM detectors are most attractive for applications in the 900–1700 nm and 120–2600 nm spectral ranges.

The design of the MSM detectors essentially combines an optical structure to capture the maximum number of incident photons and an efficient electrical structure to collect the photo-generated carriers. The important figure of merit (FOM) for photo detectors used in optical communications systems is the bandwidth efficiency product (BW $\times$ η). High quantum efficiency and the high probability of detecting incident photons with minimum loss are the most critical parameters of a photon detector or photo detector. The quantum efficiency of a conventional detector is dependent on the absorption coefficient of the semiconductor material deposited and is directly proportional to the thickness of the active regions. However, thicker active regions tend to reduce device speed or response time due to long transit times. Optimization of gain–bandwidth product (G $\times$ BW) requires higher quantum efficiency without increase in active-region thickness. The latest research and development activities on optoelectronic devices reveal that photon detector performance can be significantly increased by placing the active device structure inside an optical resonant microcavity. Optoelectronic devices based on this concept are referred to as resonant cavity enhanced (RCE) devices. The performance of most optical detectors [7] improves when operated at cryogenic temperatures. Lower cryogenic temperatures improve both the sensitivity and the quantum efficiency of the detector. Typical cryogenic operating temperatures [7] commonly used for photon detectors are summarized in Table 4.1.

The Johnson noise or thermal noise in the detector is strictly due to temperature rise at the junction of the element, which can be significantly reduced at lower cryogenic temperatures. Detector theory indicates that photon detectors have more photon-generated carriers than thermally generated carriers. Cryogenically cooled Pb:Se, Ge:Zn, and He:Cd:Te detectors offer significantly improved performance levels when operated at optimum temperatures of 195 K, 77 K, and 5 K, respectively. The above detectors yield fast response and improved detectivity (D^*), better than 10^9 cm $\sqrt{\text{Hz}}$/W. In the case of a Ge:Zn photon detector, one sees a sharp in-

TABLE 4.1 Cryogenic operating temperatures for various photon detectors

Photon detector material	Operating temperature (K)
PbS	300, 193, 77
Si	300
InAs	300, 195, 77
InSb	300, 77
Ge:Au	77
Ge:Cd	5
Ge:Zn	5
Si:As	22
Hg:Cd:Te	77

crease in detectivity if the temperature exceeds 10 K. Most sensitive photon detectors operating over wide spectral ranges and at cryogenic temperatures are shown in Table 4.2.

The extrinsic-doped silicon, germanium, and germanium–mercury detectors require operating temperatures well below 30 K for optimum sensitivity. However, He:Cd:Te and Pb:Sn detectors offer optimum performance even at 77 K. The Pb:Se detector is best suited for operations at higher optical frequencies because of its low time constant, whereas the In:Sb detector is most attractive for higher optical frequencies. The Hg:Cd:Te detector can also be used as a quantum detector. It is important to mention that the ternary alloy detectors, namely, Hg:Cd:Te and Pb:Sn:Te, are widely used over the 8–14 spectral range because of their improved responsivity and detectivity at the cryogenic temperature of 77 K. By selecting the right partial mole fraction semiconductor compound, Hg:Cd:Te, and optimum cryogenic temperature, one can achieve improved detector performance over a wide IR spectral region. These detectors have potential applications in high-performance military IR sensors such as the IR line scanner (IRLS), forward looking IR (FLIR), and IR search and track (IRST).

4.5 QUANTUM DETECTORS

Quantum detectors are widely used in the visible and near-IR spectral regions. Cryogenically cooled quantum detectors will be most suitable for fiber-optic-based systems, namely, optical data links and fiber optic ring lasers. Performance parameters of these detectors, such as sensitivity, detectivity, responsivity, response time, noise-equivalent power (NEP), and dark current, will experience significant improvement at cryogenic temperatures. Quantum well IR detectors have to be cooled down to below 77 K to achieve background-limited performance level.

4.5.1 Avalanche Photon Counting Detectors

Performance of a light detector depends on the optical signal level and its bandwidth. A photo count (PC) detector is similar to a photon detector or photo diode detector as far as operating principal is concerned. A photo diode detector is generally used for the visible portion and near-IR region of the spectrum. The optical sig-

TABLE 4.2 Performance of photon detectors at cryogenic temperatures

Detector material	Spectral range (μm)	Temperature (K)
Ge:Hg	8–14	27
Ge:Cu	8–28	5
Si:As	8–30	5
Si:Ga	8–16	27
Hg:Cd:Te	8–14	77
Pb:Sn:Te	8–14	77

nal level is typically greater than 10 nanowatt (nW) over a bandwidth not exceeding 10 Hz. As the light level decreases or the bandwidth increases, the noise from the output amplifier connected to the photo diode detector overwhelms the signal. To overcome this problem, it is desirable to use a detector with internal gain, such as an avalanche photo diode (APD) detector or a photomultiplier tube (PMT). An APD has an internal gain between 100 to 1000, while a PMT has a typical gain exceeding one million (in the 2–3 million range). Both these devices have noise-free internal gain. Because of very high gain, a PMT is roughly 1000 times more sensitive than a silicon APD detector over a spectral range from 200 to 850 nm. Typical characteristics of junction photo detectors are summarized in Table 4.3.

The signal-to-noise ratio (SNR) is the most important parameter for a detector, because the detection is strictly based on it. This ratio is a function of detector noise factor (1.4 for PMT and 2.5 for ADP), amplifier noise factor, background, dark and photo cathode currents, number of secondary electrons, detector gain (shown in last column of Table 4.3), temperature, bandwidth, resistance, and active area. The SNR is dependent on twelve different variables and its expression is most complex because of the number of variables involved. Both the APD and PMT detectors are light-to-current detectors. Conventional analog methods are used to convert the current output into voltage output, which may be amplified prior to digitization. Additional amplification may be required with an APD device but is often not needed in the case of PMT devices because of their high inherent gain. When the input light levels approach a femtowatt (10^{-15} W), the noise figure from the detectors greatly reduces the SNR, which will rule out conventional analog processing methods. In this situation, a digital approach called photon counting must be used. Because of the very large gain required, PMT is the most attractive device for a photon counting system.

4.5.2 Photomultiplier Tube (PMT) Detector

PMT is the most efficient photon counting system because of its high inherent gain and low noise factor. In nuclear imaging, it is the PMT that creates the high image quality and performance level of the imaging system, which is most desirable for medical diagnosis and for certain night vision sensors used by armed forces. A schematic diagram showing the critical elements of a PMT device along with a pulse light histogram [8] is depicted in Figure 4.5. When a photon strikes the photo

TABLE 4.3 Typical characteristics of junction photodetectors

Material	Structure	Rise time (ns)	Wavelength (nm)	Responsivity (A/W)	Current/Gain (mA/—)
Silicon	PIN	0.5	300–1100	0.5	1/1
Germanium	PIN	0.1	500–1800	0.7	200/1
InGaAs	PIN	0.3	1000–1700	0.6	10/1
Silicon	APD	0.5	400–1000	77	15/150
Germanium	APD	1.0	1000–1600	30	700/50

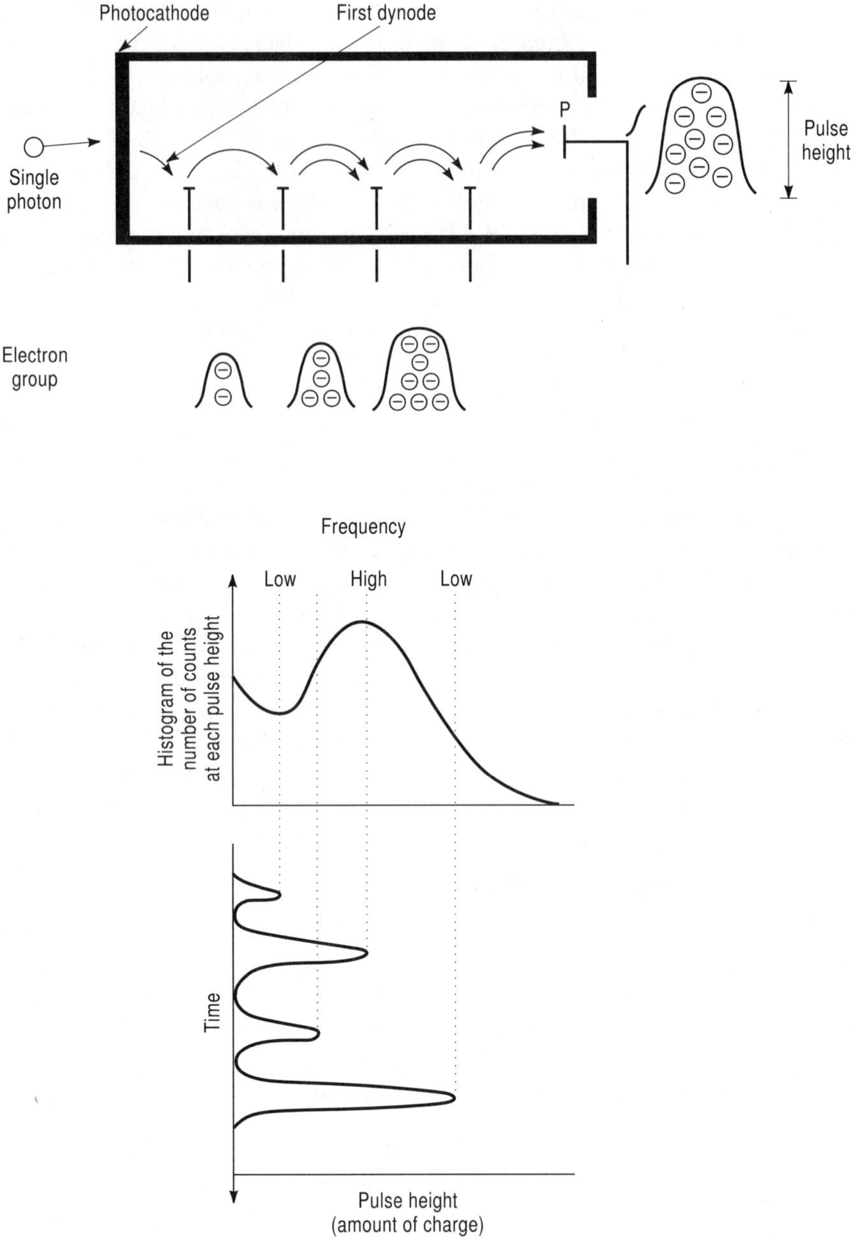

FIGURE 4.5 Critical elements and histogram of a photomultiplier tube.

cathode of the PMT, the photoelectric effect generates a single photon electron. The photon electron is collected by an electrooptic lens and then accelerated into a dynode. Secondary emission of between 3 and 5 electrons appears at the first dynode surface. These secondary electrons are then accelerated into next dynode and so on and so forth. The average number of secondary electrons generated by each collision with each dynode is four. The number of secondary electrons increases rapidly as the electron transfers from left to right towards the collector P.

In a typical standard PMT, there are 10 dynode stages with ultimate capability of generating one million or more electrons. The transit time through the PMT is in the order of 15 ns, but the pulse width can be as short as 1.5 ns. The peak output pulse voltage is typically about 5 mV across a 50 ohm load, assuming an output pulse current of 100 μA ($V = IR = 100 \times 10^{-6} \times 50 = 0.005$ V). It is important to mention that all pulses that exit the PMT do not have the same pulse amplitude, because the electron multiplication is a statistical process that generates a distribution of output pulses. The histogram of pulse sizes shown in Figure 4.5 is known as the pulse height distribution (PHD). This distribution of output pulse size generates excessive noise levels. At light levels exceeding a few femtowatts, the shot noise due to the photoelectron effect is much larger than the noise generated by the pulse height distribution, which will have very little impact on the SNR. However, it is necessary to reduce the noise generated by the pulse height distribution, which can be accomplished by using a discriminator. This discriminator circuit can be appropriately set to reject pulses that are smaller than some predetermined threshold level.

The threshold can be determined by using a PHD technique involving a LED as a stable light source and neutral density filters between the LED and PMT. The most practical approaches for the measurements of pulse height distribution are shown in Figure 4.6. After collecting millions of pulses and storing in memory, a histogram can be developed. The PHD is the one that shows the highest ratio of peak to valley. The height of the valley is used to set the discriminator threshold level, which in turn reduces the overall signal count level by 10–20 % due to thermal noise from dynodes. An alternate approach involving a discriminator followed by a counter is shown in Figure 4.6B. The lower level discriminator (LLD) is first set to zero and then the number of counts are recorded. The upper level discriminator (ULD) is increased and the counts are again measured. This particular method requires several sets of data to optimize the PHD through differentiating the count curve versus LLD level. Once the LLD level is obtained using either method, the circuit shown in Figure 4.6B can be used to count the photons. It is important to mention that the photon counting device not only eliminates the noise due to the statistical gain fluctuations, but also increases the stability of the measurement under wide variations in temperature and power supply voltage. A 2% change in power supply voltage will change the gain of the PMT by 15%, which translates directly to a 15% change in the analog signal level. However, in the photon count mode, a 2% change in power supply voltage will introduce only a 1% change in the photon count rate, provided the LLD level has been set properly.

Efficiency and reliability of a PMT used in remote or military field environments must be given serious consideration. Power supply voltage fluctuations must

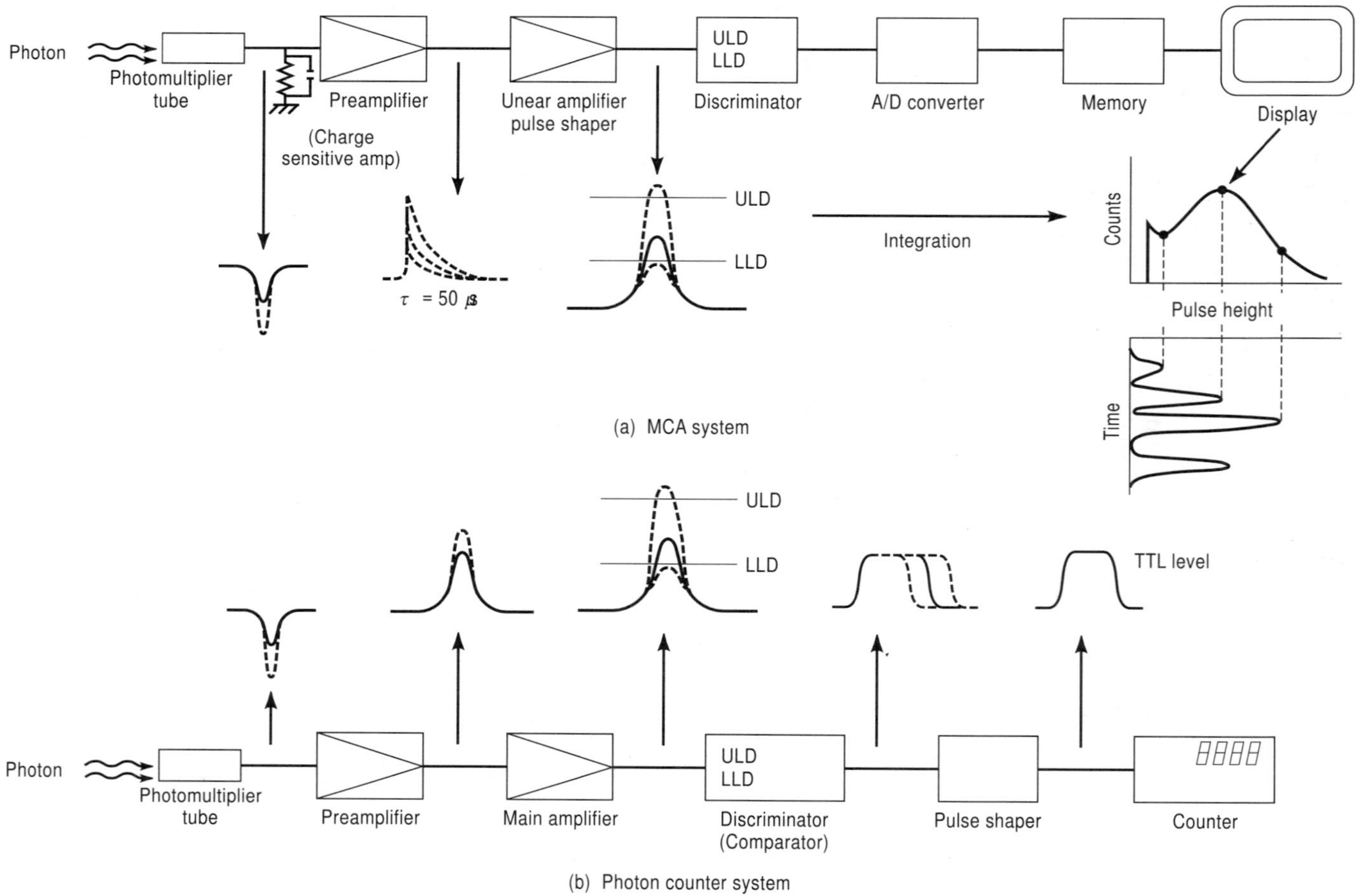

FIGURE 4.6 Schematic instrumentation diagrams to measure photon height distribution (HD) using (A) Multichannel analysis (MCA) and (B) photon counter system (PCS) methods.

be kept below 1% to maintain high accuracy in the photon count rate. A state-of-the art PMT design using a two-lens array configuration [9] demonstrated an increase in the efficiency from 25 to 65%. This lens array redirects the light onto the photomultiplier channels. Each element of the array is a telescope made out of two spherical lenses. The telescope can collect the light with incident angles less than 100 mrad (5.68 degrees), which introduces only small distortions in the image. The array not only increases the PMT detector efficiency to 65% without disturbing the positions of the incident photons, but also improves the reliability of the PMT.

4.5.3 Image Intensifiers

The photon counting technique is useful for measuring weak images with improved contrast. A multistage image intensifier is generally used to amplify the photons. However, the image intensifier reduces the spatial resolution by broadening the image of the single photon. There are several applications that require the high sensitivity, wide dynamic range, and stability of the photon counting system. Photon counting is exclusively used in the confocal microscope, which used a laser to excite a specific region within a cell. Fluorescent molecules within that cell are used to monitor local pH value or calcium concentration. Another potential application of photon counting involves the detection of bacterial contamination in food storage locations; such tests are required to meet stringent hygiene regulations established by the Food and Drug Administration (FDA). High-performance image intensifiers are widely used by front-line army units, who are required to fight in hostile territories under extremely low light levels.

4.6 OPTICAL AND QUASIOPTICAL DETECTORS

Noise-equivalent power (NEP) is the most critical performance parameter of an optical or quasioptical detector. A high-temperature, transition-edge bolometer made from superconductor YBCO-film with a critical temperature (T_c) of 91 K on strontium titanate substrate provides a NEP as low as 7×10^{-12} watt/$\sqrt{\text{Hz}}$, which is impossible to achieve from a room-temperature IR detector for measurement of radiation at 13 μm or higher wavelengths. Quasioptical detectors are simple and less expensive. However, they must provide fast response time better than 50 ps, frequency roll-off of 4000 MHz, NEP better than 5×10^{-12} W/$\sqrt{\text{Hz}}$, and wide IF bandwidth at MM-wave operations exceeding 500 GHz.

4.6.1 Superconductor Hot Electron Bolometer (HEB)

The sensitivity of a superconductor HEB made from a niobium nitride (NbN) junction with superconducting gap frequency of about 1400 GHz [10] will not deteriorate at operating frequencies up to 1400 GHz. A superconductor hot-electron transition-edge (SHETE) bolometer can be an alternative to a SIS-junction bolometer [10] at operating frequencies greater than 1000 GHz. A SHETE bolometer does not rely on electron heating and its response is independent of the gap frequency. Both the above-mentioned bolometers are widely used as quasioptical detectors in astronomical sensors and optical microscopes. Superconducting antenna-coupled microbolometers [10] have significant operational advantages in terms of heat capacity and optical efficiency. A hot-electron bolometer (HEB) employs a thin copper film connected between the antenna terminals. The RF current from antenna is dissipated into the resistive copper strips, leading to a temperature rise in the electrodes and voltage across the metallic tunnel junction deposited on the copper strips. Both the electron–electron and electron–photon relaxation rates govern the operation of the HEB with temperature as low as 300 millikelvins (mK) or 0.3 K. HEBs have demonstrated remarkable performance in the 750–970 GHz range [10] with absorption power as low as 140 pW, photon-to-signal temperature differential of 33 mK, electron-to-signal temperature differential of 36 mK, photon noise less than 1.6×10^{-31} W/$\sqrt{\text{Hz}}$, and junction noise not exceeding 2×10^{-32} W/$\sqrt{\text{Hz}}$. HEBs, when operated at a cryogenic temperature of 4.2 K, have potential applications in both ground-based and space-based astronomical photometry systems.

4.7 RESONANT-CAVITY ENHANCED (RCE) DETECTORS

High-speed RCE detectors are fabricated using molecular-beam-epitaxial (MBE) and standard photolithography to produce Fabry-Perot low-loss microcavities. Optoelectronic device performance is significantly enhanced by placing the active device structure in a Fabry–Perot resonant microcavity (FPRM) [11]. The design of a semiconductor photo detector involves an optical structure to efficiently capture the incident photons and an electrical structure to collect the photo-generated carriers. The bandwidth–efficiency product is the FOM for the photo detectors and these photo detectors are widely used in optical communications. High quantum efficiency is the most important parameter of the detector. To optimize the gain–bandwidth product, it is necessary to improve the quantum efficiency without increasing the active layer thickness. Placing an active device inside a FPRM, as shown in Figure 4.7, results in significant performance improvement. The wavelength selectivity and enhancement of the resonant optical field are provided by the microcavity. The increased optical field allows the photo detector to be made thinner and faster, while at the same time increasing the quantum efficiency at the resonant wavelength, as illustrated in Figure 4.8. Due to the rejection of the off-resonant wavelengths by the microcavity, the RCE photo detector exhibits both wavelength selectivity as well as high-speed detector response.

The quantum efficiency is defined as the current flux to photon flux ratio and is dependent on the absorption coefficient (α) and the thickness (d) of the absorbing layer. For a high-speed RCE photo detector with small depletion width ($\alpha d \ll 1$), the quantum efficiency can be defined as

$$\eta_q = (1 - R)(1 - e^{-\alpha d}) \approx (1 - R)(\alpha d) \tag{4.1}$$

where R is the reflection coefficient of the antireflection coating and d is the absorbing layer thickness.

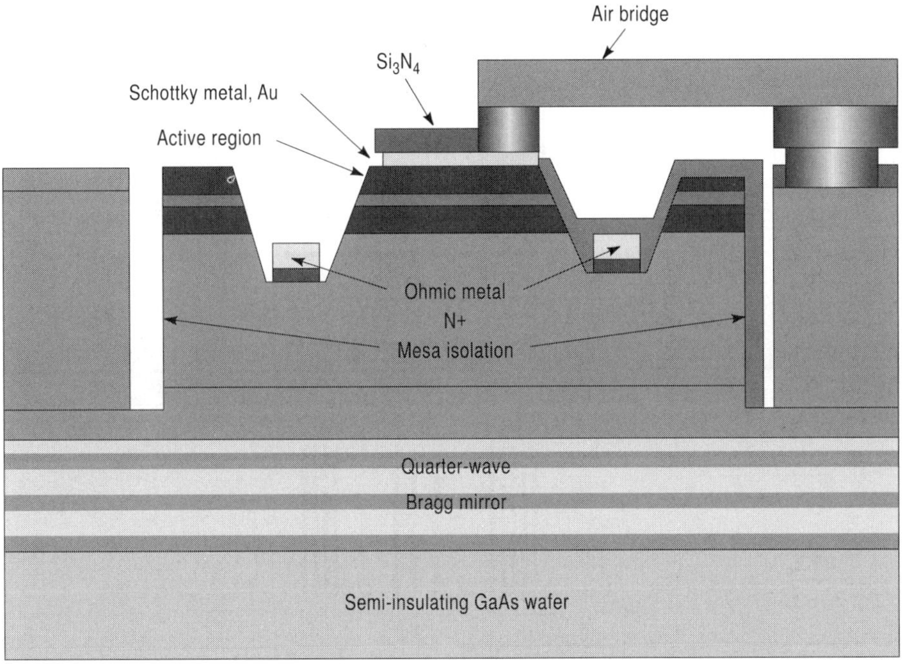

FIGURE 4.7 Fabry–Perot low-loss microcavity design using resonant-cavity-enhanced Schottky photodiodes and standard photolithography.

The speed-response limitation in a small-area photo diode is due to drift time across the depletion region. The transit-time-limited 3 dB bandwidth of a thin detector can be defined as

$$(BW)_{\text{tran}} = [(0.45)\,(v/d)] \tag{4.2}$$

where v is the carrier or hole velocity and d is the depletion region thickness. The bandwidth–efficiency product can be written as

$$(BW)_{\text{tran}}(\eta_q) = (0.45)(\alpha)(v)(1 - R) \tag{4.3}$$

For a gallium arsenide detector with zero dielectric coating ($R = 0$), the absorption coefficient is 10^4/cm and the carrier or hole velocity is (6×10^6) cm/second, the bandwidth–efficiency product comes to $(0.45)(10^4)(6 \times 10^6)(1 - 0) = 27$ GHz. This means that the bandwidth is 27 GHz at a quantum efficiency of 100%, 24.3 GHz at a quantum efficiency of 90%, and 21.6 GHz at a quantum efficiency of 80%. The quantum efficiency of a photo detector can be improved by collecting the light through the MESA edge of the optoelctronic device, which is perpendicular to the electric current direction, as shown in Figure 4.8. Simultaneously, a long, narrow absorption region and a short current path allow optimization of the bandwidth–efficiency product. The insertion loss experienced by the incoming light limits the

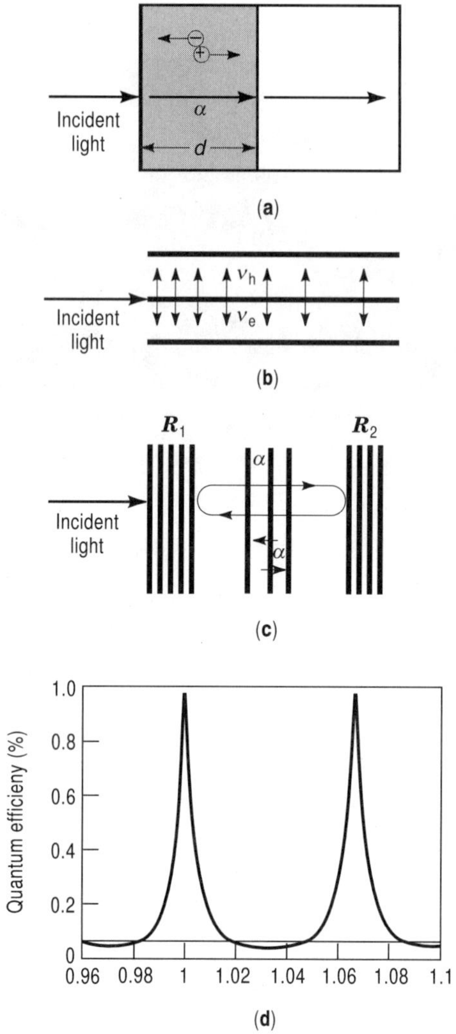

FIGURE 4.8 RCE detector design and fabrication details and quantum efficiency at 1.000 μm and 1.060 μm wavelengths.

quantum efficiency. Quantum efficiency is a periodic function of the inverse wavelength and it is enhanced periodically.

Structural details of a RCE detector are shown in Figure 4.9. The critical elements of this detector include top and bottom mirrors, top and bottom spacers, and an absorbing layer with narrow bandgap. Maximum quantum efficiency of the RCE detector [11] is

$$(\eta_q)_{max} = \frac{(1 - R_2e^{-x})(1 - R_1)(1 - e^{-x})}{[(1 - \sqrt{R_1R_2})(e^{-x})]^2} \tag{4.4}$$

where R_1 is the reflectivity of the top mirror, R_2 is the reflectivity of the bottom mirror, and the constant x is the product of the absorption coefficient (α) and the absorption region thickness (d). Calculated values of maximum quantum efficiency for a GaAs-based RCE detector as a function of mirror reflectivities and absorption region thickness are shown in Table 4.4

The computed values indicate that higher reflectivity of the bottom mirror yields progressively higher quantum efficiencies. These data also indicate that there is an optimum thickness of the absorption region for a given detector material. For a GaAs detector material, the optimum absorption region thickness appears to be around 200 nm. These data further indicate that the optimum reflectivity of the top mirror seems to be around 70%. In the case of conventional detectors, high quantum efficiency is usually observed with thin absorbing layers and the bandwidth–efficiency product no longer is limited by the material parameters. However, the overall improvement in the bandwidth–efficiency product of a RCE detector [11] is about three times of that for a conventional GaAs photo diode detector with a junction area of 10 μm by 10 μm.

High-speed photo detectors incorporating advanced technologies are best suited for gigabit-speed fiber optic communications and fiber-optic-based RF/microwave systems. Recently developed novel device structures [11] seem to improve responsivity, speed, and power-handling capacity of the high-speed detectors. Published literatures reveal that the fastest commercial photo detectors have a bandwidth of 60 GHz at wavelengths around 600 nm and 45 GHz at infrared wavelengths (950–1650 nm) with maximum responsivity of 0.2–0.4 A/W, depending on the operating wavelength. Two types of high-speed detectors, namely, resonant-cavity enhanced (RCE) detectors (Figure 4.9) and velocity-matched distributed (VMD) detectors (Figure 4.10) offer optimum performance level. The former type offers high speed, high quantum efficiency, precision wavelength selection, and low-voltage operation, whereas the later provides high speed, high power-handling capacity and excellent signal fidelity. Design of traditional high-speed photo detectors is generally based

TABLE 4.4 Maximum quantum efficiency as a function of mirror reflectivity and absorption region thickness

Reflectivity (%)		Absorption region thickness (nm)		
R_1	R_2	d = 100 nm	d = 200	d = 400
70	75	40	53	56
70	80	—	64	61
70	85	55	68	67
70	90	—	78	72
70	95	78	88	79

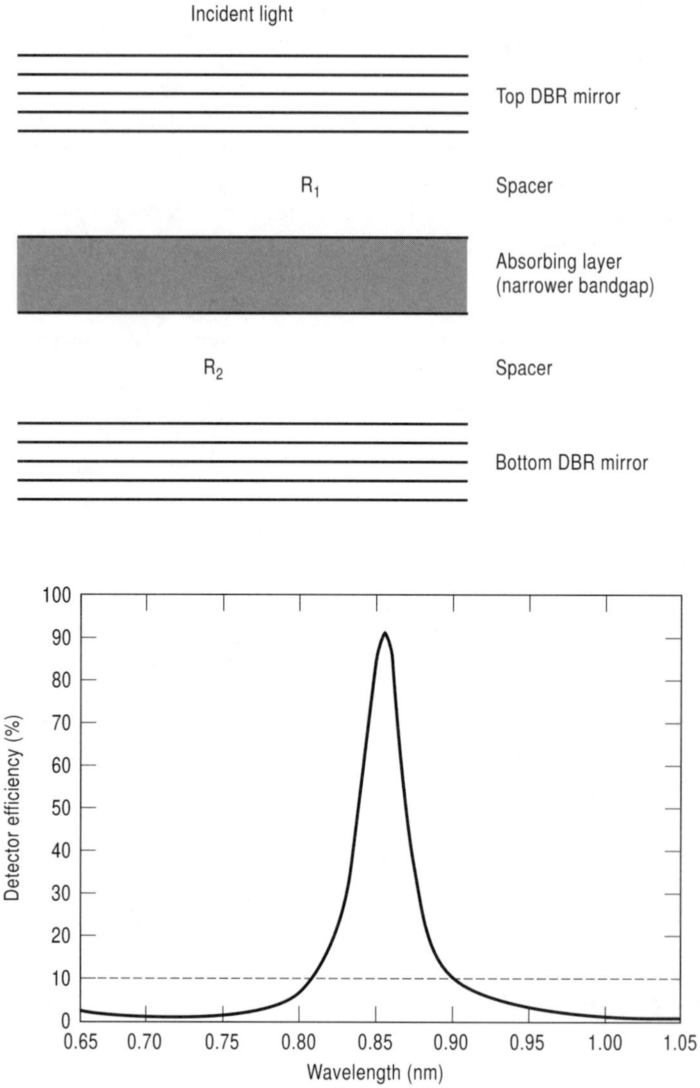

FIGURE 4.9 Structural details and quantum efficiency of RCE detectors.

on trade-offs between the bandwidth, responsivity, and quantum efficiency at the desired operating wavelength.

4.7.1 Critical Design Issues and Parameters for RCE Detectors

RCE detectors eliminate the trade-offs required by the traditional high-speed detectors by using active layers much thinner than the absorption length to absorb most

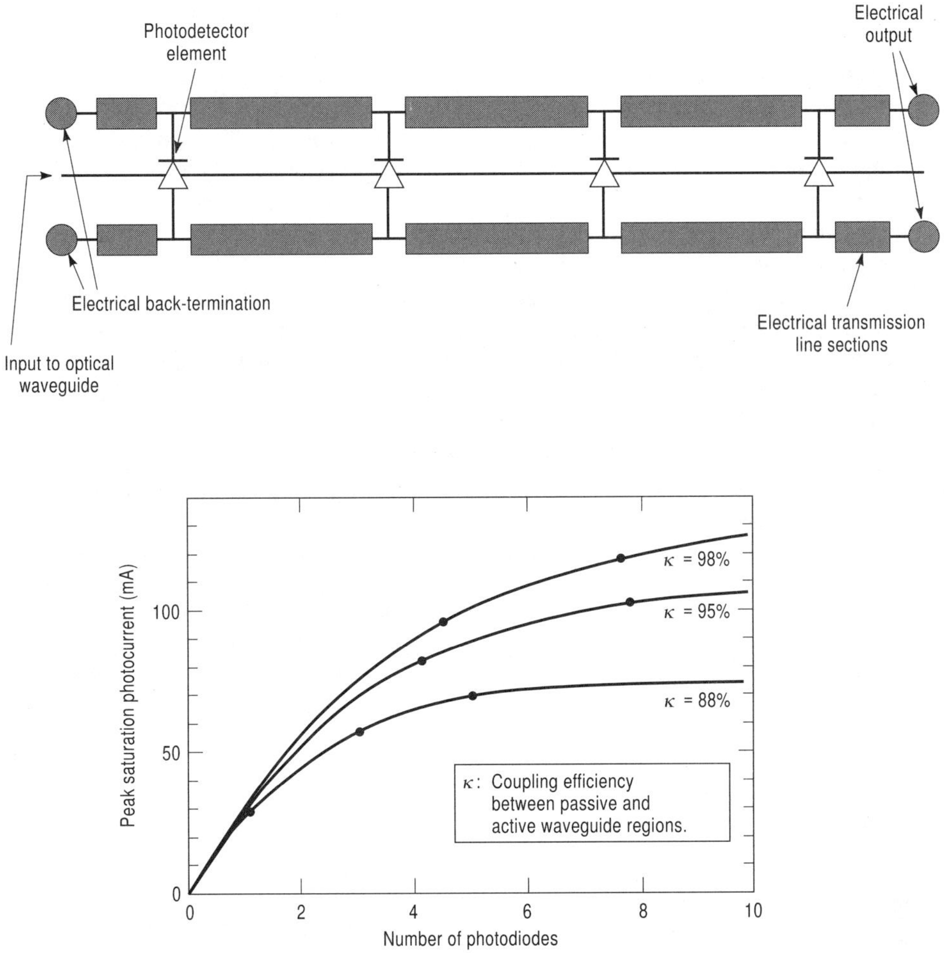

FIGURE 4.10 Structural details, coupling efficiency, and photocurrent of VMDs.

of the incident optical energy. As illustrated in Figure 4.9, a RCE device consists of a thin (compared to absorption length) narrow-bandgap absorbing layer positioned in a Fabry–Perot cavity, top and bottom mirrors, and two nonabsorbing spacers required for the adjustment of the overall cavity length. Despite several performance improvements, RCE devices suffer from low responsivity due to off-resonant reflections and increased cross-talk between the channels due to cavity insertion loss. Comprehensive examination of critical design issues and design concepts indicate that the reflections from the top and bottom cavity mirrors, positioning of the absorption layer within the cavity, effective length of the cavity, and absorption layer thickness are the most critical design parameters. The effective length of the cavity determines the resonant wavelength, whereas all other parameters determine the peak quantum efficiency and the width of the resonant cavity, as shown in Figure 4.9.

Fabrication of a RCE detector requires high-reflectivity distributed Bragg reflectors (DBRs). The mirrors must be made of low-loss materials with appropriate index of refraction compatible with high quantum efficiency requirements. Mirrors using GaAs and AlGaAs optical materials with refractive index of 3.5 and 33.0, respectively, offer optimum reflectivity. High-performance devices operating at communication wavelengths use advanced materials such as InP/InGaAs/InAlAs, AlAs/GaAs/Ge, and Si/SiGe with refractive indexes of 3.2, 3.45, and 3.2, respectively. These materials have properties close to those of GaAs/AlGaAs; nevertheless, they are best suited for RCE devices operating at wavelengths up to 1550 nm. Studies performed by the author indicate that even a small difference in indices requires more mirror layers and a thicker overall structure, which presents serious problems for epitaxial growth.

PIN, Schottky, and avalanche photo diodes (APD) all meet most performance requirements of RCE detectors. In case of a PIN diode made from InGaAs/InP material, heavily doped InGaAs layers must be used to reduce the absorption in the mirror structure. This particular mirror structure offers reflectivity of 97% and a peak quantum efficiency of 65%, which represents an improvement factor of four in responsivity over a conventional PIN diode with the same 200 nm thick InGaAs absorbing layer. A similar absorbing layer in a InGaAs/InGaP/InP RCE device structure with mirror reflectivities of 70% (top mirror) and 95%(bottom mirror) offers a quantum efficiency of 82% at the same wavelength of 1550 nm.

4.7.2 Performance Parameters of Velocity-Matched Distributed Photo (VMDP) Detectors

Analog fiber optic links, interconnects, and optical beam forming networks (BFNs) used by phased-array radar systems, require low-noise, high-power lasers to achieve good signal fidelity. Receivers in high-performance phased-array radar must detect high-power microwave and optical signals with minimum noise. Most currently available photo detectors are not able to handle the received power levels in the 10–100 mW range. The power handling capability of a photo diode detector is limited by the screening effects associated with photo-generated carriers. When the carrier density becomes too high in the absorption region, carriers are no longer collecting efficiently. For conventional photo detectors operating above 10 GHz, saturation occurs even at a few milliwatts level (around 10 mW).

To extend the saturation power to 100 mW or higher, the absorbing volume must be increased. A single device cannot have a large volume, because the increased transit time or delay will slow down the device response. A velocity-matched distributed photo (VMDP) detector solves the saturation problem. This means that a VMDP detector structure shown in Figure 4.10 simultaneously offers both high speed and high saturation power level because of higher coupling efficiency (K) and peak saturation photocurrent available when the number of photodiodes exceeds 10 or so. The structure shown in Figure 4.10 achieves the high power capability by combining the outputs of multiple photo detectors. This combination scheme maintains the speed of a single photo detector and makes the electrical and optical velocities equal, but adds the in-phase photo currents, thus leading to effective combination of RF signals. Peak saturation current increases as the coupling efficiency between the passive and active waveguide regions and the number of photo diodes increases. Increasing the number of photo diodes may not be most cost-effective, but does result in slight improvement in detector performance level. The VMDP detector [12] shown in Figure 4.10 uses a GaAs/AlGaAs device, an integrated ridge waveguide, and a coplanar strip transmission line.

If the MSM detectors are placed across the coplanar strips with an optical ridge waveguide running under them, peak combined photo current from five detectors is around 67 mA at 860 nm wavelength under pulsed operation. Using a reasonable detector spacing between, coupling efficiency of 88% is possible, which corresponds to a DC absorption optical power of 95 mW (67/0.707). By increasing the coupling efficiency from 88% to 98%, the same design can offer peak saturation current of more than 100 mA, as is evident from Figure 4.10; this corresponds to a DC optical power level of about 140 mW. In summary, VMDP detectors offer high-speed detection performance, high-power optical signals needed for achieving low noise performance, and wide dynamic range, which are most desirable for analog fiber optic systems.

4.8 FOCAL–PLANAR ARRAY (FPA) DETECTORS

4.8.1 Introduction

One-dimensional and two-dimensional FPAs are widely used for high-quality general imaging and spectroscopic imaging applications. Cryogenic cooling of FPAs will boost their performance level, but at the expense of higher cost and complexity. Fabrication of uncooled infrared FPAs based on microbolometer technology offers a low-cost thermal imaging sensor for possible commercial and military applications. An integrated CMOS silicon-readout integrated circuit with an uncooled 320 × 240 element bolometer FPA provides remarkable performance over 8–12 μm range. Amber Inc., Goleta, CA, designer of this uncooled FPA detector, claims that the images obtained with this imaging system are comparable to ones obtained with its cryogenically cooled, InSb-based Radiance I camera. According to the company, its microbolometer-based FPA has a noise-equivalent temperature difference (NETD) of 0.1 °C compared to its uncooled IR imaging system based on ferroelectric technology, which requires mechanical choppers to extract and stabilize an image signal from noise. An uncooled IR imaging sensor based on ferroelectric technology will have higher NETD (0.15–0.20 °C), increased system complexity, and poor reliability compared to an imaging sensor based on uncooled IR–FPA technology. GEC-Marconi company of England has recently developed a low-cost 100 × 100 element uncooled FPA for imaging and instrumentation applications. This FPA offers high electrooptic performance and environmental reliability with minimum cost and complexity. Imaging sensors can be designed using linear array configuration or two-dimensional array configuration, depending on the performance requirements, but both the cost and complexity increase with increasing number of elements and dimension.

A line-scanning camera based on InGaAs linear-array technology offers excellent performance at low production cost. Potential applications of the linear-array technology include paper processing (to check moisture content), food processing (to sort nuts from shells), and fruit processing (to separate fruit contents from pits). Two-dimensional FPAs are generally used in the military applications where high-quality images to identify two-dimensional features of a target are of paramount importance. A two-dimensional LWIR–FPA sensor with high sensitivity can play an important role in reliable detection of supersonic and hypersonic missile targets in the upper atmosphere (well above 100 km) known as the exoatmosphere, provided that effective discriminating techniques against decoys, debris, and space background noise are integrated into the sensor. IR line scanners incorporating two-dimensional FPA designs can be used in airborne or satellite-based reconnaissance and surveillance imaging sensors. Commercial applications of two-dimensional FPAs include deicing of leading-edge sections of the aircraft wings or deicing of roads and bridges. The choice between silica and InGaAs array technologies depends on the application, quality of image, and cost.

For optimum IR sensor performance, IR–FPAs must use detector materials and configurations capable of meeting the specific sensitivity, detectivity, and SNR over the IR spectral band of interest. It is important to mention that silicon detector arrays cannot detect light beyond 1 μm, whereas an InGaAs detector array provides excellent detection capability over the spectral ranges from 0.9 to 1.7 μm or from 1.2 to 2.2 μm, depending on the relative indium mole fraction content in the semiconductor compound.

Pixel size is considered as one of the most reliable performance parameters or FOM of a FPA-based sensor. Pixel size, pixel spacing, and quantum efficiency determine the quality of an image provided by an imaging sensor. Current fabrication technology can effectively separate each grid wavelength by five or ten pixels in a 256 or 512 element linear array with 50 μm square-pixel spacing. Linear array designs come in standard lengths of 128, 256, 512, or 1024 pixels with pixel heights from 50, 100, 200, or 1000 μm. A two-dimensional InGaAs array with read-out circuitry made of CMOS silicon structures is easier to operate than today's best designed silicon charge-coupled devices (CCDs). Silicon CMOS imaging sensors with CMOS read-out circuitry are best suited for applications in the visible range. However, two-dimensional InGaAs arrays using 128×128 elements with 60 μm-square pixels and 320×240 elements with 60 μm-square pixels are most suitable for applications where image resolution and contrast are of paramount importance. On the other hand, two-dimensional improved InGaAs CCD-based arrays offer simple design, compact packaging, faster response, and high-quality images. Such arrays have potential applications in spectroscopic imaging systems, wavelength-division multiplexing (WDM) systems, dense-WDM systems, and superdense-WDM systems because of extremely high resolution based on smaller pixel pitch (less than 50 μm).

4.8.2 Uncooled Focal–Planar Arrays (FPAs)

Uncooled FPAs with built-in CMOS read-out circuitry and intelligence networks provide improved performance, high reliability, and most cost-effective design. A new generation of solid-state cameras integrated with advanced FPA technology have potential applications in various commercial and military systems. Infrared uncooled cameras are in high demand because of good performance coupled with minimum cost and complexity. Uncooled IR sensors and cameras based on microbolometer technology respond to thermal energy instead of photon count as in case of Hg:Cd:Te and InS detectors, which require cryogenic cooling. Microbolometers have operating spectral ranges from 7.5 to 13 μm. By detecting the amount of thermal energy based on surface emissivity, IR sensors or cameras create images of objects in the field-of-view (FOV) in real time. Variation in emissivity indicates surface flaws or features that may be of significant importance in recognition and identification of objects or quality control and assurance evaluation of industrial and military products.

Continuous improvements in FPA technology over the last decade has yielded FPA detectors that offer significant improvement in camera resolution—up to four times over the previous generation of IR detectors .The FPA-based IR sensors or cameras can be designed to measure objects with temperatures varying from –25 °C to +200 °C with a measurement accuracy of as low as 0.1 °C. Advanced IR sensor architectures offer unique performance capabilities including intelligence features and software-based control capability. It is important to mention that uncooled solid-state imager design is strictly based on CMOS manufacturing process technology, which has been fully mature for a long time. Cryogenic cooling no doubt offers improved sensitivity, but it suffers from additional manufacturing and life-cycle costs, because of the need for frequent maintenance and replacement of cooling elements lost due to leakage or evaporation. On the other hand, a compact, low-cost, uncooled, solid-state imager system requires little or no maintenance over extended operations, even in severe environments.

Solid-state IR imaging sensors with built-in intelligent sensing and software control capabilities are most suitable for applications requiring temperature-based events for unattended security operations. On-line IR cameras can monitor several temperature conditions in user-defined processing areas. When specific conditions are reached in integrating manufacturing and process control systems, alarms can be triggered to warn the operators to take appropriate action. On-line IR sensors can be configured into factory automation systems to provide intelligent feedback to the process control network. Well-defined software interface programs and application-development tools provide cost-saving designs that can be beneficial in the automobile industry, canned goods manufacturing, and the tool industry by ensuring high yield, error-free processes, and high-integrity sealing techniques. If an IR sensor detects an error or yield performance below a minimum level, an error code can be sent to the process control system, which will either set off an alarm or automatically remove the product involved from the assembly line.

4.8.3 Dual-Wavelength Imaging Systems

Dual-wavelength (DW) imaging systems are receiving a great deal of attention, particularly, in the manufacturing industry [14]. By integrating dual-wavelength (visible and IR) imaging technology in machine-vision-based (MVB) analysis, improved process monitoring and controls are possible, particularly in applications where high-quality production items and high yields are involved. Infrared process control plays a key role in qualitative applications. Visible light systems do not effectively monitor the conditions of bonding or sealing processes. However, the heat generated by the chemical process during the formation of a proper bond is visible to a high-resolution IR sensor with user-configurable software, which will either demand operator intervention or will automatically reject the defective product. Uncooled DW imaging sensor technology has opened the doors for wide commercial, industrial, and military applications where high yield and high quality are desired with minimum cost. DW–IR sensors meet or exceed the performance, security, and manufacturing industry requirements while offering high product reliability, affordability, and fool-proof practical solutions for a variety of manufacturing inspection and security check problems.

4.8.4 Potential Detector Materials and Signal Processing Requirements

Detector materials capable of yielding high values of detectivity (D^*) are best suited for applications in FPAs. Some materials offer optimum performance only at cryogenic temperatures and some yield optimum performance over very narrow spectral ranges. Cryogenically cooled FPA-based IR sensors using either InSb or Hg:Cd:Te detector arrays offer significantly improved performance (Figure 4.11), but with added cost and complexity. The detectivity of an uncooled detector array rarely exceeds 10^{10} cm $\sqrt{\text{Hz}}$/W, even over a narrow spectral band, whereas the detectivity of a cryogenically cooled (85 K) InSb-based FPA with built-in signal processing module (SPM) located at the focal plane varies from 10^{11} to 10^{12} cm $\sqrt{\text{Hz}}$/W, even up to 5.5 μm wavelength, which indicates at least one order of improvement, as evident from Figure 4.11. An IR sensor using a Hg:Cd:Te-based FPA with built-in SPM and operating at 60 K temperature offers a detectivity better than 10^{11} cm $\sqrt{\text{Hz}}$/W, which again represents significant improvement in detectivity, image quality, and resolution. The SPM is a high-density, three-dimensional package that uses the Z-dimension of the focal plane to put the critical signal processing functions in proximity to the IR detector elements on the focal plane. The Z-technology offers complex signal processing on the focal plane, where the action is. Placing the signal processing on focal plane provides improved SNR, signal-to-clutter ratio (SCR), reduction in off-focal plane processing requirements, increased dynamic range, and high detection probability due to high SNR and SCR values.

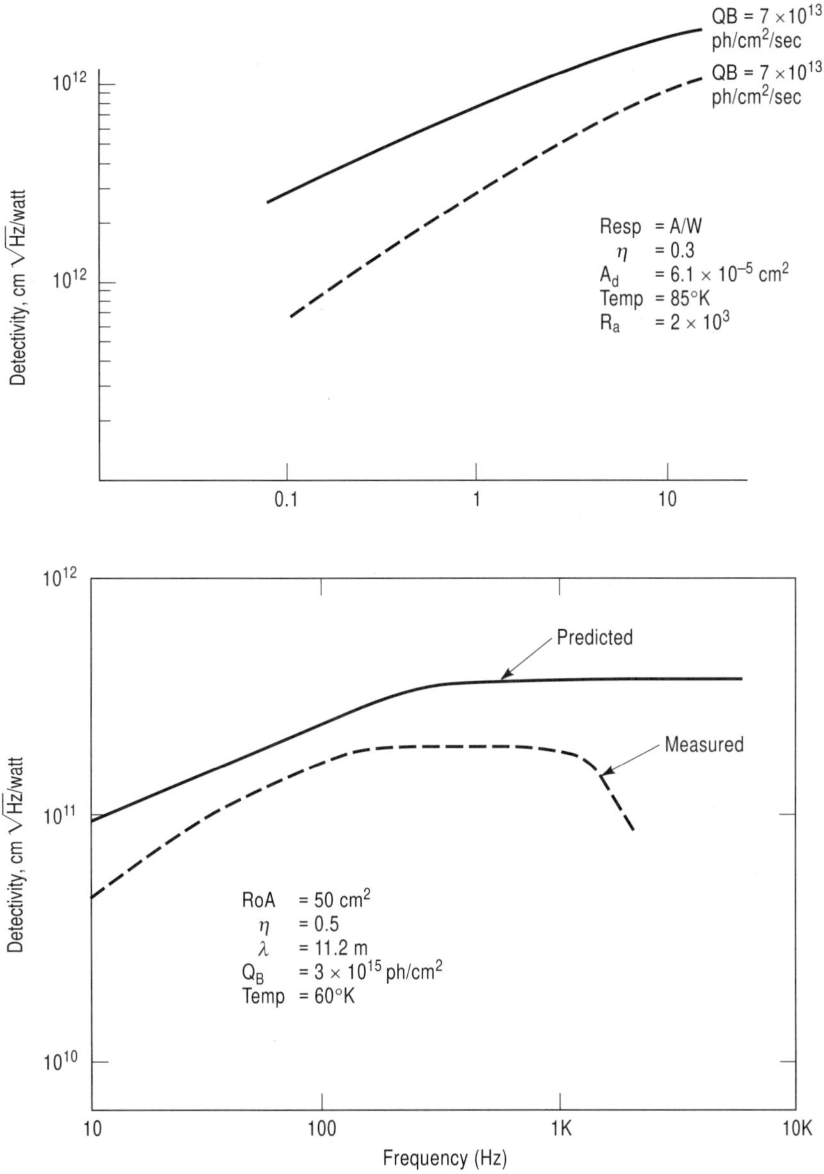

FIGURE 4.11 Detectivity of cryogenically cooled focal-planar arrays.

It is important to point out that earlier InSb-based detectors were widely used in single-detector, mechanically scanned IR sensors. This detector material offers high sensitivity because of its very high quantum efficiency (80–90%). With InSb, the wells fill in a few microseconds, but then the rest of the photons must be dumped.

The InSb detectors tend to drift over time, thereby requiring periodic two-point nonuniformity corrections in the field. As a result, FPAs using InSb detectors become more costly and complex because they require mechanical shutters, thermoelectric coolers, and additional electronics. However, an InSb-based FPA sensor can be used in applications where extreme thermal sensitivity is required, such as, long-range military imaging systems.

Quantum-well InP (QW-InP) detectors are widely used in FPA-based sensors because of their unique bandgap property, which offers optimum performance over the 9–11 μm range and reduced false alarm rates. QW-InP detectors have a quantum efficiency from 5 to 10% at 9 μm, but offer very high thermal sensitivity (better than 0.015 °C). The long-term stability and the uniformity of this material are not fully known. Under these circumstances, FPAs with QW-InP detectors have very limited applications.

It is important to mention that FPA-based IR sensors equipped with signal processing modules (SPM) will provide the capability to vary temporal discrimination filters in real time and a unique interface capability either to photoconductive or photovoltaic detectors. The SPM is the most critical component of the hybrid detector array used in a FPA. Critical elements of the SPM for a FPA comprised of 128 × 128 detector elements are shown in Figure 4.12. The SPM provides all signal processing functions, including adaptive band pass filtering, to achieve clutter rejection of 12 dB per octave, and a random access multiplication and transimpedance amplification (TIA) with dynamic range in excess of 80 dB. Note that 128 integrated circuits (ICs) are required to fabricate a focal plane read-out module. The detector array is attached to a dewar cold finger, which is connected to a cryogenic cooling source.

4.9 READ-OUT DEVICES AND THEIR REQUIREMENTS FOR FPAs

Read-out devices are the critical elements of the FPA detector arrays of uncooled or cryogenically cooled FPA-based IR sensors. These devices take the output signal from each detector element and transfer it to the signal processing unit (SPU), which is also called the signal processing module (SPM). Hybrid array design architecture using IR-sensitive detector material and a read-out multiplexer unit is shown in Figure 4.12. The read-out device can use either a charge-coupled device (CCD) detector or a complementary metal-oxide–semiconductor (CMOS) detector, depending on the processing requirements, cost, and complexity.

4.9.1 CCD Detectors as Read-out Devices

In the case of a CCD detector as a read-out device, the signal from each CCD detector is determined by transferring its electrons from one detector to the next detector down the same row until it reaches the end of the column, where it is read out. The CCD transfer process is not perfect and suffers from a phenomenon known as blooming caused by the overflow of photons into adjacent detector cells. CCD de-

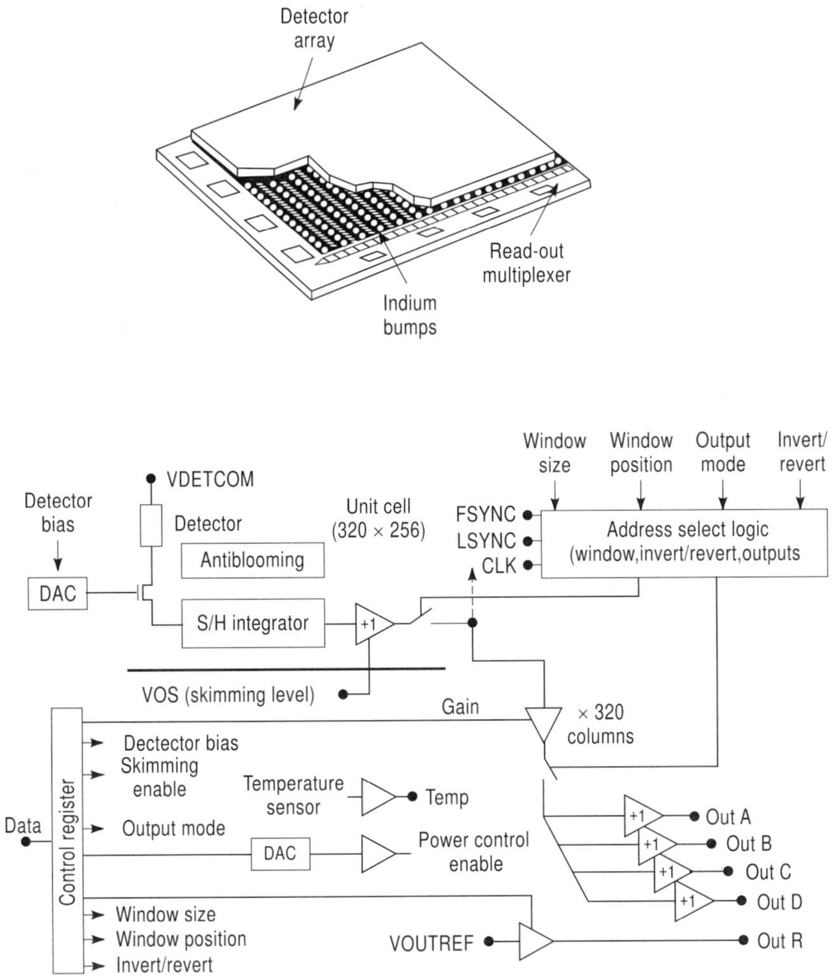

FIGURE 4.12 Design aspects of a hybrid detector array and PRIC.

vices require significantly more power than their CMOS counterparts and thus will require high-power cooling schemes. For example, a high-resolution IR sensor comprised of 512 × 512 CCD devices designed to operate over a spectral range from 400 to 1000 nm and with built-in on-chip A/D converter typically would require 50 mW of electrical input power at 5 V. When a CCD is used in a measurement IR camera, the errors must be compensated for to maintain high-quality images. CCD-based IR imaging sensors provide superior image quality because of their low dark current based on the leakage current. Low leakage current yields both low dark current as well as low overall noise. It is important to mention that the low noise and low dark current of CCD detectors are their most impressive performance parameters and are not yet matched by CMOS detectors under the same operating conditions. Typical performance specifications of CCD and CMOS devices are summarized in Table 4.5.

CCD-based imaging sensors require thermoelectric (TE) cooling to maintain low dark current. Because of this, high-performance digital cameras and scanners use only CCD devices to achieve superior image quality. However, medium-priced equipment or systems, namely, printers, scanners, and calculators, mostly use CMOS technology because of lower cost. Scientific-grade, cryogenically cooled CCD-based cameras have potential applications in biological imaging, low-noise spectroscopy, X-ray devices, and high-performance image identification. The cost of the IR sensor depends on type of detectors used, quality of image desired, size of pixel, and complexity of optics. The minimum size of a pixel is about 5 μm across, regardless of whether CMOS or CCD devices are used. Much smaller pixel size will significantly increase the cost of the pixel imaging optics. Since CCDs have a single output, they are more uniform and are easy to control.

4.9.2 CMOS Detectors as Read-out Devices

CMOS-based imaging sensors offer minimum cost, smaller pixel size, low supply voltage and wide dynamic range. However, CMOS devices suffer from very high dark current, which is at least ten times higher than that of CCD devices. CMOS devices are best suited for low-power applications such as battery-powered sensors. Power dissipation in a read-out circuit is critical and thus cooling of CMOS devices to 77 K is required to maintain acceptable performance. In a highly efficient cooler, each milliwatt of power dissipated by the read-out circuit requires about 25 mW of battery power for cooling. Optimum battery life is made possible by using CMOS multiplexed detectors for read-out circuits and high-efficiency Stirling cryogenic coolers. In the case of IR sensors, where high dark current is not a critical issue, CMOS is the best choice for the read-out device. Research and development efforts are currently being directed to reduce the dark current density to less than 100 pA/cm^2. As far as image quality is concerned, on-chip analog-to-digital converter (ADC) technology provides the clearest signal. It is possible to mix analog and digital signals on the same chip to obtain a CMOS-based sensor without signal distortion. One of the major advantages of CMOS devices is their ability to integrate the functions of several chips into one chip, resulting in a "camera on a chip." A highly integrated approach such as this will lead to application-specific sensors that could

TABLE 4.5 Performance specifications of CCD and CMOS devices

Performance specification	CMOS devices	CCD devices
Pixel size (μm)	6–9	9–20
Noise equivalent electrons	35	2–34
Dark current density (pA/cm^2)	110–1000	10–30
Operating voltage (V)	3.3 or 5.0 (single)	8 to 15
Dynamic range (dB)	120 (maximum)	60 dB (minimum)

be used in high-volume commercial applications involving digital still cameras or video conferencing sensors.

4.10 READ-OUT INTEGRATED CIRCUIT (ROIC) TECHNOLOGY

ROIC technology offers the most efficient electron interface for an IR image device, resulting in significantly lower cost and improved device performance. However, implementation of ROIC technology must be seriously considered early in the detector design phase. An off-the-shelf advanced ROIC can offer significant reduction in sensor cost and fabrication time. A ROIC with a 320 × 250 pixel design configuration using sub-micron CMOS technology and incorporating on-chip functions offers improved performance with minimum cost. Furthermore, this design configuration is fully compatible with detectors made from indium antimonide, mercury–cadmium–telluride, and indium–gallium–arsenide materials. This particular ROIC design, using state-of-the art 0.6 μm CMOS transistor technology, offers 30 × 30 μm pixels, an electron device with storage capacity of 20 million bits, dynamic range exceeding 70 dB, and read-out noise less than 500 electrons. The ROIC operates in two modes: a simplified hand-off default mode supporting a single output or a programmable array logic (PAL) mode defaulting to a single-output, full-window, normal scan order with no reference output. A schematic diagram of a control or programmable mode is depicted in Figure 4.12; it offers advanced features such as dynamic image transposition, dynamic windowing, multiple output configurations, and signal skimming. Both modes support integrate-while-read (IWR) and integrate-then-read (ITR) operations, various gains, and high-voltage QWIP bias provisions. Other ROIC design features include snapshot integration of simultaneous image acquisition capability from all pixels in the array, elimination of the need for expensive and power-hungry external frame buffers, and compatibility with various optics requirements necessary for change of image orientation. ROIC design with multiple output configurations is best suited for hand-held imagers and high-speed data acquisition systems, with each mode capable of providing an output bandwidth of 10 million pixels. Off-the-shelf ROICs, including small-format 128 × 128 pixel and large-format 640 × 480 pixel devices for cryogenically cooled IR detectors, and an optimized 640 × 480 pixel version specially for QWIRs will be available in the near future [15].

4.11 FPA DESIGN REQUIREMENTS FOR SPACE APPLICATIONS

FPA-based IR sensors for space and satellite applications are required to meet stringent radiation hardness performance specifications, depending on whether the satellite is operating in low earth orbit (LEO), medium earth orbit (MEO), or geostationary orbit (GSO). The performance of both the IR detectors and silicon CCDs are affected by radiation levels in space when operating at a specified altitude. In the case of IR detectors, the performance parameters affected are SNR, dark current, spectral response, quantum efficiency, and cross-talk. However, in the case of silicon CCD-based detectors, only the charge transfer efficiency, dark current, and flat-band voltage shift are affected by space radiation environments. Radiation-resistance capabilities of various space-based detectors are shown in Table 4.6.

The estimated radiation resistance values shown above are for Hg:Cd:Te and InSb materials generally used in the fabrication of photovoltaic cells (PV). Under space environment operations, radiation hardness is required against trapped electrons, trapped photons, cosmic rays, solar flares, and charged particles. The radiation-hardening level requirement for FPAs is dependent on severity of the radiation level from a specific source. Under nuclear weapon environments, FPAs will face extremely high levels of radiation from X-rays and gamma rays associated with prompt sharp pulses (10 ns or so), neutrons, and debris associated with delay pulses ranging from 1 μsec to 10 msec in duration, and from thermal sources over time periods exceeding 10 sec or more. The performance levels of the FPAs will be affected adversely by the environments, levels of radiation, duration of exposure, and types of radiation the detectors are exposed to.

Space radiation does affect the performance levels of other semiconductor devices such as GaAs MESFET-RF transistors, PIN switching diodes, and InGaAsP laser diodes. Published data [16] on performance levels of GaAs MESFET devices indicate that an X-band MESFET device can lose a minimum gain of 2 dB over the 8–12 GHz frequency range when operating under a gamma radiation level exceeding 10^8 rads. However, the transconductance of the same device remains unaffected when exposed to gamma radiation doses from 10^5 to 10^8 rads . Since most of the FPA-based detectors use materials such as InGaAs, InAs, InSb, or Si:As, involving either Ga or As or both elements, it is recommended that gamma radiation exposure levels exceeding 10^8 rads be avoided to ensure both the performance and the reliability of the devices involved. In brief, IR detectors and semiconductor devices using the above materials must be designed to meet the gamma radiation levels anticipated under space and nuclear weapon environments.

TABLE 4.6 Radiation resistance capabilities of various detectors, including CCD devices

Detector material	Radiation resistance with specified gamma dose
Hg:Cd:Te (PV)	10 Krad with oxides passivation
	MWIR (2–10) μm > 10 Mrad (with oxides)
	LWIR (10–12) μm > 10 Mrad
InSb (PV)	More sensitive to total dose
	More soft than Hg:Cd:Te
	Oxide passivation limits radiation hardness
Si-CCD	Much softer, < 400 rad
	With proton levels ranging from 9 to 63 MeV

4.12 SUMMARY

Performance capabilities of potential IR detectors and FLAs operating in near-IR, mid-IR, and LWIR regions are summarized with emphasis on cost, complexity, and reliability. Performance levels and limitations of uncooled and cryogenically cooled IR detectors and FPAs are revealed, pointing out the added cost, complexity, high power consumption, and frequent maintenance service required due to cryogenic cooling. Low-power, high-power, and high-energy detectors capable of operating over extended IR spectral bands are identified, with emphasis on critical performance parameters. Photovoltaic (PV) detector designs involving multiguide $GaInP_2/GaAs/Ge$ cells have demonstrated conversion efficiencies exceeding 26% under laboratory environments. GaAs metal–semiconductor–metal (MSM) photo detectors offer higher optical coupling efficiencies and are best suited for applications such as optoelectronic integrated circuits (ICs) and optical communications and telecommunications systems operating in the 900–1700 nm and 1200–2600 nm spectral regions. Performance capabilities and critical parameters of ADP detectors and photomultiplier tubes (PMTs) are summarized. Performances capabilities of potential photon count detectors to detect bacterial contamination in food storage are identified. Resonant-cavity-enhanced (RCE) detectors are described in detail because they offer the highest bandwidth–efficiency products, making them best suited for optical communications systems. Performance requirements for IR detectors and FPAs used in the sensors for commercial and industrial applications are summarized. Radiation hardness levels are specified for IR detectors for space and satellite applications. Performance benefits of one-dimensional and two-dimensional FPAs are discussed, with emphasis on cost and reliability. Critical requirements for read-out devices such as CCD detectors and CMOS detectors are identified. Design requirements for spaced-based and satellite-based FPAs are summarized, with particular emphasis on ROIC, superconductor, high-performance material, and integrated processing technologies. Potential IR detectors capable of yielding detectivities greater than 10^{10} cm $\sqrt{\text{Hz}}$/W with performance curves shown in Figure 4.13 have been identified for various applications.

NUMERICAL EXAMPLE

Compute the maximum quantum efficiency of the high-speed RCE detector made from GaAs photo diode material using equation (4.4) and the following assumed parameters:

R_1 = top mirror reflectivity of 70% for an absorption region thickness (d) of 200 μm and of 50% for a absorption region thickness (d) of 400 μm.

R_2 = bottom mirror reflectivity of 75%, 80%, 85%, 90%, and 95%, regardless of the values for top mirror reflectivity (R_1) and absorption region thickness (d).

α = 10,000/cm (absorption coefficient for GaAs material).

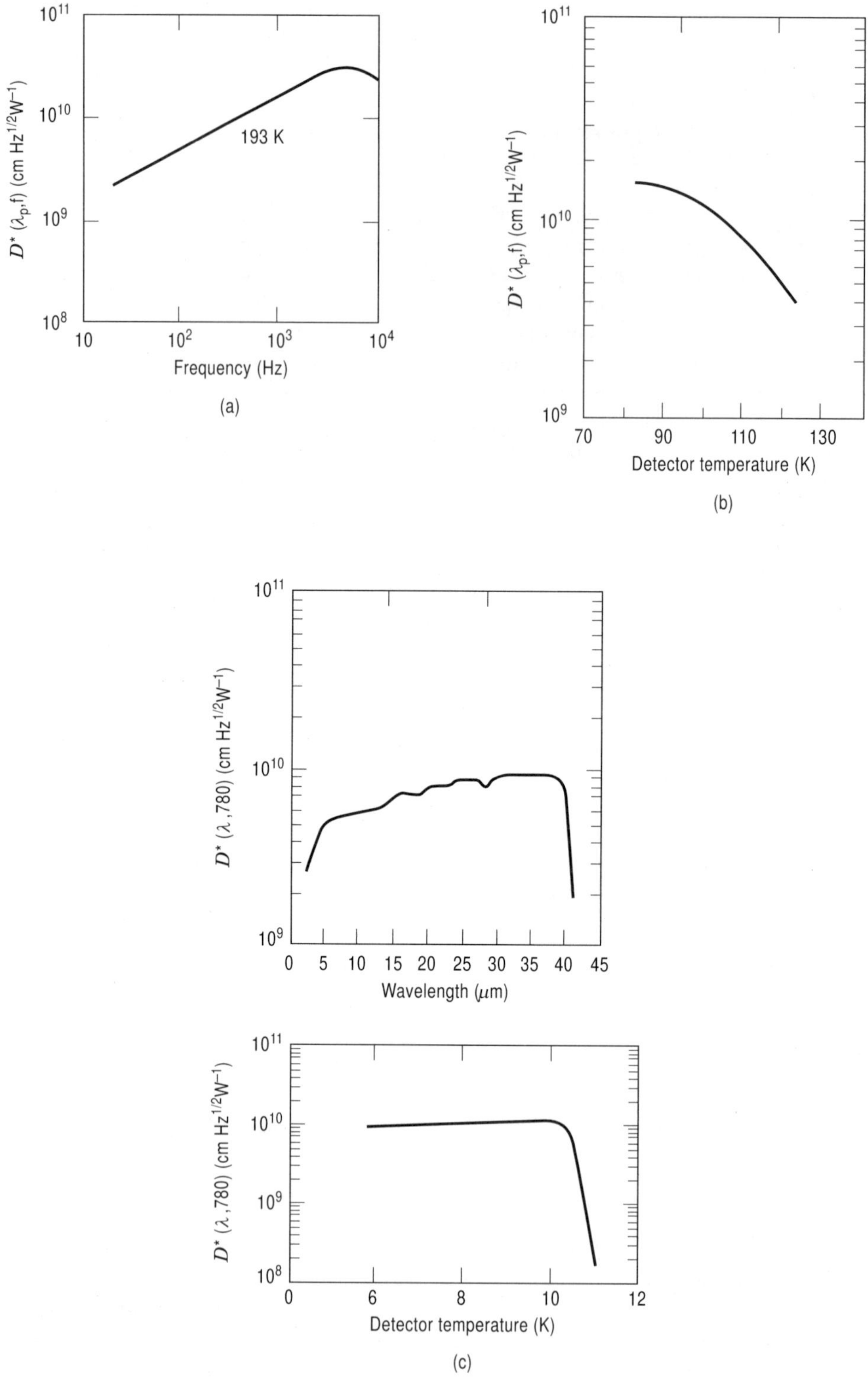

FIGURE 4.13 Detectivity of various detectors as a function of wavelength and operating temperature.

Calculated values of maximum quantum efficiency are shown in the table below.

Calculated values of quantum efficiency as a function of various parameters (%)

R_2 (%)	$d = 200$ μm, $R_1 = 70\%$	$d = 400$ μm, $R_1 = 50\%$
75	68.5	71.3
80	73.4	76.3
85	78.5	81.6
90	84.2	87.3
95	89.9	93.2

Note: The calculated data indicate that lower top mirror reflectivity around 50% and large thickness of absorption region around 400 μm offers maximum quantum efficiency for the GaAs-based RCE detector. Furthermore, higher values of bottom mirror reflectivity (R_2) yield higher quantum efficiency, irrespective of the values for top mirror reflectivity (R_1) and absorption region thickness (d).

REFERENCES

1. Newport Corp, Product Sheet, 1971 Deere Ave., Irvine, CA 92608.
2. Ibid.
3. Ibid.
4. Ibid.
5. Ibid.
6. Editorial, *Photonic Spectra,* p. 45, October 1998.
7. A, R. Jha, *Superconductor Technology: Applications to Microwave, Electrooptics, Electrical Machines and Propulsion Systems*, pp. 189–198, Wiley, New York, 1977.
8. K. Kaufman, "Photon counting extracts low-level signals," *Lasers and Optronics*, pp. 13–15, June 1996.
9. Contributor Editor, *Photonic Spectra*, pp. 56–59, September 1998.
10. A, R. Jha, *Superconductor Technology: Applications to Microwave, Electrooptics, Electrical Machines and Propulsion Systems*, pp. 196–198, Wiley, New York, 1977.
11. M. Selim, "Resonant-cavity enhanced devices improving efficiency," *Optoelectronics World*, p. 15, March 1998.
12. A. Davidson et al., "Demand for high-speed detectors drive research," *Laser Focus World*, pp. 101–108, April 1998.
13. Ibid.
14. G. McIntosh, "Imaging meets demands of on-line manufactures," *Laser Focus World*, pp. 155–158, April 1998.
15. J. D. Frank, "Off-the-shelf readout ICs standardize detector interface," *Optoelectronics World*, pp. 23–24, March 1998.
16. *Detectors for Space Applications*, viewgraph collection from JPL, Pasadena, CA, October 1989.

CHAPTER FIVE

Passive Infrared Devices and Electrooptic Components

5.0 INTRODUCTION

This chapter describes the performance capabilities and operational limitations of passive infrared (IR) devices and electrooptical (EO) components for commercial, industrial, medical, and scientific applications. Critical performance parameters and state-of-the art design concepts for passive IR and EO devices operating in near-IR, mid-IR, and long-wavelength IR (LWIR) regions will be discussed with emphasis on cost and reliability. Passive IR and EO devices and components, such as optical fibers, microlenses, optical crystals, isolators and circulators, optical bandpass filters, switches, limiters, delay lines, attenuators, optical cavities, laser pointers, IR thermometers, IR cameras, and other electrooptic and photonic devices will be described, with emphasis on their potential applications in industry, manufacturing, and scientific research. Since optical fibers and their materials are the most critical elements of electrooptical, optoelectronic, photonic devices, and infrared sensors, they will be treated first before describing any of the other devices or sensors mentioned above.

5.1 OPTICAL FIBERS, OPTICAL MATERIALS, AND THEIR PROPERTIES

Optical coplanar waveguides (CPW) or fiber optics (FO) can be used as transmission line for transmitting and distributing of optical signals with minimum loss and dispersion. Optical fiber waveguides are made from optical materials such as AgBr, AgCl, TlBr, KCl, silicate glass, fluoride glass, and chalcogenide. An optical fiber is comprised of three distinct elements, namely, core, cladding, and coating material. The core, a high refractive index material, is surrounded by a concentric cladding material with slightly higher refractive index. The coating material is intended to protect the optical fiber under severe operating environments. Fibers are generally made from silica with index modifying dopants such as GeO_2. A protective coating of cushioning material such as acrylate is used to reduce the cross-talk between the adjacent fibers and the microbending that occurs under severe mechanical environments. For greater environmental protection, fibers are incorporated into cables surrounded by an outer jacket made from steel or kevlar strands, as illustrated in Figure 5.1.

The fiber geometry and composition determine the discrete sets of electromagnetic (EM) fields that can propagate in the fiber with minimum loss. Radiation modes and guided modes are found in optical fibers. Radiation modes carry EM energy out of the core so that the energy is dissipated. Guided modes are confined to the core and propagate energy along the fiber axis, transporting information and power through the fiber. If the fiber core is large enough, it can simultaneously support many guided modes. Each guided mode has its own velocity and can be further decomposed into orthogonal, linearly polarized components. The two lowest-order guided modes are present in a circularly symmetric fiber. When light is launched into a fiber, various modes are excited. Some light is absorbed in the jacket and the rest is propagated in the fiber by the internal reflections shown in Figure 5.1. Insertion loss is one of the most critical performance parameters and is dependent on

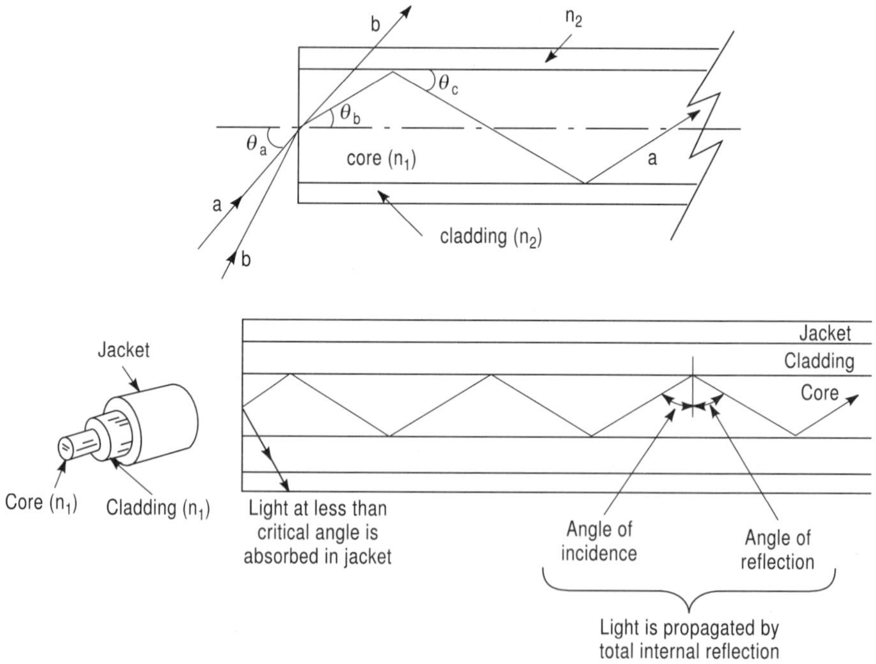

FIGURE 5.1 Critical parameters of fiber optics and their values for various types of optical fibers. *(Continued on next page)*

Ranges of Values for Common Optical Fibers

Type of fibers	Diagram of fiber	Core diameter (μm)	Cladding diameter (μm)	Numerical aperture	Nominal attenuation (dB/km)	Bandwidth (MHz-km)	Frequently encountered sizes (core/cladding, in μm)
Multimode, step index		50 to 400	125 to 500	0.15 to 0.4	<50	<25	100/140, 200/250, 400/450
Multimode, graded index		30 to 75	100 to 250	0.2 to 0.3	<10	<200	50/125, 62.5/125, 85/125
Singlemode, step index		3 to 10	50 to 125	0.10	<3 at 850 nm, <1 at 1500 nm,	<2000	9/125
Plastic-clad silica		50 to 500	125 to 800	0.2 to 0.4	<50	<25	–
Plastic		200 to 600	400 to 1000	~0.5	<1000 at 650 nm	–	–

FIGURE 5.1 *(continued)*

fiber material, operating wavelength refractive-index profile, and dimensional parameters, as shown in Figure 5.2.

5.1.1 Insertion Loss in the Optical Fiber

Typical optical fiber cable loss varies from 0.35 db/ km in a single-mode operation to 0.52 dB per km in a multimode operation at a wavelength of 1.3 μm. Theoretical insertion loss and bulk absorption for various optical fiber materials as a function of wavelength are shown in Figure 5.3. The overall insertion loss in a fiber results from

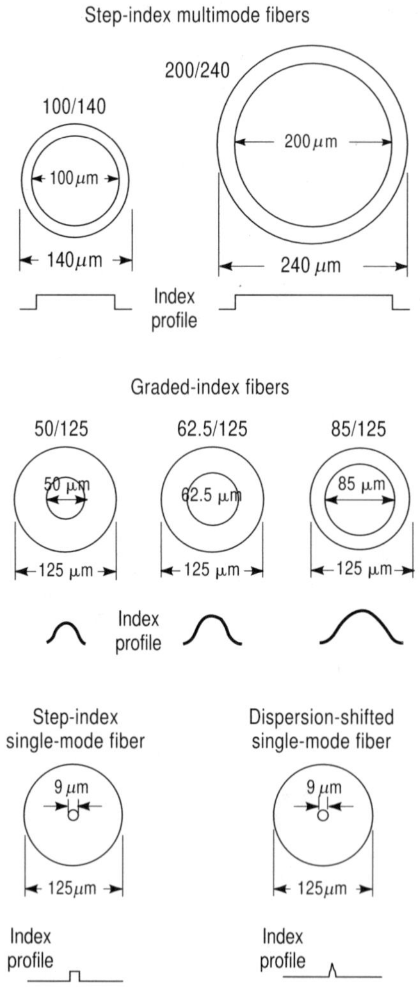

FIGURE 5.2 Typical core and cladding diameters and attenuation as a function of operating wavelength.

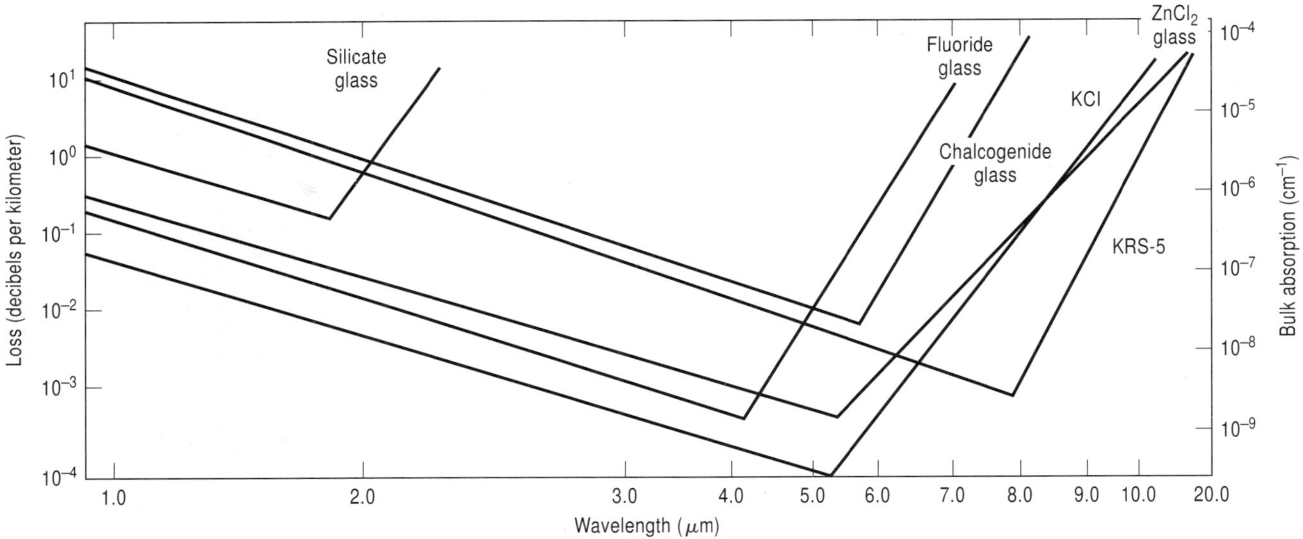

FIGURE 5.3 Minimum insertion loss and bulk absorption coefficient for infrared optical fibers identified with fiber materials.

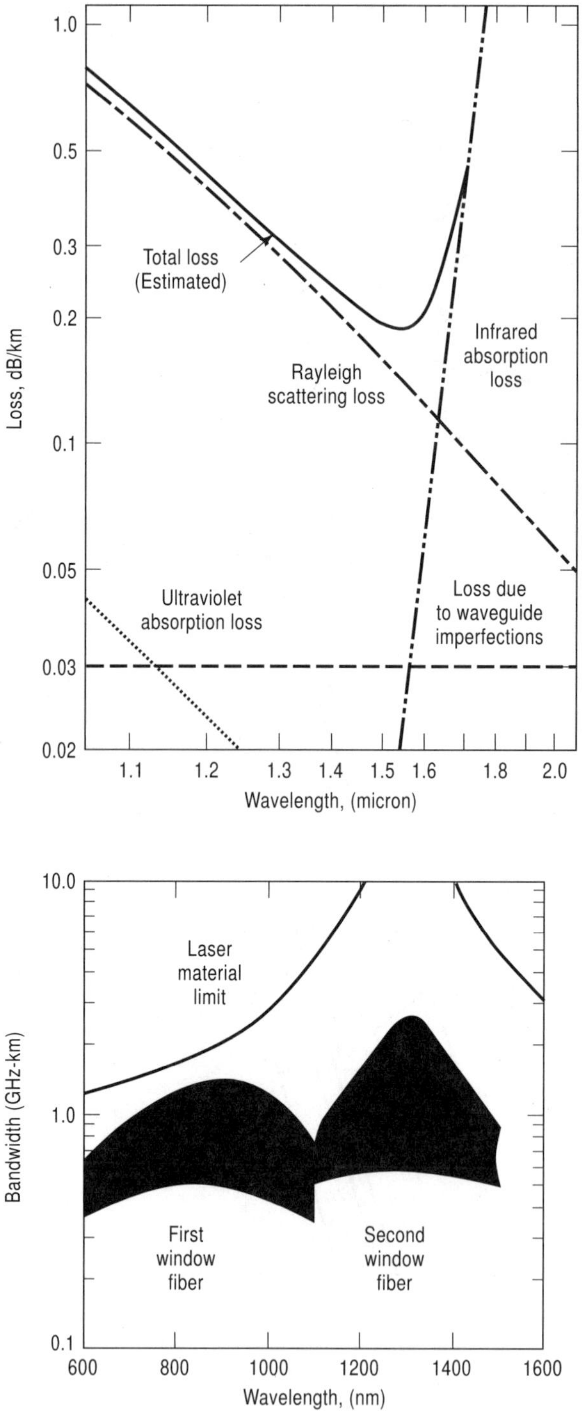

FIGURE 5.4 Various losses and bandwidth at various wavelengths.

Rayleigh scattering, infrared absorption, optical material dispersion, ultraviolet (UV) absorption, and waveguide imperfections, if any. Various factors contributing to total insertion loss and the bandwidth range for the first and second windows in the fiber as a function of wavelength are shown in Figure 5.4. Note that attenuation due to absorption and scattering losses in the optical fiber generally determine the cost of a FO delay line, FO data link, or FO telecommunications system. Scattering phenomena can couple optical energy from guided to radiation modes causing significant losses, particularly, in a multimode fiber. The latest state-of-the art optical fibers demonstrate a total insertion loss on the order of 0.1 dB/km over the 1.1 to 1.3 μm spectral range.

5.1.2 Modal and Material Dispersion in Optical Fibers

Modal dispersion is the most significant optical signal degradation mechanism. It occurs when more than one mode is excited in the fiber and causes a pulse-broadening effect proportional to the square root of the dielectric constant of the core material. Core and cladding dimensions and refractive-index profile of the optical fiber determine the modal dispersion impact on the output pulse. Core and cladding dimensions along with refractive-index profiles of various optical fibers such as step-index multimode fibers, graded-index fibers, step-index single-mode fibers, and dispersion-shifted single-mode fibers are shown in Figure 5.2. Examination of the various fiber configurations shown in Figure 5.2 reveals that the graded-index fiber exhibits minimum modal dispersion and has a wider bandwidth than that of step-index multimode fibers. Dispersion in a fiber material is caused by the refractive index of the core, which varies with the optical frequency components. This means that different frequency components of a time-varying signal propagate at different velocities, thereby causing a bandwidth limitation in the fiber length. Material dispersion is proportional to source bandwidth and is present in both single- and multimode fibers, but it is much smaller than modal dispersion in multimode fibers. In optical fibers made from silica glass, material dispersion is extremely low around an operating wavelength of 1.3 μm. The spectral bandwidth for minimum material dispersion is generally very narrow.

5.2 REFRACTIVE-INDEX-DEPENDENT PARAMETERS OF OPTICAL FIBERS

5.2.1 Critical Angle

The critical angle determines the maximum acceptance angle, θ_a, which is written as

$$\sin\theta_a = \frac{n_2}{n_1} \tag{5.1}$$

where n_2 is the refractive index of the cladding, n_1 is the refractive index of the core, and θ_a is the maximum acceptable angle for total internal reflections between core and cladding regions.

5.2.2 Numerical Aperture

Numerical aperture (NA) defines the refractive-index profiles of optical fibers and can be expressed as

$$NA = \sqrt{n_1^2 - n_2^2} \tag{5.2}$$

Higher values of numerical aperture (typical values 0.25 to 0.37), offer greater light coupling efficiencies. High light coupling efficiency is highly desirable in applications where the output of an optical source is connected to a detector.

5.2.3 Total Number of Modes (*M*)

The parameter *M* defines the total number of modes that can be sustained in an optical fiber and can be expressed as

$$M = (k)(r)(NA) \tag{5.3}$$

where, k is a propagation constant, r is the core radius and NA is the numerical aperture. Studies performed by the author on various optical fibers indicate that a step-index fiber will propagate only a single dominant mode if M is less than 2.405. The studies further indicate that for graded-index fibers, the value of parameter M denoting single-mode cutoff is increased by a factor on the order of 1.5, depending on the refractive-index profiles of the optical fibers involved.

5.2.4 Material Dispersion (*D*)

Material dispersion is also known as chromatic dispersion and is defined as

$$D = \left(\frac{\lambda}{c}\right)\left(\frac{d^2 n}{d\lambda^2}\right) \tag{5.4}$$

where λ is the wavelength, c is the velocity of light, and n is the refractive index. Material dispersion is expressed in picosecond/nanometer-kilometer (ps/nm-km) units. As stated earlier, material dispersion can reduce the optical coupling efficiency of the light entering into the optical fiber.

5.3 STRUCTURAL ASPECTS AND CONSTRUCTION CONCEPTS FOR OPTICAL FIBERS

Potential construction concepts for optical fibers, including standard communication fibers , high-temperature communication fibers, hermetically sealed communication fibers, high-power density fibers, and low-loss fibers are illustrated in Figure 5.5. Appropriate optical fibers must be selected to meet optical performance requirements under specified operating environments.

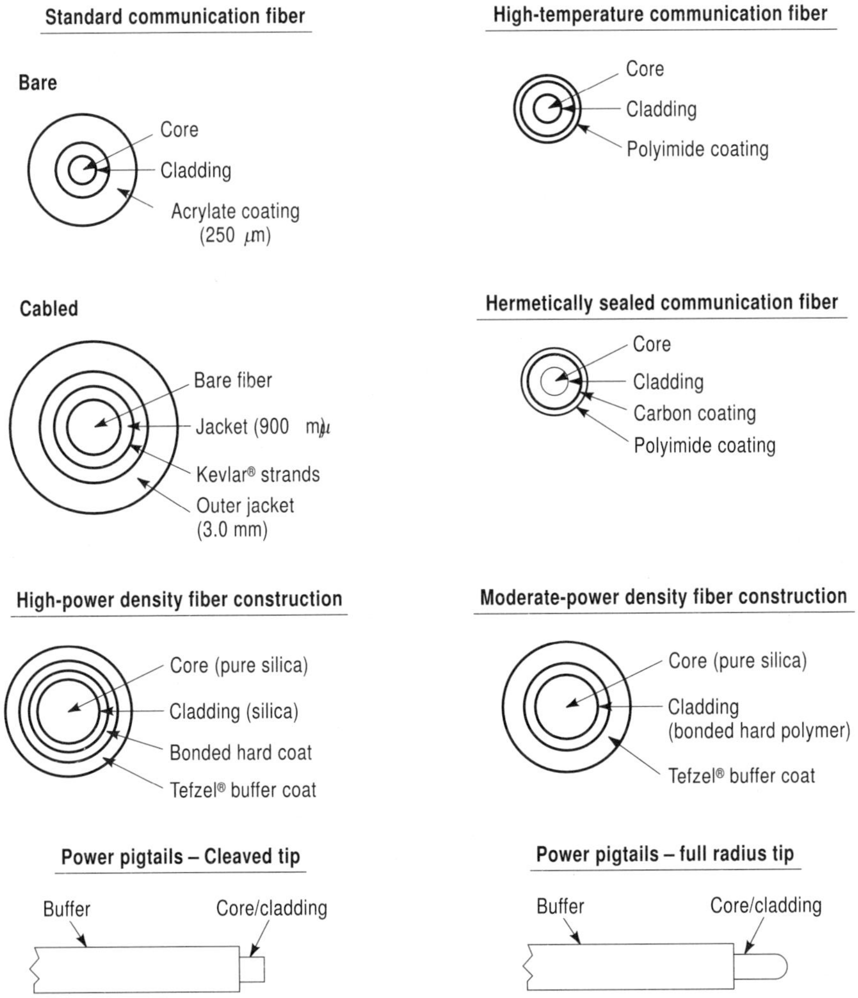

FIGURE 5.5 Construction techniques for optical fiber configurations for various applications.

Standard communication-grade fibers generally have a 250 µm acrylate buffer coating to allow continuous operation over the –65 °C to +135 °C temperature range without performance degradation. High-temperature silicon single-mode fibers optimized for operation over the 1310 to 1550 nm spectral range require a core and cladding diameter of 93 µm and 125 µm, respectively. Multimode graded-index fibers available with core-to-cladding diameter ratios of 50/125, 62.5/125, and 100/140 (all dimensions expressed in µm units) offer large bandwidth and low attenuation the over 850 to 1300 nm spectral range. High-temperature multimode communication-grade fibers are available with polyamide buffer coatings to extend the operating temperature limit to +375 °C and to increase the chemical and abrasive resistance needed under harsh industrial environments. High-temperature, all-silicon semi-rigid optical fiber are best suited for medical and industrial applications where optimum performance over wide temperature range is of paramount importance. Hermetically sealed single-mode and multimode communication grade fibers are widely available with a carbon coating directly deposited on the cladding followed by a polyamide coating to provide safe operation in the harshest humidity and temperature environments. Single-mode sensor-grade and bend-insensitive multimode step-index sensor-grade optical fibers have been designed with large pure silica core diameters ranging from 110 to 1000 μm. These optical fibers are most suitable for medical, industrial, aerospace, and precision scientific measurement applications. Optical fibers with small diameters are bend-insensitive and are most ideal for fiber optic gyroscope and space-restricted sensor applications.

Polarization-preserving single-mode (PPSM) fibers are available with optimum performance over 820 to 870 nm and 1300 to 1550 nm spectral ranges. PPSM fibers [1] with beat-length as small as 2 mm are readily available at moderate cost. The beat-length is defined as the length of the fiber over which the fiber polarization rotates through 360 degrees. Studies performed by the author indicate that the shorter the beat-length, the better the polarization-preservation properties of the fiber. The studies further indicate that a "bow-tie" cladding architecture offers high polarization-preservation capability with stress-induced, intrinsic birefringence. Typical applications of PPSM fibers include polarization state sensors, coherent detection systems, and transmission of optical power with minimum polarization loss.

5.3.1 Optimum Fiber Coupling Configurations

Various fiber coupling configurations with associated insertion losses are illustrated in Figure 5.6. The higher the coupling efficiency, the lower the fiber-end loss. Coupling efficiency is an important parameter for all electrooptical systems, where optical power has to be transmitted to other components or the output needs to be measured with high accuracy. Higher coupling efficiency is essential to minimize power loss during the transmission. Coupling is dependent on acceptance angle (the larger the acceptance angle, the higher the coupling efficiency), light loss from the core-to-cladding interface, and the number of fibers involved in the optical cable. Coupling configurations in Figure 5.6 indicate that a coupling efficiency of 98.5% causes a fiber-end loss of 0.066 dB (configuration A), which is the lowest loss among the various configurations considered. Coupling efficiencies of 97.5% , 64.6%, and 50% cause fiber-end losses of 0.110 dB, 1.90 dB, and 3.01 dB, respectively. Minimum fiber-end loss requires high coupling efficiency, which is directly proportional to the numerical aperture of the fiber.

5.3.2 Side-Polished Optical Fibers (SPOFs)

Removing enough cladding to create an optical window into the fiber core provides functionality and transparency, which are essential for precision variable optical attenuators, modulators, optical amplifiers, tunable optical filters, and optical switch-

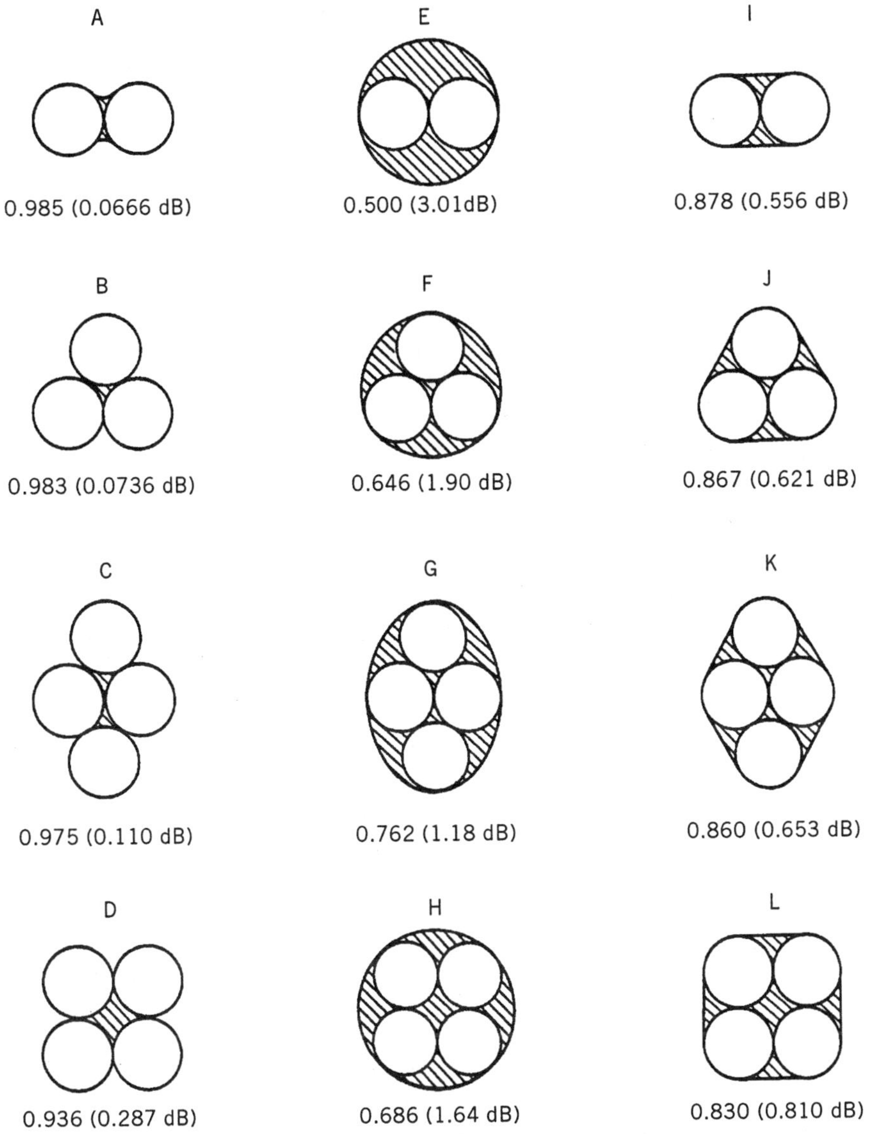

FIGURE 5.6 Estimated end insertion losses for various coupled optical fiber configurations.

es. SPOFs offer full compatibility with practically any passive or active optical material, thereby providing a versatile tool to design high-performance fiber optic devices with minimum cost and complexity. Use of different optical materials gives rise to different optical functions in devices that feature near-zero insertion loss and back reflection. Integrated optical components such as phase modulators and thermooptic switches generally use single-mode optical fibers. A fiber optic variable attenuator using the SPFO technique can be electronically controlled with no moving parts and with zero loss from attenuation and back reflection [2].

Most of the free-space microoptic devices including wavelength-division multiplexers (WDMs) using interference filters, and polarization-insensitive isolators, circulators, and switches, are made and assembled on an optical table. However, components and devices made in this manner suffer from excessive losses due to optical misalignment and multiple surface reflections. Integrated optical devices using SPFO technology can be monolithically cascaded with minimum cost and lower device-to-device losses. This technology offers a cost-effective approach for the design and development of critical optical components.

5.3.3 High-Power Fiber Optic (HPFO) Cables

Fiber optic (FO) cables, which demand high-power handling capability, diffraction-limited beam quality, high efficiency, improved reliability, and safe operation, are receiving much attention. HPFO cables have potential applications in high-power industrial laser fiber delivery systems, medical diagnostic equipment, and high-power IR tracking and detection systems. An optical fiber with AsS cladding of 250 μm and AsSe core of 150 μm demonstared a peak power density of 50 MW/cm^2 at 2 μm wavelength output from the Ho:YLF laser [3]. One meter of the same cable with core diameter of 200 μm and cladding diameter of 330 μm demonstrated a peak power density of 126 kW/cm^2 and a transmission efficiency exceeding 60% at a wavelength of 10.6 μm.

Graphic arts industry, material processing, LIDAR systems, and specialized alignment applications demand high-power, diffraction-limited beams for optimum performance with minimum cost. Research studies performed on high-power optical cables reveal that double-clad (DC) fibers are inherently capable of handling high power [4]. Optical fibers using rare-earth dopants such as erbium (Eb) and ytterbium (Yb) are widely used for high-power optical systems operating at 10.6 μm. In the fiber using Yb as dopant, Yb absorbs the pump energy and the absorbed energy transfers the energy to neighboring Eb atoms, leading to realization of a gain at 1550 nm. Optical fibers doped with Yb offer broad absorption bands around 920 nm and 980 nm, improved reliability, and high conversion efficiency. Integration of DC fibers in optical amplifiers has created a high-power, high-speed optical source with high efficiency and improved beam performance. A 2.5 Gbit/sec, Yb doped-fiber amplifier based transmitter delivered more than 5 W (CW) of power at 1070 nm. Optical fibers made from pure fused quartz demonstrated a damage threshold of 1 to 10 GW/cm^2, which would correspond to a fiber laser with output levels between 10 and 100 kW.

5.4 IR WINDOW AND DOME TECHNOLOGY AND MATERIALS

Optical windows and domes are critical components of high power optical systems. Recent research activities on IR window and dome materials have identified several materials for operations over the 1 to 14 μm range. Advanced window material technology will have an impact electrooptic sensors, optoelectronic devices, and airborne, spaced-based, and sea-based IR systems. Computer modeling can provide reliable information on the critical performance parameters for high power IR windows, including impact resistance, rain and sand-based erosion resistance, thermal shock resistance, mechanical integrity under high power and high temperature environments, and thermal efficiency at elevated temperatures. Potential optical materials for IR window and dome applications are sapphire, diamond, optical ceramics, nitrides, sulfides (ZnS), phosphides, and germanium (Ge). Critical materials for IR window and dome applications are shown in Table 5.1.

Examination of these window materials indicates that diamond is the most appropriate window material for high-power IR lasers and for airborne multispectral sensors. The optical material ZnSe is widely used for low-power CO_2 lasers. Because of its poor thermal conductivity, ZnSe is not recommended for high-power laser applications. Despite its high absorption coefficient, diamond is best suited for multikilowatt CO_2 laser window applications.

Laser scientists have concluded that a 1 mm thick, 2 inch diameter CVD diamond window will be capable of handling CO_2 laser power exceeding 100 kW (CW) with high reliability. Diamond windows offer low insertion loss, highest thermal conductivity, high thermal dissipation capability, and high resistance against thermal shock.

Studies performed by the author on high-power IR window materials indicate that the fundamental lattice absorbing frequency is directly proportional to force constant and inversely proportional to the atomic mass of the material. Dome or window materials that are made up of strongly bonded atoms generally have fundamental lattice absorption frequencies that are very critical in the 8–12 μm region. Stress concentration at preexisting flaws determines the strength of dome materials like ZnS and ZnSe. In some materials, application of surface compressive layers

TABLE 5.1 Properties of high-power window and dome materials

	Window material				
Properties	Diamond	ZnSe	ZnS	Ge	Al_2O_3
Band gap (eV)	5.48	2.7	3.9	0.66	9.9
Cutoff wavelength (μm)	—	20	14	23	5.5
Absorption coefficient					
at 1060 nm	0.1–0.3	0.0005	0.2	—	—
at 535 nm	—	—	—	0.02	1.9
Refractive index					
at 1060nm	2.38	2.40	2.19	—	—
at 500 nm	—	—	—	4.00	1.63
dn/dT (1/K)	0.001	0.0064	0.0041	0.040	0.0013
Thermal conductivity (W/cm·K)	20–23	0.19	0.27	0.59	0.35
Thermal coefficient (ppm/K)	1.3	7.6	7.9	5.9	5.8
Hardness factor (kg/mm²)	8300	137	230	780	1800

will lead to significant increases in strength and ductibility, thus avoiding catastrophic failure. The threshold velocity (V_{th}) for observable impact damage in a window material is directly proportional to the damage parameter (D), which is the product of fracture toughness (K_f) and elastic wave velocity (V_{elas}). Now the threshold velocity can be written as

$$V_{th} = [(A)\,(K_f)^{2/3}][V_{elas}]^{1/3} \tag{5.5}$$

where A is a constant dependent on flow size distribution.

Studies performed on window materials indicate that all LWIR materials have relatively low damage threshold velocity. The studies further indicate that threshold velocity decreases with increase in exposure time. The threshold velocity is typically less than 1 Mach (1117 ft/sec) for most dome materials. The lack of a durable , rain erosion resistant LWIR optical materials presents a serious problem. LWIR window materials with high values of fracture toughness are necessary to obtain high modulis of elasticity of the surface coating. It is important to mention that the protective coating must provide improved optical efficiency.

5.5 OPTICAL CRYSTALS

Lamp-pumping of a laser rod in a pump cavity generates tremendous amounts of heat outside the crystal's narrow absorption band. Efficient thermal management is very important in the design of a solid-state laser source. With the emission wavelength of the diode pump source matched to the absorbing band of the laser crystal, the thermal problem is significantly reduced. Because of their excellent thermal properties, such as thermal conductivity, coefficient of thermal expansion, specific heat capacity, and thermal shock, Nd:YAG optical crystals are best suited for near-IR applications. However, at wavelength operations over the 1 to 3 μm range, other optical crystals such as Nd:$YLiF_4$ and Nd:YVO_4 are widely used. Rare-earth laser crystals are available for specific wavelength operations. The properties of a few state-of-the art optical crystals are summarized in Table 5.2 [5].

Tunable laser crystals are used in applications where tuning capability over wide spectral regions is required. Cobalt- and chromium-based optical crystals are widely used where tuning capability of 1750 to 2400 nm and 1200 to 1500 nm, respectively, is required with high conversion efficiency. Optical crystals shown in Table 5.2 are generally used for fixed wavelength operations. Nonlinear crystals used in the design of harmonic-generator lasers are shown in Table 5.3. Critical performance parameters such as emission wavelength, operating mode (CW or pulsed), and operating temperature of the multidopant laser crystals are summarized in Table 5.4.

It is interesting that some doped crystals such as Er:YLF, Nd:YAG, and Nd:YLF operate very efficiently at fixed wavelength or over a very narrow spectral band. Nd:YAG and Nd:YLF crystals are widely used in solid-state lasers because of their abilities to generate higher output powers with improved efficiency over the 1–2

TABLE 5.2 Optical and physical properties of crystals for 1 μm lasers

Properties	YAG	YLF	YVO	GSGG
Emission wavelength (nm)	1064	1047/1053	1064	1061
Nd concentration (%)	0.7–1.1	1.0–2.0	1.0–3.0	2.0
Crystal structure	cubic	tetragonal	tetragonal	cubic
Melting point (°C)	1950	825	1825	1800
Thermal conductivity (W/cm·K)	13	6.3	5.2	0.7
Thermal expansion ($\times 10^{-6}$/°C)	7	8, 13	7, 11	7.5
Fluorescence lifetime (μsec)	240	520	110	240
Hardness factor	8.5	4-5	4-5	8
Absorption coefficient (1/cm)	15–20	32, 8.5	40, 10	—
Slope efficiency (%)	38	—	45	—
Absorption peak (nm)	808	792	809	460, 645
Absorption bandwidth (nm)	5	12	20–30	—

μm range. Crystal sizes of the order of 100 mm^3 are required for most diode-pumped solid-state (DPSS) lasers. Optical crystals such as YVO_4, $GdVO_4$, and $Sr_5(VO_4)_3F$ have demonstrated gain and conversion efficiency higher than that of Nd:YAG crystals. The above vanadium-based crystals have laser cross-sections ranging from 1.8 to 6.5 times higher than Nd:YAG. In addition, these particular crystals maintain a strong single-line emission with nearly the same peak emission wavelength as Nd:YAG, around 1067 nm. The vanadium-based crystals are uniaxial

TABLE 5.3 Nonlinear optical materials generally used in the design of harmonic-generator laser systems

Material	Wavelength (μm)	Nonlinear coefficients
$Ba_2NaNb_5O_{15}$	1.06	$d_{15}, d_{24}, d_{31}, d_{32}, d_{33} = 15 \times 10^{12}$ m/V
SiO_2	1.06	$d_{11} = 0.4 \times 10^{12}$ m/V
$LiNbO_3$	1.06	$d_{33} = 40 \times 10^{12}$ m/V
$LiTaO_3$	1.06	$d_{33} = 20 \times 10^{12}$ m/V
KDP	1.06	$d_{14}, d_{36} = 0.50 \times 10^{12}$ m/V
	0.6943	$d_{14}, d_{36} = 0.48 \times 10^{12}$ m/V
ADP	1.06	$d_{14}, d_{36} = 0.55 \times 10^{12}$ m/V
	0.6943	$d_{14}, d_{36} = 0.48 \times 10^{12}$ m/V
$LiIO_3$	1.06	$d_{31}, d_{33} = 5 \times 10^{12}$ m/V
Te	10.6	$d_{11} = 2000\text{–}5000 \times 10^{12}$ m/V
GaAs	0.8435→0.845	$d_{14} = 137 \times 10^{12}$ m/V
	1.06	$d_{14}, d_{36} = 250 \times 10^{12}$ m/V
	10.6	$d_{14} = 250 \times 10^{12}$ m/V
Ag_3AsS_3 (proustite)	1.15	$d_{31} = 15 \times 10^{12}$ m/V
	10.6	$d_{15}, d_{22} = 28 \times 10^{12}$ m/V

TABLE 5.4 Mode of operation, emission wavelength, and operating temperature for various crystalline laser systems

Host/Dopant	Wavelength (μm)	Mode of operation	Operating temperature (K)
$A1_2O_3/Cr^{3+}$	0.6929	p	350
	0.6934	p	350
$CaWO_4/Nd^{3+}$	1.0584	p, CW	300
	0.9145	p	77
	1.3392	p	300
$CaMoO_4/Nd^{3+}$	1.0673	p, CW	300
$YAlO_3/Nd^{3+}$ (polarized output)	1.0795	p, CW	300
	1.0645	p, CW	300
$Y_3Al_5O_{12}/Nd^{3+}$	0.946	p, CW	230
	1.0519	p, CW	300
$Y_3Al_5O_{12}/Nd^{3+}(Cr^{3+})$	1.0613	p, CW	300
	1.0640	p, CW	300
	1.0736	p, CW	300
	1.319	p, CW	300
	1.338	p, CW	300
	1.358	p, CW	300
$Y_3Al_5O_{12}/Er^{3+}$ (Yb^{3+})	1.6459	p	300
$YAlO_3/Er^{3+}$	i.663	p	300
$Y_3Al_5O_{12}/Tm^{3+}$ (Cr^{3+})	2.019	p	300
Ca_2MoO_4/Ho^{3+}	2.059	p	300
$Y_3Al_5O_{12}/Ho^{3+}$ (Er^{3+}, Tm^{3+})	2.1288	p	300
CaF_2/U^{3+}	2.57	p	300
	2.613	p	300
CaF_2/Er^{3+}	2.69	p	300

and produce only polarized laser output without undesirable thermally induced birefringence. They have nearly identical spectral features to that of Nd:YAG. The peak pump wavelength for all vanadium-based crystals is around 808 nm, so they are widely used in the manufacturing of high-power diode arrays for pumping applications. These new crystals of high optical quality must have low scattering and strain losses with uniform doping. These crystals yield improved laser performance because of large emission and absorption cross-sections. The major disadvantages of these crystals include shorter excited-state lifetime, about 45% of the lifetime of the Nd:YAG crystals.

Codoped and multidoped crystals can be tailored either for pumping or diode-pumping applications. The multidoped crystals, namely, ErTmHo:YLF (at 2060 nm), CrEr:YSG (2796 nm), and TmHo:$GdVO_4$ yield optimum performance at the stated wavelengths. Multidoped CrTmHo:YAG laser crystal represents a whole family of materials that can be tailored either for pumping or diode-pumping with high conversion efficiency. In the case of lamp-pumped designs, the chromium ions absorb lamp energy effectively and then transfer that to the thulium ions with an efficiency greater than 90%. The crystal can be optimized for holmium (Ho) emission at 2080 nm or thulium emission at 2010 nm. In contrast, diode-pumping at 780 nm can eliminate the need for chromium codoping because thulium has a strong absorption band at 780 nm. The latest Nd-doped crystals offer twice the energy of Nd:YAG crystals with an emission wavelength of 1067 nm and self-Raman-shifting in the picosecond-pulse mode to operating wavelengths of 940, 1060, 1180, and 1320 nm. DPSS lasers have potential applications in interferometry, material processing, and IR surveillance functions in military, environmental, and telecommunications fields.

5.5.1 Inorganic Nonlinear Optical Crystals (NOCs)

Nonlinear optical crystals are mostly used for converting a fundamental laser frequency to another higher or lower frequency. Conversion to higher frequency is called harmonic generation, whereas conversion to lower frequency is known as parametric oscillation. It is possible to combine tunable dye lasers, tunable color-center solid-state lasers, and tunable semiconductor injection lasers in crystals to cover almost continuously the spectral range from 170 nm to 18 μm [6].

Beta-barium borate (BO; B-BaO_4) has played an important role in the development of nonlinear optics [6]. This unique nonlinear optical (NLO) material possesses both the low-temperature β-phase and high-temperature α-phase, thereby giving it the unique phase-matching capability critically needed in the design of second harmonic generators (SHG). A nonlinear BBO crystal offers [6]:

- A wide transmission region from 200 nm to 3500 nm
- A broadband phase-matching capability from 409 to 3500 nm
- A large effective SHG coefficient (roughly 3 to 6 times greater than that of a KDP crystal), depending on the operating wavelength
- Significantly high damage-threshold—close to 10 GW/cm^2 for 100 ps pulses and 5 GW/cm^2 for 10 ns pulses, measured at the Nd:YAG laser wavelength of 1064 nm
- Extremely low thermal birefringence (*dn/dT*) compatible with the refractive index of ordinary rays and extraordinary rays
- High optical homogeneity of 1×10^{-6}/cm
- High chemical stability and high hardness factor

5.5.2 Second Harmonic Generator (SHG) Efficiency

The conversion efficiency of a SHG is of paramount importance and can be expressed as

$$\eta_{SHG} = \frac{P(2W)}{P(W)} = \tanh^2(x) \tag{5.6}$$

where $P(2W)$ = output power at twice the input frequency, W
$P(W)$ = input power at frequency, W
x = crystal parameter function of crystal length, pump power level, and the effective nonlinear coefficient at the phase-matching angle. The nonlinear coefficient is dependent on the symmetry of the crystal. A BBO crystal has a small acceptance angle and, thus, may not provide higher conversion efficiency than other NLO crystals.

Based on equation (5.6), the conversion efficiency is 29.8%, 46.8%, and 70.9% when parameter x is equal to 0.5, 0.6, and 0.7, respectively. However, high damage threshold and small thermal detuning due to small birefringence of the BBO crystal are major advantages over other NLO crystals. Dispersion curves and phase-matching characteristics for BBO crystals as a function of wavelength are shown in Figure 5.7. Critical optical parameters of BBO and D-KDP nonlinear crystals used by SHGs are summarized in Table 5.5 [6].

BBO is sensitive to depolarization effects. The high damage threshold and large temperature acceptance of BBO make it most ideal for doubling the output of high-power lasers; for example, producing 100 W (CW) output from a Nd:YAG laser. The laser-induced heating does not significantly affect the conversion efficiency of the BBO, thereby resulting in improved stability of the crystal.

5.6 COMPACT OPTICAL LENSES OR MICROLENSES

Compact optical lenses are known as microlenses or miniaturized lenses. Microlenses are the backbone of many optical components and systems. Self-focussing lenses of high optical quality are used to focus or collimate light in a wide variety of optical signal transmission applications. Their gradient-index composition, single-piece construction, and rod shape offer several advantages over conventional lenses. Microlenses are widely used fiber optic and electrooptic applications, including optically steerable phased-array antennas, imaging sensors, communications systems, scientific measurements, and control and data processing equipment. Self-focussing lens arrays are the biggest users of microlenses and are widely used in copiers, printers, facsimile machines, electronic whiteboards, and other industrial imaging products. These microlenses are widely used in miniaturized cameras and instruments for industrial and medical applications, where high resolution and precision focusing are of great importance. Microlens arrays are made from high-quality optical materials to meet stringent focusing requirements.

5.7 OPTICAL RESONATORS AND SUPPORT STRUCTURES

High-power laser systems used in scientific research, medical diagnosis, manufacturing, and the entertainment industry are extremely sensitive to temperature variations and vibrations produced by support structures. Even small temperature varia-

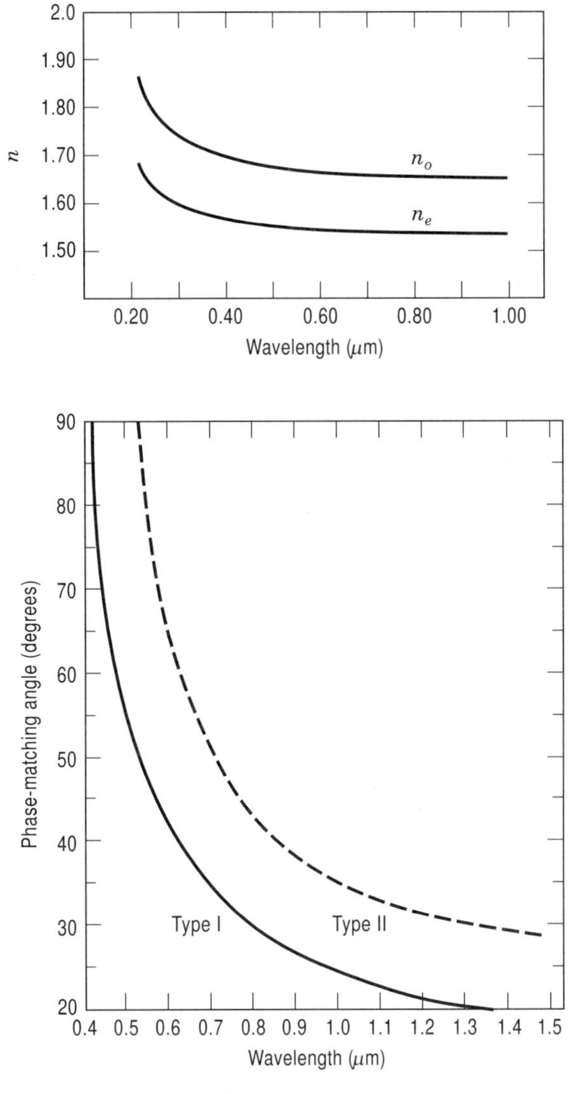

FIGURE 5.7 Dispersion characteristics and phase-matching angle as a function of operating wavelength.

tion or slight optical misalignment due to vibration can cause unacceptable performance degradation in some applications. That is why high-performance, high-power, lasers require optical cavities or resonators with excellent thermal stability. The resonator maintains critical alignment of the mirror that establishes the performance parameters of the laser cavity. The laser cavity is constructed with a flat mirror and a long radius-concave, curved mirror to maintain the optimum cavity alignment needed for high beam-pointing accuracy. Optical alignment of a laser cavity is very

TABLE 5.5 Optical parameters of BBO and D-KDP nonlinear crystals

Type	Crystal wavelength (nm)	Range (nm)	acceptance angle (mrad)	walk-off angle (deg)	Damage threshold (GW/cm^2)
BBO	1064, 532	200–3500	1.4	2.74	14 @ 1060 nm
D-KDP	1060, 532	200–1500	2.2	1.30	6 @ 1060 nm

critical and requires extremely tight angular tolerances. Even a misalignment of 20 to 30 microradians can cause the laser output power to drop to 50% of the full power rating. An angular change of 1 microradian can introduce a positional error of 0.1 mm at a distance of 100 m or an error of 1 mm at a distance of 1 km.

5.7.1 Material Requirements for Support Structures and Optical Resonators

The key components of a high-power laser are generally located in a structure made from Super Invar 32 alloy rods with appropriate dimensions. Because of the high electrical power dissipation into heat energy, the laser cavity is challenged to provide high mechanical integrity under intense thermal environments. If an optical mirror in the laser cavity were to become misaligned even by 10 microradians, both the laser efficiency and the output would be seriously impaired. When a resonator or cavity goes out of alignment, several things can happen simultaneously, namely, reduced output power, increased output noise, deterioration of transverse mode profile, and defocusing of the laser beam. All these problems can be eliminated or minimized by selecting optimum cavity materials and supporting mirrors by rods made from material with high mechanical integrity, high thermal mass, and low thermal expansion coefficient. The cavity design must include appropriate stabilization mechanisms to negate the adverse effects of environmental disturbances.

5.7.2 Resonator Performance Requirements

Generally, a Nd:YAG resonator has right circular cylinder geometry, but slab, rectangular, and other shapes are also used, depending on space availability and operating mode requirements. The rod end faces are polished flat and parallel for low transmission loss without optical coatings. The resonator is comprised of a rod with a highly reflective dielectric coating at one end and a reflector at the other. Separate mirrors are attached at each end of the resonator. Most scientific lasers use graded reflecting mirrors to produce low beam divergence, high optical energy, and improved brightness. The lowest-order stable transverse mode (TEM_{00}) has a smooth, Gaussian spatial beam profile in both the near and far fields. Most stable resonators are operated in multimode configurations. Comprehensive review of optical properties of cavity materials and various cavity configurations reveals that Nd:YAG resonators have the narrowest line width, between 4 to 5 Å, which results in the maximum spectral purity, excellent phase-matching capability, high optical pumping efficiency, and high thermal conductivity (0.14 W/cm·K) most desirable for low birefringence loss and low fluorescence lifetime. In brief, the material selected for the resonator must meet all the desired optical, thermal, and mechanical strength requirements.

5.8 LASER POINTERS

Laser pointers with red light (650–685 nm) are mostly used for highlighting special aspects of slide presentations in technical conferences, management review meetings, or industrial seminars. Laser pointers are also available with green light, which appears to be much brighter than the dull red beam because its 532 nm wavelength is closest to the human eye's most sensitive wavelength of 550 nm. Red laser pointers are up in the 650 to 685 nm range [7], which make them appear to be less bright. Studies performed by the author on laser pointers indicate that a battery-operated laser pointer costing around $50 can create a hologram capable of creating the coherent depth or linear polarization necessary to produce a high-quality three-dimensional image.

5.9 OPTICAL DISPLAYS USING LIQUID CRYSTAL DISPLAY (LCD) AND LIGHT EMITTING DIODE (LED) TECHNOLOGIES

Optical displays or monitoring systems can deploy either LCD technology or LED technology, depending on the display performance requirements and procurement costs. In the opinion of the display designers, LCD technology offers the most compact display for small handheld devices or appliances, where size, weight, cost, ruggedness, and high viewability are the pressing requirements. Advanced scientific calculators, desk calculators, and other commercial displays use LEDs as light sources. The quality of display depends on viewing area dimensions, size of each pixel, and the number of pixels that are required to present a suitable display to the user.

Glass manufacturers can fabricate displays with pixel sizes that are smaller than can actually be viewed by a human eye. When pixel size is below a threshold dimension of about 0.25 mm, the effectiveness of the display in the handheld devices begins to degrade. Character size is another important issue in a single-pixel format. Below the 0.25 mm threshold, multiple pixels must be used to create alphanumeric characters, which will increase the cost and complexity of both the hardware and software.

Most of the handheld and desk calculators have a small back light comprised of a light pipe, diffuser material, and LED illumination source. The light pipe serves both as an efficient LED light dispenser to illuminate the display and as a mechanical structure for the display glass and printed circuit board. The light pipe is generally made from a polycarbonate material, which serves two distinct functions, namely , efficient dispersion of the light from the LED sources to illuminate the display, and providing a mechanical structure for the display glass. The light from the LED sources must be efficiently dispersed throughout the light pipe to provide uniform illumination at the surface of the light pipe. Microstructures are usually molded into the light pipe to diffuse the light into a uniform pattern.

Liquid crystal displays offer good performance, but suffer from high cost and complexity. LCDs are mostly used in the cockpits of fighter and reconnaissance aircraft. Applications of LCDs are limited in commercial, medical, and industrial applications. However, some applied scientific research programs have used LCDs, because LED-based displays are not able to meet their stringent resolution requirements.

5.10 ANALOG-TO-DIGITAL CONVERTER (ADC) DESIGN USING OPTICAL TECHNOLOGY

Optical technology offers the most effective way to speed up the ADC process that currently limits the use of the available signal processing capacity for high-performance radar, electronic countermeasures, and communications systems. The new method based on optical technology developed for telecommunications in the area of wavelength-division-multiplexing (WDM) have demonstrated improved SNR performance. The fastest ADC with 10-bit resolution that one can now buy is about 100 MBPS using conventional technology. An ADC using optical technology can offer a 8-bit ADC device with sampling rate exceeding 40 GS/sec. It is possible to design an ADC device with an architecture that uses a method of time-stretching of incoming analog signals based on stretching and compressing of Gaussian signals. The method involves a three-step process (Figure 5.8) of dispersing the incoming signal,, chirping it, and dispersing it again. The signal-stretching process could be simplified by using a chirp bandwidth much larger than the bandwidth of the stretched signal. In other words, using a short optical pulse of few femtoseconds (say, 10 fs) wide, one can achieve a bandwidth of several terahertz (THz) compared to a stretched signal of several several hundred gigahertz, which represents a ratio of 1000:1. This indicates that by performing a chirp operation using ultrashort pulses, one can achieve much higher chirp bandwidth [8]. The optoelectronic time-stretching technique illustrated in Figure 5.8 uses a dispersal optical pulse to impose a wide-band pulse on the incoming analog signal. The use of optics also allows the carrier, along with the modulated signal, to be stretched prior to A/D conversion and ultimate digital signal processing. The stretching process maintains the signal pulse shape but spreads it over a long duration. One can obtain a bandwidth of 7.5 THz using a 160 fs pulse from a from a modulated erbium-doped fiber ring laser with a 1.1 km single-mode optical fiber. Intensity modulation of the chirp pulse by the analog input signal is accomplished through an electrooptic modulator using lithium niobate material. The signal is time-stretched by passing through the second stage of a single-mode optical fiber of 7.6 km length. This ADC device based on

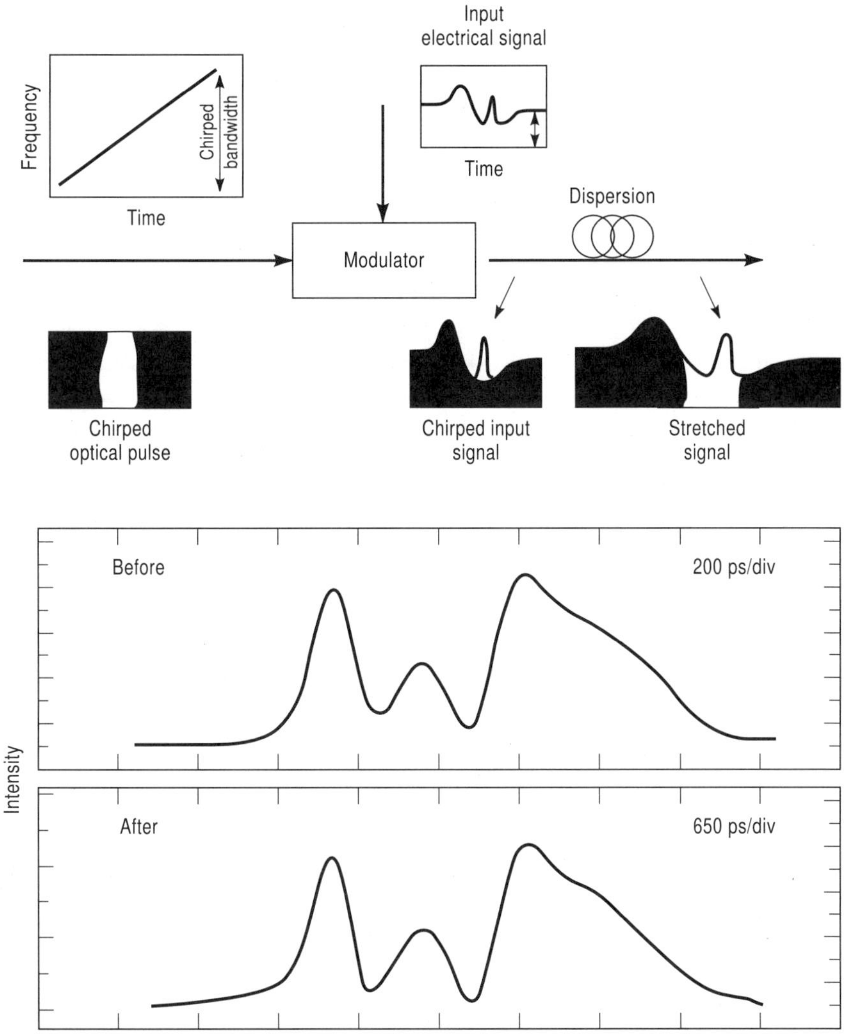

FIGURE 5.8 Analog-to-digital converter (ADC) architecture using optical technology for time-stretching of incoming analog signals.

electrooptic technology provides sampling rates exceeding 50 GS/sec with a 8-bit device.

5.11 LASER ALTIMETERS

Microwave radar altimeters operating in the S and C bands are widely used where a moderate accuracy of 3–5 ft is acceptable in height measurements. When ultrahigh resolution and accuracy are required, laser altimeters are the only sensors capable meeting the stringent performance requirements. The laser altimeter makes surface height measurements by measuring the time of the flight from an extremely narrow pulse, typically from a 10-mJ Q-switched, 1060 nm Nd:YAG laser source. The Mars orbiter laser (MOL) altimeter [9] abroad the Global Surveyor spacecraft generated the first three-dimensional map of the Martian North Pole using its 50 cm diameter optical telescope. This telescope gathered more than 2.5 million optical pulses reflected from the Mars surface over 206 orbits. The Mars mission scientists reported a surface profile showing ice 3.5 km thick (about 1.5 million cubic kilometer of ice). The Nd:YAG laser altimeter beam provided a surface spot size between 70 and 330 m [9], depending on the spacecraft altitude and pointing angle. The optical telescope of the system routes the reflected light to a photo diode, which is electronically gated to select the desired range window. The time-of-flight measurements are correlated with the orbital position to generate topographical maps with a spatial resolution of 0.6 km and vertical accuracy between 5 and 30 m. This laser demonstrated other Martian features such as impact craters and clouds.

5.12 OPTICAL LIMITERS

Optical limiters are used to protect the most expensive and sensitive optical amplifiers and receivers from unexpected high-power optical pulses or signals. They only allow the transmission of low-intensity optical signals, while blocking the transmission of high-intensity light signals. These passive optical devices use microlens arrays and porous optical substrates. The new generation of optical limiter designs will be based on sol–gel glass fabrication technology capable of providing ultra-low-threshold levels. Limiters with low optical threshold levels will have potential applications in the protection of delicate high-precision optical sensors used by military or space scientists. In addition, the optical limiters might have potential applications in telecommunications, cable TV (CATV), aerospace, and scientific research.

5.13 OPTICAL ISOLATORS AND CIRCULATORS

Optical isolators and circulators are performance-improving devices. Optical reflections can degrade the performance of lasers or optical amplifiers unless optical isolators block the reflected energy. Optical circulators are three-port devices that allow the use of a single fiber for both transmit and receive operations. Circulators are widely used in applications where backward reflection is to be detected, compared, and measured. An isolator is a passive, unidirectional optical device that allows the optical beam to transmit in the forward direction with minimum insertion loss, while blocking the optical signal transmission in the reversed direction with an attenuation of 40–70 dB.

Several different design configurations of optical isolators are available such as all-fiber isolators, fiber-embedded isolators, fiber Faraday rotator isolators, and waveguide-based isolators. Most commercially available isolators include Faraday rotator isolators with 45 degree rotation and a pair of birefringent crystals or pair of polarizers. A pair of microlenses such as gradient-index lenses are used to couple the optical beam from one fiber to the other. The birefringent crystal-based structure is called a polarization-independent isolator, whereas the polarization-based design is called a polarization-dependent isolator.

Most semiconductor diode lasers emit linearly polarized optical beams with extinction ratios of about 20 dB, but the single-mode fibers do not maintain the input polarization. Polarization-independent isolators are widely used in applications where a minimum isolation of 40 dB and maximum insertion loss of 0.05 dB are the performance requirements. In a polarization-independent isolator, the first wedged crystal splits the input beam into two parallel beams that are then refocused by a lens into an optical fiber in the forward direction. The optical beams, after passing through a 45 degree Faraday rotator, enter the second wedge crystal. The two beams exit the surface of the second wedge crystal parallel to each other. The lens focuses them both into the output fiber. In the reverse direction, the second wedge again splits the incoming beam into two beams. After passing through the 45 degree rotator, both beams are rotated by 45 degrees in the same direction; the beam directions are interchanged in the first crystal. The first wedge separates the two beams, blocking of the backward propagation from coupling into the input fiber.

5.13.1 Cost and Performance Parameters of Isolators

A single-stage high-performance isolator costs between \$600 and \$800 depending on the isolation requirements. A single-stage isolator [10] provides a minimum isolation of 32 dB, maximum insertion loss of 0.5 dB, return loss better than 60 dB, polarization-dependent loss less than 0.05 dB, dispersion less than 0.2 ps, instantaneous bandwidth of ± 20 nm and maximum power handling capability of 300 mW. Single-stage isolators are widely used in optical transmission systems using distributed-feedback (DFB) lasers. A DFB laser frequency is most sensitive to back reflections, which have adverse effects on laser gain profile and power stability. Erbium-doped fiber amplifiers (EDFAs) with gain ranging from 20 to 40 dB are more sensitive to the reflections and often use more than one isolator to eliminate reflections from various joints and surfaces in the circuit. Fiber optic isolators have potential applications in telecommunications, biotechnology, fiber optic gyros, and other optical systems requiring multipath-free and cross-talk free operations.

Multistage isolators offer the higher isolation values, but at the expense of higher insertion loss and cost. A typical two-stage isolator capable of operating over telecommunication spectral bands can cost between \$800 and \$1400. However, a two-stage device offers isolation greater than 55 dB and insertion loss around 0.75 dB, which is roughly 0.25 dB more loss due to the second stage. Most of the least-expensive isolators are designed for standard telecommunication wavelengths of 1310 to 1550 nm and moderate isolation requirements. However, fiber optic isolators operating in the UV or near-IR regions are bulky and more expensive. Isolators must be operated at power levels and temperatures recommended by the manufactures to ensure specified optical performance and reliability. Regardless of the isolator type, devices with low polarization-dependent loss (LPD) must be used to preserve overall system performance.

5.13.2 Optical Circulators

As stated earlier, circulators are three-port devices that allow the use of a single fiber to transmit and receive optical signals. They are mostly used where reverse reflection signals are to be detected, compared, and measured with high accuracy and reliability. These devices are available for operation either at 1310 or 1550 nm and come with 1 m pigtails. Circulators are also available for four- or eight-port operations. The polarization-dependent loss (PDL) and polarization-mode dispersion (PMD) for circulators are higher than those for isolators due to device complexity. Circulators are available with typical room-temperature maximum insertion loss of 1.0 dB, minimum isolation of 50 dB, minimum bandwidth of 50 nm, maximum PDL of 0.2 dB, maximum PML of 0.3 dB, and return loss better than 50 dB. High-performance circulators are relatively more bulky and expensive than isolators. Unit cost varied between \$3000 to \$4200 in 1999, depending on isolation, bandwidth, and other critical performance requirements. Circulators are used in specific applications where high isolation and minimum insertion loss over a wide bandwidth are the principal requirements.

5.14 FIBER OPTIC COUPLERS AND MULTIPLEXERS

Single-mode fiber optic (SMFO) couplers and wavelength-division-multiplexers (WDMs) generally employ fused-fiber technology and are widely used in telecommunications, aerospace, defense, and scientific applications.

5.14.1 Couplers

Fiber optic (FO) couplers are available with 1×2, 2×2, 1×4, and 1×8 configurations and are capable of operating over narrow as well over wide spectral bandwidths. Single-mode fused-fiber couplers are bidirectional devices that can be used to either split or combine optical signals with minimum loss. Single-wavelength coupler designs are optimized to provide best performance at center wavelengths of 633, 780, 850, 1310, and 1550 nm because of great demand at these wavelengths. Couplers operating either at 1310 or 1550 nm wavelengths are available with bandwidth of ± 20 nm. Optical couplers with 1×2 and 2×2 configurations and coupling values of 3 dB are generally available at minimum cost because of straightforward design procedures. Single-mode couplers are best suited for power monitoring in power EDFAs and repeaters, where minimum loss and wavelength-independent coupling ratio are of paramount importance.

5.14.2 Wavelength-Division Multiplexers (WDMs)

Ultra-low-PDL fused-fiber couplers with insertion loss well below 0.5 dB across the 1550 nm spectrum are widely used for WDM and dense-WDM applications and other polarization-sensitive telecommunications systems. Specific details of a WDM coupler capable accepting both 1.31 μm and 1.55 μm signals containing voice and video data, respectively, are shown in Figure 5.9. It is evident from Figure 5.9 that the 1.55 μm video signal is split in two by a WDM coupler, whereas the 1.31 μm voice signal is received by a photo detector. Note that the functional diagram of a two-wavelength splitter/combiner uses a cascaded series of diffractive optics to alternately deflect, collimate, split, and focus the light from the input fibers and from the laser diode operating at 1.31 μm. The 1.31 μm laser light is split by the diffractive grating, forming two beams. One of these beams is focused by an off-axis diffractive lens (not shown in Figure 5.9) onto the photo detector, while the other laser beam is collimated by a diffractive lens and sent back along the return path.

Single-mode WDM couplers are generally used to separate or combine optical signals at telecommunication wavelengths, thereby allowing bidirectional communications on a single fiber. WDM couplers offer high isolation between the two wavelengths with minimum cost and complexity. Critical performance parameters of a WDM coupler include isolation level, excess loss, instantaneous bandwidth, extinction ratio, coupling ratio, directivity, return loss, and polarization-dependent loss.

5.15 FIBER OPTIC RING LASER GYROS

Fiber optic ring lasers use high-performance optical fibers to achieve optimum gyro accuracy and stability under severe operating environments. These operating environments include extreme temperatures, high magnitudes of shock, and nonlinear, multifrequency vibrations. Optical fibers for use in ring lasers must meet stringent performance requirements to ensure ultra-high gryo accuracy and stability under severe environments. Fiber optic ring laser gyros have potential applications in missile guidance systems, sophisticated military systems, and airborne surveillance sensors. The ring laser-based gyroscopes offer optimum accuracy in measurements of linear and angular displacement under roll, pitch, and yaw conditions, which are difficult to achieve from other conventional devices. During the launch phase of a missile, the missile undergoes severe shock and vibration. It is of critical importance to maintain the desired missile guidance accuracy during the launch phase, cruise phase, and terminal phase to accomplish the mission objectives successfully. Performance of missile guidance strictly depends on the gyros deployed to provide the most accurate information, minute-by-minute, on range to target, missile velocity, and angular positions during various phases of missile fight.

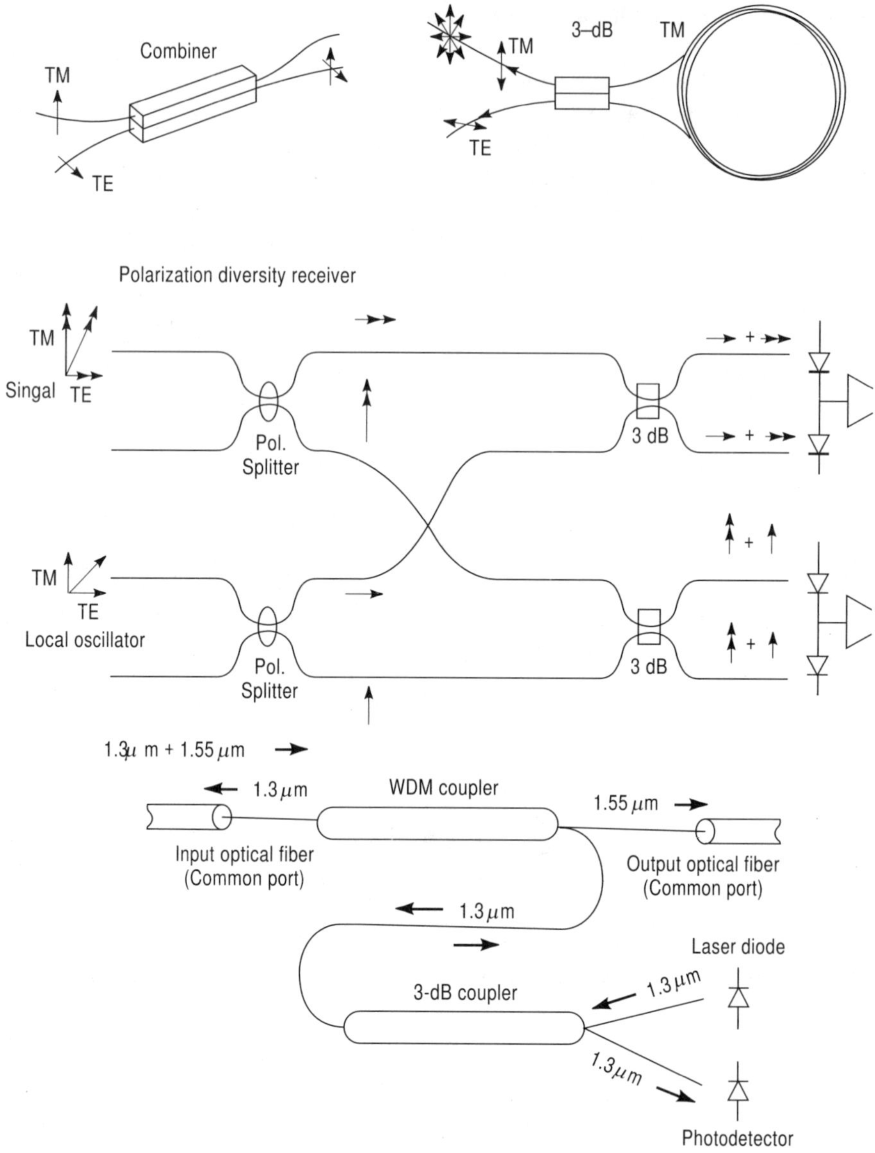

FIGURE 5.9 Schematic diagrams of optical coupler, combiner, and polarization splitter for optical communications systems.

5.16 FIBER OPTIC COMMUNICATION LINKS

Communication links using fiber optic technology offer transfers of data with maximum security, minimum insertion loss, and no RF interference. The optical link typically involves a LED transmitter, modulator, preamplifier, and photo diode detector. The output of the LED transmitter is modulated by the intelligence information to be transmitted and is coupled to an optical fiber transmission line. Modulation of the source is accomplished by modulation of input current. At the other end of the optical fiber, the signal intelligence is recovered through conversion of the optical power, which is amplified to a desired level by a preamplifier. Total system loss is equal to the sum of losses contributed by the fiber loss, coupler loss, and photo diode inefficiency. Two-way transmission can be achieved by providing both a transmitter and a receiver at each end of the fiber. The data transmission capability is dependent on the transmitter power, receiver sensitivity, power loss in the link, and link margin meeting specific bit-error-rate (BER) requirements. Data transmission over an optical link can be achieved with minimum cost and complexity without being susceptible to any intentional or nonintentional jamming.

5.17 OPTICAL SWITCHES

Optical switches are designed to switch signals between a pair of alternate output fibers. Optical switching promises to cure telecommunications logjam problems. A 16×16 cross-connect switch requires 256 switching nodes. Optical signals go straight through the "off" state condition but are switched in the "on state" condition. Advanced optical switches are currently in development, including complete cross-connect configuration with multiple input and output ports and add/drop switches that can switch individual wavelengths into and out of WDM signals. Various techniques are being investigated to meet diversified switching requirements. They include optomechanical switches that move fibers or mirrors to redirect optical signals, and solid-state devices that might shift light signals between fibers or planar waveguides. Research scientists are working on other technologies, such as nonlinear devices for wavelength conversion needed to switch signals from one wavelength to another, as in the case of wavelength-division-multiplexed systems. High-speed optical logic gate circuits must be developed for implementation in fast time-division-multiplexed (TDM) systems.

Regardless of switch types, optical control of signals remains an engineering challenge. Optical logic gates use the signals' contents to control switching. Designs of optical AND and INVERT logic gates operating at 100 Gbits/sec data rate have been optimized. The telecommunications industry is the major user of optical switches. The rapid spread of WDM is creating an urgent need for an add/drop capability that can switch wavelengths into or out of a WDM signal. Dense-WDM (D-WDM) systems are capable of carrying dozens of wavelengths [11]. The latest system developed by Lucent Technology [11] can transmit up to 80 wavelengths with high optical efficiency. With effective use of this tremendous capability, telecommunications operators can add or drop certain wavelengths at system nodes without completely separating all wavelengths. Switching speed, isolation between the fibers, and insertion loss are the most critical performance parameters of the optical switch.

5.17.1 Other Switch Technologies for Optical Switching

Optomechanical switches are simple and inexpensive but suffer from high switching times. Such switches have switching times between 5 and 10 ms for the fastest two-channel switches up to 500 ms for 16 × 16 matrix switches. Switching speeds under 2 ms are possible with thermooptical switches, which shift optical signals by changing the temperature of a waveguide that is one arm of an interferometer. Thermooptical switches can be cascaded to form complete arrays with rise times of 100 μs and fall times of 180 μs in silica-on-silicon devices

Acoustooptic (AO) switches change transmission characteristics of one arm of a waveguide interferometer by application of a voltage, which changes the refractive index of lithium niobate medium, providing the switching function. These AO switches are relatively fast, but they are most expensive and complex. Microelectromechanical (MEM) switches represent the emerging switching technology and are based on tiny optical and micromechanical components deposited on a semiconductor substrate. Electronic input signals control the motion of the components, thereby rapidly redirecting the optical signals. MEM switches have demonstrated switching speeds better than 500 μs, which is expected to go down to below 100 μs as the MEM switch technology matures.

Cross-connect (CC) switches provide logical circuit connections in telephone networks by routing signals from any input channel to any output channel. CC switches can move optical signals from one wavelength to another as well as between optical fibers. CC switch technology has the potential to hook up to as many as 640 × 640 elements (or 409,600 lines).

An all-optical switch approach uses nonlinear interaction in a semiconductor-based optical amplifier. Two signals enter the optical amplifier, namely, the modulated input and a CW signal at the desired output wavelength. Nonlinear wavelength conversion is also possible in optical fibers, but it will require more than one kilometer of optical fiber, thereby making it economically infeasible.

Comprehensive review of various optical switching schemes in terms of overall performance, switching speed, cost, and complexity shows that an optical switching concept using fast AND and INVERT logic gates seems to transmit data at rates exceeding 100 Gbits/second, which will lead to potential applications in WDM and dense-WDM telecommunications systems.

5.18 PLOTTERS, SCANNERS, AND PRINTERS USING ELECTROOPTIC TECHNOLOGY

5.18.1 Laser-Based Plotters

Electrooptic technology makes possible a large-format, high-resolution raster laser plotter with a maximum photo printing area of 72″ × 40″. This laser-based plotter provides a variable resolution of 100 to 4000 lines per inch with an accuracy repetition of 20 μm. Plotting can be achieved either with negative or positive polarity with emulsion capability depending on customer requirements. These devices can offer reductions onto high-resolution emulsions on film or glass.

5.18.2 Optical Scanners

Optical scanners provides a continuous light-line scan using two input fiber bundles of appropriate length, each with an active diameter of 0.5 inches, that plug into two light sources.The process allows light from each input to be distributed over the entire length of the line, resulting in evenly reduced intensity over the line. The light line can be calibrated to ensure uniformity of ± 15 gray scales at a mean of 200 using a gamma (parameter of the scanner) setting of 1.0. A cylindrical lens, if necessary, can be attached to produce a concentrated line of light with intensity levels exceeding 90%, which can be directed onto any inspection surface. Optical line scanners are the critical elements of machine vision and inspection system applications

5.18.3 Laser Printers

Laser-based printers capable of producing 1200 dots per inch (DPI) output are in great demand because of high print quality. A high-performance laser printer is comprised of a controller, laser diode source, collimating lens, nonsymmetric lens, polygonal mirror, scanning lens assembly, and photoconducting drive. Some applications may require even higher resolution than stated above to reduce jaggedness in oblique lines and to allow quality image enlargement during a reproduction process. According to laser printer designers, each time the resolution doubles, the number of dots increases by fourfold, indicating that the number of dots is proportional to the square of the resolution. The laser-beam size must increase as the focused-spot size decreases or as resolution improves to avoid differentiation effects needed for high quality print match. This requires large-sized mirror facets. This means that to maintain a specified printing speed, the mirrors must rotate relatively faster to produce more scans per second.

Greater processing rates or more pages per minute increase demand on the modulation frequency of the laser diode necessary to generate crisply separated spots. For example, a laser printer capable printing two pages per second or 120 pages per minute at 1200 dots per inch requires 200,000 scans per minute and a laser modulation frequency of 400 MHz. Such a high modulation frequency is extremely difficult and costly to achieve in actual practice. An alternate solution to this is to use lower printing speeds such as 10 or 20 pages per minute.

Varying the spot size across the span will produce a smoother outline without increasing the number of dots per inch, scan rate, or modulation frequency. Changing the near-field spot size of the laser diode translates to variable spot size at the photo conductor. A 2:1 variation in spot size offers considerable smoothness (similar to 1200 DPI) of the printed line even at 600 DPI and reduction of the modulation frequency by a factor of four.

Several color printers use the facet tracking approach, which deflects a beam from a HeNe laser to maintain the beam at the center of the rotating mirror facet. This allows the use of small facets as well as more scan lines for a given revolution rate. A liquid crystal attenuator controls the beam intensity to maintain proper exposure of the photo conductor. These laser printers offer high quality laser image recording and medical hard copy with computer-based diagnostic imagery. In medical applications, the laser prints directly onto photographic film to prepare hard copies of digital images. This permits physicians and surgeons to have instant access to dozens of computer-based diagnostic images. Computer-based diagnostic imaging is also known as computed tomography (CT), position emission tomography (PET), magnetic resonance imaging (MRI), and digital angiography (DA).

Computer-to-plate (CTP) technology is gaining popularity in commercial printing. In the CTP technique, a laser transfers the image directly from the computer to the plate, which is chemically processed using thermally cross-linking polymers. The threshold exposure is generally twice that for ablation of thin layers of material. The material is optimized either for 830 nm laser diodes or 1064 nm DPSS lasers for improved image quality. Multiple-beam systems can be used to gain speed, but they increase the cost and complexity. However, laser printers using high-power diode arrays can overcome the above problems.

5.19 INFRARED THERMOMETERS

Infrared radiation is a part of the electromagnetic (EM) spectrum, which includes radio waves, microwaves, visible light, UV light, the IR spectrum (extending from 0.7 μm to 1000 μm), gamma rays, and X-rays. However, for IR measurements the spectral range from 0.7 μm to 14 μm is widely used. Noncoherent thermometers can measure temperature accurately over the above spectrum using advanced optical detectors and electrooptic components. Since every object emits maximum IR energy at a specific wavelength, each measurement requires a unique sensor equipped with appropriate optics and detectors. For example, a sensor operating at 3.43 μm can be optimized to measure the surface temperature of polyethylene and related materials. Similarly, a sensor operating at 5 μm is best suited to measure glass surface temperatures. An IR sensor operating at 1 μm is used to measure surface temperatures of metals and foils. A sensor capable of measuring lower surface temperatures is best suited for paper, board, plywood, and composites. In general, military and industrial targets reflect, transmit, and emit IR energy at appropriate wavelengths. The energy emitted by a target is dependent on its emissivity, surface condition, and peak radiation wavelength, which indicates the temperature of the object. IR sensors have adjustable and calibrated emissivity settings over the 0.1–1.0 μm spectral region, which will permit rapid and accurate temperature measurements of different surface types and structural materials.

High-speed linearization, exotic signal processing techniques, and surface mount technology are integrated with standard IR detectors to achieve fast temperature response. High-performance InGaAs photon detectors are widely used where response less than 1 ms is desired at temperatures as low as 250 °C. IR thermometers are most attractive for applications in which the sensor is used to provide a temperature signal for a control loop. High-quality optics, advanced microelectronics, and embedded software have led to improved performance of IR thermometers for making single- and dual-wavelength measurements. Single-wavelength IR thermometers with variable focus optics offer high optical resolutions with response times as low as 1 ms over wide dynamic temperature ranges. Such IR thermometers are well suited for rapidly moving manufacturing processes such as steel rolling, where quick measurement and fast response are of critical importance. Dual-wavelength thermometers can simultaneously measure IR radiation at two distinct wavelengths and calculate temperatures through appropriate means. These IR thermometers can measure targets obscured by other particles or elements such as dust, smoke, and atmospheric gases.

Fiber-optic-based IR thermometers are most appealing because of their integration of compact electronics and embedded software, high reliability, and lower costs. The optical fibers transmit IR energy to the IR detectors the same way that light is transmitted over fiber-optic-based phone lines. The optical fibers can survive under harsh operating environments such as high temperature, nuclear radiation, and high EM fields. Because of small diameter and high mechanical integrity, optical fibers can be routed to a target location that may not be in a clear line of sight (LOS), a situation in which use of standard IR thermometers is not advisable.

This new class of IR thermometer technology offers unique features, namely, fast response time, smaller spot size, wide temperature range, remote sensor operating capability, integrated microelectronic package, and embedded software programs for remote monitoring and data analysis. These advanced IR thermometers are most suitable for applications requiring statistical process control, monitoring of critical process temperatures, and remote-control adjustment (without having to walk near a process control center operating at 1500 °C or higher).

5.20 FIBER OPTIC DELAY LINES

Single-mode (SM) optical fiber is the most attractive delay medium for signal processing applications because of low loss (less than 0.1 dB per μs delay) and large time–bandwidth product (greater than 10^5). Recent developments leading to efficient tapping of light from SM fibers have helped in designing the recirculating and nonrecirculating delay-line structures. These delay lines provides a variety of important signal processing functions, namely, coded sequence generation, convolution, correlation, matrix–vector multiplication, and frequency filtering. SM optical fibers [12] are capable of transmitting modulated signals with bandwidths exceeding 100 GHz over a 1 km distance with insertion loss not exceeding 0.5 dB. The low loss and low dispersion of SM optical fibers allow the signals to propagate over large distances with minimum loss and distortion.

The real-time signal processing of broad band radar signals, ECM signals, and optical communications signals will benefit the most from optical fiber delay lines. SM fiber and SM fiber components such as couplers, switches, and isolators are readily available to build recirculating and nonrecirculating delay lines and filters. Even though standard delay lines using conventional devices such as coax cables, charge-coupled devices (CCDs), and surface acoustic wave (SAW) devices are available, they suffer from high cost, excessive weight, and limited bandwidth. Delay lines using magnetostatic wave (MSW) devices involving ferromagnetic materials offer very wide bandwidth, but they also suffer from very high insertion loss. Superconducting delay lines using niobium transmission lines and proximity coupled taps offer ultra-low loss with bandwidth as large as 20 GHz, but with high cost, complexity, and frequency of maintenance problems that are serious drawbacks.

5.20.1 Critical Performance Parameters of Optical Fibers

Several critical performance parameters of optical fibers, such as attenuation, scattering, dispersion, and linearity, must be given serious consideration for specific applications. Attenuation results from a number of wavelength-dependent mechanisms, including absorption, scattering, modal dispersion, microbending, and Rayleigh scattering. Rayleigh scattering is most dominant at operations below 1.6 μm and is inversely proportional to the fourth power of wavelength. Preliminary calculations indicate that the insertion loss in a SM fiber is about 3 dB/km at 0.83 μm, compared to 0.5 dB/km at 1.3 μm. Propagation loss per μs of delay is lowest around 0.2 dB for SM fiber, compared to 0.4 dB for a superconducting niobium-based microstrip delay line, 3.5 dB for a SAW delay line, and about 18 dB for a copper microstrip delay line, all operating at 2 GHz. The loss for the superconducting delay line remains practically constant at 0.2 dB over the 2–100 GHz bandwidth, whereas the losses in other delay lines increase as the bandwidth increases.

Studies performed by the author indicate that the SM fiber has the maximum modulation bandwidth, whereas the modulation bandwidth of a multimode (MM) fiber is limited by the modal dispersion due to the presence of several modes in the fiber. In a MM fiber, a number of transverse modes that exist do not travel at the same group velocity, thereby imposing a bandwidth limitation. Modal dispersion can be reduced in MM fibers by using graded index profiles. On the other hand, the modulation bandwidth of a SM fiber delay line is only limited by the refractive index of the material, which is a function of temperature and wavelength.

The modulation bandwidth is dependent on half-width of the spectrum of the optical source, optical fiber length, and differential dispersion (expressed in ps/nm·km). Low values of RMS half-width of the spectrum and differential dispersion are necessary to achieve high modulation bandwidth. An operating wavelength of 1.3 μm has a near-zero differential dispersion, which offers a bandwidth–distance product greater than 100 GHz·km.

Large modulation bandwidths and long delays are possible with pump-modulated fibers because of large time–bandwidth products. Assuming a bandwidth–distance product of 100 GHz·km and a delay of 5 μs per km, the computed time–bandwidth product exceeds 5×10^5, which is not possible with other conventional delay lines. The linear power handling capability of a fiber is set by the stimulated Raman and stimulated Brillouin scattering phenomena. The input power above which the stimulated Raman back-scattering becomes critical is given as

$$P_{\text{crit}} = \frac{(21)(A)}{(L)(\gamma_0)} \tag{5.7}$$

where A is the cross-sectional area of the fiber core [13], L is the fiber length, and the parameter γ_0 is the Raman gain coefficient function of the operating wavelength.

Assuming a SM fiber of 200 m length with a 6 μm core diameter and using a value of 5×10^{-10} cm/W, the critical input power comes to 0.594 W for a 1.31 μs delay. In other words, a SM fiber delay line 152 m in length and with a core diameter of 6 μm offers a delay of 1 μs with the same critical input power of 0.594 W. Furthermore, for a 10 μs delay line with length of 1520 m and core diameter of 6 μm, the critical power comes down to 0.059 W (59 mW).

The inherent dynamic range of a SM fiber is inversely proportional to the signal bandwidth, but proportional to Raman back-scattering . Raman scattering is independent of signal modulation frequency, but the quantum noise is linearly proportional to the signal bandwidth. This means that the fundamental optical dynamic range (FODR) of a fiber optic delay line decreases with increasing signal bandwidth. Based on published data, the FODR for signal transmission for a 152 m (with 1 μs delay) delay line comes to about 80 dB at 10 GHz and 130 dB at 5 GHz.

Single-mode optical fiber delay lines have several excellent features, including high fiber flexibility, highly precise time intervals between taps, lower group velocity of light in the fiber (around 2×10^8 m/s), ultra-low time delay spacing of 5 ps or lower between the fiber taps, high-temperature operation capability, insensitivity to environmental effects, and built-in immunity against electromagnetic interference (EMI) and high-intensity electromagnetic pulses.

5.21 DIGITAL STILL CAMERAS USING OPTICAL TECHNOLOGY

Digital still cameras (DSCs) are widely used by consumers because of relatively low costs and high-quality prints. The megapixel capacity of CCD-based DSCs affords small physical pixel areas that meet consumer photographic requirements. However, for industrial, commercial, scientific, and medical applications, large pixel sizes [13] are desirable. A high-performance megapixel CCD-based DSC provides about 1400 × 1000 active elements with square pixels of 4.65 μm per side. CCD devices with the same pixel count but different pixel sizes perform quite differently because the physical size of each pixel is the key performance indicator. Furthermore, the pixel size affects the responsivity, dynamic range, frame rate, and cost of the optical system. The DSC performance is considered not acceptable when the pixel size drops below 5 μm or so.

5.21.1 Impact of Pixel Size and Area

The quantum efficiency, also known as the photon-to-electron conversion efficiency, of most CCDs is similar, but size differences of the photosensitive area makes their performance levels unequal [13]. Responsivity from each pixel is directly proportional to the pixel area. As the responsivity increases, the same amount of signal can be collected in a shorter period of time with high SNR. In other words, increased responsivity results in higher SNR, lower illumination, and increased frame rates, leading to higher throughput. As stated earlier, pixel size also affects the dynamic range of the camera. Parametric tradeoff calculations indicate that a pixel size of 4 μm square results in a SNR of 12 dB, 8 μm square a SNR of 18 dB, and 12 μm square a SNR of 22 dB. The larger the pixel size, the larger the charge density and higher the signal levels. Wide dynamic range permits areas of high brightness. A DSC camera can easily distinguish between objects in shadow and those located near a bright object.

Dynamic range is normally characterized by the number of gray levels in a camera and it is a measure of image contrast on a display. An 8-bit DSC can produce 256 gray scales with 48 dB dynamic range, whereas a camera with a 10-bit design offers 1024 gray scales with 60 dB dynamic range. Machine-vision applications require a camera with minimum CCD dynamic range of 60 dB that will cost more than an 8-bit design. Dynamic range drops as the camera frame rate (which is typically 30 or 60 frames/sec) or the total noise level increases.

The overall performance of a CCD-based camera depends on the response of the photon transfer function (PTF) of the CCD device, which relates the overall noise level as a function of input light intensity. There are three distinct noise regimes, namely, the noise floor representing the level under totally dark conditions; shot noise, which is dependent on the amount of illumination impinging on the CCD surface; and the fixed pattern noise, which is only visible at high light levels, and is proportional to the input signal. It is important to mention that a CCD-based camera eliminates the need for a frame grabber, but its higher cost can limit its applications. A digital color microcamera developed by Panasonic Corp. offers both high pixel density and real-time imagery capability at reasonable cost. This particular high-resolution, color, single-chip, real-time, video camera uses a 900,000-pixel CCD chip that provides 560 horizontal lines of resolution and a SNR exceeding 54 dB, which is unmatched by other similar cameras.

5.22 HIGH-SPEED IR CAMERAS

High-speed IR (HSIR) cameras equipped with FPA technology differ from CCD-based cameras and offer high sensitivity, improved IR imagery, powerful windowing functions, compact packaging, and improved reliability. High quantum efficiency (close to 90%) of the InSb-based FPA provides unparalleled performance with crisp and clear images. With variable frame rate control capability and large video memory, these cameras deliver 512×512 image quality. The windowing capability makes it possible to view high-speed thermal events at frame rates as high as 6100

Hz. This high-speed thermal event capability makes it possible to record the trajectory of a maneuverable airborne target such as a military jet aircraft or missile. Switching between the three FPA integration times yields an effective dynamic range exceeding 130 dB. This unique feature is ideal for applications involving multispectral imaging, nondestructive testing, and investigation of thermally dynamic events. A choice of 32,000 integration times is available that can meet virtually any IR application. Appropriate wavelength filters are available for sub-band spectral analysis in the 3 to 5 μm range. This FPA-based IR imaging device is capable of processing images of exceptional quality from 20,000 ft or higher, which indicates that it has potential application in battlefield sensors.

Conventional CCD-based cameras use detectors that are about 10 μm in size and normally operate in the visible region around 500 nm. Since these cameras operate in the detector-limiting region, the equivalent resolution does not improve if the f-number of the camera lens is below 5. If the f-number is greater than 8, the optics resolution affects the minimum discernible target size. The overall imaging system equivalent resolution is given as

$$R_{eq} = \sqrt{R_o^2 + R_d^2} \tag{5.8}$$

where R_o is the resolution of the optics, equal to 1.845 $\lambda \cdot F$, and R_d is the resolution of the detector, equal to detector width d. The parameter F is the lens f-number and λ is the operating wavelength. Cameras designed for LWIR operations use detector size d of 10 μm; such cameras are called optics-limited devices. In the optics-limited region, the minimum target size is given as

$$[\text{Target size}]_{\min} = (1.845)(\lambda/D_o)(R) \tag{5.9}$$

where D_o is the diameter of the optical system and R is range to target.

5.22.1 IR Camera Design Requirements

IR camera design requirements depend on the operational scenario. The overall design requirements must be determined based on the detector–optical combined performance requirements. Optimum camera design can be achieved either for diffraction-limited optics or for detector-limited performance. Sometimes, even the minimum aberrations are not acceptable. Detector size and the optic size must be selected to meet the overall performance requirements of the IR camera. The IR camera design requirements must focus on optics/detector package operational capabilities and limitations.

5.23 OPTICAL FILTERS

Practically all optical systems deploy optical filters to optimize their performance. Fixed-frequency filters (low-pass, high-pass, and band-pass filters) and tunable filters will be discussed, with emphasis on performance capability, cost, and complexity. In some applications, tunable filters reduce the number of detectors from three to one, thereby realizing substantial reduction in cost and significant improvement in reliability. Performance requirements of the filters for possible applications in optical amplifiers (i.e., EDFAs), optical parametric oscillators (OPOs), WDM communications systems, and dense-WDM telecommunications systems will be briefly described, with emphasis on insertion loss, skirt selectivity, and stop-band isolation.

5.23.1 Band-Pass Filters (BPFs)

Band-pass filters are generally used for noise rejection over a specific spectral region, for gain-flattening purposes, and for separation of spectral bands over a wide spectral region. BPFs with stringent performance requirements are used in demanding applications, such as undersea, airborne, or space-based systems, to meet the critical system performance requirements.

BPFs with low insertion loss, minimum dispersion, and sharp skirt selectivity are widely used in the WDM and dense-WDM telecommunications systems for reducing cross-channel noise. BPFs with appropriate band-pass regions are available for use in WDM and dense-WDM systems with 16, 32, 64 or higher channels. A single-mode BPF operating at 1550 nm wavelength with 1 nm band-pass region typically has an insertion loss of about 1.0 db, a 3-dB bandwidth of 65 nm, back-reflection of –50 dB, polarization-dependent loss (PDL) of 0.03 dB, and a center wavelength stability of 0.005 nm per °C.

5.23.2 Low-Pass Filters (LPFs) and High-Pass Filters (HPFs)

LPFs and HPFs are available for systems applications where signals at specified lower wavelengths are required to pass through the filter with minimum loss, while attenuating all the optical signals at higher wavelengths, and vice-versa. There are certain applications where LPFs or HPFs can provide optimum filter performance levels at minimum cost. However, such applications are very limited. Thus, optical filters with band-pass characteristics have wide applications, including commercial, industrial, medical, military, and space applications.

5.23.3 Add–Drop Filters

Studies performed by the author indicate that a two-dimensional digital micromirror (TDM) can act as a fault-tolerant add–drop filter for potential applications in dense-WDM communications [14] systems. The TDM can be coupled by free-space optics to fiber-connected array-waveguide-grating (AWG) multiplexers that form a two-dimensional in–out feed. Currently, most add–drop filters used in the telecommunications industry rely solely on on-chip planar technology integrated with optical waveguides and switches. This technology offers very high space–bandwidth products. A TDM is comprised of more than a million mirrors, and involves macropixels of several hundred micromirrors, each having switching capability at individual wavelengths. This design architecture is fault-tolerant, because failure of individual mirrors will have an insignificant impact on the optical-alignment sensitivity, and because of the overall robustness of the device. Graded-index fibers seem to provide improved performance. Add–drop filters need to be developed at 1550 nm, the most popular telecommunications wavelength, because of their potential for application in WDM and dense-WDM telecommunications systems.

5.23.4 IR Absorbing Filters

Entertainment lighting, medicine, optical communications, UV curing and heating, and fiber optic based applications are the areas in which precise control of unwanted heat energy is of critical importance. Excessive heat can degrade the performance of EO sensors. Thin-film coatings of specific optical materials can separate the light from excessive thermal energy, thereby minimizing the heat-related performance degradation. Optical filters using such heat-absorbing coatings are called heat-absorbing filters. They offer improved transmission at visible wavelengths, while rejecting the thermal energy at IR wavelengths. When using such filters, it is recommended to remove the heat energy absorbed by circulating cold air in the vicinity of the filter. In the case of a system with intense focused energy, air cooling may not be sufficient, leading to a catastrophic failure of the optical system.

5.23.5 Dielectric Heat Rejection Filters (DHRFs)

A DHRF offers an alternative to critical cooling requirements for an IR-absorbing filter. DHRFs are made by depositing alternating layers of high- and low-refractive-index materials onto a substrate using a specified deposition technique [15] for optimum performance. This deposition technique increases the refractive index and reduces the porosity of the films, leading to increased stability and durability of the coating. The number and thickness of the individual layers depend on the optical signal wavelength to be reflected.

Optical and thermal properties of the coating materials determine the transmission efficiency and durability of the heat filter. The coating materials must have minimum absorption properties needed to reduce heating of the individual layers and thermal stress on the coatings. The design of a multistage DHRF must maximize both the visible-wavelength transmission and the IR reflection. An acrylic-based DHRF shows significant absorption below 375 nm and above 1000 nm, but offers a transmission of about 90% over the 400 to 800 nm range.

5.23.6 Dielectric Hot-Mirror Filters (DHMFs)

Combination of an absorbing filter and reflective dielectric coating filter produces a new type of filter called the DHM filter. This filter is comprised of a mirror coated onto a substrate material with low thermal expansion coefficient, such as borosilicate or fused silica. The filter can be placed very close to the source in order to reflect the near-IR energy, typically over 750 to 1200 nm range, with minimum reflected energy of 90%. The dielectric hot mirror, which is coated on a substrate of excellent thermal performance, is not an absorptive filter, so it can withstand more thermal energy than a pure absorptive filter. Hot mirrors reduce transmission efficiency at IR wavelengths, while maximizing it at visible wavelengths (400–750 nm). In summary, a combination of a DHM filter and IR absorbing filter can withstand much more heat energy than an IR absorbing filter alone. The combination filter design has potential applications in night-vision displays and other military night-vision sensors.

5.23.7 Tunable Filters

Tunable filters have potential applications in military imaging sensors and telecommunications systems. Several categories of tunable filters are available in the market; the most popular ones will be described here, with emphasis on performance and cost..

5.23.7.1 Fiber Optic Tunable Filters (FOTFs)

Fiber-optic-based tunable filters offer good performance at minimum cost and are widely used at telecommunication wavelengths. These filters come with typical bandwidths of 1 nm and 3 nm over the tuning range from 1290 to 1320 nm, and with 30 nm pass band over a specified spectral region at either 1310 nm or 1550 nm wavelength. These filters have maximum insertion loss of 1.5 dB, tuning resolution of 0.05 nm, back-reflection better than –50 dB, PDL of 0.03 dB at 1568 nm, center-wavelength stability of 0.005 nm/°C and operating temperature range of 0 to + 50 °C. FO-based tunable filters have potential applications in EDFAs for sideband noise suppression, tunable lasers, and wavelength selecting devices for optical tests and measurements.

5.23.7.2 Tunable Filters for Military and Commercial Applications

WDM is the clear choice for increasing the information-carrying capacity of an optical fiber. A tunable filter can select individual wavelength channels from a stream. Conventional demultiplexing schemes require several fixed-wavelength filters, leading to a heavy and costly filter package. A FO-based tunable filter for commercial or military application costs around $2000 in 1999 dollars, but will be available at much lower cost within next 5 years, if produced on a mass scale. A low-cost tunable filter will be found to most cost-effective for demultiplexing schemes.

The most critical use of this filter is in military high-power laser system applications. A tunable notch filter provides maximum eye and sensor protection from nearby operating high-power tunable lasers. A tunable notch filter will block out unwanted laser radiation, but will allow the sensor to conduct the necessary battlefield operations. Tunable filters will protect airline pilots and fighter aircraft pilots from enemy high power lasers intended to blind them. Performance capabilities and type of tuning filter determine the cost of tunable lasers.

Liquid crystal filters (LCFs) are getting more popular and are best suited for image detection and display applications because of low cost and lower power requirements. Fabry–Perot-based tunable filters will be used in WDM applications, when and if they move to the marketplace. Tuning is also possible through piezoelectric materials or electrooptic elements with a liquid or solid crystal in the optical cavity; however, the costs of such filters will be much higher than other filters discussed above. Acoustooptic (AO) tunable filters have demonstrated remarkable performance capabilities for microscopy and remote sensing applications, but they suffer from high cost, excessive weight, and very high power consumption. State-of-the art AO tunable filters have demonstrated transmission efficiency greater than 88%, tuning period less than 1 second, and tuning resolution less than 0.4 nm over the 400–500 nm spectral region. Their potential applications include underwater detection systems, optical imaging sensors, and astronomical equipment.

5.24 SUMMARY

State-of-the art performance parameters of IR passive devices and sensors using electrooptic technology are summarized. Passive devices discussed include optical fibers, nonlinear optical crystals, optical cavities and resonators, microlenses, fiber optic cables, liquid crystal displays, laser pointers, A/D converters using electrooptic technology, laser altimeters, optical isolators and circulators, switches, fixed-wavelength and tunable filters, fiber optic links, laser-based scanners and printers, delay lines using optical fiber technology, IR thermometers, and high-speed IR/digital cameras. Potential applications of these passive devices or sensors have been identified for military, commercial, or industrial applications. Degradation in performance parameters such as quantum efficiency, conversion efficiency, power consumption, and reliability under severe operating environments are indicated wherever deemed necessary. Tunable filters are described in greater detail because of their critical importance in telecommunications systems, high-power military sensors, and industrial equipment requiring high sensitivity, resolution, and reliability.

NUMERICAL EXAMPLE

Compute the conversion efficiency of a second harmonic generator (SHG) using the equation (5.6) and assuming various values of parameter x from 0.5 to 0.75, with an interval of 0.05.

The conversion efficiency of a SHG is given by equation (5.6), which can be written as $\eta_c = \tanh x$. The calculated values of conversion efficiency as a function of parameter x are shown in the table below.

Parameter x	Conversion efficiency (%)
0.50	29.85
0.55	37.60
0.60	46.81

Parameter x	Conversion efficiency (%)
0.65	57.80
0.70	70.96
0.75	86.81

Remarks: Typical conversion efficiency of a SHG varies from 40 to 75%, depending on optical crystal length, effective nonlinear coefficient at the phase matching angle, and pump intensity or optical pumping power.

REFERENCES

1. Newport Corp., Irvine, CA, Product Catalog, pp. 4–27.
2. K. J. McCallion, "Side-polished fibers provide functionally and transparency," *Optoelectronics World,* pp. S-19 and S-24, September 1998.
3. Contributing Editor, *Laser Focus World,* p. 13, December 1995.
4. G. Mitchard et al., "Double-clad fibers enable lasers to handle high power," *Laser Focus World*, pp. 113–115, January 1999.
5. Contributing Editor, *Laser and Optronics*, pp. 13–15, December 1994.
6. R. S. Adhav et al., "BBOs nonlinear optical phase-matching properties," *Laser Focus and Electrooptics*, pp. 88–100, September 1987.
7. *Electro-Optics Handbook*, p. 149, RCA, Commercial Engineering Division, Harrison, NJ, 1974.
8. Contributing Editor, "Digital-signal properties," *Laser Focus World*, pp. 37–40, January 1999.
9. Contributing Editor, "Altimeter Maps Mars' North Pole," p. 36, February 1999.
10. "Isolators protect fiberoptic lasers and optical amplifiers," *Laser Focus World*, pp. 147–151, November 1998.
11. Contributing Editor, "Optical switching promotes cure for telecommunications applications," *Laser Focus World*, pp. 69–71, September 1998.
12. K. P. Jackson et al., "Optical fiber delay-line signal processing," *IEEE Trans. on MTT, 39*, 7, 193–197, 1985.
13. K. Wetzel et al., "When digital cameras need large area pixels," *Optoelectronics World*, p. S-11, March 1999.
14. Contributing Editor, "Add/drop filters for IR applications," *Laser Focus World*, pp. 11–12, November 1998.
15. R. Cabrera, "Coating separates heat from light," *Laser Focus World*, pp. 177–180, January 1999.

CHAPTER SIX

IR Active Devices and Components

6.0 INTRODUCTION

This chapter describes the performance capabilities and limitations of active IR, photonic, and optoelectronic components and devices. State-of-the-art performance parameters of active IR devices and sensors for possible commercial, industrial, and space applications are summarized. Components and devices for military applications will be described in subsequent chapters. Critical performance requirements are identified for active IR components and devices including: electrooptic modulators; high-speed fiber optic communication links; optically controlled-phased array antennas; optical surveillance receivers; optical correlators; spectrometers; optical transmitters; optical time-stretchers for A/D applications; IR devices for micromachining involving drilling, cutting, soldering, and brazing; optical spectrum analyzers; and high-power devices for heavy industrial applications. Performance degradation of IR devices or sensors used by telecommunications and space systems under severe operating environments will be identified, with emphasis on reliability.

6.1 ELECTROOPTICAL MODULATORS

Electrooptical (EO) modulators are the critical elements of high-speed, long-distance optical data transmission systems, high-speed A/D converters (ADCs) for complex signal processing in EW and radar systems, high-data-rate telecommunications systems, space-based sensors, and sophisticated high-technology military systems. EO modulators employ birefringent optical crystals, polarizers, and other optical devices needed for polarization of the propagating electric wave signal. EO modulators can operate based on electrooptic effect, acoustooptic effect, or magnetooptic effect, depending on the materials used. Various modulation architectures will be described, but EO modulators using electrooptic materials will be described in greater detail.

6.1.1 Modulator Designs Based on Operating Principle and Device Architecture

Optical modulators are classified based on their operating principle and device architecture. Acoustooptic modulators, magnetooptic modulators, and electrooptic modulators are classified based on operating principles or technologies used in their designs, whereas waveguide modulators, bulk rod modulators, traveling-wave (TW) modulators, fiberoptic coupler modulators, semiconductor modulators, and interferometric modulators are classified based on the type of structures used in their architectures. However, selection of a modulator depends on application, performance specifications (including bandwidth and drive voltage requirements), cost, and complexity.

Ti:$LiNbO_3$-based optical modulators using traveling-wave electrodes are very popular. The modulators can be designed with various electrode configurations such as coplanar waveguide (CPW) electrodes with wider center conductor and narrow gap, asymmetric coplanar strip electrodes, and inverted slot-line electrodes, all shown in Figure 6.1. The shielded velocity-matched (SVM) design [1] with ridge electrodes is shown in 6.2. The bulk rod modulator design shown in Figure 6.2 is not attractive for many applications because of high voltage drive requirement and limited bandwidth capability. Semiconductor-based multiple quantum well (MQW) modulators using waveguide strips are best suited for applications involving large signal operating conditions. Regardless of modulator types, low drive voltage and large instantaneous bandwidth are considered the most critical requirements of a modulator. In addition, all efficient modulators require excellent microwave–optical velocity matching and with minimum electrical leakage to suppress undesirable optical modes.

6.1.2 Traveling-Wave (TW) Modulators

Waveguide modulators were designed several years ago and have not used the latest technologies. Among all the modulators, EO modulators are best suited for applications where high-speed data transmission over long distances is the principal requirement, such as transoceanic lines carrying data at ultra-high rates. However, these lines occasionally suffer from jitter, dispersive waves, and spontaneous emission noise. These shortcomings can be overcome with in-line synchronous modulation techniques. When a transmission of 40 G bit/sec signal or higher is required in a single channel over long distances exceeding 5,000 km, the signal must be treated with solitron control and narrowband filtering to reduce jitter and bit error rate (BER). The solitron control function is provided by an electrooptical modulator using a lithium niobate rod with minimum cost but limited bandwidth.

The TW modulator overcomes these problems with minimum cost and complex-

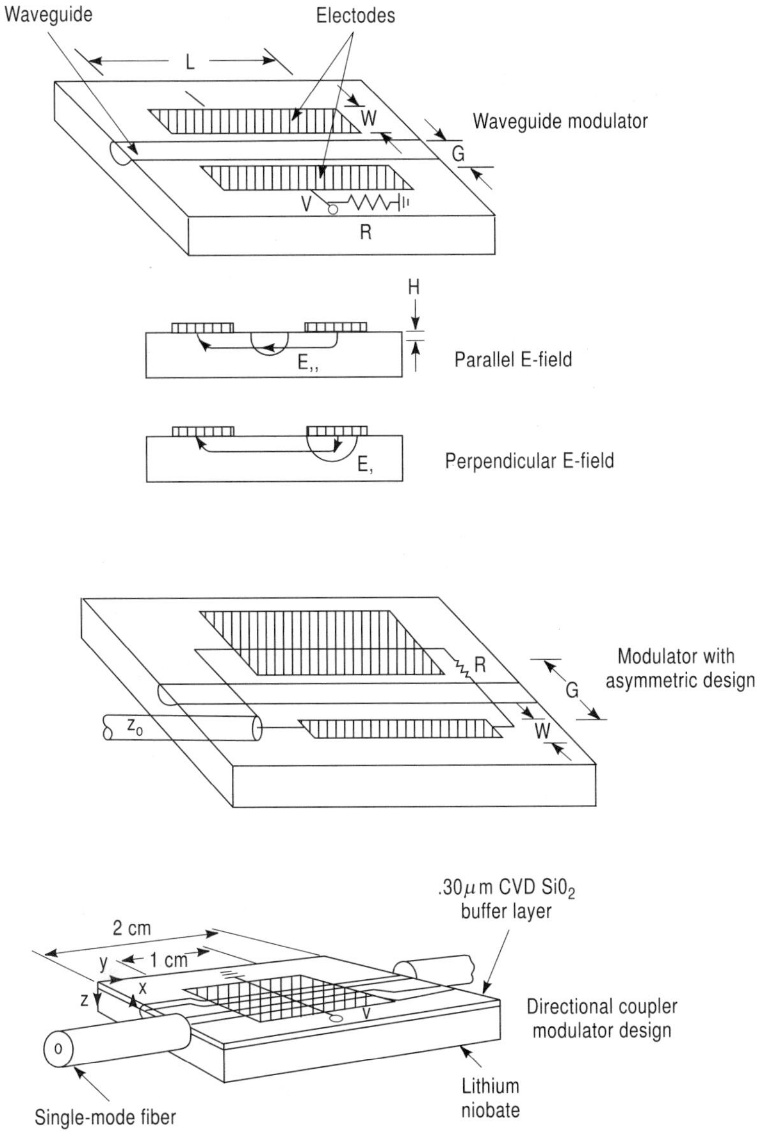

FIGURE 6.1 Electrooptic modulator configurations using coplanar strip traveling-wave electrodes and fiber optic directional couplers.

ity because the latest structure of a TW modulator provides both velocity matching and impedance matching simultaneously over the modulation bandwidth [2]. Modulation bandwidth is the most critical performance parameter; it is limited by impedance matching and velocity dispersion. Different power levels are required to modulate the signals from various laser wavelengths [3]. Bandwidths as high as 40

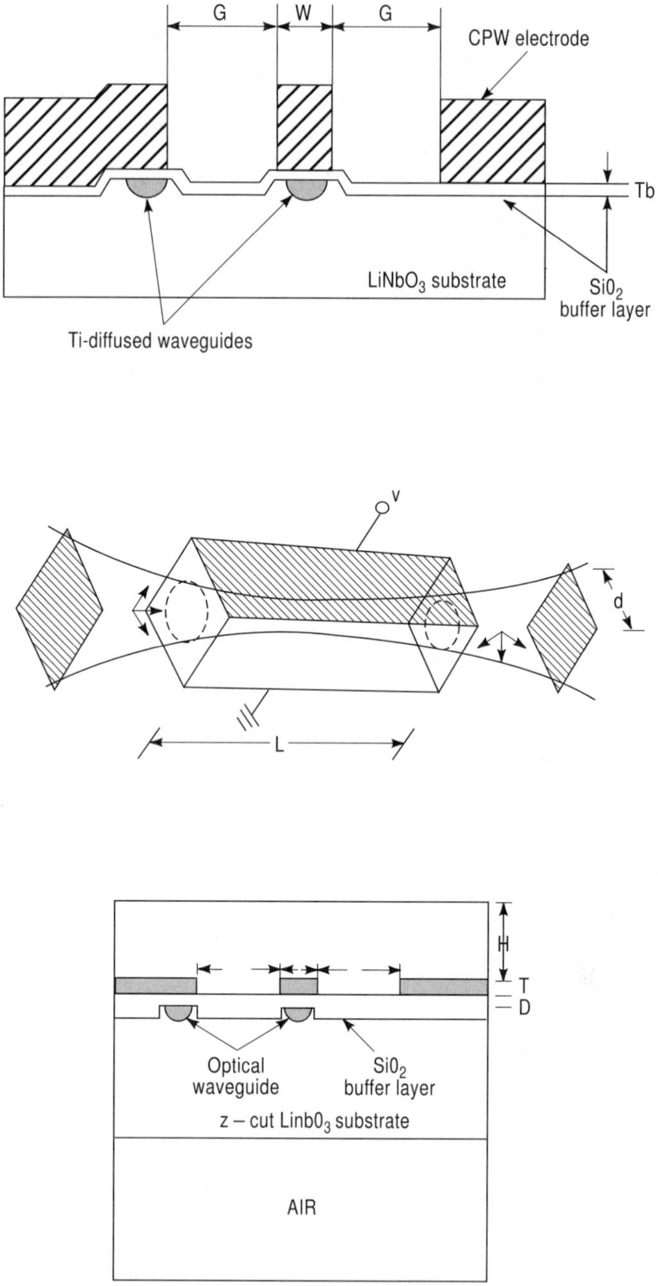

FIGURE 6.2 EO modulator designs using bulk rod and Ti:$LiNbO_3$.

GHz can be achieved with 1 cm modulator length using small cross-sectional dimensions and parallel-strip line structure.

The performance of a velocity-matched, traveling-wave modulator is dependent on microwave effective index, spacing between the electrodes (known as electrode gap, G), and electrode width (W). Computed values of capacitance per unit length (pF/cm) and bandwidth–length product (GHz·cm) as a function of gap-to-width (G/W) ratio for a symmetric strip electrode on lithium niobate material are shown in Figure 6.3. Capacitance per unit length (pF/cm) and characteristic impedance of the

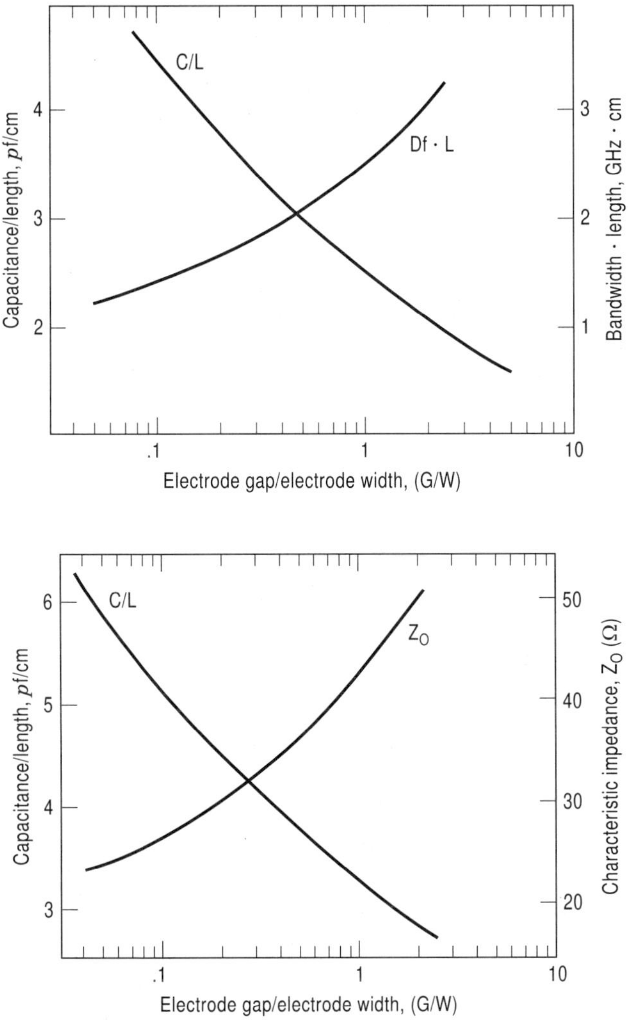

FIGURE 6.3 Capacitance, characteristic impedance, and bandwidth–length product as a function of electrode gap-to-width ratio.

electrode (Z_0) as a function of (G/W) ratio are also depicted in Figure 6.3. The curves shown in Figure 6.3 indicate that both high impedance and large bandwidth require a high (G/W) ratio. A high-impedance electrode would require high drive voltage, leading to high power consumption, which may be not be acceptable in some applications.

Other techniques are available for velocity matching without compromising the bandwidth capability. Optical modulators using inverted slot line (ISL) structure configuration offer a unique method for perfect velocity matching by reducing the index diameter to the optical refractive index equal to that of $LiNbO_3$, which is around 2.2. In the case of ISL electrode configuration used by the TW modulator, the effective refractive index decreases with decrease in spacing between the electrode and the ground plane. The spacing is dependent on modulation frequency and the cut of the substrate material. For a 5 GHz modulator, the spacings required for perfect velocity matching are 25.6 μm and 17.2 μm for Y-cut and Z-cut $LiNbO_3$ substrates, respectively.

The characteristic impedance (Z_0) of the ISL electrode on Y-cut lithium niobate substrate increases monotonously with increase in modulation frequency. Its value as a function of frequency is about 32 ohms at 40 GHz, 35 ohms at 50 GHz, and 37 ohms at 60 GHz modulation frequency. Low insertion loss, high modulation efficiency at MM wave frequencies, and perfect velocity matching between optical and modulation signals are possible with ISL configuration because of larger inherent spacing between the electrode and ground plane [4]. The velocity matching technique has been also demonstrated in a TW wave, direct coupled, intensity modulator design.

6.1.3 Semiconductor-Based TW Modulators

The semiconductor-based TW modulator (SBTWM) uses a thin coating of tantalum pentaoxide (Ta_2O_5) of high dielectric constant ($e_r = 27$) at microwave frequencies and low refractive index at near-IR wavelengths ($n = 2.03$ at 830 nm) to achieve good velocity matching between the optical and microwave signals. This particular modulator can operate at 830 nm with a maximum derive voltage of 5 V with a figure-of-merit (FOM) exceeding 100 GHz/V, which represents the state-of-the art performance to date of a TW modulator.

6.1.4 Shielded Velocity-Matched (SVM) EO Modulators

Ti:NbO_3-based SVM EO modulators using ridge electrode structure [5] have demonstrated significantly improved performance over other modulator designs. High-speed optical transmissions systems require large modulation bandwidth and low drive power, which are possible with external optical modulators that use titanium-diffused lithium niobate substrates and TW electrodes. As stated earlier, the modulation bandwidth is strictly dependent on velocity matching between the optical and microwave signals. However, in this modulator design, microwave channels of the TW electrodes determine the modulator bandwidth performance. Wide-band modulation capability can be achieved either by using the grooves between the optical guides, by employing thick electrodes, or by incorporating ridges into the SVM-CPW structure. The bandwidth performance of a SVM optical modulator design depends on electrode thickness. Because a very thick overlaid layer corresponds to a conventional CPW structure with a thick electrode and without a shielded plane, one will observe that a shielded plane has the same effect as a thick electrode. The overlaid layer thickness reduces the microwave effective index and the Z_0 of the TW electrodes, leading to a significant increase in modulation bandwidth, as illustrated in Figure 6.4. In summary, it can be concluded that a SVM EO modulator configuration incorporating a ridge into a SVM structure and using CPW TW electrodes offers lower microwave effective index, lower drive voltage (less than 3.5 V), and significantly wider modulation bandwidth.

6.1.5 EO Modulators Using Bulk Rod Concepts

Critical elements of an EO modulator using a bulk rod are shown in Figure 6.2. The optical beam passes through the rod volume of dimensions $h \times w \times L$, as illustrated in Figure 6.2. For a focused Gaussian beam, the minimum geometrical factor is limited by the optical diffraction to a value defined below:

$$d^2L = (S)^2(4\lambda/\pi n) \tag{6.1}$$

where S is a factor with a unity value for a crystal having the Gaussian diameter at each end of the crystal, equal to d, and n is the refractive index of the crystal material. Geometrical factors can be reduced without limiting the optical waveguide structure because the diffractive limitation does not apply. A good modulator design requires efficient interaction between the optical and microwave fields. Preliminary calculations using equation (6.1) indicate that relatively higher drive power levels are required because of poor coupling efficiency. A TW bulk electrooptical modulator using 1 cm Z-cut $Li{:}NbO_3$ material demonstrated a bandwidth of 4 GHz with a drive voltage of 0.5 V. This modulator offered a low FOM of 8 GHz/V with a power consumption greater than 5 mW. Because of high power consumption and limited bandwidth capability, this particular EO modulator is used in applications where power consumption and bandwidth are not the critical performance requirements. However, bulk modulators employing miniature guiding structures and integrated optics can operate with much less modulating power but with higher cost and complexity.

6.1.6 Acoustooptic (AO) Modulators

The operating principle of an AO modulator is based on the phase grating created by an acoustic wave through a photoelastic effect. This effect can either diffract a light beam by several orders, as in the Raman–Nath regime of operation, or deflect a light beam into a single order, as in the Bragg regime. In either regime, intensity modulation of moderate bandwidth is easily achieved without regard to the polar-

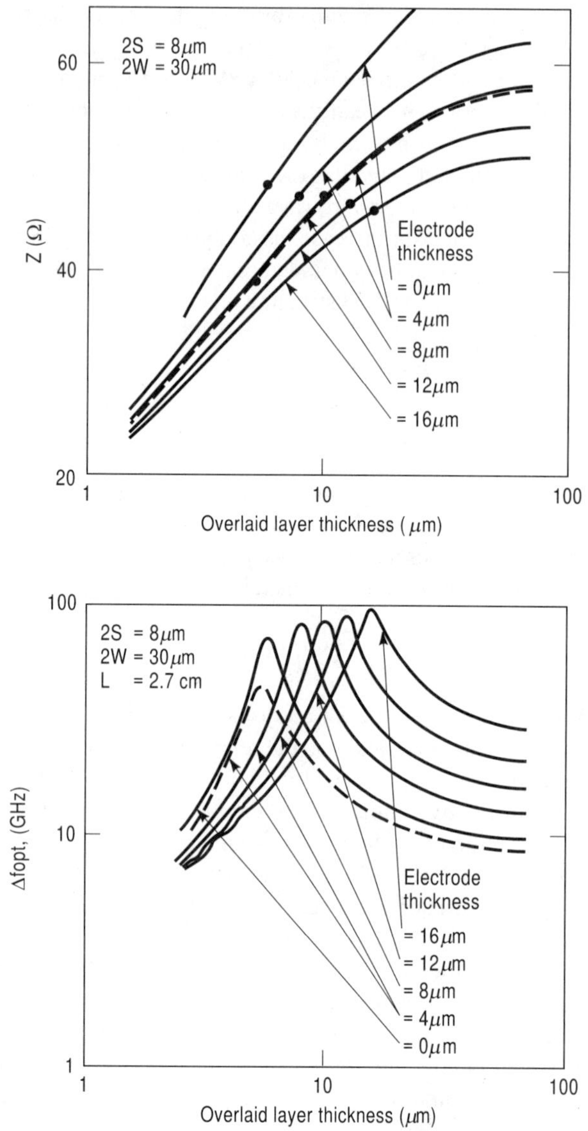

FIGURE 6.4 Characteristic impedance and optical 3 dB bandwidth as a function of electrode and overlaid layer thickness.

ization of the incident light. The modulation bandwidths of the AO modulator is limited to about 2 or 3 GHz. Lithium niobate is considered the most suitable acoustic material for the design of an AO modulator. The surface-wave acoustooptic has potential application in the development of guided-wave devices. Deflection and diffraction of optically guided waves by surface-acoustic waves (SAW) have been achieved in a lithium niobate thin film. High-speed AO deflectors are particularly suitable for time division multiplexing (TDM) and demultiplexing of optical digital signals.

6.1.7 Magnetooptic (MO) Modulators

Iron garnet crystals are most suitable for the design of efficient MO modulators. MO modulators are particularly useful at 1.52 μm operation because of low loss at 1.52 μm compared to high optical loss at 1.06 μm at room temperature. Because of high optical losses, MO modulators using Nd:YAG laser sources at 1.06 μm operation are not recommended. Comprehensive investigation of several magnetic iron garnet materials indicate that certain materials containing praseodymium or neodymium will allow modulation operations at 1.06 μm with reduced loss less than 3 dB and with a power-to-bandwidth ratio as low as 0.01 mW/MHz or 10 μW/Hz, which appears too high compared to other modulators. However, MO modulators using a thin film of high-performance iron garnet offer possibilities for high-speed switching and modulator applications using a light source at 1.15 μm.

6.1.8 Semiconductor Waveguide (W/G) Modulators

A semiconductor W/G modulator design involves a waveguide electrode geometry appropriate for a GaAs or InP semiconductor-based phase modulator. The electrical field is developed across either a diffused PN junction or a Schottky barrier. Its efficient design requires the voltage–length product to be as small possible, approaching the depletion region dimension of 1 μm needed for optimum performance. By optimizing the geometrical parameters, a GaAs W/G phase and polarization modulation operation have been demonstrated with a voltage–length product of 1.1 V·cm at an operating wavelength of 1.06 μm.

Because of the rectifying junction required for a semiconductor-based W/G modulator, the bandwidth analysis is more complicated than for insulated substrates. The depletion layer thickness depends on the applied voltage for both the PN junction and Schottky barrier. As a result, the capacitance depends on the drive voltage and time-varying signal parametric effects, which results in harmonic distortion of the optical modulation. In a GaAs-based modulator with proper W/G fabrication, the second harmonic distortion has been reported below 5% for modulation depth. Because of small velocity mismatch for both GaAs and InP, semiconductor-based TW modulators using these materials seem to offer improved performance. Relatively lower dielectric constants of these materials will allow smaller gap-to-width (G/W) ratio for a given impedance level in the lithium niobate material. Difficulty to realize high-quality strip transmission lines on these substrates and large optical losses in semiconductor waveguides preclude the use of a long modulator to take full advantage of the small velocity mismatch of this modulator architecture.

6.1.9 Magnetostrictive (MS) Light Modulators

In these modulators, the light is conducted via a multiple-reflector assembly through a germanium internal reflection element, which is optically transparent to infrared radiation wavelengths exceeding 2 μm. The reflectivity can be continuously controlled by controlling the separation between the interfaces. A magnetostrictive device controls the separation of two media. The absorbing medium consists of a MS ferrite block with a hole through the middle and solenoids wound around the resulting legs. A DC bias current is applied to the solenoid for fine adjustment of the distance separating the germanium reflector element and the ferrite block. An AC current is applied to the solenoid to expand and contract the ferrite block, causing modulation of the light absorption. The performance data on this modulator indicate a modulation depth of 15%, 25%, and 35% at an operating wavelength of 2.5 μm, 5.0 μm, and 6.5 μm, respectively. This particular modulator is best suited for low-frequency chopper modulation applications, where mechanical choppers not desirable. Modulation amplitudes close to 100% are possible with MS modulators using high-quality flat surfaces.

6.1.10 Properties of Electrooptic Materials Used for Modulators

Table 6.1 summarizes the important properties of potential electrooptic materials for possible applications in electrooptical modulators. The parameters n^3/r and Δn are of critical importance. Neglecting the higher-order terms, the change in the refractive index (Δn) is proportional to the cube of the refractive index (n), inversely proportional to the tensor coefficient (r), and directly proportional to the applied voltage (V). For an applied voltage along the z-axis or the longitudinal axis of the EO modulator, the electrooptically induced refractive index approaches the extraordinary index (n_e) and strong diagonal coefficient (r_{33}). Only room temperature properties are shown in Table 6.1. Note that n is the refractive index of the material, r is the linear tensor coefficient, and Δn is the change in refractive index due to application of voltage. The other parameters were defined earlier.

In the case of acoustic materials, the coupling coefficient and insertion loss per unit length are the most critical parameters that determine the performance of AO modulators. Both these parameters are functions of propagation direction or cut (X-cut, Y-cut, or Z-cut) and acoustic mode (longitudinal wave or shear wave). High values of coupling coefficient and low values of insertion loss are most desirable for AO modulators. Examination of acoustic materials such as lithium niobate, lithium

TABLE 6.1 Room temperature properties of electrooptic materials

Materials	n^3/r, 10^{-12}mV	EO coefficient, r_{ij}	λ, nm	n_3	Δn	e_r	Loss, dB/cm
Lithium niobate	328	$r_{11} = 16$	4000	2.05	0.010	43	0.3
Lithium titanate	—	$r_{33} = 27$	4000	2.03	0.002	—	—
Gallium arsenide	49	$r_{41} = 1.2$	1250	3.42	0.010	12.5	0.4

titanate, and zinc oxide indicates that lithium niobate offers the highest coupling coefficients and minimum insertion loss at 1 GHz. It is important to mention that the electric-to-acoustic conversion efficiency is proportional to coupling coefficient. Dominant properties of AO materials are summarized in Table 6.2.

6.2 OPTICAL RECEIVERS

Optical receiver performance requirements are nearly similar to those for microwave or MM-wave receivers (Table 6.3). The ultimate performance of an optical receiver is limited by the noise fluctuations that are present at the receiver input. The noise degrades the S/N ratio, which adversely affects the receiver performance. Critical performance parameters of an optical receiver include sensitivity, dynamic range, detection criterion, amplifier noise figure, coherence of the input signal, shot noise, dark-current noise, quantum noise, and the surface-leakage current noise. Some of these parameters are of critical importance because they largely determine the overall receiver performance level. However, photo detectors and low-noise amplifiers are the major components that can limit the performance of an optical receiver.

6.2.1 Optical Receiver Using the Direct-Detection Technique

Receiver noise is the most critical performance parameter; it depends strictly on the method of demodulation and the electronic components and circuits used by the receiver. Two methods of demodulation of optical signals are generally used, namely, direct detection method and heterodyne detection method. The direct detection method is simple to implement and least expensive, because it does not depend on signal coherence and polarization state. In a heterodyne detection system, an incoming optical signal is mixed in the detector along with a coherent signal from a local oscillator (LO) to produce a difference frequency from which information is extracted. The heterodyne detection method, which is applicable to single-mode transmission systems, is not cost effective because stable single-frequency laser

TABLE 6.2 Properties of AO materials

Material	Acoustic mode (001 axis)*	Coupling coefficient (K_c)	Attenuation (dB/cm) @ 1 GHz
$LiNbO_3$	BAW (L)	0.49	0.10
	BAW (S)	0.68	0.15
	SAW	0.21	3.00
$LiTaO_3$	BAW (L)	—	0.10
	SAW	0.12	1.10
ZnO	BAW (L)	0.28	7.50

*BAW stands for backward acoustic wave, SAW for surface acoustic wave, L for longitudinal mode, and S for shear mode.

TABLE 6.3 Performance comparison of integrated optical receivers (or spectrum analyzers) with other receiver designs for RF spectrum surveillance application

	Sensitivity	Dynamic range	Frequency measurement accuracy	Signal separation capability	Acquisition time	Processing complexity	Simultaneous emitter capability	Probability of detection	Dense environment capability	Production cost
IO RF spectrum analyzer	High (–80 dBm)	Moderate (40 dB)	Excellent (1–5 MHz)	Good	Excellent (1–10 μsec)	Simple	Good	High	Excellent	Low ($2 K)
IFM	Low (–75 dBm)	Fair (30 db)	Fair (1 MHz)	Poor	Moderate (2–10 μsec)	Moderate	Poor	High	Poor	Medium ($18 K)
Channelized	High (–79 mm)	Moderate (40 dB)	Fair (50 MHz)	Fair	Fair (20–100 μsec)	Simple	Good	High	Fair	High ($30 K)
Superheterodyne	High (–82 dBm)	Good (55 dB)	Excellent (0.2%)	Good	Poor (4 sec)	Moderate	Moderate	Low	Good	Medium ($15 K)
Compressive	High (–89 dBm)	Fair (30 dB)	Excellent (1 MHz)	Good	Excellent (9 μsec)	Complex	Moderate	High	Good	High ($25 K)

sources are required. On the other hand, the direct detection method shown in Figure 6.5 is widely deployed in fiber-based optical communications systems because of minimum cost and complexity.

Various noise sources associated with avalanche photo detectors (APDs) and amplifiers are identified in Figure 6.5. The background radiation noise is negligible in a fiber system. The beat noise from various spectral components of an incident carrier is insignificant when a large number of modes are transmitted and received. The quantum noise, dark-current noise, and surface-leakage current noise all contribute to shot noise. The quantum noise due to fluctuations in the photo excitations of the carriers imposes an ultimate limit on the optical receiver sensitivity. Despite the var-

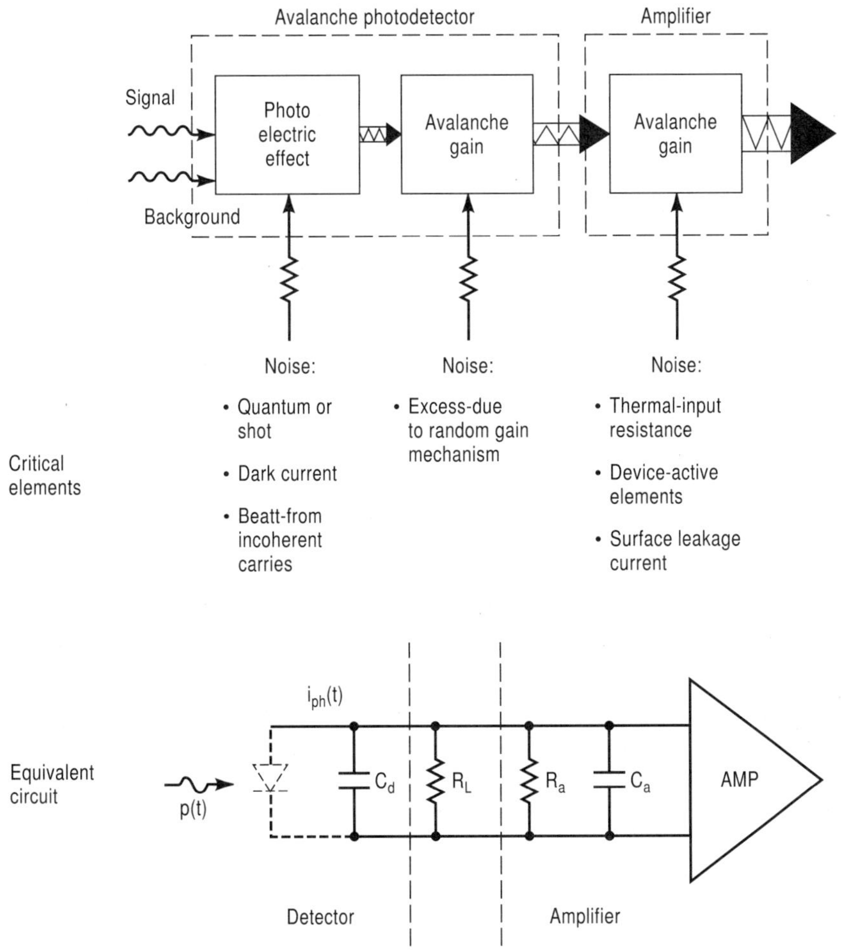

FIGURE 6.5 Block diagram and equivalent circuit of an optical receiver based on the direct detection concept.

ious noise contributions, avalanche multiplication of the photo carriers does improve the sensitivity of the receiver.

6.2.1.1 *Effects of Detector Materials on Receiver Sensitivity*

The absorption coefficient of the semiconductor detector materials is strictly dependent on the wavelength, whereas the absorption of a large percentage of the incident light is a function of the detector materials and the operating wavelength. It is important to mention that the avalanche photo diode (APD) combines the detection of optical signals with internal amplification of the photo current and provides a gain–bandwidth product close to 100 for various silicon and germanium photo diodes. Silicon APDs with rise times of 0.32 ns or 320 ps and quantum efficiency of 85% are widely used for operations at 1060 nm. On the other hand, a germanium APD exhibits its fastest response time of 0.1 ns or 100 ps, but requires moderately high voltage (200–400 V) and thermal stabilization circuitry to maintain constant gain. These requirements plus the high cost of the APD itself must be seriously considered in the design phase of an optical receiver where cost is the most demanding requirement.

6.2.2 Channelizer Optical Receiver Using Acoustooptic Technology

Block diagrams of channelizer receivers using acoustooptic (AO) technologies such as SAW, BAW, and BRAGG are shown in Figure 6.6. An AO-based channelizer receiver is capable of detecting and sorting out optical signals in bands of specific spectral bandwidths. This receiver may employ various transducer devices, such as BRAGG cell devices, surface-acoustic-wave (SAW) devices, or bulk-acoustic-wave (BAW) devices, and an array of several optical detectors, depending on the performance requirements of the optical receiver. Performance capabilities of various channelizer receivers are summarized in Table 6.4.

Critical elements of an AO-based channelizer receiver using BRAGG cell technology are depicted in Figure 6.6. The incoming RF signal is converted down to a convenient IF signal and transduced to acoustic signals in a wide-band acoustic wave device called a BRAGG cell. The acoustic signals modify the refractive index of the BRAGG cell and diffraction gratings are formed that diffract the collimated incident laser beam. Acoustic waves propagate in the BRAGG cell. The laser beam is expanded and collimated to achieve the beam dimensions necessary for sufficient interaction time between the acoustic signal and collimated optical signal. The beam is then passed through a spatial filter for channel passband shaping. The passive optics are designed to prevent scattering and defocussing optical signals, leading to an acceptable dynamic range.

The acoustic wavefronts are orientated to interact with the collimated light at the BRAGG angle [6] defined as follows:

$$\sin \theta_B = \frac{\lambda_0}{2\lambda_a n} \qquad (6.2)$$

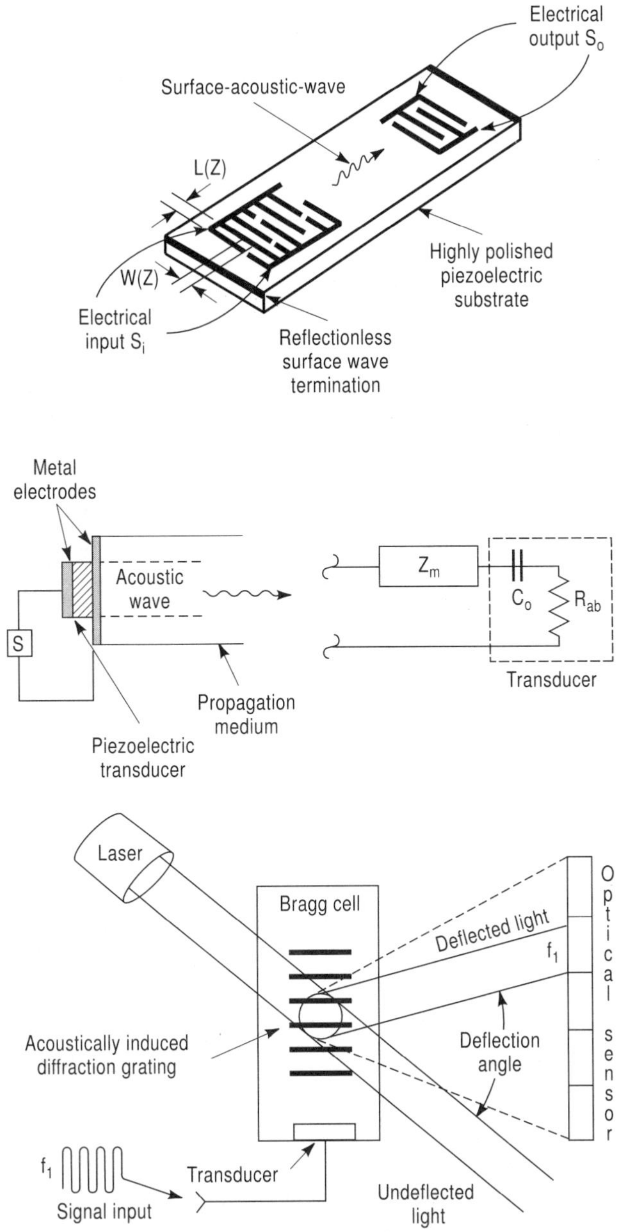

FIGURE 6.6 Block diagram of channelizer receivers using SAW, BAW, and BRAGG cell technologies.

TABLE 6.4 Performance capabilities of various channelizer receivers

Device and technology	Bandwidth (GHz)	Time–BW	Dynamic range (dB)	Size (in^3)	Cost
Filter	2	100	60	100	High
SAW	0.8	25	50	100	Medium
BAW	1	50	20	10	Medium
AO	2	1000	60	10	Medium

where λ_0 stands for optical wavelength, λ_a stands for acoustic wavelength, and n stands for refractive index of the material. Computed values of BRAGG angle for a $LiNbO_3$ ($n = 2.2$) cell as a function of acoustic wavelength and optical wavelength are summarized in Table 6.5.

It is important to mention that lower values of IF frequency are required to achieve small values of Bragg angle and vice-versa. The refractive coefficient of the acoustic material is dependent on temperature and optical wavelength. Here the assumed value of refractive index equal to 2.2 at both wavelengths of 0.85 and 1.06 μm may not introduce large errors in BRAGG angles because of narrow separation between them. However, if the two optical wavelengths are wide part, then the values of refractive index must be used at their respective wavelengths. The optical energy resulting from a given acoustic signal is diffracted at an angle that is twice the Bragg angle from the collimated beam. Since the Bragg angle is a function of acoustic wavelength and optical wavelength, signals applied at frequencies result in spatially diverse optical deflectors. The angle by which the optical beam is diffracted for a given IF acoustic signal is proportional to the frequency of the IF acoustic signal. However, the pixel size of the photo detector array is determined by the acoustic IF frequency. The deflected beam carries the frequency, amplitude, and time-of-arrival (TOA) of the RF signal. The Bragg angle provides the precise information on signal frequency, TOA (time-of-arrival), and AOA (angle-of-arrival), which are critical in radar and EW systems.

The optical signal at the photo diode array is the instantaneous Fourier transform (FT) of the acoustic signal because of the FT lens between the acoustic wave and the photo diode array. Frequency channelization is achieved because of the frequency dependence of the diffraction angle. This method provides simultaneous process-

TABLE 6.5 Magnitudes of BRAGG angle (degrees)

		Optical wavelength (μm)	
Frequency (MHz)	λ_a (μm)	0.85	1.06
200	17.5	0.63	0.79
400	8.75	1.26	1.58
600	5.83	1.90	2.37
800	4.38	2.53	3.16

ing of incoming signals without loss of information because of the AO channelization effect.

The phase and frequency of the input IF signals are also modulated onto the diffracted light; this can be used for subsequent coherent signal processing to achieve high resolution, which is of critical importance for side-looking radar and electronic warfare systems. This type of receiver is both complex and costly because it requires a stable coherent laser source and well-designed acoustic transducer. The lateral acoustic beam spread is determined by acoustic transducer aperture size. Since the angular beam spread of the light source is extremely narrow (on the order of few microradians), the range of angles and corresponding frequencies over which the conditions for Bragg angle are satisfied is determined strictly by the acoustic beam spread.

If the acoustic beam spread increases, the diffraction efficiency decreases because the percentage of acoustic power at the precise angle for Bragg interaction is totally beam-spread dependent. A trade-off may be required between the diffraction efficiency and the bandwidth of the Bragg cell to optimize the receiver performance. This trade-off must use the microwave-based acoustic material's parameters, such as bandwidth capability, sensitivity, transducer dynamic range, and resolution, as shown in Table 6.2. Transducer materials [7] must have low insertion loss, wide dynamic range, high electric-to-acoustic conversion efficiency (which is proportional to effective coupling coefficient), high power handling capacity, and low-level spurious responses. Furthermore, detector parameters must be included in the trade-off studies. Detector sensitivity depends on signal interaction time, detector material, operating temperature and wavelength, and the input signal level. Studies performed by the author indicate that the quantum-limited noise floor of an optical receiver using a BRAGG cell is about 20 dB higher than the thermal noise floor of an electronic channelized receiver.

The optical signal processing concept based on BRAGG cell technology is best suited for applications where high resolution, wide dynamic range [7], high probability of intercept (POI), high-density electromagnetic environments, and instantaneous signal intercept characteristics are the principal requirements. In summary, such an optical receiver is best suited for applications in high-performance side-looking radar surveillance sensors and EW systems.

6.2.3 Surveillance Receivers Using Integrated Electrooptic Technology

A microwave surveillance receiver (MSR) using integrated electrooptic technology is similar to an integrated optics version of an optical RF spectrum analyzer. The MSR uses a coherent laser source at 0.85 μm, SAW transducer, detector array, and a reference cell. The incoming RF signal is down-converted, amplified, and applied to the SAW transducer. The laser beam is collimated and the guided coherent optical beam intersects the acoustic beam at the Bragg angle. A portion of the peak is deflected at angle θ_d equal to twice the Bragg angle (θ_B), which is proportional to the incoming RF frequency. The 3 dB bandwidth of the electrooptic-based RF surveillance receiver can be express as

$$\left(\frac{\Delta f}{f}\right)_{3\text{dB}} = \frac{(2n)(\lambda_a)}{(\alpha)(\lambda_o)} \tag{6.3}$$

where α is a beamwidth-related factor in terms of acoustic wavelength with a typical value of about 48, λ_a is the acoustic wavelength, λ_o is the optical wavelength, n is the refractive index of the AO material, f is the down-converted frequency, and Δf is the 3 dB RF bandwidth. Assuming $LiNbO_3$ ($n = 2.2$) as the AO material, factor α of 48, f of 600 MHz, optical source wavelength of 0.85 μm, and acoustic velocity of 3.5×10^5 cm/sec, the calculated value for the 3 dB bandwidth for the MSR comes to 62.9%, which is better than an octave bandwidth. It is evident from equation (6.3) that higher 3-dB bandwidths are possible with high refractive index materials and coherent laser sources operating at lower wavelengths.

The diffraction efficiency of the MSR is a function of input acoustic power level, acoustic-to-optical FOM, overlap integral parameter (I_{mn}), acoustic aperture size (L), refractive index of the AO material, and optical wavelength (λ_o). The diffraction efficiency for a MSR receiver with given FOM and acoustic power can be given as

$$\eta_d = \sin^2[(\pi/2\lambda_o)(n)^3(I_{mn})(L)] \tag{6.4}$$

It is evident from this expression that higher diffraction efficiency for a given material is possible with larger acoustic apertures and lower optical wavelengths.

6.2.3.1 *RF and Optical Resolution Capabilities*

It is important to mention that optical detector size (S) and the optical beam diameter (D) determine the RF resolution (Δf), whereas the maximum permissible crosstalk between the adjacent detector cells determines the optical resolution limit. The RF resolution and optical resolution limits require that a complicated variable V satisfies the following conditions:

$$V = \frac{SN_{\text{eff}}}{\lambda_o F} < \frac{\Delta f}{V_{\text{SAW}}} \quad \text{for RF resolution limit} \tag{6.5}$$

$$> \frac{g}{D} \quad \text{for optical resolution limit} \tag{6.6}$$

where N_{eff} is the effective refractive index of the material, g is the parameter depending on optical spot size (typical values between 1.21 and 2.48), F is the optic constant, S is the detector cell size, V_{SAW} is the SAW phase velocity, and λ_o is the optical source wavelength. A graphical display of the parameter V, defined by equations (6.5) and (6.6) as a function of RF resolution (Δf) and optical beam diameter (D) for an optics-based MSR system, is shown in Figure 6.7. The RF resolution can be obtained from the straight line, whereas the optical beam resolution can be determined from three hyperbolic lines designated by three distinct values of parameter g. Good RF resolution is possible over large bandwidth with the BRAGG modula-

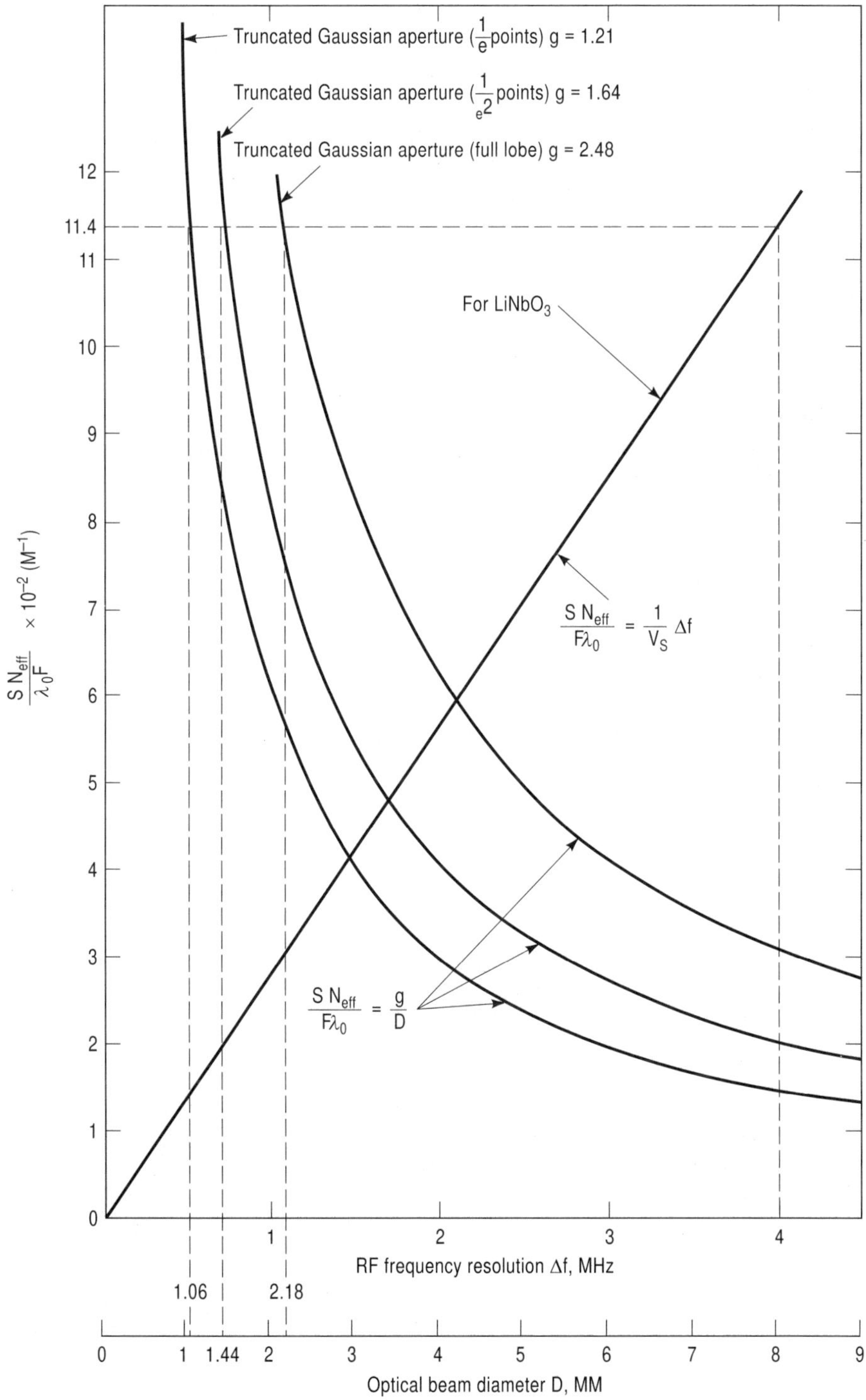

FIGURE 6.7 Graphical representation of relationship between detector cell size (S), RF frequency resolution (df) and beam diameter (D).

tion technique using several tilted transducers designed for optimum Bragg angle. Preliminary computations indicate that for a 4 MHz RF resolution, one requires a minimum beam diameter (D_{min}) of 1.06, 1.44, and 2.18 mm, respectively, corresponding to $(1/e)$, $(1/e)^2$, and full lobe spot size. The corresponding adjacent channel cross-talk (not shown Figure 6.7) will be about –24.5 dB, –26.5 dB, and –27.5 dB, respectively, for a (D/D_{min}) ratio of unity. However, if D is 6.54 mm and D_{min} is 2.18 mm, the adjacent channel cross-talk level will be roughly –33 dB for the full main beam case.

Estimated values of optical beam diameter for various RF resolutions are summarized in Table 6.6 and these values are valid for a MSR system using a coherent laser source operating at 0.85 μm and lithium niobate Z-cut substrate. This particular MSR will require about 500 mW at the output of the IF amplifier with a net acoustic power level of about 17 mW, assuming an overall insertion loss of 15 dB. The IF amplifier in the MSR can be easily adjusted to meet RF input drive power and LO power levels.

6.2.4 Optical Spectrum Analyzer (OSA)

An optical spectrum analyzer (OSA) is identical to a RF spectrum analyzer as far as its principal function is concerned, except that it analyzes only the optical signals over the specified optical spectrum. The OSA has potential applications in wavelength-division-multiplexing (WDM) systems, dense-WDM systems, telecommunications systems, optical receivers, and other optical sensors requiring optical spectral analysis. The OSA is particularly best suited for WDM system analysis to obtain complete information on wavelength, optical power level, and S/N ratio for each channel with high wavelength accuracy and low polarization error needed for accurate measurement of optical amplifier gain and other critical parameters. This analyzer can directly measure optical power levels up to 100 mW without using an optical attenuator. It has a built-in high-speed printer and large color LCD display with several interface provisions for keyboard, mouse, video, printer, and so on. The OSA calculates and displays the optical parameters on a WDM device's peak wavelength, channel spacing, peak power level, and S/N ratio for up to a hundred channels from the spectrum of the transmitted signal. Critical performance parameters such as gain and noise figure of the optical fiber amplifiers can be evaluated with high accuracy, reliability, and repeatability. This instrument can evaluate the notch width of the transmitted spectra produced by the fiber grating with high accuracy.

Comprehensive review of current OSA technology indicates that it is possible to

TABLE 6.6 Optical beam diameter requirements (mm) for various RF resolutions

RF resolution (MHz)	V (meter)	$g = 1.21$ $(1/e)$	$g = 1.64$ $(1/e)^2$	$g = 2.48$ (full lobe)
1	285	4.24	5.75	8.70
2	575	2.10	2.85	4.31
4	1140	1.06	1.44	2.18

measure wavelengths over the 1200 to 1700 nm range with an accuracy better than ± 0.3 nm, wavelength linearity of ± 0.02 nm, wavelength reproducibility of ± 0.05 nm, wavelength resolutions of 0.2, 0.5, and 1.0 nm, optical power level range of –90 to + 20 dBm (1200–1600 nm) and –80 to + 20 dBm (1600–1700 nm), accuracy level of ± 0.3 dB (1310–1550 nm), polarization dependency of ± 0.5 dB (1310–1550 nm), level flatness of ± 0.1 dB (1500–1570 nm), dynamic range of 40 dB at 1523 nm with a peak of ± 1 nm, return loss of 30 dB over the 1310 to 1550 nm range, and sweep time less than one second. In summary, the OSA provides accurate measurement of optical parameters, three individual trace displays, WDM analysis with minimum cost, EDFA analysis, spectral width search, notch width search, and graphic display of long-term monitoring results.

6.3 SEMICONDUCTOR OPTICAL TRANSMITTERS

High-power optical sources and transmitters, including lasers and lamps, were extensively discussed in Chapter 2. In this section, discussion will be limited to quasi-solid-state and semiconductor optical transmitters with low to moderate power levels. Review of the state-of-the art semiconductor sources indicates that optical sources using epitaxial heterojunction devices are the most attractive IR semiconductor optical sources, with power levels around 1 W (CW) or less. InSb and InAlSb semiconductor diodes provide reasonable power outputs with 100 mA drive current at peak wavelengths of 6.2 μm and 4.1 μm, respectively. These mid-IR sources can be used in applications such as gas staring imagers, radiation shields for large IR detector arrays, dynamic IR scene projectors, remote pollution detection devices, general spectroscopy, and secured communications. However, such semiconductor IR diode sources require cryogenic cooling around 77 K for optimum performance. Studies performed by the author [8] on semiconductor diode sources indicate that diodes with highly strained InGaAs quantum well (QW) active regions imbedded within the InGaAsP heterostructures provide maximum IR power with high quantum efficiency over the 1.5 to 1.8 μm range. Diode designs using the InGaAs/InGaAsP heterojunction structures provides the following performance capabilities:

- Improved thermal performance due to high thermal conductivity of the binary compound InP
- High-power CW capability because of instant heat removal capability
- Lower power dissipation due to reduced series resistance
- Improved differential quantum and conversion efficiencies due to lower thresholds
- Low production cost due to matured fabrication technology

CW power levels between 100 and 500 mW over the 1.5 to 1.8 μm region have been demonstrated at room temperature using a diode array pumping scheme comprised of several InGaAs/ InGaAsP QW laser diodes operating at 980 nm. Such an optical transmitter will require diode arrays with ratings of 10 to 20 W, optical cavity of appropriate length, low-loss single-mode output cable, and optics with minimum loss. Optical cavity dimensions and strip width must selected to provide maximum optical power output with high total differential quantum efficiency. This efficiency is inversely proportional to the cavity length. Computer modeling indicates that a cavity length of 1 mm and strip width more than 200 μm are required to achieve 500 mW (CW) output over the 1.5 to 1.8 μm spectral region at 10 °C with minimum drive.

6.3.1 Optoelectronic Oscillators (OEOs)

The operating principle of an OEO is similar to that of a Van Der Pol vacuum-tube oscillator. In the OEO, the power source and cathode are replaced by the Nd:YAG-based diode-pumped solid-state (DPSS) laser [9]. The grid is replaced by an electrooptic modulator and the plate is replaced by a photo diode. The LC circuit is replaced by a long length of low-loss optical fiber acting like a delay line. Replacing the mechanical and electrical elements with optical ones eliminates the dispersion losses that limit the oscillator frequency. The fiber is dispersive to the laser but not to the modulation. Increase in dispersive losses with increasing frequency will improve the overall performance of the OPO shown in Figure 6.8 at higher frequencies.

The overall performance of an OPO is much better than other counterparts. For example, the operating frequency of an electromechanical quartz oscillator can reach the 30 to 50 MHz range compared to the operating frequency of an OPO of as high as 100 GHz. The operating frequency of an OPO is only limited by the electrooptic modulator architectural design. A low-power (5–8 mW) three-port MESFET oscillator design using CPW technology and operating at 3 V or less will be most attractive for modulating optical signals with minimum power. Such an oscillator topology on a GaAs or any other electrooptic substrate will be best suited for high-speed electrooptic modulators operating around 20 GHz and up. This MESFET oscillator design can offer efficient operation even at 1 V, which can significantly reduce the power requirements for the modulation [10]. Despite its several advantages, it cannot match the speed performance provided by an OPO.

6.4 OPTICAL CORRELATORS

Automatic target recognition is limited by the processing speed of the signal processor. Studies performed by the author indicate that even a miniature optical correlator can boost the processing speed by a factor of ten. The performance capability of the optical correlator is strictly dependent on the computational capability of the processor in terms of data output per second. Electronic processors have limited computational capabilities and are too slow to be effective in real-time situations such as automatic recognition under rapidly changing battlefield environments.

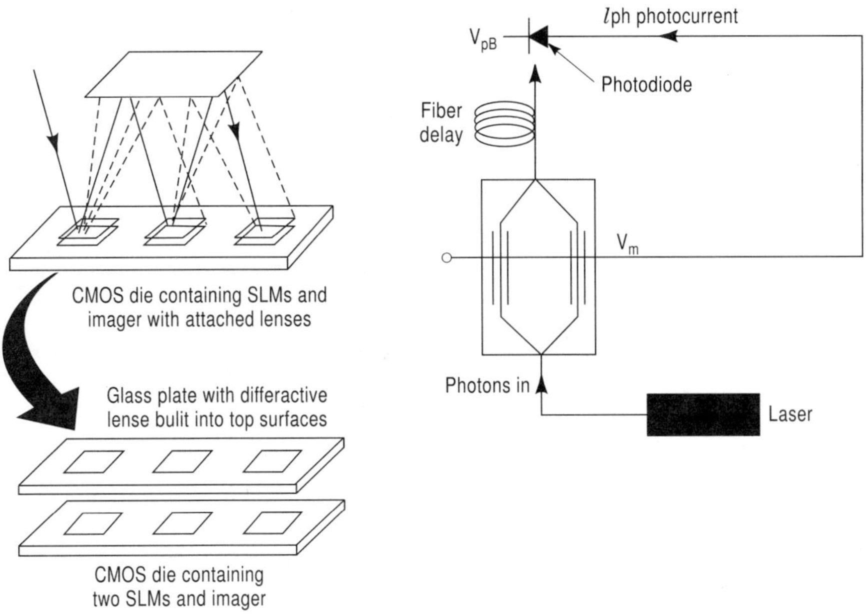

FIGURE 6.8 Block diagrams of an electrooptic oscillator and integrated optical correlator using SLM technology incorporating liquid crystals.

Optical correlators rely on optical computation speed and are capable of making 8 to 10 comparisons per frame, which is only possible with Spatial Light Modulator (SLM) technology. This particular device exhibits low coefficient of thermal expansion, which makes it most attractive for a wide variety of operating environments. An optical correlator architecture incorporating CMOS technology and SLM design with liquid crystal technology is shown in Figure 6.8. This optical correlator design concept can play a key role in real-time, complex signal processing schemes capable of providing speed and resolution unmatched by any other technology. The integrated optical correlator design shown in Figure 6.8 uses the CMOS die, two SLMs using liquid crystal technology, and a high-resolution camera with built-in diffractive lenses (not shown in Figure 6.8) on the cover glass of a single chip.

The ability to achieve 1000 comparisons or more per second in gray scale is called the figure-of-merit (FOM) of an optical correlator. A correlator with high FOM is necessary for automatic detection and recognition of moving targets under heavy clutter environments. Optical correlators with high FOM capabilities are best suited for fingerprinting identification by law-enforcement agencies, high-resolution airborne and space-based reconnaissance sensors and sophisticated medical diagnostic applications from tumor location and identification to ophthalmology [11].

The optical correlator design shown in Figure 6.8 requires alignment of two mirrors only. The incoming image is displayed on the SLM, which is "read out" via a laser beam (not shown in Figure 6.8). A lens Fourier transforms the image, and the transforms meet the second SLM. This modulation contains the Fourier transform (FT) of an image that was stored and processed digitally, which represents a target. When the transforms match, light continues through the system to be detected at the camera plane, creating a correlation peak at the location of the matching object. In the integrated design, the correlator can be folded back on itself via a mirror with diffractive lenses on top of each element, performing the necessary FT. Because the SLMs consist of small pixels, the liquid-crystal-on-silicon devices cause diffractive errors, which can be reduced by replacing the one big mirror by two small mirrors. A coherent laser source with precision control of light sensitivity can further reduce the diffraction errors. This particular optical correlator design offers superior performance with minimum weight, size, weight, and complexity.

6.5 ANALOG-TO-DIGITAL CONVERTER (ADC) DEVICES USING ELECTROOPTIC- AND PHOTONIC-BASED TECHNOLOGIES

ADC speed is of critical importance to the high-speed digital signal processors used in advanced radar systems, high-resolution surveillance sensors, and sophisticated, complex electronic warfare equipment. The speed of an ADC device can be significantly increased by reducing the incoming RF signal bandwidth involving the carrier and its modulation. Two ADC design architectures using photonic-based and electrooptical technologies are shown in Figure 6.9. The upper design shows the critical elements including the fiber optic cable and modulator using EO technology, and the lower architecture describes the locations of various ADC elements such as EO modulator electronics, polarization separators, photodetectors, amplifiers, and analog comparators.

6.5.1 ADC Design Using the Photonic-Based Time-Stretch Technique

The first ADC design using the photonic-based time-stretching concept [12] seems to offer high speed with minimum cost, size, power consumption, and complexity. This design employs time-stretching using fiber optic cables and a modulator incorporating electrooptic technology. The electrical signal to be stretched is intensity modulated onto a linearly optical chirped pulse using the EO modulator. The optically chirped signal waveform is then dispersed in an optical device with linear dispersive characteristics. Dispersion imposes different delays on different wavelength components, leading to signal stretching or compression in time, depending on the sign of the dispersion and of the chirp signal.

After the dispersion, the signal is detected by the photodetector and amplified, if required. The time-stretching concept has been demonstrated [12] by dispersing a narrow pulse of 150 fs over a 50 ns spectral range originating from an erbium-doped fiber laser (EDFL) in a spool of a low-loss, single-mode optical fiber of length L_1 with a dispersion coefficient (D_1) of about 17 ps/km/nm. The total dispersion, which is the product of fiber length and dispersion coefficient, in the fiber

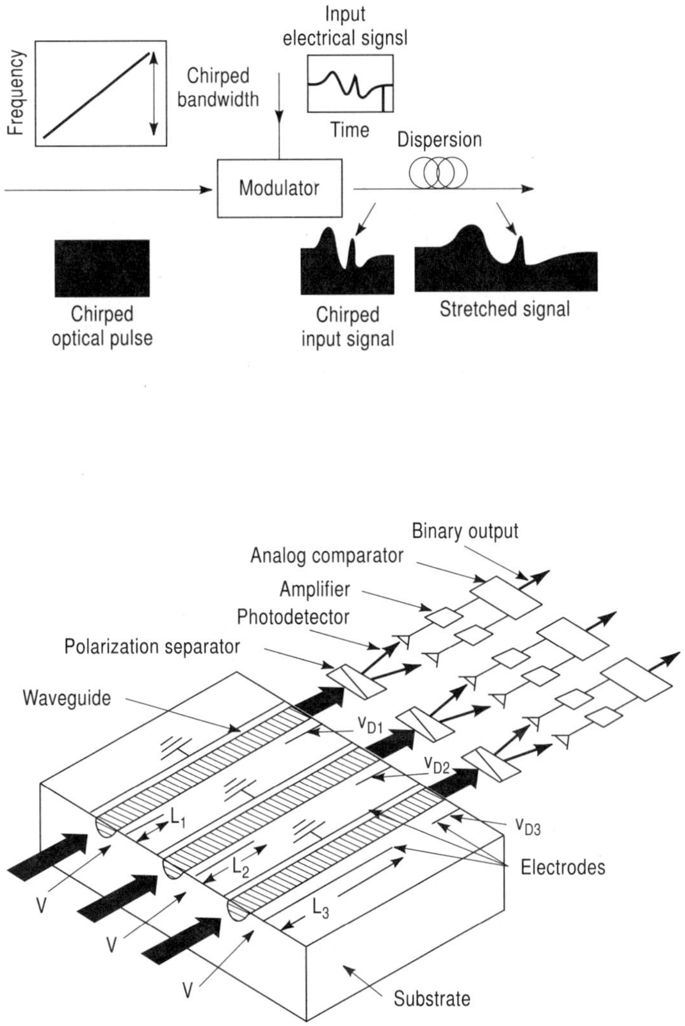

FIGURE 6.9 Design concepts for A/D converters using EO technology.

generates the chirped waveform. The electrical signal then modulates the intensity of the chirped waveform in the lithium-niobate-based electrooptic modulator leading to a direct time-to-wavelength correspondence [12]. The modulated waveform is then further dispersed in a second spool of single-mode fiber of length L_2, with dispersion coefficient of D_2. The envelope of the modulated pulse is stretched in time, while propagating through the second dispersive element. The photonic-based time-stretching technique is defined by a stretch factor (M), which is a function of two fiber lengths and their dispersion coefficients. The stretch factor can be written as

$$M = 1 + \frac{D_2 L_2}{D_1 L_1} \tag{6.7}$$

where, D and L indicate the dispersion coefficients and lengths of the two optical fiber elements, and subscripts 1 and 2 refer to fiber elements 1 and 2, respectively. The above equation indicates that the stretch factor M increases with increase in dispersion coefficient and length of the output or second optical fiber and with the decrease in the dispersion coefficient and length of the first fiber element. Stretch factors and corresponding sampling rates available from such an ADC device are summarized in Table 6.7. Stretch factor computations use output optical fiber with different lengths and input optical fiber of 1.1 km, but with same dispersion coefficient for both fiber elements.

The photonic-based time-stretch preprocessing technology can revolutionize the ADC design with significant increase in both the input signal bandwidth and the effective sampling rate (shown in Table 6.7) by a factor known as time-stretch factor (M). The photonic-based time-stretch technology has demonstrated a sampling rate of 1 GSPS or 1000 MSPS, capable of computing a high-speed analog signal with 100 ps transition times. This level of performance is not available from conventional ADC devices. It is important to mention that the ADC speed can be improved using over-sampling techniques, but the resolution is obtained at the expense of bandwidth and power consumption. In addition, a digital decimation filter is required to suppressed the quantization noise. However, in the case of an ADC designed with photonic-based time-stretch technology, the sampling rate is significantly improved, with no impact on power consumption, and the S/R ratio is improved by a factor of (M^{2n+1}), where M is the stretch factor and n is the number of loops deployed by the EO modulator using optical fibers.

6.5.2 ADC Design Using Electrooptic Technology

Critical elements of a 3 bit ADC design using EO modulator technology are shown in Figure 6.9. This ADC design uses a polarized CW input light source, lithium-niobate-based EO modulator complete with electrodes, polarization separators, photodetectors, amplifiers, and analog comparators.

TABLE 6.7 Stretch factors and sampling rates for ADC devices using the photonic-based time-stretching preprocessing technique and EO modulator technology

Output fiber length L_2 (km)	Stretch factor (M)	Sampling rate (MSPS)
2.2	3	300
5.5	6	600
6.6	7	700
7.7	8	800
8.8	9	900
9.9	10	1000

An array of identical dielectric channel waveguides is fabricated in a single-crystal substrate of a linear electrooptical material such as lithium niobate. Each waveguide, while supporting predominately one TE and one TM waveguide mode, is excited by a linearly polarized light from a CW laser operating at optimum wavelength. The performance of this EO modulator depends on the intensity modulation function of linear EO phase retardation. The phase retardation of the light in the TE mode with respect to the TM mode is a function of the electrooptical coefficient of the material, applied voltage, waveguide parameters, and electrode spacing. The light from each optical waveguide is passed through a polarization separator and intensities of two orthogonally polarized components are detected separately. The magnitudes of these intensity components are dependent on static phase shift, modulation depth or amplitude, and the DC terms detected in the signals. A binary representation of the applied voltage (V) is obtained by electronically comparing the two intensity components. As the applied voltage changes, conversion errors are most likely to occur near the intensity cross-over points. It is important to mention that both the resolution and dynamic range of the ADC increase with increase in number of bits, at the expense of higher power consumption and added complexity.

6.6 OPTICALLY CONTROLLED PHASED ARRAY (OCPA) ANTENNAS

Conventional phased array (CPA) antennas suffer from disadvantages including narrow-band operation, high component losses, excessive weight, high cost, high power consumption, and poor reliability. CPA antennas are extremely complex because of several hundred microwave delay lines and associated components. CPA antennas have great difficulty in providing wide instantaneous bandwidth, high S/N ratio, large dynamic range, high clutter rejection performance, uniform antenna patterns at wide steering angles, and deep-null steering capability. Most CPA antenna architectures require a series of true-time delays or phase delays at the antenna elements.

6.6.1 Coherent OCPA Antenna Technology

OCPA antenna technology not only overcomes most of these problems, but also offers superior performance in terms of fast response time, simultaneous multiradar functions, squint-free antenna beams, antenna patterns with high integrity, high reliability, and minimum power consumption. OCPA antenna technology can be applied to passive systems such as adaptive phased array receiving antennas or multiplexing with improved performance at minimum cost. Critical optical and fiber-optic-based components such as erbium-doped fiber amplifiers (EDFAs), optical couplers, multiplexers, and de-multiplexers, antenna elements, and laser sources used in such passive systems are shown in Figure 6.10. Note EDFAs are required to compensate for high losses in various optical components. All optical fibers that carry the microwave and optical signals from input to the antenna terminals must be of the same length to avoid extraneous delays. Furthermore, these passive systems can use noncoherent sources for acceptable performance levels. How-

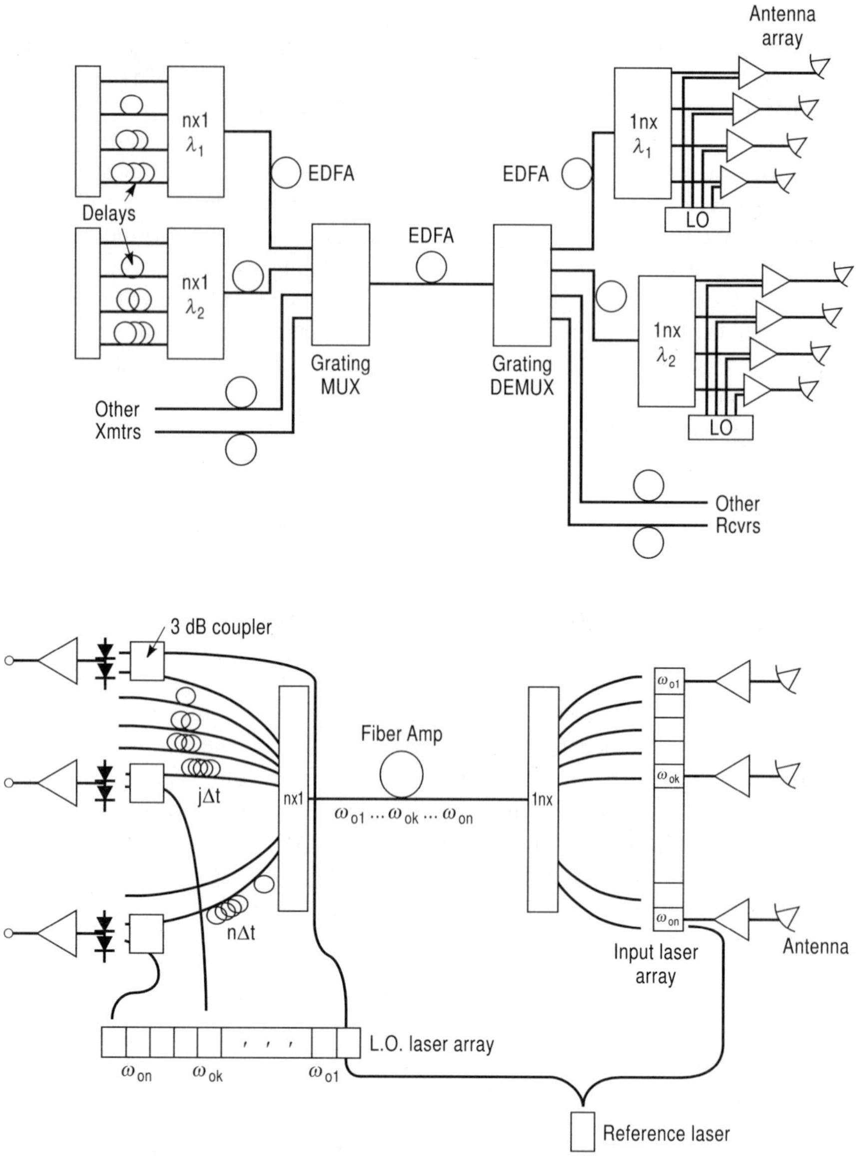

FIGURE 6.10 Multiplexer and phased array antenna using EO technology and fiber-based components.

ever, OCPA antennas using coherent sources will provide improved antenna performance, but at high cost and complexity.

Coherent OCPA antenna technology is best suited for applications where high pointing accuracy, very high clutter rejection capability, antenna patterns with ultrawide lobe levels (–50 db or better), high probability of intercept, multiradar function capability with high S/N ratios and low false alarm rates, and detection of targets with very low radar cross-section (RCS). Block diagrams of optically fed phased array antennas with optical control elements and coherent laser sources are shown in Figure 6.11. The coherent OCPA antenna system [13] is based on the coherent optical transmission of signals from input to antenna array and vice-versa. Phase or frequency modulation can alternately be employed to encode microwave signals onto optical signals. Both microwave amplifier gain and LO gain can be tuned at each antenna element to optimize the antenna performance, including the beam shape. Phase locking is necessary to maintain constant amplitude. Wide-band antenna operation requires continuously tunable laser arrays.

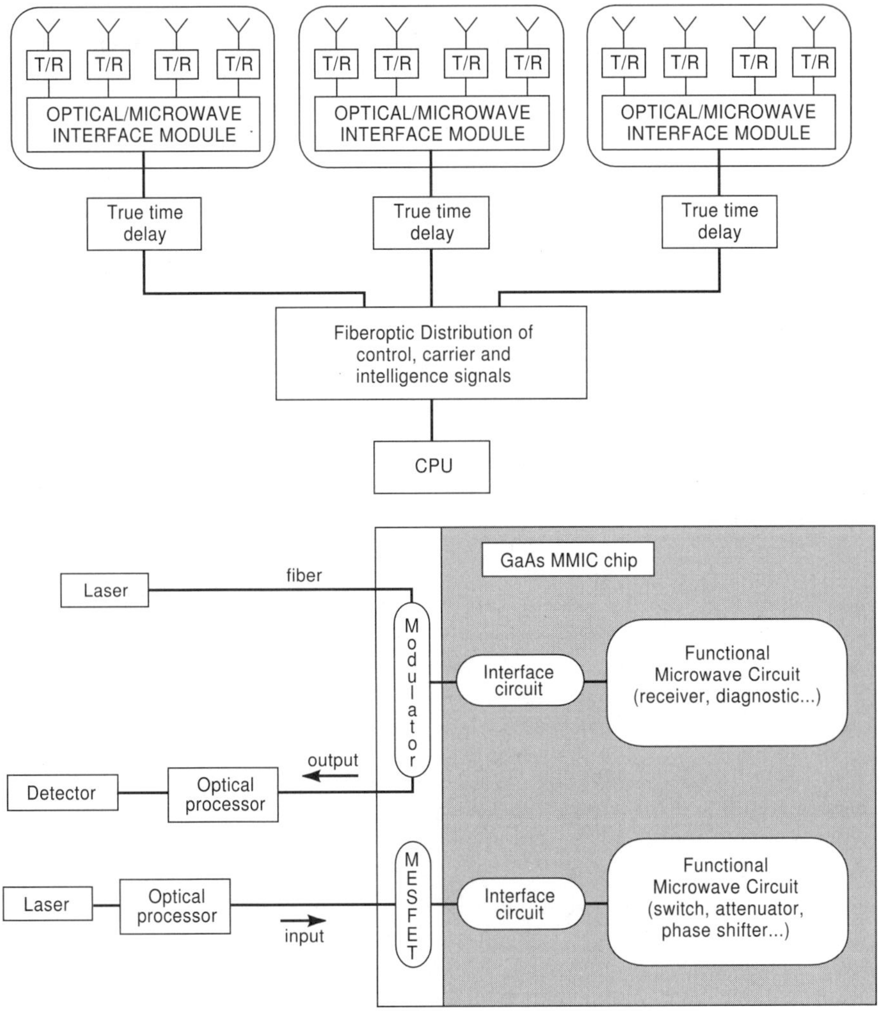

FIGURE 6.11 Block diagrams for an optically fed phased array antenna and for control signals and data inputs via fiber optics.

A 128 element, multichannel, broadband coherent OCPA antenna system for telecommunications applications demonstrated a S/N ratio better than 40 dB with an optical loss of only 3 dB at a wavelength of 1.55 μm and over an instantaneous bandwidth of 1 GHz. This level of performance is not possible with a conventional 128-element, microwave phased array antenna. An OCPA antenna is comprised of optical combiners/splitters, optical fibers, true time delay elements, optical/microwave interface modules, antenna elements, EDFAs, transmit/receive (T/R) modules comprised of amplifiers and other microwave integrated circuit devices, laser sources, detectors, optical processors, and the central processing unit (CPU), as shown in Figure 6.11. A 1 × 8 combiner has a typical loss of about 9 dB, whereas a 1 x 16 combiner has a loss of about 12 dB. The optical fiber loss will be less than 1 dB and the erbium-doped fiber amplifier (EDFA) typical gain and noise figures will be about 20 dB and 4 dB, respectively.

Assuming an optical tuning range of 100 Å or 0.1 nm, 1 Å representing a bandwidth of 12 GHz at 1.55 μm, and each channel having an instantaneous bandwidth of 1 GHz, the OCPA-based WDM system will have a channel capacity of 100, capable of occupying a 1200 GHz bandwidth (12 × 100). If the spectral width of the EDFA is increased to greater than 100 nm, a significant number of channels can be achieved in a multiple beam antenna system. Assuming a spectral bandwidth of 100 nm, an instantaneous bandwidth of 1 GHz, and a channel spacing of 5 GHz needed to eliminate interchannel interference, a WDM system will have a channel capacity of 2500 channels.

In summary, the coherent OCPA antenna system offers an accommodation of more than 100 channels in a single fiber and more than 2500 channels in a multiwavelength system. The coherent OCPA antenna has potential applications in null steering, multiple beam formation with minimum loss, wide-band surveillance, and squint-free multiple frequency operations. Furthermore, coherent OCPA architecture offers optimum design flexibility because of high gain from the combination of optical and electronic amplification, local oscillator gain, tuning capability of radiating elements, tunable lasers, T/R modules using GaAs MMIC technology, true time delay devices, optical processors, and subarray techniques, as shown in Figure 6.11.

6.7 OPTICAL TRANSLATORS

Translators have potential applications in microwave and optical systems and are used to translate the microwave or optical operating frequency to a specific value with minimum loss or phase and amplitude distortion. The critical elements, input–output responses, and BER penalty for an optical translator as a function of number of cascade wavelength translators are shown in Figure 6.12. The penalty curves shown in Figure 12 indicate that translators using optoelectronic technology suffer maximum losses from 3.3 to 5.5 dB, depending on the frequency, while the all-optical translators have a maximum loss of about 1.5 dB, even when the number of cascade wavelength translators approaches 100.

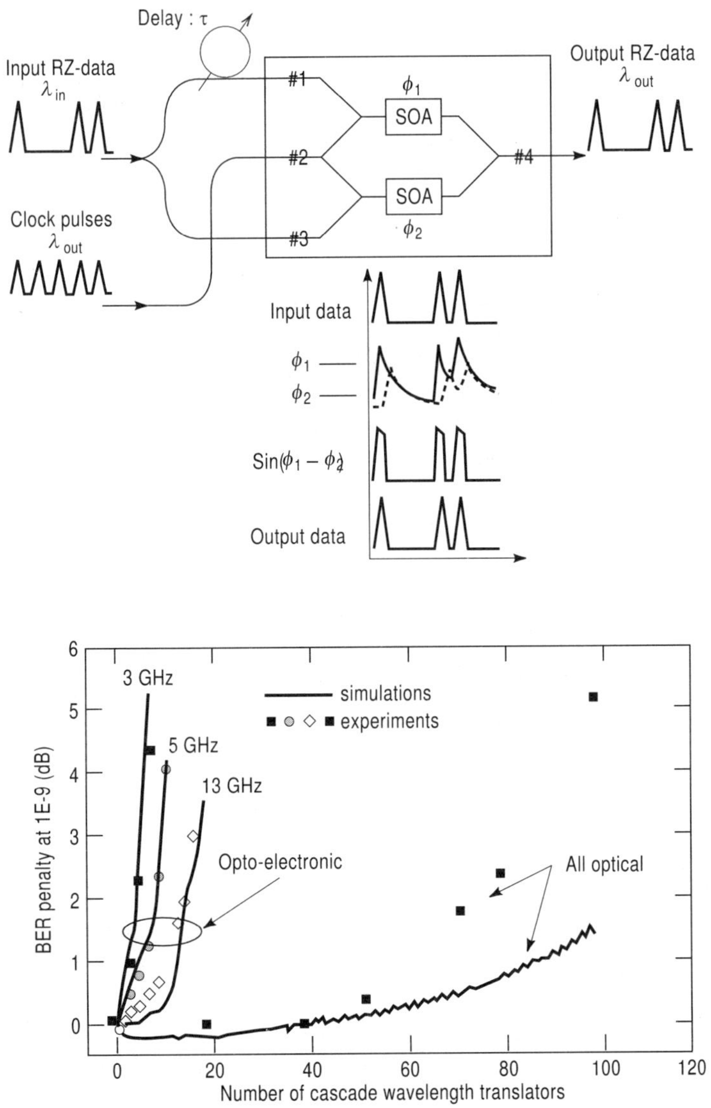

FIGURE 6.12 Block diagram of an optical translator using optical technology and performance comparison with translator using optoelectronic technology.

6.8 SUMMARY

Potential design aspects and performance capabilities of electrooptical modulators, including bulk modulators, waveguide modulators, shielded velocity-matched EO modulators, acoustooptic modulators, magnetooptic modulators, semiconductor modulators, and magnetorestrictive light modulators, are described, with emphasis on cost, performance, and complexity. Performance characteristics of various optical receivers, including receivers based on direct-detection methods, acoustooptic channelizer receivers, and surveillance receivers using electrooptic technology, are summarized. Operational capabilities of the latest optical oscillators using semiconductor and optical technologies are discussed, with emphasis on noncryogenic performance and cost. Performance capabilities and limitations of optical correlators and photonic-based time-stretch ADC devices are briefly summarized and their potential applications identified. Potential applications of photonic-based time-stretch preprocessors are identified for high-performance radar systems, high-resolution airborne and space-based side-looking radar systems, and EW systems operating under severe electromagnetic environments. Performance capabilities of coherent OCPA antennas are described, with emphasis on multibeam formation features, fast response time, squint-free antenna patterns, and multimode radar capability. Potential applications of coherent OCPA antennas include surveillance receivers with deep-null steering, wide-band reconnaissance sensors, and squint-free multiple-frequency systems.

NUMERICAL EXAMPLE

Compute the time-stretch factors and sampling rates for a photonic-based time-stretch preprocessor using the following assumed parameters:

Initial sampling rate or clock frequency: 100 MHz
Input line length (L_1) : 1.1 km
Output line lengths (L_2) : 2.2 km, 5.5 km, 6.6 km, and 8.8 km
Dispersion coefficient : Equal for both lines

Using the equation (6.7), the following values for the time-stretch factor are obtained:

Output line length (km)	Time-stretch factor	Sampling rate (MSPS)
2.2	3	300
5.5	6	600
6.6	7	700
8.8	9	900

REFERENCES

1. T. Yoneyama et al. "Characteristics of inverted slotline for traveling wave optical modulator." *IEICE Trans. Electronics*, pp. 229–237, February 1993.
2. K. Kawhno. "High speed Ti:LiNbO$_3$ and semiconductor modulators." *Trans. Electronics*, pp. 183–190, February 1993.

3. E. Yamashita et al. "Some experiments on design of broadband electro-optical modulators." *IEEE Trans. on MTT, 1*, 3, 703, September 1978.
4. M. N. Khan et al. "Technique for velocity-matched TW-electro-optic modulator in AlGaAs/GaAs." *IEEE Trans. on MTT, 41*, 2, 244–249, February 1993.
5. K. Kawano. "Improvement of performance of shielded velocity-matched Ti:NbO_3 optical modulator by using a ridge structure." *IEICE Trans. Electrons E76-C*, 2, 238–243, February 1993.
6. G. W. Anderson et al. "Advanced characterization technique for RF, Microwaves and MM-waves."'"*Proceedings of IEEE, 79,* 3, 355–372, March 1991.
7. M. N. Khan. "Transducer geometry, materials for acuostooptic (AO) channelizer receiver." *Proceedings of IEEE, 79,* 33, 362–364, March 1991.
8. A. R. Jha. "Development of high power 1.5 to 1.89 micron semiconductor laser diodes." Technical Report, Jha Technical Consulting Services, Cerritos, CA, pp. 6–16, January 2, 1996.
9. Contributing Editor, "Diode-pumped solid state lasers." *Laser Focus World*, 48, January 1999.
10. Z. V. Radisic et al. "CPW oscillator configuration for an EO modulator." *IEEE Trans. on MTT, 41,* 9, 1645–1647, September 1993.
11. J. L. Morey. "Optical systems speed automatic target recognition." *Photonics Spectra,* pp. 46, January 1999.
12. B. Jalali et al. "Time-stretch preprocessor overcomes aDC Limitation." *Microwaves and RF*, pp. 57–66, March 1999.
13. P. M. Freitag et al. "A coherent optically controlled phased array antenna system." *IEEE Microwave and Guided Wave Letters, 3,* 9, 293–95, September 1993.

CHAPTER SEVEN

Application of Infrared and Photonic Technologies in Commercial and Industrial Devices and Systems

7.1 INTRODUCTION

This chapter will focus on the application of infrared (IR) and photonic technologies in commercial and industrial devices and systems. Potential applications of these technologies will be identified for the development of key components such as diode-pumped solid state (DPSS) lasers, semiconductor tunable lasers, optical fiber lasers, and fiber-based optical amplifiers such as erbium-doped fiber amplifiers (EDFAs). Performance capabilities of sensors and systems incorporating these IR devices and components such as high-definition TV (HDTV), optical projectors, commercial printers, laser wafer marking systems, spectrometers used to control manufacturing processes, fiber optic (FO) data links, optical spectrum analyzers, optical displacement sensors, imaging sensors, communication systems, IR cameras,weapon detection sensors, optical storage systems, FO interferometer sensors, smoke/fire detection sensors, and other industrial sensors are described, with emphasis on cost and reliability. Attempts will be made wherever possible to distinguish between the industrial and commercial applications.

7.2 APPLICATIONS IN COMMERCIAL AND INDUSTRIAL DEVICES AND SENSORS

7.2.1 High-Definition TV (HDTV)

Consumers are showing great interest in HDTV because they are impressed with the high-quality imaging that results from the improved resolution and contrast offered by the optical technology. Recently, a spatial-light-modulator (SLM) design was used to develop a HDTV projector based on a scanned linear array incorporating laser technology. High resolution is the most impressive performance parameter of HDTV. It is achieved by using a grating light valve (GLV)-based SLM technique shown in Figure 7.1 The scanning configuration [1] deployed is comprised of three arrays, each with a column having 1080 pixels. Each SLM device represents a red, blue, or green pixel in the HDTV projection display, with 1920 × 1080 resolution capability. The projection device separates red (656 nm), green (532 nm), and blue (457 nm) components from a white light laser source and then directs the distinct color beams through line-generating optics onto the three linear arrays. The modulated light then combines through a diachronic assembly and projects onto the

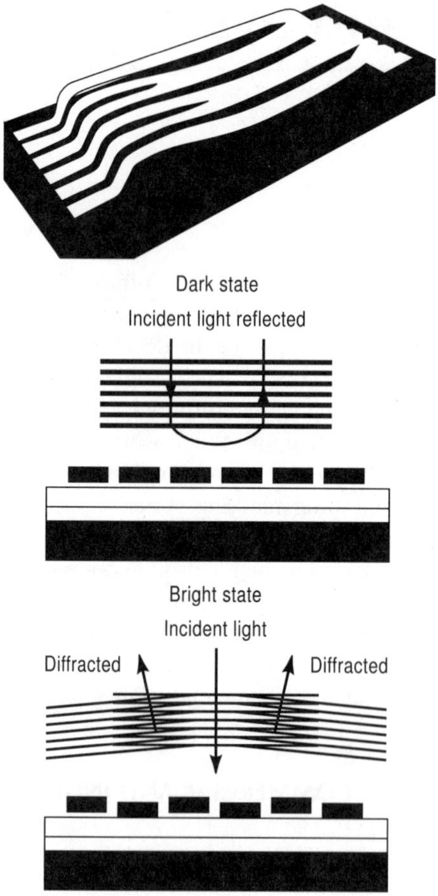

FIGURE 7.1 Architectural design of a grating light valve (GLV) spatial light modulator using aluminum-coated silicon ribbons.

screen via an optical fiber. A galvanometric mirror scans across the image horizontally, so that each pixel paints one row of color channel data for each scan refresh.

This particular projector [1], with 1920 × 1080-pixel resolution offers a contrast ratio greater than 200:1, convergence of ± 0.25 pixel, refresh rates up to 96 Hz, diagonal image size of 110 inch, and gray scale of 8 bits per channel. Such high-definition performance requires SLMs with speeds better than 150 ns and optical power capacity of about 20 mW/pixel for a 10,000 lumen projector. This kind of performance level is only possible by integrating GLV technology into each SLM modulator device.

Each SLM modulator comprises of a row of parallel, reflective silicon ribbons (Figure 7.1). The silicon ribbons are extended above a thin air gap and can be displaced differentially with respect to each other with the aid of applied voltage across the gap.The differential displacement of adjacent ribbons effectively switches each modulator from its reflective (unaddressed) state to its diffractive (addressed) state.

Conventional HDTV projector designs based on either area arrays or scanned laser beam (SLB) systems are more complex and expensive. The HDTV projectors using GLV-based SLMs offer a cost-effective approach compared to the other two conventional design concepts. Area array architecture would require more than 2 million pixels, as opposed to 1080 pixels in the linear array system, because of poor yield in the manufacturing process. With 200 times fewer pixels, the scanned GLV-based architecture has an impressive yield relative to any other fabrication technology. A SLB system modulates fewer active line elements simultaneously. Furthermore, the SLB system requires a costly diffraction-limited light source as opposed to a low-cost source using diode bars. Even the current state-of-the art scanning technology would have a hard time in meeting the present HDTV performance requirements. A SLB system with a refresh rate of 96 Hz and a 24-facet polygon would require a rotating speed of 260,000 revolutions per minute, which could present a serious reliability problem. Many-sided polygons are very inefficient because of the blanking requirements as the beams cross the boundaries between the facets.

7.2.2 Portable Microlaser Projectors

Existing laser projectors using either gas lasers or lamp-pumped solid state (LPSS) lasers are complex, expensive, and bulky and require sophisticated cooling systems. Low-cost mass production for commercial applications is not possible for projectors using these laser sources. Diode-pumped microlasers [2] emitting at red, green, and blue wavelengths are most attractive for possible applications in laser-based projectors because of high electrical-to-optical efficiency, compact size, high reliability, lower cooling power requirements, and relatively minimum cost. A block diagram of a microlaser projector using liquid-crystal-display (LCD) technology is shown in Figure 7.2. The portable microlaser projector offers long life, high reliability, high resolution, uniform image quality, large color gamut (or entire range), and high color contrast and efficiency. Reflective active-matrix LCD devices provide high throughput efficiency and full video frame rates with high pixel resolution of 1280 × 1024 pixels. A prototype portable projector [2] with 500 lumen rating using

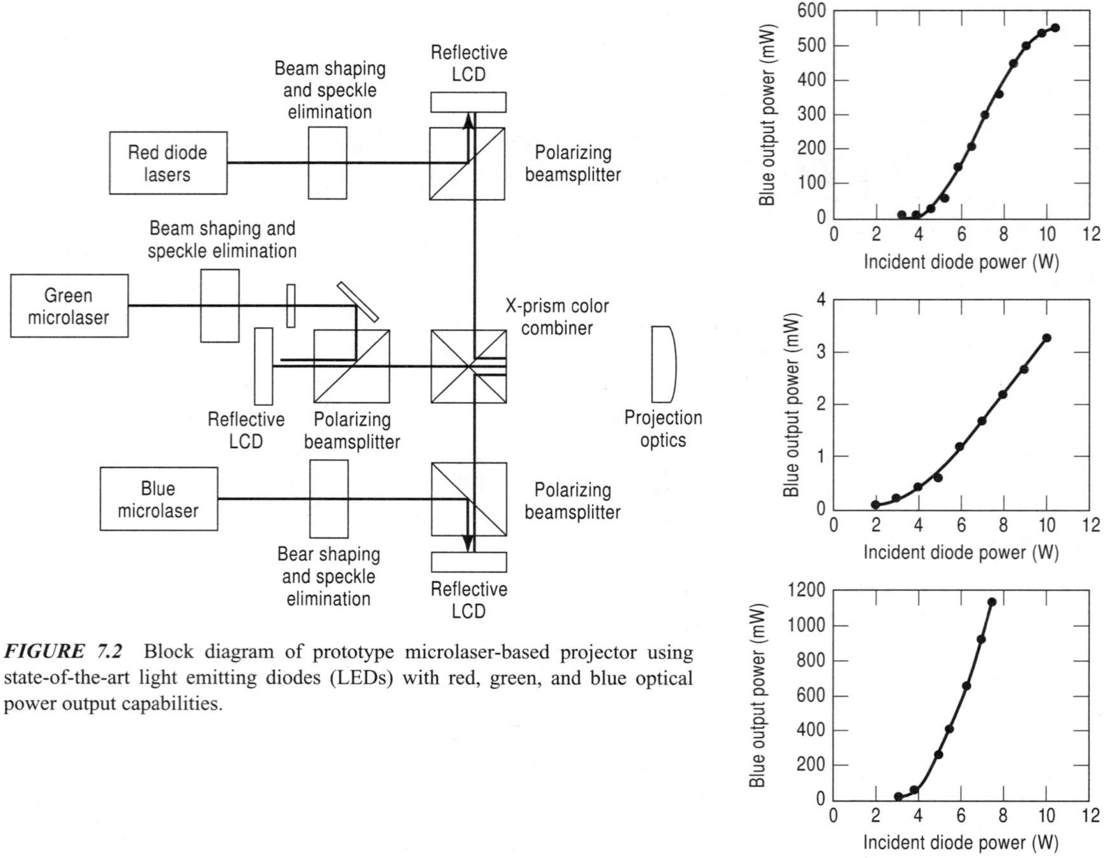

FIGURE 7.2 Block diagram of prototype microlaser-based projector using state-of-the-art light emitting diodes (LEDs) with red, green, and blue optical power output capabilities.

a single green output microlaser, three blue microlasers, and an array of red diode lasers demonstrated optical power output of 1.35 W at 457 nm (blue), 1.87 W at 532 nm (green), and 4.87 W at 650 nm (red), with a total optical system throughput efficiency exceeding 33%. Performance characteristics of the prototype LCD-based microlaser projector are summarized in Table 7.1.

Direct-emitting visible laser diodes are currently being developed, but high-efficiency, multimode, broad-stripe, watt-level AlGaInP red output diodes are commercially available at reasonable costs. Blue and green diode lasers with high-power outputs and efficiency are still years away from commercialization at acceptable cost. In addition, these diode lasers suffer from poor beam quality, which is of critical importance in high-resolution projection applications. It is important to mention those DPSS lasers with miniature red, green, and blue outputs appear to be most suitable for high-resolution portable projectors. These microlasers use low-cost, long-lifetime, near-IR AlGaAs semiconductor lasers as pump sources for rare-earth, ion-doped crystals in conjunction with intercavity nonlinear processing to yield visible wavelengths. Proper selection of nonlinear laser crystals and thin-film coatings of appropriate dielectric materials can produce a variety of visible wavelengths. Air-cooled microlasers with electrical-to-optical efficiency up to 5% are available with minimum size and weight, low power consumption, and minimum cost if produced in high volume. To generate visible light from a microlaser, the near-IR emission of a diode-pumped nonlinear-doped crystal is required for the frequency doubling process. The microlaser consists of two crystals—a neodymium-doped linear crystal for gain and a nonlinear crystal for frequency doubling. The fundamental lasing occurs at 1313 nm, 1064 nm, and 914 nm, which will lead to generation of visible light at 656 nm, 532 nm, and 457 nm, respectively, through an intercavity doubling procedure. The AlGaAs laser–diode array with nominal power rating of 10 W at 800 nm, microlaser crystals, and the integral thermoelectric (TE) cooler can be housed in an enclosure not to exceed the dimensions of 5″ × 1″ × 1″, which can be easily mounted on an air-cooled heat sink. The beams from the multiple microlasers can be easily combined to achieve any desired brightness level, which is not possible with optical lamp sources. The overall electrical power required for a microlaser, in-

TABLE 7.1 Performance characteristics of a portable microlaser projector

Performance characteristics	Parametric values
Pixel resolution (pixels)	1280 × 1024
Luminous output (lumens)	500 (minimum)
Estimated reliability (hours)	10,000 (minimum)
Contrast ratio	100:1 (minimum)
Image uniformity	± 1.5% (across image field)
Image size (feet)	2 to 10 (diagonals)
Prototype unit size* (inch)	15 × 15.5 × 8.5

*With further engineering developments involving design refinements and with improvements in microlaser technology, significant reductions in weight, size, and cooling requirements are expected.

cluding power for cooling, varies from 30 to 50 W, depending on the ambient temperature. Typical characteristics of red, green, and blue microlasers are summarized in Table 7.2.

According to the latest research and development data, the blue and green microlasers are very close to meeting all the specification requirements for both backlit and direct-write displays. The output power from a red microlaser (656 nm) is currently marginal. High-power, red output diode microlasers are best suited for backlit displays. However, the poor spatial beam quality and high numerical aperture inherent in diode lasers will require sophisticated beam conditioning and expensive optics for the display. Even slight variations in optical crystal quality, surface finish, and thin-film coating quality can lead to variations in output power. The ultimate lifetime of the microlaser will be limited by the near-IR diode pump laser source. As mentioned earlier, AlGaAs diode laser devices have demonstrated lifetimes between 10,000 and 15,000 hours, due to slow degradation in the lattice structure.

7.3 FULL-COLOR VIRTUAL (FCV) DISPLAYS

Virtual displays use light-emitting diodes (LEDs), which when integrated with appropriate optics form an image of desired quality on a screen. Virtual displays are currently used in camcorders and headsets. These displays use tiny liquid crystal displays (LCDs) that are about 0.7 inch along the diagonal axis; optics are placed in front of the LCDs to make the viewed image appear much larger than actually it is. New emerging technologies such as scanned LED arrays offer full-color minidisplay capability, which will have potential applications in digital cameras, cellular phones, smart-card readers, video games, and portable viewers.

In a scanned linear-array (SLA) technology based on LEDs as the image source, a LED array is placed vertically between the focal point of the lens and the lens itself. The LED array is driven to produce one column of data that is magnified and reflected into the viewer's eye via a mirror. The next column of data is then written to the LED array and the mirror is moved to spatially displace the light. A full image is generated by sweeping the scanning mirror at 60 frames per second. A full color-

TABLE 7.2 Typical characteristics of current red, green, and blue microlasers

Characteristic	Blue	Green	Red
Wavelength (nm)	457	532	656
Maximum CW output power (W)	0.78	4.50	1.50
Typical output power (W)	0.50	3.00	1.00
Peak noise (over 10 Hz to 1 MHz)	< 3%	< 3%	50%
Beam radius (mm)	0.125	0.150	0.310
Beam divergence (half angle)* mrad	10	6.3	3.0
Beam product (mm·mrad)	1.25	0.94	0.93
Electrical power consumption (W)	(30–60 W)	(30–60 W)	(30–60 W)
Laser head size (inches)	(1 × 1 × 4.5)	(1 × 1 × 4.5)	(1 × 1 × 4.5)

virtual (FCV) display with 640 × 480-pixel resolution capability can produce 4096 colors. Such a FCV display provides brightness of 12 fL and a contrast ratio of 5000:1 with power consumption less than 300 mW. It is important to mention that a FCV display requires LED emission in each of the three distinct colors, namely, red, green, and blue. A least expensive FCV display can be designed comprised of two separate LED arrays, one array using AlGaAs diodes for the red color and the other using of side-by-side blue–green GaN-based LED arrays. Both blue–green arrays are fabricated on the same substrate and share the same driver. Reflection technology used by the FCV display offers outstanding performance, small size, low power consumption, high color purity, rugged design, and a display architecture that can be easily modified for higher resolution.

7.4 DIGITAL VERSATILE DISC (DVD)

The performance of a DVD depends on the storage capacity of the diode lasers. Studies performed by the author on diode lasers indicate that for DVD applications, shorter wavelength laser diodes are most suitable to achieve the desired data storage capacity and greater information density. Successful application of optical storage technology has been demonstrated in the audio compact disc (CD), CD-ROM, personal computers (PCs), laptops, and video games. Most of the current CD laser heads operate at higher wavelength, around 780 nm for data reading and writing, which yields a maximum data storage capacity of about 650 megabytes (MB) from a standard 12 cm compact disk.

7.4.1 Laser Diode Requirements for DVD Applications

Studies performed on laser diodes indicate that for DVD applications, shorter wavelengths are highly desirable to meet high data storage requirements. The focussed spot size from a diffraction-limited light source is proportional to the optical wavelength of the laser source. This indicates that the shorter the emission wavelength of the laser diode, the smaller the data pits written on the optical disc, leading to significant increase in data density. Preliminary calculations reveal that a 12 cm optical disk is limited to 0.65 GB of storage capacity at 635–650 nm (red) laser diode wavelength, but limited to 13.5 GB of capacity at 400 nm (blue) laser diode wavelength, based on a single-side, single-layer format of the 12 cm disk. These storage capacities can be further increased with both sides accessible and two layers on each side of the 12 cm disk. The studies further reveal that red laser diodes with emission wavelengths over the 635 to 650 nm spectral region are best suited for laser heads of current DVD technology.

Performance stability and reliability at higher operating temperature, around 70 °C or so, are of paramount importance, particularly, for red laser diodes. Good temperature characteristics of red diodes are necessary for DVD applications [3]. Power requirements vary for various DVD applications including DVD-ROM (read-only), DVD-Video (read-only disc for high-quality video images), DVD-audio (read-only disc for high-quality audio signals), DVD-R (write once), and DVD-RAM (read–write). For read-only and write-once DVD applications, laser diodes with power ratings greater than 30 mW are required for optimum performance. Since, the smallest spot size from a laser source is dependent on its beam quality, it is absolutely necessary to have a near-diffraction-limited beam. However, small aspect ratio and beam astigmatism of the laser diodes are necessary to reduce focussing optics cost and complexity.

The most critical laser diode performance parameter for the DVD application is the relative intensity noise (RIN). DVD designers indicate that a RIN level better than –120 dB from a laser diode is necessary to reduce the reading and writing errors.

Both the theory and experimental data indicate that the RIN is attributed to mode-hopping of the laser diodes caused by the optical feedback from the optics and the optical disk. The optical feedback can be reduced by using special optics incorporating a polarizer and a quarter-wave plate. Deployment of a mode-selective cavity configuration such as a distributed feedback (DFB) or a distributed BRAGG reflector (DBR) cavity is the most effective way to reduce the mode-hopping noise, but at the expense of higher cost and complexity. Reduction of the RIN is also possible by using the self-sustained pulsation technique. A RIN level as low as –140 dB has been achieved over the temperature range from + 20 to + 60 °C using this particular technique [3], in addition to remarkably stable DVD operation even at elevated temperatures.

7.4.2 Application of Blue Laser Diodes in DVD

The latest research and development activities indicate that replacing the red laser diode with a blue-emitting laser diode in a DVD player increases the storage capacity by about 2.5 times [4]. A blue–violet output semiconductor laser offers extremely narrow line width with higher stability, lower fabrication costs, significantly higher storage capacity, and improved lifetimes (in excess of 10,000 hours). This semiconductor laser diode opens up new opportunities not only in the DVD, but also in spectroscopy, interferometry, and microscopy. A Japanese chemical company [4] has developed a 400-nm blue laser by fabricating an epitaxial laterally overgrown (ELOG) GaN layer on top of a sapphire substrate. This device, operating at 400 nm, demonstrated a minimum output power of 5 mW (CW) at room temperature, lifetime exceeding 100,000 hours, and significant improvement in heat transfer efficiency. The maximum output of this semiconductor laser diode is about 30 mW with significant reduction in lifetime if operated above 5 mW. This blue free-running diode laser delivered 5 mW of output power with an input power of 250 mW leading to an efficiency of 2%.

A free-running laser diode is a high-gain device with a large inherent fundamental line width, and the line width of the laser source is inversely proportional to the photon lifetime in the cavity. The line width of the laser diode can be decreased by increasing the photon lifetime, which is possible with longer cavity and resonator mirrors with high reflectivity. Furthermore, improved optics for beam collimation and focussing will provide high stability and narrow line width. Besides applications to DVD, blue laser light is extremely useful in detecting chemical elements that have resonant lines in the blue region of the spectrum. Blue light sources have potential applications in DVD, Raman spectroscopy, laser microscopy, interferometry, holography, multiple-wavelength printers, and telecommunications because of lower cost.

7.5 LASER-BASED COMMERCIAL PRINTERS

Lasers are at the heart of three-fastest-growing industries, namely, computers, printers, and telecommunications. Diode lasers are widely used in the high-quality printers that revolutionized computer-based desktop publishing. All diode lasers are based on the LED operating principle, with the exception of the direction of applied bias voltage. The efficiency of a LED is typically less than 1%, and the emission is emitted equally in all directions, as in the case of an incandescent lamp. In case of diode lasers, the light is generated by stimulated rather than spontaneous emission as in the case of LEDs, resulting in higher conversion efficiency. Furthermore, diode lasers require higher drive current compared to LEDs. Quantum efficiency as high as 78% and overall device efficiency as high as 45 % are possible with diode lasers. The small size and low cost of diode lasers makes them most attractive for commercial printers. Current laser printer technology is focused to move towards shorter and shorter wavelength to realize high data storage densities, because smaller spot sizes are only possible with shorter wavelengths. Laser printers use diode lasers to scan images of high quality onto photoconductive material with minimum cost and high reliability. Low cost and high reliability are the major advantages of diode lasers for the use in commercial printers.

7.5.1 Laser Printers Using Computer-to-Plate Technology (CTP)

CTP technology eliminates a complete stage in the production of printing plates, leading to significant saving in commercial printing applications. This technique allows imaging directly onto the plate surface by exposing it with a laser beam. Current state-of-the art technology uses frequency-doubled, DPSS lasers emitting at 532 nm to expose the photopolymer plates. Integration of thermal plates into the CTP system offers a very sharp exposure threshold, very hard "halftone" dots, and high-quality printing. Commercial printing companies say that their customers are highly impressed with operational simplicity, fast speed, and high quality of platemaking, when they use the CTP systems.

Offset-lithographic-printing plates based on laser ablation of thin films offer exposure thresholds of less than 100 mJ/cm^2. Plates using thermal cross-linking technology of polymers have threshold exposures of 200 mJ/cm^2 and are less suitable for very short pulse exposures, which require longer reaction times. Nevertheless, such printing plates offer better printing quality than the ablated plates.

7.5.2 Power Requirements for Laser Printers

Printing plates use different materials optimized for absorption of emissions from either 830 nm laser diodes or 1064 nm DPSS lasers. The 830 nm wavelength is a good compromise between the shorter wavelength (560 nm), which permits more efficient absorption at the plate surface, and the longer wavelength (1064 nm), which allows more powerful diode lasers to be deployed. The power requirement for a laser printer is strictly determined by the plate material exposure threshold level, required imaging speed, optical efficiency, and scanning geometry involved. The laser power required can be expressed as

$$P_{\text{laser}} = \frac{(K)(E_{th})(A_p)}{(\eta_o)(T_{pe})} \tag{7.1}$$

where K is a constant (2 to 3), E_{th} is plate exposure threshold (typically from 100 to 200 mJ/cm^2), A_p is the plate area, η_o is the optical efficiency of the scanning system (typically 60%), and T_{pe} is the plate exposure time. Assuming a constant of 2.5, image size or area of 5000 cm^2, a plate exposure threshold of 200 mJ/cm^2, an exposure time of 120 seconds, and optical scanning system efficiency of 60%, the power requirement for this laser printer is about 35 W at room temperature. Calculated laser power requirements for a laser printer as a function of plate size, exposure time and plate threshold level are shown in Table 7.3.

It is important to mention that image-resolution requirements are very demanding because of halftone dot structure needed to achieve tone gradations in the offset lithography printing method. Halftone dot requirements vary from 800 dpi (dots per inch) for newspapers to 4000 dpi for high-quality art reproduction. This corresponds to a laser spot size of 10 to 50 μm. An experienced plate maker can expose a wide range of plate sizes with specific resolution requirements, but at the expense of higher cost.

7.5.3 Performance Requirements for Laser-Based Commercial Printers

The spot size, image resolution, printer speed or copies per minute, and cost per copy are the critical requirements of a commercial printer. Single-spatial mode operation and high beam-pointing stability are absolutely essential for good image

TABLE 7.3 Laser printer power requirements as a function of plate size, threshold level, and exposure time (W)

	100 mJ/cm^2		200 mJ/cm^2	
Area (cm^2)	6 sec	30 sec	60 sec	120 sec
5000	360	72	72	36
2500	180	36	36	18
1250	90	18	18	9

performance in a single-beam laser scanning system. The rotation rate of the scanning mirror, which scans the laser beam around the internal drum surface, limits the scanning speed and the reliability of the printer using conventional bearings. However, recent advances in hydrodynamic bearings offer scanning mirror rotating speeds as high as 40,000 rpm without any adverse impact on reliability.

The type of modulation required to produce the image must be carefully selected. In the case of an AO modulator using a BRAGG cell device, extremely high optical-power density inside the modulator crystal is required to obtain high modulation bandwidth. This limits the power handling capability to less than 20 W to avoid catastrophic failure or permanent damage to the crystal.

Single-mode output power levels exceeding 30 W or so can be obtained from lamp-pumped 1060 nm Nd:YAG lasers, but the power supply rating and cooling requirements can present a reliability problem. Power limitation problems can be overcome by deploying the latest DPSS lasers specifically designed for commercial applications. These DPSS lasers use more efficient host crystals such as YLF and YVO4 and offer power output levels greater than 30 W, higher fiber coupling efficiency, reduced thermal effects, and compact packaging [5]. These laser sources are not only more efficient and compact, but also exhibit excellent pointing accuracy and can be mounted on a moving carriage within the plates. These laser sources are most attractive for high power commercial laser printers due to their performance capabilities.

7.5.4 Multiple-Beam Exposure Technology for Commercial Printers

The practical problems of modulation and stringent laser specification requirements for a single-beam system can be significantly reduced, if not completely eliminated, by using a multiple-beam laser source. The lowest cost per watt of a diode laser makes the multiple-beam exposure (MBE) concept most attractive for commercial printers. In addition, mechanical and modulation speed requirements can be scaled down in proportion to the number of optical beams used. Most MBE systems use external drive configurations in which the printing plate is attached to a large cylindrical drum that rotates while the optical system is on a line slide. Image quality can be significantly improved using several individual fiber-coupled diode lasers with equal spacing along the length of the cylinder. An array assembly compromising of 32 or 64 fiber-coupled lasers can be used to meet the desired laser output requirements. Superimposing several single-mode diode laser outputs with a diffractive optic element offers built-in redundancy, automatic compensation for failure of any one diode, and overall improved reliability of the system. A 16-channel printing system with 12 fiber lasers per channel with each diode of 200 mW power rating provides a total laser output of 38.4 W.

7.5.5 Monolithic Linear Arrays (MLAs) for Commercial Printers

Uniform illumination of a linear spatial light modulator (SLM) offers an alternate approach to a multiple-beam exposure concept. Multimode laser bars comprised of monolithic linear arrays can provide the uniform illumination. Laser bars with output levels of 20 W, 30 W, and 40 W are currently available. Laser bar technology can assist in developing a thermal plate setter incorporating a light-valve-array (LVA) technology comprised of 500 channels. Such a system scans the plate with x–y linear translation of the optical head, instead of an external drum configuration. This approach provides efficient exposure of plates of different sizes and an automatic plate-handling system, leading to significant improvement in the overall efficiency for commercial printers.

7.6 HIGH-VOLTAGE SENSOR USING ELECTROOPTIC TECHNOLOGY

The step-down transformer (SDT) has been a standard device for decades for measuring line voltages ranging from five to hundreds of kilovolts. The primary terminal SDT must be connected to the high-voltage line to convert the line voltage to a safe level (say, 115 V) for measurement on a standard volt meter. An optic probe containing a Pockets cell with a folded light path offers a safe and reliable method of high-voltage measurements. The optical fibers that link the probe to transmit and receive optoelectronic signals can be 100 m or more in length with minimum insertion loss (typical fiber loss is about 0.25 dB per kilometer at 635 nm wavelength). In a typical Pockets crystal cell, the birefringence of an electrooptic crystal increases in proportion to an electrical potential applied directly across it. The potential arises from the external electric field emanating from a high-voltage power line. A ground plane must be provided near the probe for safety reasons, but the probe itself does not contain any electric wires. Three optical fibers link this electrooptic high-voltage (EOHV) device to a remotely located box containing the measurement circuitry and a laser diode emitting at an optimum wavelength of 635 nm. The optical light from the laser source is delivered to the probe via a low-loss single-mode optical fiber capable of keeping the beam size as small as possible within the optical crystal, which is made of magnesium oxide-doped lithium niobate material. This crystal material is durable and the doping significantly reduces the birefringence. Durability is of critical importance, because a strong electromagnetic field mechanically stresses the optical crystal, which can damage the crystal structure if the stress produced exceeds the safe limit of the optical crystal. Two single fibers exit the probe adjacent to the input fibers. Signals of identical strength at zero external electric field become unbalanced in proportion to the electric field. The entire optical system is enclosed in a rugged waterproof ceramic case. The EOHV device is capable of measuring line voltages ranging from 1 kV to 150 kV with maximum reliability and safety.

A high-voltage insulator with ideal geometry must surround the hermetically sealed EOHV device to ensure additional safety and reliability for continuous monitoring of the line voltage. This device is highly recommended for the use by electric power utilities companies, regardless of power generating schemes used, such as steam or nuclear power.

7.7 IR SENSORS FOR FIRE AND SMOKE DETECTION

Firefighters often face serious problems from the smoke produced by fires. Smoke inhalation not only pose a suffocation threat to both the victims and firefighters, but also significantly reduces visibility for the firefighters, resulting in huge damage to property and casualties.

Portable long-wavelength viewers using uncooled focal planar array (FPA) detectors are available to pinpoint the source of a blaze through heavy smoke and prevent damage to the property and occupants. The latest IR cameras operating over the 8 to 12 μm range and using uncooled FPA detector technology offer firefighters far better smoke penetration capability. Various advanced techniques are being investigated for smoke detection devices. Solid-state pyroelectric array technology is available for possible application in fire detection devices. IR thermal imaging devices comprised of 510 × 492 element ferroelectric FPA detectors operating over the 8 to 12 μm spectral range offer efficient and reliable operation in a compact package weighing less than 4 pounds. This hand-held device offers superimposition of the visible and infrared images with direct temperature read-out capabilities, which will allow firefighters to minimize damage to property and its contents and occupants. Compact, light-weight, helmet-mounted IR thermal viewers using uncooled pryoelectric FPA detector technology are specially designed for firefighters, enabling them to control the fire damage with minimum time, effort, and damage. The binocular display incorporated in the IR viewer provides the user with a clear visual path through the flame and smoke. The compact, fireproof processing unit along with batteries is belt-mounted, and the video port of the device can be displayed onto a remote monitor for authorities to take further appropriate action. The sensitivity of this IR imaging device is better than 0.5 F for room-temperature targets. This IR device has been demonstrated to spot a man within 20 seconds after entering a house under fire. According to firefighters, without the devices they do not see flames but dense smoke, and under such conditions, the IR thermal viewer can play a critical role in the detection and location of fire.

The latest IR cameras [6] involve no moving parts and incorporate antireflection coated germanium optics and an uncooled 320 × 240 element microbolometric FPA detector capable of operating over the 8–14 μm range. The bolometer-based FPA detector requires minimum electric power and a thermoelectric (TE) cooling device to maintain a stable reference temperature near the ambient temperature. The thermal sensitivity of this particular FPA detector is better than 80 mK or 0.080 K. The camera lens and a 4″ × 4″ video screen are angled up to provide a comfortable view of the scene under fire. This IR device is powered by standard cam-corder batteries with two-hour continuous operating life between charges.

7.8 OPTICAL CONTROL OF PHASED ARRAY ANTENNAS (PAAs)

Optical control of modern phased array antennas is necessary in advanced radar and EW systems, where fast response, high beam pointing accuracy, symmetrical antenna patterns in AZ and EL planes, ultra-low side-lobe level (SLL), squint-free beam, multiple beam forming (MBF) capability, and fast beam steering with minimum loss and cost are the principal requirements. The integration of monolithic microwave integrated circuit (MMIC) technique, transmit/receive (T/R) modular concept, and photonic technology will play a key role in future PAA systems. Maturing of photonic technology and availability of optoelectronics are necessary to integrate the optical control technology in the T/R modules for PAA systems involving a large number of individually controlled MMIC-based T/R modules. The size and complexity of future PAA systems demand unique concepts for distribution of RF energy and processing of received signals over wide band with minimum time. In case of a PAA system, one must distinguish between three types of signals to be routed, namely, reference signals, coding signals, and control signals, all of which can be transmitted with minimum loss through optical fibers.

7.8.1 Synchronization of T/R Modules

Large PAA systems comprised of several hundreds of MMIC-based T/R modules must be synchronized to the master oscillator to assure signal coherence at the antenna aperture. One approach is to provide a local oscillator (LO) at each T/R or subarray level and then optically injection-lock all the LOs with the master oscillator. This approach will be expensive and complex. Another approach is to transmit the master oscillator signal via the high-speed, low-loss fiber optic (FO) links shown in Figure 7.3, detect the optical at the T/R module, and amplify it. This concept is fast, reliable, and inexpensive.

7.8.2 Signal Coding Techniques

Modulation of the carrier is required during the transmit mode, whether in a radar or communication system. In case of a communication system, an analog or digital signal is superimposed on the carrier using either AM or FM modulation techniques. Radar applications generally require direct pulse coding of the carrier by linear FM direct on–off keying or phase shift keying (PSK) using a Barker code or another pseudorandom code as the carrier envelope. Two distinct coding configurations are illustrated in Figure 7.3. The central coding (CC)configuration executes the coding at the central processing unit (CPU), while in the case of the remote coding (RC) technique, coding is performed at each T/R module. The CC configuration requires one FO link between the CPU and the T/R modules, while the RC technique requires two FO links for each T/R module.

The sensitivity, noise figure, and dynamic range are the critical performance parameters of a FO link. The FO link generally consists of a AlGaAs optical transmitter, optical receiver, and an optical fiber of suitable length. Laboratory measurements indicate that the RC method has a dynamic range at least 10 dB higher than the CC technique. Reactive matching of laser diodes yields an additional 10 dB improvement in the sensitivity of the link. Reactive matching of the optical transmitter or laser (Figure 7.3) can be achieved by using a laser chip embedded into a reactive-

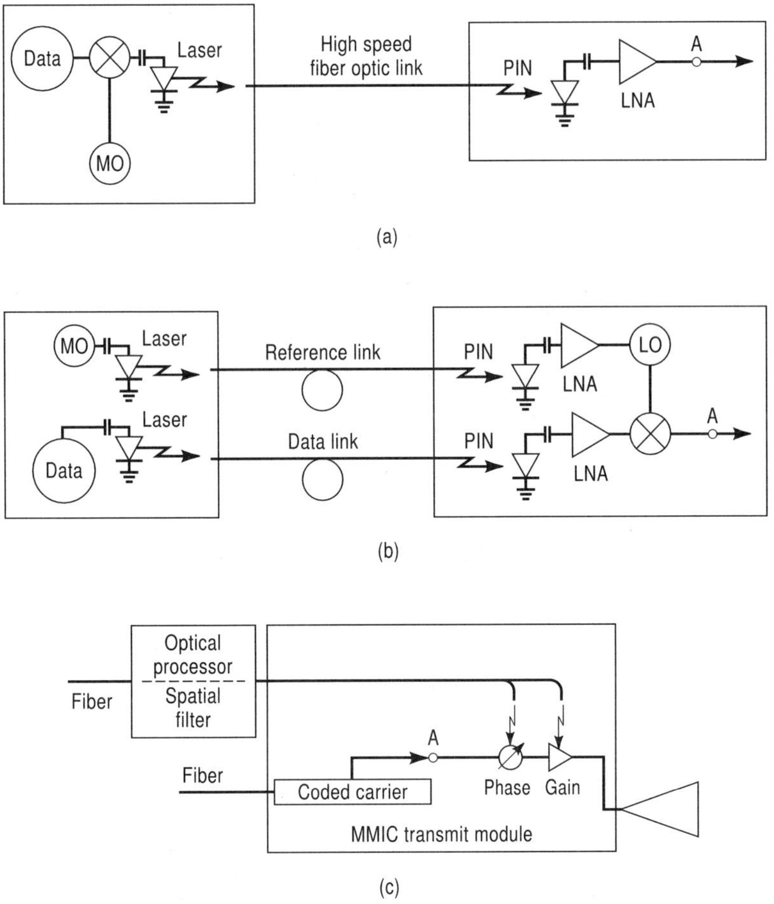

FIGURE 7.3 High-speed fiber optic link configurations (A and B) and optical control of transmit module using MMIC technology (C).

ly matched circuit. The RC technique outperforms the CC scheme in avionics systems operating below 10 GHz.

7.8.2.1 Optical Beam-Forming Technique

Optical beam-forming (OBF) capability in a MMIC-based PAA system is of paramount importance because it allows the steering of the radar beam in the desired direction with required adjustment of the beam shape to achieve low side-lobe suppression and select nulling in specific directions. OBF capability is obtained through optical control of the phase shift setting and amplifier gain. In an active PAA, the excitation of each antenna element must be controlled to allow amplitude tapering and beam steering with optimum accuracy. In an optically controlled PAA system, the gain and phase values of each radiating element are controlled through optical source intensity levels, as shown in Figure 7.4.Two distinct beam forming schemes are available depending on whether the phase shift and attenuation are implemented in the frequency domain or in the time domain. Typical performance capabilities of an optically gated reflective-phase shifter and optically gated switch-type variable attenuator as a function of microwave frequency and laser illumination level are shown in Figure 7.5. All microwave components such as phase shifters, attenuators, and other control devices use optical control signals [7] transmitted via optical fibers. The control functions generated by the beam steering computer (BSC) are used to control the gain and phase of each T/R module. Since all the com-

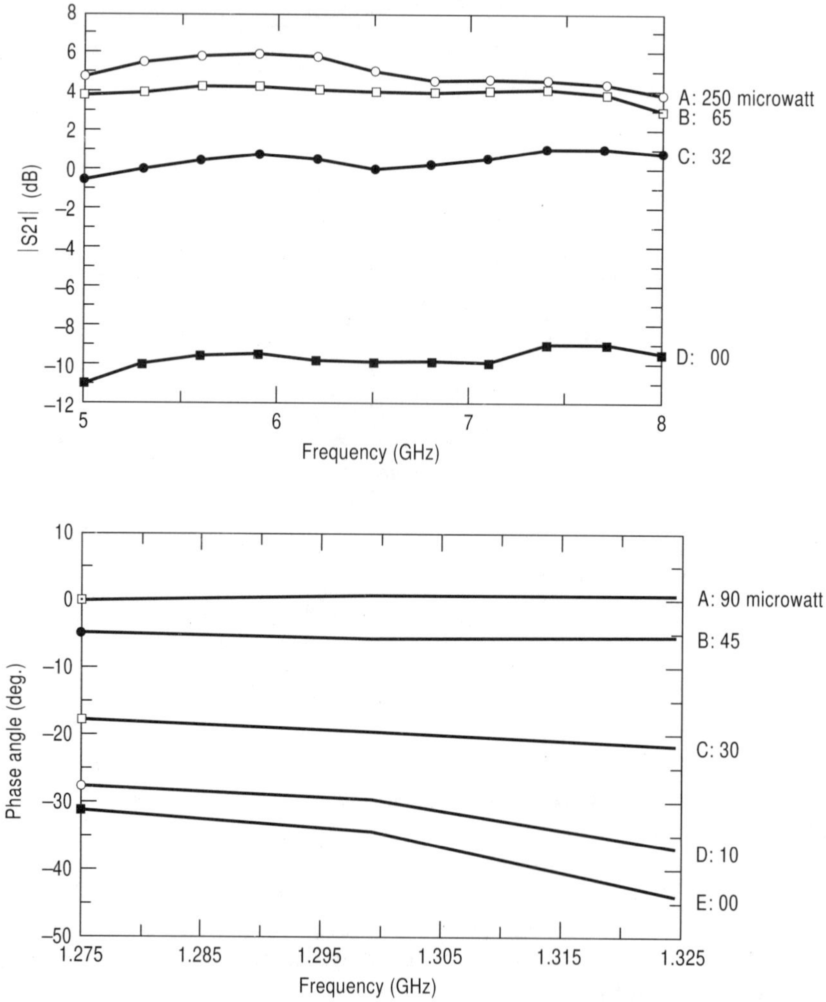

FIGURE 7.4 Transmission coefficient and phase values of an electrooptical phase shifter as a function of optical intensity.

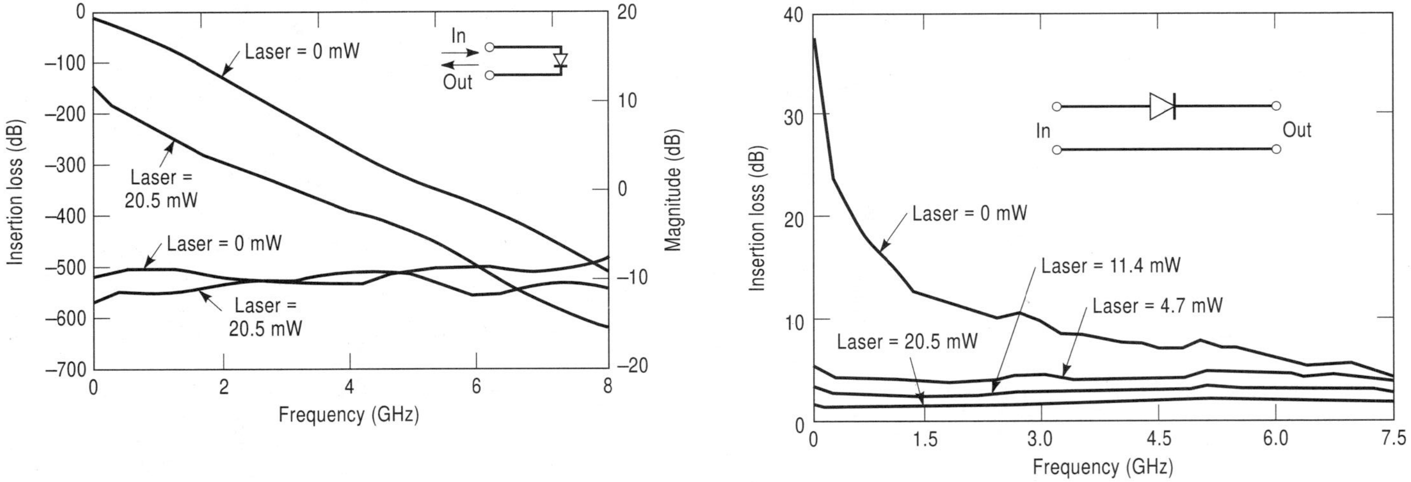

FIGURE 7.5 Reflection coefficient of a phase shifter and insertion loss for a switch/variable attenuator as a function of optical power using electrooptical technology.

ponents used in fabrication of a T/R module are compatible with GaAs-MMIC technology, they could be integrated on a single MMIC chip with minimum cost, complexity, loss, and power consumption.

The frequency response of a MESFET amplifier remains practically flat, as illustrated in Figure 7.4. It is possible to control the amplifier linearly over a 20 dB range using only 60 μW of optical power from an inexpensive LED source. Similarly, optical control has demonstrated a 360-degree continuously variable phase shift capability with a phase shifter using MMIC technology. L-band, S-band, C-band, and X-band MMIC-based phase shifters demonstrate excellent performance in terms of accuracy, speed, and optical power consumption. In brief, optical control techniques offer:

- Phase and amplitude control independent of frequency band
- Complete compatibility with MMIC technology
- This technology can be extended to other MMIC functions including switching, limiting, and modulation
- This technology is fully compatible with emerging optical signal processing technology such as spatial filtering.
- It is most cost-effective, because it uses low-cost, low-power LED devices and optical fibers.
- Optical control schemes require only microwatts instead of milliwatts, thereby affording significant reduction in power consumption.

7.8.3 Direct Optical Control (DOC) Technique

The direct optical control technique is a sophisticated approach that uses optically controlled microwave devices in conjunction with optical processing devices involving spatial filtering that are incorporated into the T/R modules. This method is referred to as direct, because the microwave control device such as a PIN diode or FET acts both as a detector and as a control element. Lateral PIN diodes are used in phase shifters, switches, and attenuators and their structures are amenable to low-cost MMIC processing. The optically controlled PIN-diodes have great versatility. These diodes are used in several applications including frequency tuning, dual-frequency antennas, tunable filters, modulators, attenuators, switches, and phase shifters. Direct optical control of MMIC-based gain circuits and oscillator configurations using state-of-the art solid state devices such as MESFETs, HEMTs, p-HEMTs, and HBTs have been successfully developed [7]. Recent research and development activities indicate that heterojunction devices exhibit better performance than the homogeneous devices under similar operating conditions. Studies performed by the author indicate that HEMT devices offer higher sensitivity than MESFET devices. The studies further indicate that multiple-quantum-well (MQW) devices offer even more sensitivity to optical control power than HEMT devices. Regardless of the devices deployed as control elements, the optical coupling to the light-sensitive regions limits the performance level. Optical control of passive microwave components such as dielectric resonators, printed circuit antenna elements, and microstrip circuits were demonstrated about ten years ago [7]. This is made possible by incorporating a photosensitive material into a microwave circuit or device and then activating the photoconductor by LED illumination. High optical intensities are required to generate high densities of electron-hole pair.

7.8.4 Multiple Beam Forming (MBF) Capability Using Optical Technology

The MBF technique using two stable laser sources suffers from high cost, poor reliability, and operational complexity. The latest MBF technology with FO links is based on true time-delay (TTD) phase shifters using optical FO links connecting the CPU with the T/R modules. As stated earlier, MMIC-based PAA systems are receiving great attention because of several advantages offered by the optical control technique.

TTD-beam-forming technique offers significant reduction in hardware requirements, particularly in case of large phased array antennas. The use of TTD devices at the subarray level using conventional phase shifters offers the most practical and cost-effective design. A two-dimensional PAA using a TTD scheme incorporating polarization switching spatial light modulators can significantly enhance the operational capability of the antenna system. TTD devices incorporating optical fibers with low loss (0.0005 dB per meter at 1.55 μm) and dispersion (0.02 nm per meter) characteristics makes the multiple beam-forming scheme [8] most attractive for several military applications.

Another beam-forming technique, illustrated in Figure 7.6, offers independent control of microwave frequency and phase using two distinct networks. Two networks of equal lengths, one dispersive and the other nondispersive, are used. A length increment is used to provide the phase control frequency, which determines the interelement phase shift. A significant reduction in array element numbers is possible when optoelectronic mixers are used for signal recovery. The electrooptic technique for the frequency-independent beam formation scheme provides the multiple beam-formation capability by simultaneously sweeping two frequency sources. However, this technique is most attractive for electronic systems, where arbitrary phase setting at each radiating element is not required.

It is important to mention that in both noncoherent beam forming schemes described here, phase shifts have been determined by the relationship between the path length and modulation frequency. Furthermore, exact frequency of the optical source is not important, because the direct detection method has been used. Beam-forming techniques employing coherent optical sources offer improved beam pointing accuracy, but with high cost and complexity.

Integration of optical control techniques and MMIC technology could greatly expand the capabilities of future radar, communication, and electronic warfare systems. Distribution of high-frequency and high-data-rate signals via optical fibers has been successfully demonstrated. As stated earlier, optical control offers TTD ca-

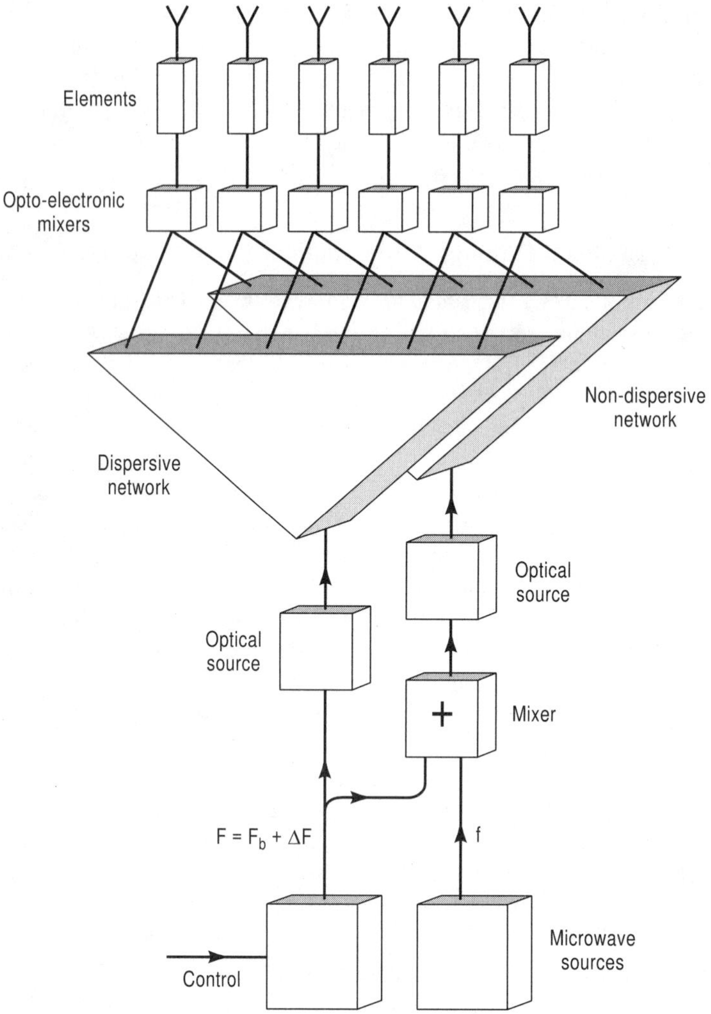

FIGURE 7.6 Electrooptical technique for frequency-independent beam-forming.

pability over wide band, larger dynamic range, high immunity against radiation and EMI, and improved antenna performance with minimum power consumption and cost. Future research activities must be directed towards hybrid optical–microwave antenna system involving chip-level integration of microwave and optoelectronic components, improved FE link configurations, and system-level integration of optical and microwave components and devices with particular emphasis on optical signal processing.

7.9 OPTOELECTONIC DEVICES FOR DETECTION OF DRUGS AND EXPLOSIVES

7.9.1 Introduction

Portable, supersensitive optoelectronic devices, when integrated with computed-tomography (CT) equipment, are best suited to check luggage and shipping containers for traces of explosives, plastics explosives, chemical and biological weapons, and other organic compounds. Detection of weapons carried by terrorists is of paramount importance. Passive IR devices can sense emissions from concealed weapons or explosives. Even a small amount of plastic explosive can do tremendous damage to property and serious injury to persons living nearby. Current research efforts are being focused on quadrupole-resonance (QR) technology, specifically for airport security applications. QR technology is most effective in detecting explosives and illegal drugs concealed in passenger baggage. Today, crime-fighting agencies face a host of new challenges from sophisticated criminals who have no regard for human life or property. International and domestic terrorists have access to newer and more effective explosives. Drug trafficikng involves other activities such as prostitution, murder, theft, and sexual assault. Detection of drugs is the major problem facing law enforcement agencies. Sophisticated technologies and techniques such as Fourier transform infrared (FTIR) spectroscopy and Fourier transform Raman (FTR) spectroscopy are currently being investigated for rapid identification of drug compounds.

7.9.2 FTIR Spectroscopic Technique

FTIR spectroscopy is the most sophisticated technology. It allows scientists and investigators to determine the origin of an exploded bomb based on a trace of residue or to determine the make and year of the car or truck left by terrorists at the scene of an explosion based on a retrieved paint chip. This technique can provide DNA evidence from a drop blood that can either rule out a suspect in a sexual assault case or strengthen the prosecution's argument. Experienced chemists can analyze bags of drugs and can even identify types of drugs using the FTIR spectroscopic technique. This technique offers the right tools to chief medical examiners, forensic crime laboratory directors, and DNA specialists for performing their tasks with high accuracy and reliability.

FTIR spectroscopy is one of the leading photonic techniques and is widely used by major crime laboratories because it offers investigators a quick and reliable method to determine the identity of the suspected drug dealer or a criminal. Furthermore, FTIR sensors can easily distinguish between the various isomers that may be hard to distinguish using gas chromatography–mass (GCM) spectrometry technology. In the case of methamphetamine and phentermine drugs, the investigators would prefer the FTIR technique to achieve a clear and foolproof reading. However, the time-consuming preparation required for each sample is primarily responsible for the decline in the popularity of FTIR technology. Both the heating of the sample in

GCM spectrometry and the time-consuming sample preparation required in the FTIR technique can be eliminated by using the FTR spectroscopic technique.

7.9.3 Fourier Transform Raman Spectroscopic (FTRS) Technology

The FTRS technique provides rapid identification of chemical compounds without complicated pretreatment. This nondestructive method has several other advantages, including the ability to sample through glass or the thin plastic bags widely used by the drug dealers. Availability of libraries of Raman spectra, expertise in this field, and familiarity with Raman spectra are of paramount importance for all investigators and forensic scientists. Field-portable Raman spectrometers are commercially available; they can be used to test a jar or glass tube of clear liquid suspected to be poison. A Raman spectrometer has provided positive identification of a substance that was cyanide.

7.9.4 Low-Cost Devices for Detection of Drugs

Cities and counties with financial constraints will generally prefer low-cost, low-tech devices to combat drug problems. However, some crime labs use FTIR techniques in situations where GCM spectrometry does not provide conclusive distinction between crack cocaine and cocaine hydrochloride. Other laboratories use UV spectrometric techniques because of their ability to screen the samples with minimum time in the 220 to 360 nm wavelength range, a spectral region particularly useful for positive detection of both cocaine and methamphetamine. In the case of mixtures of drugs or mixtures of contaminants with overlapping UV spectra, it is difficult to distinguish between vitamin C, with a high absorption at 250 nm, from cocaine, methamphetamine, or nicotine, all with absorption in the 220 to 270 nm range. In such situations, one must consider an alternate technique known as capillary chromatography.

Crime lab scientists have indicated their preference for the gas and liquid chromatography techniques coupled with mass spectrometry particularly for analyzing a wide range of residues and powders left by the explosives. The high-pressure liquid chromatographic technique is most suitable for compounds that are not stable when heated. When attached to such a system, a diode array detector with a spectral range between 190 and 460 nm makes it possible to analyze the compounds, because an organic molecule absorbs at a particular wavelength.

Gas chromatographic systems equipped with chemiluminescene detectors can sense IR pulses that explosive molecules emit after reacting with atmospheric ozone. Such a system is known as a thermal energy analyzer and its portable version is widely used at airports for on-site testing passenger belongings.

Hand-held ion mobility spectrometers (IMSs) can be successfully used to trace residues of tear gas. These devices are best suited for riot control environments and to maintain civil order. Low-tech devices such as polarized light microscopes (PLMs) are widely used to identify hair and clothing fibers. In certain cases, these devices can identify explosives, but the operator requires extensive experience to recognize specific crystal structures. Identification of hairs or fibers is a leading method of linking a potential suspect to a crime. A more obvious link is possible when a perpetrator leaves behind at tiny sample of blood or semen. Forensic scientists or biologists can generate a DNA "fingerprint" of the perpetrator even from the tiniest sample. DNA typing is accomplished by determining the specific number of tandem repeats on a chromosome [9]. Forensic scientists consider these DNA tests most reliable; they can either rule out a suspect or strengthen the prosecutor's case.

7.9.5 Laser-Based DNA Technique to Fight Crime

The laser-based DNA (LBDNA) technique, also known as the capillary electrophoresic process, is getting a lot of attention today. In this process, forensic scientists extract the DNA and create a sort of carbon copy of each strand with a specific number of tandem repeats; these are then treated with fluorescent tags to highlight their unique number needed for conclusive and meaningful DNA analysis. An argon ion laser operating either at 488 nm or 514 nm wavelength shines directly on the tiny glass capillaries, causing the DNA to fluoresce. A CCD array with 640 × 512 pixel resolution capability automatically records each sample with high accuracy. This system allows a lab technician to perform reliable analysis on smaller samples of genetic material. Furthermore, the LBDNA equipment is capable of analyzing samples that have been degraded or are few in number. State-of-the art LBDNA equipment incorporates a 20 mW laser source emitting at 532 nm that can record the fluorescence of DNA bands with two photomultiplier tubes with much higher gain than conventional ADP detectors. Lab technicians can switch optical filters and fluorophores to mark characteristics on the DNA strands with high resolution. Capillary electrophoresis requires high a degree of skill in operating the system and considerable experience in proper loading of each sample. The LBDNA technique represents the latest implementation of photonic technology in taking a bite out of crime.

7.9.6 Laser-Based Fingerprinting (LBFP) System

The LBFP technique has become a widely accepted procedure capable of yielding fingerprinting of high quality and very high resolution. High-power lasers have picked up the faintest fingerprints and tiniest fibers. The LBFP system uses the latest photonic technology and enables the police to crack the most difficult criminal cases, leading to successful convictions. The system uses argon-ion laser emitting bright green light at 574 nm to unveil latent prints with the highest clarity and resolution. Under the argon ion laser light, the human palm print shows more clearly distinct lines and ridges. This level of fingerprint performance can positively identify and track a criminal responsible for murder or rape.

Despite high cost and occasional maintenance problems, law enforcement agencies have found LBFP equipment most reliable and effective in fighting crimes. High procurement cost problems can be solved by switching to a DPSS source. However, maintenance problems remain the same because of the additional cooling requirement needed for optimum system performance. The latest LBFP devices deploy portable light sources that are tunable over wide spectral ranges to provide more operational flexibility. Low-cost light sources such as xenon lamps or metal halide arc lamps fitted with several optical filters offer lower procurement cost and still allow investigators to obtain other trace evidence such as hair or clothing fibers linking a suspect to specific crime. A tunable light source can pick up fingerprints left on many surfaces, including coffee cups or bags of potato chips, because of its ability to tune to the optimum wavelength needed to create the best contrast for retrieving a print.

7.10 LASER SENSOR TO DETECT CLEAR AIR TURBULENCE (CAT)

Clear air turbulence (CAT) presents a serious safety problem both to passengers and the aircraft and can occur anytime with very little warning. Turbulence can be both a nuisance and an unexpected danger. The latest Doppler LIDAR (Laser-induced Doppler and ranging) sensors provide a first defense against the dangerous patches of swirling wind. The LIDAR sensor sends laser beams several miles (up to 100 miles) ahead of the aircraft. When the laser beams hit airborne particles such as dust or aerosols, they reflect back the laser energy to the plane. The reflected optical signal amplitude is determined by the scattering, diffraction, and absorption coefficients of the particles, which are functions of altitude and operating wavelength. This sensor with built-in computer will characterize the turbulent air motions ahead in time for the pilot to warn the passengers and crew to fasten their seat belts. The laser-based sensor saves the passengers and crew from getting injured by the sudden reduction of altitude needed to avoid the turbulence region. LIDAR provides the pilot with 1 to 15 minutes of warning time of turbulence ahead of the aircraft, which seems sufficient to take appropriate safety precautions, thereby avoiding any injury to the passengers or crew.

7.11 PHOTONIC-BASED SENSORS TO COUNTER TERRORIST THREATS

Unexpected and isolated terrorist attacks involving biological, chemical, or small nuclear weapons and rapid proliferation of weapons of mass destruction (WMD) have compelled the international scientific community to investigate photonics technology as a first line of defense. Nuclear threats require a global-scale effort to analyze intelligent data and satellite photos for initial clues. On the domestic front, the twin threats of biological and chemical weapons present a more immediate danger without any warning. Early detection of weapons and terrorist locations remains one of the best defenses. Studies performed on photonic devices by the author indicate that GCM spectroscopy, ion mobility spectroscopy, LIDARs, and other advanced IR sensors are available for detection and location of the weapons used by the terrorists. These systems are bulky, costly, and not easy to operate. However, photonic-based sensors are being considered as an integral part of the current and future generation of weapon detection technology.

7.11.1 Photonic Sensors to Counter Biological Threats

An anthrax attack can kill thousands of persons within a few hours; victims exhibit symptoms of influenza involving cough and fever. Death can be expected from massive hemorrhaging, edema, and seizures. Evidence of foolproof detection technology is not currently available. Scientists are developing photonic technology that will assist emergency personnel in rapid identification and quantification of biological toxins, minimizing death and injury.

Fourier transform Raman (FTR)spectroscopy and surface-enhanced resonant Raman (SERR)spectroscopy techniques, which are widely used for detection of lethal traces of explosives including Semtex, RDX, and TNT are being considered to counter biological terrorism. The above-mentioned techniques use lasers operating at wavelengths that are sensitive enough to detect biological agents. At present, reliable spectral signatures to detect or identify biological agents in air are not fully known. Detection and identification of biological agents requires knowledge of the specific luminescence characteristics of anthrax, botulism, and plague, all composed of amino acids, the building blocks of life, to differentiate between the hazard levels of certain particles in the biological agents from those that occur naturally in the environment.

Research scientists think that LIDAR is the most promising sensor for the detection of pathogens in distant clouds. The LIDAR is equipped with an Nd:YAG laser operating at 1064 nm with a pulse power up to 450 mJ. Backscattered optical signals from the clouds are collected by a telescope and detected with a sensitive silicon avalanche photo diode (APD). However, the sensitivity of this laser wavelength is not impressive. Therefore, scientists have developed a more sensitive UV–LIDAR using a lamp-pumped Nd:YAG laser that is frequency-tripled to 355 nm wavelength (1065/3 = 355 nm). This particular sensor incorporates a double-grating monochromator coupled to a photomultiplier tube (PMT), which detects the fluorescence levels associated with aerosol clouds with impressive sensitivity due to very high gain of the PMT (about 1000 times higher gain than offered by a APD). This sensor will allow gathering of more information on suspended aerosol clouds, including the types of molecules present, whether dead or alive. The LIDAR sensor operating at UV wavelengths is most effective as an initial warning about the possible nearby danger. The short-range (5 to 10 km) LIDAR, when mounted on a pickup truck or military vehicle could provide a first line of defense against biological agents such as anthrax and botulism.

The latest research has focussed on a detector configuration that combines the transducer and probe components. The new detector system takes advantage of crystallized rare-earth compounds that emit a photon in response to absorbing two or three photons at a specific IR wavelength. This process is known as phosphor up-conversion. When photons are used in conjunction with an antibody coating [10] and a diode laser emitting at 980 nm as an excitation source, the sensor emits a unique, narrow band between 500 and 600 nm in the presence of certain biological agents. The emission spectra can be recorded using either a PMT or an APD, depending on the sensitivity requirement. This laser-based systems can be fine-tuned to achieve highest sensitivity, and has a capability of detecting of several dozens toxins simultaneously with high reliability. However, research efforts must be continued to upgrade sensor performance for quick, accurate detection of toxins in real time with high probability of success.

7.11.2 Low-Cost Photonic Devices to Counter Chemical Threats

The 1995 Tokyo subway incident illustrates the vulnerability of large cities to chemical agent attacks. The most dangerous chemical agents include deadly nerve gases sarin, cyanide, phosgene, and mustard gas, all of which can cause mass casualties. Because these attacks occur in stealth with slow release of a vapor or aerosol, detection of chemical agents or gas will be rather slow and difficult. Due to the small size and unique chemical makeup of each particle of the agent, it is more sensitive to spectroscopic detection. Detection of chemical agents is affected by false readouts.

A wide variety of detection techniques are available including disposable paper tickets that operate like litmus paper, mobile ion-based spectrometry sensors, photoionization detectors, FTIR spectrometers, and Raman IR spectrometers. However, these detection systems suffer from high cost and weight and require extensive time-consuming analysis procedures. A sensor centered on improving real-time sensitivity and portability aspects will be most attractive for this particular application. A polymer-based lanthanide luminescent sensor recently developed by The Johns Hopkins University Applied Laboratory (Laurel, MD) offers reliable detection of soman and sarin nerve gas. The fiber optic probe used by the sensor is covered with a polymer containing the metal europium, which gives off a strong orange-red glow when the coating is exposed to blue light at 457 nm from an inorganic laser. Incorporation of a smaller, inexpensive light sources such as LEDs emitting at 470 nm or the latest blue diode lasers will further reduce the sensor procurement cost, weight, size and power consumption. When the coating encounters certain chemical derivatives of either of the two above gases, the color changes perceptibly, indicating the presence of toxic gases. To prevent false readings from chemically similar pesticides, the europium must be embedded in a plastic film that binds to the hydrolysis product of soman.

Photonic-based up-converting phosphor detectors may be suitable for detecting biological agents that may elude fluorescence detection techniques. The polymer-based lanthanide sensor will better differentiate between a nerve gas molecule and a common pesticide. Because of the operating environments of such a sensor, it must meet high reliability, false-alarm-free operational capability, high sensitivity, cost-effective design, real-time detection, and portability requirements.

7.12 IR SENSORS FOR INDUSTRIAL APPLICATIONS

Under this section, IR sensors capable of providing the real-time data needed to develop, optimize, monitor, and control industrial processes will be described. Gas chromatography (GC) and mass spectrometry (MS) are widely used in gas analysis, but these techniques do not provide real-time process monitoring and control. Even though the high-speed GC systems have reduced the analysis time from hours to minutes, still they suffer from other limitations leading to a costly approach.

7.12.1 IR Sensors to Monitor Semiconductor Fabrication Processes

Optical emission spectrometers (OESs) are generally used in semiconductor fabrication procedures. However, they require low pressure and lack qualitative accuracy because of instrumentation variations and pressure fluctuations. True qualitative analysis is impossible with plasma variations, window-material degradation, and limited field of view (FOV). These drawbacks can be overcome with an IR absorption spectrometer (IRAS).

IR spectra of a molecule are fairly constant but dependent on chemical structure and temperature variations. A small IR spectral absorption collected from a combustion source can be compared with calibration spectra. The calibration and combustion spectra collected from different IR spectrometers indicate no variations in spectral characteristics, thereby providing highly reliable and fairly accurate qualitative analysis data. However, reliable monitoring of the processes during the etching of a complex, multilayer, patterned wafer requires a versatile sensor such as a FTIR spectrometer. These spectrometers provide a reliable method of optimizing the plasma power settings to maximize the active etch constituents. No other sensor can meet or beat the performance capabilities of a FTIR spectrometer in the etching process. However, if the FTIR spectrometer is found to be too expensive for production monitoring involving one or few species, a low-cost, nondispersive IR (NDIR) spectrometer can be used to minimize the product monitoring costs.

7.12.2 IR Sensors for Environmental Control

Exhaust gases from semiconductor processes generally contain toxic and greenhouse gases not suitable to release into the atmosphere. Abatement systems are too expensive and unreliable. An FTIR multigas analyzer samples the exhaust contents, such as CO, NO, SiF_4, C_2F_6, CF_4, etc., before and after the abatement system and thus provides a reliable method to analyze and identify gaseous constituents present in the exhaust. This sensor allows an environmental engineer to select, quantify, control, and periodically validate the performance and effectiveness of the system. FTIR analyzers can also be used to measure the destructive effectiveness of the abatement system as well as the potential production rate of secondary pollutants. A trained operator is required to interpret and quantify species present in the exhaust system, but spectral analysis software is needed to provide immediate interpretation and quantification required for complete real-time answers. FTIR spectrometers provide larger spectral throughputs, thereby transmitting more IR energy to the detector. This will direct all IR frequencies on the detector, distribute detector noise uniformly across the entire spectrum, and maintain better frequency accuracy.

The data are collected via a single-mode laser acting as a spectral reference, which allows data collection from multiple instruments and eliminates the need for recalibration. These advantages of a FTIR spectrometer allows the collection of high-resolution data with high signal-to-noise (S/N) ratio and rapid real-time analysis of most gas compositions with improved sensitivity. The use of multiple optical filters permit simultaneous analysis of several gases by a single FTIR sensor. In summary, a FTIR-based gas monitoring system provides rapid, reproducible, qualitative and quantitative measurements of processes from abatement system to combustion exhaust to semiconductor processing chamber.

7.12.3 FTIR Spectrometer for Monitoring Turbine Engine Performance

Monitoring of combustion gases from commercial and military turbine engines is possible with FTIR spectrometers. The high-temperature, high-pressure combustion gaseous products can be monitored with high accuracy and reliability without continual calibration. Since the FTIR spectrometer offers full spectral coverage capability, only one sensor is needed to measure a number of combustion chamber gases concurrently, unlike other techniques that would require numerous sensors, each dedicated to monitor one specific compound or a group of compounds. In addition, the FTIR spectrometer provides rapid and reliable monitoring of other organic gaseous mixtures, which further identifies and characterizes combustion source products.

In the case of high-performance supersonic fighters, accurate monitoring of temperature and pressure of various gases associated with gas turbine emissions are of paramount importance for tactical reasons. Multigas FTIR analyzers are widely used by gas turbine engine manufacturers and military maintenance depots for real-time monitoring the gas turbine emissions. This sensor can simultaneously monitor the gas emissions as well as the hydrocarbon contents over a wide concentration range with minimum cost and high reliability.

7.12.4 Raman Spectroscope for Real-Time Monitoring of Manufacturing Processes

The low-resolution Raman spectroscope (LRRS) realizes significant reduction in hardware costs while providing both effective real-time manufacturing process quality control and environmental control capabilities. The LRRS uses a multimode diode laser with typical output power levels in excess of 300 mW at 785 nm, a low-resolution monochromator matched to a CCD-detector, optical filters with specified skirt selectivity, and a 400 μm optical fiber. This portable LRRS system measures approximately 10 × 20 × 28 cm, weighs less than 6 pounds, costs about $10,000 in 1999 dollars, and can be interfaced to a desktop or notebook computer.

The diode laser is coupled to a high-sensitive spectrometer capable of detecting and analyzing optical signals over a spectral range from 785 to 1000 nm. The optical signals are collected by a sensitive probe. The detector is a linear silicon–CCD array comprised of 2048 elements. The sensitivity of the system is 86 photons per count for a one-second integration period, which can be varied from 4 ms to 60 seconds with a 500 kHz A/D converter card or from 20 ms to 60 seconds with a 100-kHz A/D converter card. This spectroscope offers grating density of 1200 lines per mm and a resolution of 25 to 30 lines per cm.

7.12.4.1 LRRS for Pharmaceutical Manufacturing Applications

Recent tests performed by a LRRS system [11] at a pharmaceutical manufacturing facility making steroid pills demonstrated significant improvement in quality control of the product under the production schedule. This LRRS offers real-time monitoring of quality control processes for manufacturing medications in pill format or solid oral-dosage forms. In pharmaceutical manufacturing, it is critical to measure and control crystalline morphology—the geometry involving the crystalline lattice. The LRRS can also be used to monitor a lamination process. During the two-part lamination of packaging films, accurate measurements and precision control of the blend ratio of the urethane adhesive to the curative agent are absolutely necessary. The ultimate performance of lamination [11] depends on a full and proper cure to produce full bond strength and required heat resistance. This miniature fiber optic Raman spectrometer can obtain spectra every 60 seconds from a methylene biphenyl polymer (MBP) adhesive and a thermoplastic polymer curative. The peak reading at 1619 per cm indicates the concentration of the adhesive, whereas a peak at 1731 per cm shows the concentration of the curative. These peaks indicate the relative strengths of the adhesive and the curative material in a lamination application and can be accurately controlled by the LRRS system with minimum cost and high reliability.

7.13 IR CAMERAS FOR IR MACHINE VISION (IRMV) SYSTEMS

A new generation of uncooled IR cameras has been developed by several companies for potential application in IRMV system. These new, low-cost cameras involve solid-state uncooled detector arrays and built-in intelligent networks to deliver the performance and cost-effectiveness that are urgently needed by manufacturing engineers. The IRMV system uses image data to automatically control the production rate of a specific product, monitor the quality control and integrity of the process, and guide the precision movement of the machine involved. An IRMV system consists of an imaging sensor that gathers the data, a high-speed digital computer that processes and intreprets the information, and a machine-controlled interface unit that provides the feedback. The broad range of applications satisfied by the IRMV system is based on the thermal image. The IRMV system can monitor the moisture content in a product, detect the burn pattern produced by a jet engine turbine (which can identify plugged fuel nozzles), and examine a rubber melting process to ensure that the thickness of the material in automobile tires is uniform (based on uneven temperature distribution).

The latest IRMV systems use an uncooled solid-state detection sensor operating over the 8–12 μm range and is capable of monitoring the thermal characteristics of the materials and processes involved. Two outstanding features of this sensor are the ability to detect the events that appear only in the IR range of the spectrum and the ability to eliminate visible noise that prevents the analysis of the problem. Surface roughness can indicate noise over the 4.5–5.6 μm region of visible light cameras, because the optical light can reflect off tiny surface features. Potential applications of IRMV systems to various manufacturing processes will be discussed.

IR sensors play critical roles in the lumber industry. These sensors are capable of revealing buried knots in logs, because the dense composition of the knots exhibit thermal properties different from those of the surrounding wood. Thus, cutting tools equipped with IRMV guidance systems can evaluate sawlogs, thereby optimizing the amount of salvageable lumber produced by a mill.

In the automobile industry, machinists and engineers using standard machine-vision systems are not able to find dents in bare metal surfaces because of the visible-light noise arising from the highly reflective surfaces. However, IRMV systems are able to ignore the visible-light noise and disclose imperfections on shiny surfaces with high accuracy and reliability.

In the case of steel industry applications, the middle of the surface of cast steel is more exposed to ambient air and thus cools more rapidly compared to those surfaces near the edges. The opposite condition exists in rolled steel operations. The convection process is slow in cooling the middle surface of the roll, but cools more rapidly at the edges due to direct exposure to ambient air. The operator might wish to circulate more air across the center of the steel band before rolling to promote even cooling. IRMV technology allows steel workers to monitor the cooling process with high efficiency in minimum time.

7.13.1 IRMV System Performance Requirements for Optimum Results

Selecting the right IRMV imaging sensor is of critical importance in manufacturing process control depending on the material used and the geometry and surface conditions of the product involved. The imaging sensor should provide the reliability of uncooled solid-state detector design, the accuracy of radiometric performance, the power of intelligent sensing features, and the flexibility of software control. The uncooled solid-state design involving a microbolometer detector is the key element of the IRMV. Cryogenically-cooled IR sensors offer improved performance, but at the expense of higher cost and poor reliability. The IR imaging camera used by the IRMV system offers maintenance-free operation and provides thermal sensitivity better than 0.02 °C, which makes it most desirable for the thermal detection of moisture. An IR camera provides a noninvasive technique and a thermal map of all the moisture in the wall of a house within a few hours. When the rest of the wall cools down, the moist areas are still relatively warm and appear as "hot spots" in the IR image. Although the thermal images could not reveal the amount or the depth of the moisture, they clearly identify the locations of high moisture. The IRMV system will help in determining the effectiveness of the moisture barriers placed in the wall within the last ten years, and therefore, will be found most suitable for use in housing construction.

Thermal IR analysis is a useful tool in automation or mass-production cases. Monitoring the thermal images and surface temperature through IRMV can:

- Detect a thermal defect in a product
- Ensure high quality via thermal uniformity
- Indicate the presence or absence of materials based on emissivity differences
- Reveal chemical consistency based on thermal profile
- Ensure that food and packaging reach the desired processing temperatures during the automatic food processing

It is important to mention that radiometric imaging using IR camera sensors in combination with temperature calibration tables and algorithms to provide environmental feedback can alter the performance of the imager, allowing for appropriate compensation. Software-based control provides monitoring of several temperature conditions and automatic instructions for any change in the process control mechanism. The widespread use of thermographic instrumentation based on IRMV technology has demonstrated the value of thermal imaging in various automated manufacturing process controls.

7.14 LASER-BASED SOLDERING PROCESS

A diode-based laser provides a noncontact reflow soldering process specifically designed for terminating extreme fine, insulated copper magnet wire to an electronic contact. High-power laser diodes are critical elements in noncontact flash soldering. The laser-based soldering (LBS) system is a combination of a 15 W diode laser and focussing optics required to achieve the desired spot size and to avoid spillover. The laser diode acts like a precision, localized, noncontact heat source for flash soldering. It offers a clean process resulting in reliable connections while avoiding damage to the wires and surrounding components or circuits. The flash-soldering process simplifies wire termination, leading to reduced labor costs and improved productivity.

Flash-soldering incorporating laser technology is widely used in manufacturing of electronic devices such as DC–DC converters, LAN filters, and miniature toroid transformers, which are critical components in telecommunications and other commercial and military electronic systems. The spot size can be less than 0.5 mm and the wire is heated for 200 ms using the laser CW power of 10 W at 1060 nm wavelength. The laser energy strikes the electric contact surface immediately adjacent to the wire with heat transfer through conduction. This prevents damage to the wire and other sensitive circuits in the vicinity. The flash-soldering process is a clean one, because it does not require added solder or wire stripping using mechanical or chemical methods. LBS flash soldering offers a high degree of precision control with minimum cost and time and requires a spot size equal to half the width of the part to be soldered to avoid spillover. Conventional soldering methods such as solder-dipping or thermocompression bonding suffer from unreliable contact, spillover, and higher failure rates and labor costs.

7.15 APPLICATION OF LASER TECHNOLOGY IN RICE GROWING

Japanese researchers [12] have demonstrated that rice grows faster under red and blue laser light than under sodium lamps. Japanese research scientist have used 30–40 W red diode lasers emitting at 680 nm and 660 W blue fluorescent lamps to grow rice seedlings in hydroponics containers. The laser diodes emitting at 680 nm produce a barely visible reddish light that plants readily use in photosynthesis. Rice growth is dependent on the photosynthetic photon flux density (PPFD). The seedlings are exposed to PPFD of 350-μm-mole/m^2/sec/day over a period of 19 hours. In June sunlight this PPFD averages to about 380 PPFD units. The scientists gave the operation 12 hours in the dark to breathe. After three months of this regimen, the laser-illuminated rice was ready for the harvest.

Photonics scientists predict that rice factories based on the laser-based hydroponics concept could produce five crops a year in controlled environments, that are not affected by the harsh outside climatic conditions. Mass-scale rice production could be achieved by starting with seedlings at a leaf age of 5 to 6. Leaf age is an indication of seedling growth and is determined by the number of leaves [12] on the plant. Rice production can be significantly increased by using seedlings with leaf age of 8 or 10 to meet rice demands of the increasing population of the world. This laser-based technology could be of significant benefit to countries with high population but with limited land for rice production, such as Japan.

7.16 LASERS FOR TRACKING OF IC PRODUCTION RATES

Lasers are widely used by semiconductor fabrication facilities for automatic tracking of the wafer through its manufacturing stages leading to production of IC chips. A typical wafer marking system consists of a pulsed solid-state laser, two scanning galvanometers, a focussing lens, associated electronics, and software necessary to make the system function. The galvanometers steer the beams angularly while the lens focuses the spot on the wafer. The position of the focussed spot is linearly proportional to the angle of the beam entering the lens, which produces a spot with a diameter ranging from 10 to 15 μm. A wafer identification mark can be produced through various methods such as optical characteristic recognition (OCR), barcode, or two-dimensional matrix of 18 × 18 pixels comprised of 22 code characters.

Nd:YAG laser sources are preferred in such operations, because its 1064 nm light is completely absorbed by the silicon. Arc-lamp-pumped, Q-switched Nd:YAG lasers currently being used are slowly giving way to diode-pumped solid-state (DPSS) lasers.

The DPSS lasers not only offer higher beam quality and reliability, but also eliminate the water-cooling and large power supplies used by the lamp-pumped lasers. The absorbing coefficient for the silicon varies with both temperature and wavelength. At room temperature (300 K), the 1064-nm laser light penetrates deeper into silicon than green light at 532 nm. Penetration depth is also dependent on temperature and laser wavelength. Computed penetration depths or distances into a silicon wafer as a function of temperature at two distinct wavelengths are shown in Table 7.4

Some companies are developing wafer-making lines based on Nd:YLF (yttrium lithium fluoride) lasers, which emit at 1053 nm and have certain advantages over Nd:YAG laser for diode pumping. This Nd:YLF laser offers high pulse-to-pulse uniformity, leading to improved process control for semiconductor fabrication. Some lasers provide superior marks on certain wafer material. For example, a frequency-doubled version of the Nd:YLF laser produces superior marks on GaAs and InP wafers. A CO_2 gas laser (10.6 μm) is used to mark sapphire wafers, which are transparent over a wide spectral region and cannot be marked by semiconductor lasers emitting in the visible or near-IR wavelength regions. Diode-pumped version of Nd:YAG lasers provide marking rates of 25,000 to 30,000 per hour. A silicon wafer transmits enough of Nd:YAG laser light at 1064 nm fundamental wavelength that even a small amount of light can reach all the way to the underside of the die, causing irreversible damage to the IC chips. Performance degradation from the underside irradiation can be avoided using some prescribed precautions.

7.17 LASER-BASED ROBOTIC GUIDANCE (LBRG)

LBRG systems [13] are widely used by automobile and aircraft manufacturers to achieve precision body alignment, mating of components, and high production rates with high reliability and minimum cost. High fuel efficiency, passenger safety and comfort, long operational life, and low maintenance are of paramount importance to the transport industry. However, high fuel efficiency or low fuel consumption is the principle requirement for an aircraft. High fuel efficiency, passenger comfort, and aircraft safety rely heavily on the precision alignment of the aircraft body and its

TABLE 7.4 Penetration distance in silicon wafer as a function of temperature and laser wavelength (μm)

Temperature (°C)	1064 nm	532 nm
1600	4	2
1200	17	5
800	34	7
400	56	1

critical components. Laser sensors have become the integral part of the robotic guidance system that facilitate and optimize production rates, particularly for automobile and aircraft bodies, where several parts need to be assembled with great precision and reliability.

7.17.1 LBRG System for the Auto Industry

LBRG systems comprised of laser-based sensors and placement devices are deployed by the auto industry to realize significant reduction in labor costs and warranty repairs. U.S. auto makers have incorporated laser technology into the body assembly process as early as the 1980s. Noncontact, laser-based machine-vision (LBMV) can track [13] body-to-body variation within few seconds from several key dimensional control points. LBRG systems can pick up a windshield, locate the windshield opening on the vehicle, and insert it with great precision in few seconds; which normally takes 15 to 20 minutes manually. LBRG systems are widely used for automobile glass decking, rear glass insertion, seam sealing, and stud welding, with significant reduction in fixturing, material costs, and labor costs. This system is also used to balance and align wheels on newly assembled cars in minimum time and attach doors to the vehicle frame on the assembly line. A three-dimensional laser-based camera finds the door and communicates its position to the robot, which picks up the door. As the robot moves the door towards the opening, x, y, and z coordinates are calculated by the computer along the contour lines. The actual location is compared with the programmed location, and the deviations are calculated in six degrees of freedom and are merged into the robot's dimensional frame of reference. The adjusted values are used to calculate the best path for the robot to use to insert the door into the auto frame. The robot uses the same approach to place windshields, quarter panels, gas tanks, instruments panels, and fenders and to locate studs for body welding. Laser-based visual fixing is used to inspect and analyze three-dimensional components, including plastic moldings, foam seats, fabric cover trims and other soft parts with impressive accuracy.

7.17.2 LBRG Systems for Aircraft Industry Applications

The assembly of wings and fuselage sections on an airframe requires huge fixturing far beyond the scope of automobile body assembly. Part alignment involves compensation for large structural errors caused by flexure and thermal properties. Boeing's laser alignment and leveling system aligns the wings to the center fuselage section and the four other fuselage sections (forward, aft, center, and wing) to each other. Laser alignment makes the fuselage as straight as possible to meet stringent flight performance and structural requirements. For example, the tail-to-tip centerline on the first Boeing 777 aircraft was within 0.023 inches. For assembly, the completed sections are placed on an electric servo-controlled jacking/positioning system incorporating as many as fifteen 20-ton jacks. The jacks use the inputs from the alignment lasers to maintain correct position as they rivet the wings and fuselage sections into the prescribed place. Two rotating lasers are used to establish horizontal and vertical reference planes. The LBRG alignment system improves the speed of assembly and offers better quality, tighter dimensional control, improved worker efficiency, safety on the assembly floor, and lower assembly costs.

7.18 HIGH-SPEED FIBER OPTIC (HSFO) LINKS

Fiber-optic (FO) links offer several advantages over conventional coaxial and waveguide transmission lines. FO links are widely used in medium-haul and long-haul telecommunications systems and local area networks (LANs). Their major advantages include light weight, small size, low attenuation, high immunity against EMI, large bandwidth capacity, low probability of intercept (LPI), and excellent compatibility with optical processing schemes. FO links have potential applications to interconnect microwave components in phased-array antennas, RF discrete Fourier systems, and radar and communication systems with optical beam steering requirements. Their applications over short microwave distances (less than 100 m) is limited by the optoelectronic transducers. A directly modulated analog RF laser FO link offers a cost-effective approach to transport received signals to a location greater than 15 km over a single optical fiber with no equalizer or repeater. A HSFO link can transfer data from a digital still camera, notebook computer, hand-held PC, or digital cellular phone to nearby buildings and central locations at short distances. The critical components of a FO link include modulator, LED, low-loss optical fiber, photo diode, and preamplifier.

7.18.1 Critical Performance Parameters of FO Links

Performance requirements of a high-speed FO (HSFO) link operating at microwave frequencies may be slightly different than those for operating at infrared wavelengths. However, performance parameters such as dynamic range, bandwidth, link gain, and S/N ratio are equally important regardless of whether microwave or optical frequencies are used. In the case of an FO link operating at IR wavelengths, two parameters are of significant importance, namely, relative intensity noise (RIN) and carrier-to-noise (C/N) ratio (similar to S/N ratio). Fiber optic links operating at millimeter-wave frequencies and optical frequencies are in great demand because of their high data rate capabilities. The optical link offers the widest bandwidth, reduced weight, immunity against RFI and nuclear radiation, and low probability of intercept (LPI). However, its performance can be limited by modulation restrictions, pointing accuracy, atmospheric conditions, and signal acquisition capability due to its narrow beam.

7.18.2 Link Performance as a Function of Laser Wavelength

As stated earlier, a directly modulated, analog RF laser-based fiber optic (LBFO) link offers a cost-effective means to transmit or to receive signals over wide bandwidth in a single optical fiber without any repeater or equalizer over a distance greater than 15 km. LBFO links can be operated at any wavelength, but most links are operated at 1 μm (GaAs/GaAlAs-DH laser) or 10.6 μm (CO_2). It is evident from the performance data summarized in Table 7.5 that the S/N ratio and noise power for the 10.6 μm laser are 40 dB and –133 dBW, respectively, compared to 16 dB and –109 dBW at 1 μm.

It is important to mention that the tracking accuracy of the 1 μm laser-based FO link is much better than the 10.6 μm laser-based FO link. The tracking accuracy is dependent on the FOV, which is defined as the ratio of wavelength to optic size. The FOV of a 20 cm optical aperture at an operating wavelength of 1 μm is 5 microradians. If a communication satellite has a stability of ±0.1 degree or ± 1.75 milliradians, tracking becomes extremely difficult for the 1-μm optical receiver. This means that the satellite stability accuracy must be close to that of the optical tracking receiver. Computed values of the tracking accuracy for an optical receiver as a function of optic size and laser wavelength are summarized in Table 7.6.

Lasers operating in the near-IR region offer FOV compatible with the stability of communication satellites, which is about ± 1.75microradians. The high directivity of the optical communication link allows for an optical transmitter with much lower output power, leading to significant reduction in the optical transmitter cost. Certain defects in the active region of the laser diode (GaAs/GaAlAs-DH) can cause reduction in the internal quantum efficiency and, consequently, output power. If the semiconductor is sensitive to a wide variety of failure mechanisms, its normal life of 10,000 to 50,000 hours will be significantly reduced.

7.18.3 Critical Performance Parameters of Various Elements of a Satellite-Based Optical Link

Doppler shift, which is dependent on microwave frequency, optical wavelength, and relative velocity, has a big impact on the performance level of a satellite-based optical link. Relative velocities in the case of low-earth-orbit (LEO) satellites can approach as high as 28,080 km/hour or 17,444 miles per hour. At a wavelength of 1 μm, the Doppler frequency will be 7.8 GHz, which is beyond the bandwidth capa-

TABLE 7.5 Performance comparison for terrestrial communication optical links operating different wavelengths

Performance parameters	CO_2 (10.6 μm)	GaAs/GaAlAs (1 μm)
Transmitter power (dBW)	0	–17
Transmit losses (dB)	1.5	1.5
Transmit antenna gain (dB)	96.0	115.6
Propagation loss (dB) over 44,000 nm	280	300
Receive antenna gain (dB)	96.0	115.6
Conical scan loss (dB)	1.0	1.0
Total received power (dBW)	–93.0	–93.3
Noise power (dBW)	–133.3	–109.3
S/N ratio (dB)	40.0	16.0

TABLE 7.6 Optical receiver tracking accuracy (microradians) as a function of operating wavelength and optics size

Wavelength (μm)	20 cm	30 cm	40 cm
0.84	4.2	2.80	2.10
1.00	5.0	3.33	2.50
10.6	53	35.3	26.5

bility of most optical mixers used in heterodyne receivers. However, in the case of a 10.6-μm laser, the Doppler shift will be about 827 MHz, which is compatible with the bandwidth capability of existing optical mixers. Long operating distances favor optical transmission lines. For terrestrial applications, the fiber optic link offers several advantages. Commercial optical fibers are available with insertion losses from 0.0003 dB per foot to 0.002 dB per foot at a wavelength of 1 μm. Low cost and low attenuation are the most impressive parameters of the optical fibers for satellite-to-satellite optical links.

Gain–bandwidth product is the next important performance parameter of an optical link. The maximum operating frequency of a directly modulated FO link is limited by the relaxation oscillator frequency of the laser, which is in the vicinity of 10 GHz for the commercially available semiconductor heterojunction GaAs/GaAlAs laser diodes.

The relative intensity noise (RIN) of the semiconductor laser increases drastically at the proximity of the relaxation frequency and above it the attenuation becomes prohibitive. Since the frequency response of the commercially available PIN photodetectors now exceed 20 GHz, the FO link operating frequency of the laser can be reduced to achieve lower RIN.

The gain of a FO link at frequencies below the 3 dB bandwidth of laser and photodetector is a function of the laser and detector coupling efficiency (about 7.5%), optical fiber insertion loss (about 13dB at 1 μm), connector losses (about 2 dB), responsivity of the photodetector, laser and photodetector input resistance, and the reflection coefficient looking into the laser. The fiber loss has two components, namely, insertion loss and the mismatch loss. The insertion loss is proportional to link length, but dependent on the operating wavelength. The mismatch loss is dependent on the circuit configuration, operating frequency, and bandwidth, which is roughly 13.5 dB for this FO link configuration. A 50 m FO link comprised of a 1 μm semiconductor laser and a photodetector modulated by a 2 GHz analog RF signal will experience a total loss of about 36 dB.

The dynamic range of a FO link is strictly determined by the difference between the 1 dB compression point of the input power and the minimum detectable signal (MDS). The dynamic range (DR) of the link can be expressed as

$$DR = (P_{1\ \mathrm{dB}} - MDS)\ \mathrm{dB} \tag{7.2}$$

where $P_{1\ \mathrm{dB}}$ is the 1 dB compression point of the input power (dB) and MDS is given as

$$MDS = [-174 + NF + 10 \log (BW)] \text{ dB} \tag{7.3}$$

where NF is the noise figure of the link dependent on the laser noise and the detector noise levels and BW is the link or RF signal bandwidth in MHz. The 1 dB compression point is typically 10 dB below the third-order intercept point of the laser diode. For a FO link modulated with an analog microwave signal at 6 GHz and a signal bandwidth of 500 MHz, the estimated dynamic range is about 22 dB, which is not adequate for many applications. Improvement in overall FO link losses and dynamic range is possible through optimized design configuration of the link. A remote data mixing (RDM) link configuration offers significant improvement both in systems losses (10dB, minimum) and dynamic range (about 15 dB), because it uses two FO links: one for the data transmission and the other for the carrier, each of which can be independently optimized for performance. The dynamic range is limited by the noise figure (NF), which is dominated by the RIN parameter. Since the RDM link configuration operates at lower frequency, the RIN level will be lower and the dynamic range will be higher accordingly.

The carrier-to-noise ratio (CNR) of a subcarrier is dependent on the optical modulation index (m), RIN level, signal bandwidth, intermodulation noise level, DC photo current, and input noise current spectral density. The overall noise includes the RIN of the laser diode, shot noise, thermal noise, and intermodulation distortion noise.

The signal power is proportional to the square of the modulation index, while the third- order intermodulation distortion is proportional to the sixth power of modulation index (m^6). The major contribution to the CNR comes from thermal noise. Rapid improvement in the CNR is possible with higher values of the modulation index parameter (m). Calculated values of CNR and CIR (carrier-to-intermodulation ratio) as a function of modulation index (m) are summarized in Table 7.7.

Fading of received signal degrades the quality of the voice communication. In mobile radio environments, the magnitude of the signal varies over distance due to atmospheric characteristics and multipath effects. The rapid variation in the received signal due to multipath propagation is known as fading. Transmission quality of the optical link will be degraded because the RF signals suffer from the fading effects. Under fading conditions, CNR degradation is inversely proportional to the square of modulation indices. The optical modulation indices m_1, m_2, and m_3 associated with three carriers are mutually independent random variables.

TABLE 7.7 Calculated values of CNR and CIR as a function of modulation index

Modulation index (m) dB	CNR (dB)	CIR (dB)
–40	45	118
–30	54	98
–20	63	74
–10	73	56

Fast fading occurs in a situation where the rapid fading (characterized by maximum Doppler frequency) is fast compared with the signal interval. In fast fading cases, the CNR represents an average value. The maximum value of the modulation index (m) has the optimum value that maximizes the mean value of the CNR, similar to the case of nonfaded signals [14]. Data shown in Table 7.7 indicate that when the modulation index increases, CNR increases while CIR decreases. The third-order intermodulation is the product of three different RF signals in case of three carriers. Computer simulation yields the optimum value of (m) as –16 dB, corresponding to an average CNR value of 66.5 dB. The simulation further indicates that the CNR is not degraded in the average point of view and the net CNR is nearly equal to that of a nonfaded case with a difference between them is within 1 dB.

Slow fading occurs in a situation where the rapid fading is slow compared with signal interval. System degradation due to slow fading is generally measured by means of a statistical probability, which is the probability of failing to achieve required CNR. From the outage probability (P_o), one can estimate the bit error rate (BER). Maximum value of average CNR is dependent on the number of carriers (N_c) and decreases accordingly as $N^2{}_c$. The maximum value of CNR is about 65 dB, 61 dB, 59 dB, and 56 dB when the number of carriers (N_c) are 3, 5, 6, and 10, respectively. Optical nonlinearity limits the number of carriers that can be accommodated per microcell, the smallest zone of which is reduced to several hundred meters, and one can reuse the same frequency as that used several distances away. Microcellular mobile communication systems using FO link technology offers an effective frequency utilization and can accommodate large numbers of subcarriers.

7.19 SUMMARY

Potential applications of infrared, photonics, and electrooptics technologies in commercial and industrial systems are discussed, with particular emphasis on cost, reliability, and efficiency. Performance capabilities and limitations of IR laser-based systems for applications in manufacturing processes are summarized, with emphasis on speed and performance level. Methods to realize significant improvement in performance level, reliability, and power consumption are identified wherever possible. Performance characteristics and capabilities of HDTV, laser marking devices, IR laser-based high-volume monitoring devices for product manufacturing, CMOS imaging sensors for commercial applications, Fourier transform Raman spectrometers, laser-based sensors for controlling manufacturing processes, IR lasers for rapid growth of rice, FO links for microcellular mobile communications and phased-array antenna control, high-speed digital cameras, laser-based airport security devices, laser-based alignment systems for automobile and aircraft industries, IR smoke and fire detectors integrated with high-resolution cameras, photonic devices for drugs and biological/chemical weapons, full-color virtual displays and laser-based systems for reliable detection of biological, chemical, and small nuclear weapons used by terrorists are described, with emphasis on cost, reliability, and performance.

REFERENCES

1. H. JonesBey. "Scanning linear array produces HDTV images. *Laser Focus World,* pp. 31–34, July 1998.
2. D. Hargis et al. "Diode pumpled microlenses promise portable lasers." *Laser Focus World,* pp. 243–250, May 1998.
3. G. T. Burnlucr. Red laser diodes for DVD applications." *Laser & Optronics,* pp. 13–14, March 1998.
4. Contributing editor. "Markey-ready blue diodes excite spectrum copist." *Laser Focus World,* pp. 69–75, April 1999.
5. R. Gibbs. "Diode lasers fuel plate making advances." *Laser Focus World,* pp. 135–144, January, 1998.
6. H. Kaplan. "Smoke penetration vision systems save lives." *Photonics Spectra,* pp.63–64, July 1998.
7. A. Seeds. "Optical technologies for phased array attenna." *IEICE Trans. Electronics, 76,* 198–206, 1993.
8. P. R. Herczfeld et al. "Optical control of MMIC-based T/R Modules." *Microwave Journal,* 309–320, 1998.
9. M. Wheeler. "Photonics takes a bite out of crime." *Photonics Spectra,* pp. 93–110, November 1998.
10. M. D. Wheeler. "Changing the face of warfare countering a terrorist threat." *Photonics Spectra,* pp. 124–132, April 1999.
11. M. W. Womble et al. "Low resolution Raman method offers low cost and portability." *Laser Focus World,* p. 132, April 1999.
12. Contributing editor. "Rice growth faster under red and blue light." *Photonics Spectra,* July 1998.
13. H. Kaplan. "Laser alignment helps appearance, safety and efficiency." *Photonics Spectra,* pp. 96–100, July 1998.
14. H. Mizugutu et al. "Performance analysis of optical fiber link for microcellular mobile communications systems. *IEICE Trans. Electronics, E76,* 2, 271–277, 1993.

CHAPTER EIGHT

Application of Infrared and Photonic Technologies in Medicine, Telecommunications, and Space

8.1 INTRODUCTION

This chapter describes the applications of infrared and photonic technologies in medical, telecommunications, and space systems. Performance capabilities and limitations of IR laser-based sensors and photonic devices for applications in medicine, space surveillance, telecommunications, and communications are summarized, with emphasis on cost and performance. Performance capabilities of IR sensors such as IR digital cameras for biotechnology image analysis, photodynamic therapy (PDP) involving light and chemistry, noninvasive optical tomography, optical endoscopy, erbium-based lasers for transmyocardial revascularization (TMR), lumpectomy, somatology, ophthalmology, hyperspectral imaging spectrometers, photonics remote sensing devices, IR sensors for environmental research, space surveillance sensors, and optical communications and telecommunications equipment, including wavelength-division-multiplexing (WDM) and dense WDM systems, are discussed, with emphasis on cost, reliability, and performance.

8.2 IR AND PHOTONIC SENSORS FOR MEDICAL APPLICATIONS

IR lasers and photonic sensors are playing key roles in medical treatment and surgical procedures. FDA-certified therapeutic applications of lasers now include general and cosmetic surgery, ophthalmology, and transmyocardial revascularization. Worldwide sales of nondiode lasers for medical therapy exceeded $500 million by the end of 1999. Ophthalmology is one of the largest medical applications; it involves photorefractive surgery for cornea reshaping and permanent vision correction. Another popular application is the PDP procedure, which is widely used for treating various diseases with great success. Photosensitive dye lasers are used to localize cancer treatment. The short-pulse ultraviolet (UV) output of an excimer laser is best suited for specific medical applications. Lasers have gained wide acceptance in medical diagnostic procedures and in chemical laboratories for routine testing of compounds. Raman spectroscopy seems to offer early detection and diagnosis of certain life-threatening diseases.

Ultrashort laser pulses (USLs) have potential applications in several critical surgical procedures. USL pulses used for tissue ablation cause minimum collateral mechanical and thermal damage because of the short energy deposition time and the high efficiency of the ablation process. Minimum energy deposition into the secondary tissue means less discomfort to the patient. The USL could be of significant importance in very sensitive procedures such as spinal surgery, where collateral damage could have devastating consequences. USL pulses are being investigated for TCR invasive procedures to improve patient's pain tolerance threshold and to provide relief from angina pain. Nd:YAG lasers with CW output levels of 36 W at 0.5 μm to 100 W at 1.06 mμ , Holmium (Ho) lasers with CW power output of 30 W at 2.1 μm, and Thulium(Th) lasers with CW output greater than 20 W at 1.96 μm are available for surgical procedures with short recovery times.

8.2.1 IR Sensors for Biotechnology Image Analysis

IR digital cameras have produced high-resolution images of cellular and subcellular configurations, such as a living and unstained Drosophila preparations [1]. High-speed digital cameras with enhanced software detect faulty cells by capturing cell material in 24 bit color with 1200 × 1600-megapixel resolution. The captured cell images can be compared to a cell signature analysis results, revealing patterns in healthy and diseased cells. Because the images are digital, one can process them as statistical entities to detect patterns in faulty cell development based on molecular structures within the cells. Using a IR digital microscopic camera with enhanced software, one can collect, compare, and process images of faulty cell development. The IR digital camera is a leading tool in biotechnology image analysis.

8.2.2 Photodynamic Therapy (PDT)

PDT technology offers the most promising and effective medical treatment for various diseases such as heart disease, breast cancer, bladder cancer, lung cancer, skin cancer, and deep malignant tumors. PDT uses a photosensitive drug that creates a toxic form of oxygen, also known as singlet oxygen, when exposed to specific laser light. Since, the photosensitive drug remains in the cancerous tissue longer than in healthier tissue, the PDT treatment can destroy the cancerous tissue selectively. The photosensitive drug is administered to a patient intravenously and collects in the body tissues. When the drug is cleared out of the healthy tissue after a day or two, a surgeon uses an appropriate laser source to activate the drug, leading to destruction of the diseased tissue responsible for cancer.

8.2.2.1 Benefits and Risks of PDT Technology

Unlike chemotherapy, which can make patients sick, PDT's main side effect is that the patient's skin may be sensitive to sunlight for a month or so after the treatment, [2] which is not that serious. PDT has a major limitation: The surgeon must have clear access to the tissue in order to focus the laser beam directly to the tissue area involved. This means that PDT treatment will be somewhat difficult to implement on very large tumors or tumors buried deep inside the body.

8.2.2.2 Alternate illumination for PDT procedure

Research scientists are exploring an alternate illumination technique known as multiphoto excitation, which uses two or more photons from a longer wavelength laser source operating in the near-IR region to simulate the same reaction as one photon from a short wavelength source. However, the use of high-power laser sources is absolutely necessary to ensure that multiple photons strike a molecule almost simultaneously. Multiphoton excitation offers both the improved spatial control and greater tissue penetration, making it ideal for treatment of large tumors. Furthermore, with two-photon excitation, some photosensitive molecules can be activated from 600 to 800 nm wavelengths. Surgeons have observed that multiphoton excitation in the 740–920 nm spectral range causes minimum tissue damage compared to short-wavelength light sources operating at 300–400 nm.

Laser sources for PDT applications include xenon arc lamps and dye lasers pumped by either an argon ion or frequency-doubled Nd:YAG laser. LEDs and diode lasers are beginning to replace traditional light sources because of lower cost, low power consumption, and high reliability. Clinical tests to date indicate that LEDs are most likely to play a significant role in providing cost-effective light sources for emerging cancer therapies using PDT technology. The tests further indicate that PDT can be extremely effective in treating the cancers at early stages.

8.2.2.3 Laser Source Requirements for PDT

Laser source requirements are strictly dependent on the properties of the photosensitive drugs used, type and size of tumor involved, and tissue penetration requirement. Some photosensitive drugs activate at 632 nm and some at 652 nm wavelength; one can use KTP-based frequency-doubled Nd: YAG pumped dye lasers or argon ion pumped dye lasers. However, the dye lasers, despite their high power capability, are bulky, contain toxic dyes, consume huge amounts of electrical power, and need frequent maintenance and cost over $100,000. Drugs companies are aggressively investigating other drugs that activate at longer wavelengths ranging from 740 to 920 nm and have reduced sensitivity.

Semiconductor diode lasers and DPSS lasers for PDT applications are being evaluated, with emphasis on cost and reliability. These laser sources are much smaller, more reliable and cheaper. Semiconductor AlGaInP diode lasers deliver between 1 and 2 W of power to the fiber port and work well with some photosensitive drugs. However, a laser with output power exceeding 3 W at the end of the optical fibers offers best results for PDT applications, according to clinical scientists. A new semiconductor is laser comprised of a AlGaInP strained quantum well (SQW) active region sandwiched between two AlGaInP waveguide regions with AlInP cladding layers. A monolithic laser array comprised of several SQW devices demonstrates a maximum output power of 15 W at 630 nm with maximum power-conversion efficiency of 25%. Such lasers have output power levels of 10–15 W at 630 nm with built-in feedback circuits to control the laser intensity to the cancerous tissue, making suitable for illumination the next generation of photosensitizers. Some clinical researchers are using LED devices with longer wavelengths (752 nm) to activate specific photosensitive drugs in treating metastatic cutaneous breast cancer tumors [2]. Other researchers are using 732 nm LED devices to test the ability of a special drug to destroy the plaques that cause heart disease . Clinical researchers indicate that longer wavelengths are most suited for blood, because blood has a low absorption at 732-nm wavelength. However, some researchers selected a blue fluorescent lamp to activate a specific photosensitive drug in treating acute actinic keratosis, because this light source does not require precise dosage requirements [2]

8.2.2.4 Red "Diode" for PDT Applications

Since longer wavelengths are best suited for blood because of low absorption characteristics, red diode lasers using AlGaInP/GaAs SQW diodes operating at 732 nm will be most attractive for PDT applications. These diodes use compressively strained structures by adding excess indium content to reduce threshold current. Arrays of red diodes have delivered CW power in the 50–100 W range. One of the most impressive applications of such diodes is in PDT of cancerous tumors. A patient is administered a photosensitive drug that concentrates in the tumor after few hours. The tumor is exposed to red light at 732 nm, killing all the diseased tissue. Clinical tests performed by scientists indicate that the use of red diode lasers makes it possible to use a wider range of photosensitive drugs with maximum effectiveness.

It is important to mention that the narrow bandwidth of 2–3 nm does not take full advantage of PDT drug absorption characteristics. However, the LED-based probe operating at 732 nm or 688 nm offers a wider bandwidth in the order of 30–40 nm, leading to increased energy absorption by the photosensitive drug. Photosensitive drugs exhibit high peak absorption at wavelengths over 670 to 690 nm spectral range, resulting in deeper penetration of light to the affected area. A probe with a variable LED wavelength capability to match the absorption wavelengths of the second-generation PDT drugs will be most suitable for PDT treatment. Lower cost and wider bandwidth of LED devices are most likely to play a critical role in the development of new light-based equipment for cancer therapies.

8.2.3 Hyperspectral-Imaging (HSI) Spectroscopic Systems for Medical Applications

HSI is a unique imaging technique by which life science scientists can identify and quantify the relationship between biologically active elements, thereby leading to acquisition of spectral information of high reliability. Traditional spectrometers [3] acquire the spectra of homogeneous materials. If the sample is a liquid mixture, then the research scientist uses chromatography or any appropriate separation technique to purify the components and then analyze the results. If a sample is a tissue or cell specimen, it is possible to separate the constituents without changing their molecular structure, which is necessary for a thorough and reliable analysis.

In the case of the HSI technique, HSI instrument acquires the individual spectrum of each cell component and assumes that the material under analysis is heterogeneous. The HSI then identifies spatially resolved objects based on their spectral signatures. The HSI technique requires up to 200 spectral data points per spectral object, whereas multispectral imaging (MSI) system require only a maximum of 20 data points per spectral object. Using the HSI technique, the research scientist can simultaneously obtain spectra from complex multiple overlapping fluorophores and separate them through deconvolution algorithms.

In both the HSI and MSI techniques, scientists need to obtain both the spectral and spatial information with high reliability. Spatially resolved spectral-acquisition devices can be designed based on two operating concepts, namely, spectral cube and image cube, as illustrated in Figure 8.1. An image cube acquires the entire spectrum from 365 to 750 nm simultaneously and generates images sequentially, one slice at a time across the field of view (FOV). In this method, the sample scans across the entrance slit of the spectrometer. However, a spectral cube acquires a fixed FOV layered in sequentially acquired wavelength slices. Each slice is identified by its own specific wavelength, ranging from λ_1 to λ_n over the 365 to 750 nm spectral range (Figure 8.1). This means that the HIS instrument [2] builds one spectrum at a time. Each pixel in the IR CCD-based camera builds a complete spectrum defined by the number of wavelengths acquired. The cubes resulting from the two distinct methods present spatial and spectral data in different ways. Spectral cubes can be created by various tunable filters, including acoustooptic filters, dielectric filters, and liquid-crystal filters. The spectral cube method is best suited for locating fluorophores dispersed in cells or tissue sections. The spectral cube method takes a long time. If the spectra within the sample change under various environmental conditions such as temperature, pressure or humidity, then information from the same sample may be different at the first and last wavelengths. Furthermore, the huge size of a spectral cube file will slow the processing speed and thus delay the presentation of data to the researchers. One can estimate the file size per wavelength and the overall file size for specific numbers of wavelengths involved. Assuming a spectral coverage of 400 nm (from 365 to 765 nm) and 80 wavelengths needed to enable a deconvolution algorithm of multiple adjacent color centers, the image requires a minimum of 80 acquisitions. If a CCD device has 180,000 pixels (240 × 750 or 300 × 600), their each file will have a capacity of 14.4 Mbytes (180,000 × 80). Note fu-

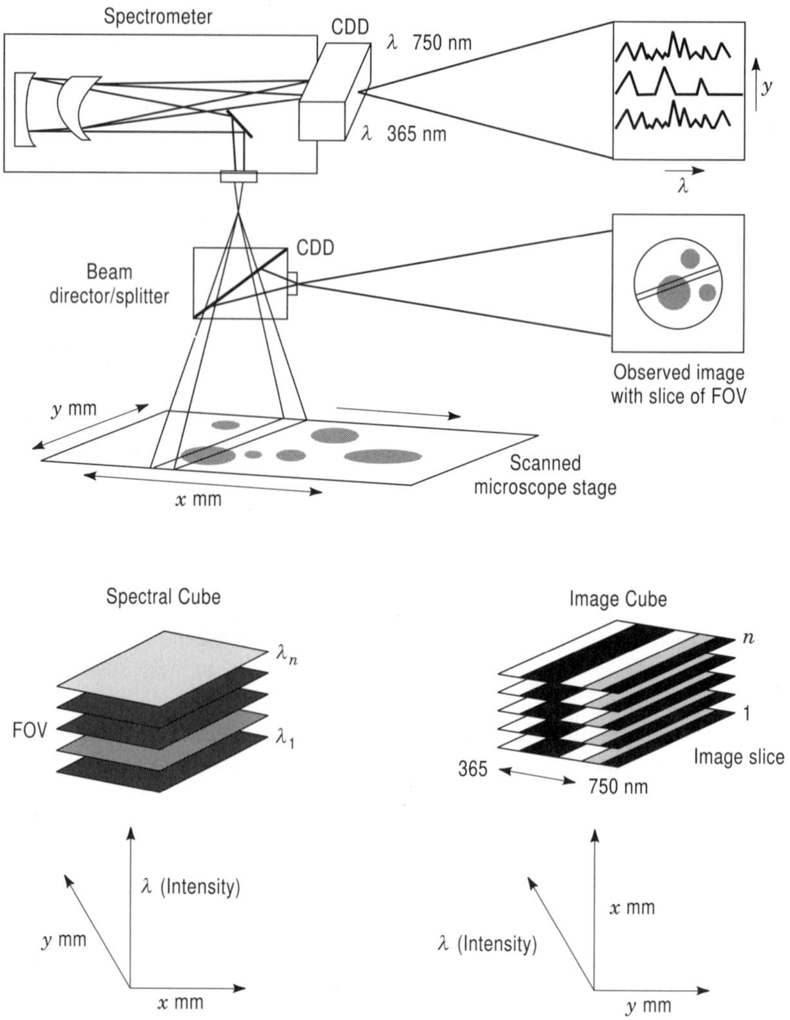

FIGURE 8.1 Block diagram showing hyperspectral imaging for life-science applications.

ture system developments involving high throughput for screening of new drugs, medical diagnostics, and cell physiology, may require far more than 80 data points in 400 nm spectral increments. For an application requiring 200 data points, the file size will grow to at least 36 Mbytes. Massive data acquisition and storage will be required over several minutes before the first computation is completed. The spectral cube approach will limit the number of acquisitions, thereby compromising the quality of the research.

As stated earlier, the image cube method captures all wavelength data simultaneously in a 180 K file. An image cube acquires 240 spectra, each with 750 data points, assuming the same CCD pixels. This method allows a scientist to view a specimen in the microscope, visually locate it, target a feature of special interest, and then simultaneously acquire full spectral evaluation of the sample. This method has potential application to epitope tagging, cell smears, microtiter wells, and high-throughput scenes. Armed with new imaging methods and sophisticated software technology, life science researchers will be able to identify the spectral morphology of cells, examine spectral characteristics of cells to indicate a specific disease, and identify appropriate treatment.

8.2.4 Near-IR Spectroscopic Method for Epilepsy Treatment

The latest medical research [3] has developed laser-based technique for pinpointing the location of an epileptic focus in the human brain. The intensity-modulated energy from a near-IR laser diode operating at 780 nm or 830 nm wavelength passes through the scalp and the reflected laser energy from the cerebrum of the brain is collected. The reflected laser beam is collected, detected, and processed to produce a real-time image of the blood flow to the brain during an epileptic seizure. This technique involves noninvasive optical tomography to obtain two-dimensional images of the brain surfaces, leading to detection of brain activity [4] under epileptic seizure conditions. This method locates radical increases in blood flow that occur during epileptic seizures. This laser-based spectroscopic technique makes it possible to carry out necessary examinations without subjecting a patient to excessive stress and discomfort during diagnostic procedures.

Current methods to determine the location of an epileptic focus include a surgical method to permit the measurements for discharges during the seizures and single-photon-emission (SPE). The computer-based tomography method involves injection of radioactive isotope into the blood to determine the blood-flow increase during the epileptic seizures. Both these methods subject patients considerable stress, discomfort, and radiation hazard. This intensity-modulated near-IR (IMNIR) spectroscopic method provides the differences in the absorption index of hemoglobin in blood, measures the changes in blood flow on both sides of the brain during the seizures, and displays the information to neurosurgeons in real-time.

Multichannel version of IMNIR spectroscopes offer more detailed information on brain activity to the neurosurgeons. A multichannel version of this sensor uses eight channels and two diode lasers emitting at 780 and 830 nm wavelengths. This particular system is capable of determining whether an epileptic focus is on right side or left side of the brain and it can locate the focus within the brain hemisphere with high accuracy and reliability. The ability of the IMNIR spectroscope [4] to observe time-based variations in cerebral blood flow during the epileptic seizure offers an important diagnostic tool to neurosurgeons.

8.2.5 Laser-Based Transmyocardial Revascularization (TMR) Procedure for Treating Heart Disease

TMR is a laser-based surgical procedure in which tiny holes are drilled in the heart to increase the blood flow to the oxygen-starved heart tissue, thereby relieving the patient of severe angina pain and other symptoms of severe coronary disease. These holes act like new vessels capable of delivering oxygen-rich blood to the heart muscle.

Holmium-based Ho:YAG lasers and Thulium-based Tm:YAG lasers emit around a 2-μm, eye-safe wavelength that does not reach the retina, and are considered most suitable for TMR procedures. The 2-μm wavelength is strongly absorbed by water and affects only the outer layers of the tissue; it can be easily transmitted by silica optical fibers with minimum loss. Critical elements of the DPSS holmium laser are shown in Figure 8.2. The Ho-based laser for the TMR procedure is less costly, less traumatic to patients, and requires no transfusion or heart–lung machine. After the procedure, the recovery is faster, more cost-effective, and more free from postoperative complications than open-heart surgical procedures.

TMR can be considered as a replacement therapy to bypass surgery and angioplasty and could become the third major medical treatment in the near future. Clinical scientists reveal that a percutaneous (a procedure performed through the skin) TMR procedure, which relies on the fiber optic catheters to deliver the laser energy to the heart involves a minimally invasive procedure. Clinical evidence of the safety and effectiveness of the percutaneous approach suggests that the medical community will prefer percutaneous myocardial revascularization (PMR) over the TMR procedure.

PMR is based on the same principle as TMR but it uses a fiber optic catheter delivery system. The surgeon places the laser probe against the ventricle wall for drilling holes in the myocardium. The PMR requires only a few holes (15 to 30) compared to the TMR procedure and takes only one or two hours to complete. The entire procedure can be done under local anesthesia in a cardiac-catherterization laboratory rather than a surgical suite and the patient goes home the same day as opposed to four days, following the TMR procedure.

8.2.6 Laser-Based Spectrometers for Dental Treatment

Laser-based spectrometers are playing an important role dental surgery and diagnosis of dental disease. Erbium-doped YAG (Er:YAG) lasers emitting at 2.94 μm and semiconductor diode lasers are best suited for diagnosis of dental diseases. Nd:YAG laser-based Fourier transform Raman spectrometry (FTRS) is most effective in the detection of tooth decay. Dentists indicate that the areas of decay with significant bacterial activity are highly fluorescent. This means that Raman scattering from a Nd:YAG laser-based FTRS could prove most effective for locating hard-to-detect infection beneath the tooth surface. Whereas advanced tooth decay is easy to distinguish on healthy tooth enamel, it is much more difficult to detect in infected dentin [5] beneath the health enamel. Researchers believe that the ability of the Raman spectroscopy to locate underlying infections and distinguish them from the healthy tissue could result in a powerful diagnostic tool for both dentists and dental surgeons. A Raman spectroscopic microscope system shown in Figure 8.2 will be found most effective in treating dental diseases. This system is comprised of a argon ion laser pumped with a Ti:sapphire laser and a dispersive spectrometer. The near-IR system will be most suitable for precise diagnosis of various dental diseases. The

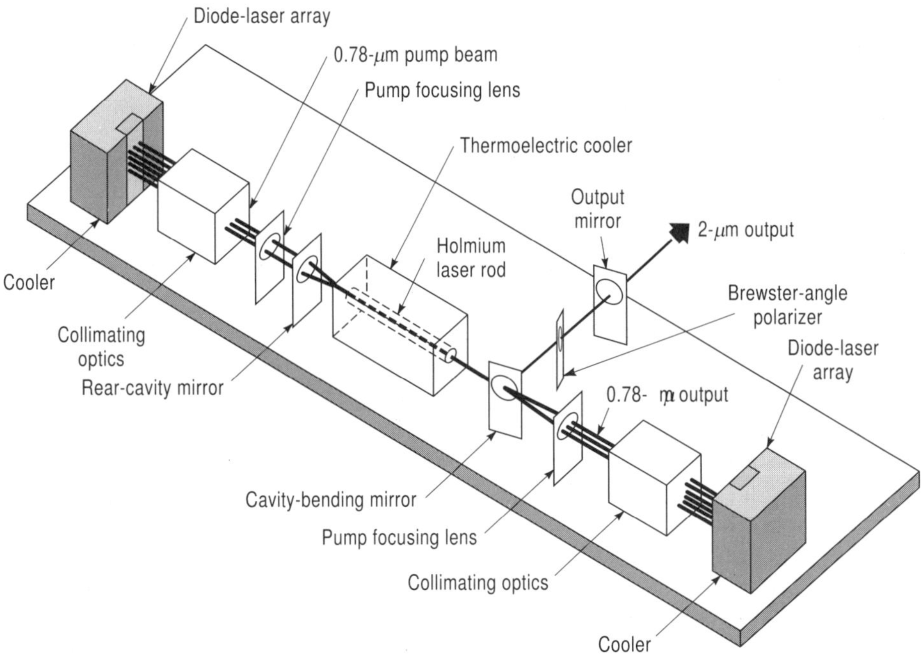

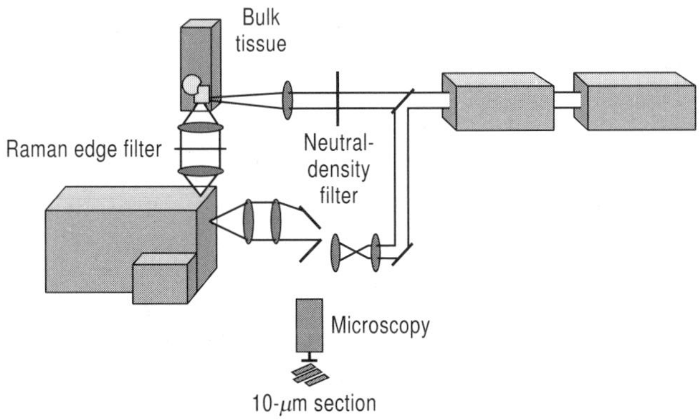

FIGURE 8.2 (A) Critical elements of a DPSS holmium laser and (B) Raman spectroscopic microscope using tunable Ti:sapphire laser.

near-IR tunable laser-based system operating over the 750–850 nm spectral range is able to collect Raman spectral signals much more rapidly and has reduced the collection times from more than 30 minutes to less than 5 seconds [5].

8.2.6.1 Various Lasers for Dental Applications

It is important to mention that water absorption is an order of magnitude at the 3 μm peak compared to the 10.6 μm wavelength. Dentists feel that the 3 μm lasers are best suited for cutting bone and drilling teeth. However, special fibers are needed for transmission 3 μm optical signals with minimum loss. The standard silica fibers are not suitable for this wavelength operation due to high losses.

Erbium-doped lasers are most ideal for dentists. These lasers include Erbium-doped lasers such as Er:YAG and Er:YSGG lasers and have been used in many medical diagnostic applications. Note the 2.94 μm Er:YAG laser is closer to the nominal water-absorption peak than the 2.794 μm Er:YSGG absorption line. Er:YAG lasers are generally flash-lamp-pumped and the Er:YSGG lasers can be codoped with chromium to enhance pump light absorption performance. Flash-lamp-pumped Er:YAG lasers can generate average power levels up to 30 W, whereas a singly doped erbium diode-pumped laser at 970 nm can provide optimum power levels exceeding 500 mW, which is more than adequate for hard-tissue dental applications.

A new laser technique could end dental fillings, thereby relieving the patient of pain and dental expenses. This new technique, developed by the University of Glasgow [6], is based on two-photon absorption from a solid state Cr:LiSAF laser, which monitors changes in laser-excited fluorescence that occur as decay attacks the enamel coating of healthy teeth. Dental tissue contains a natural fluophore that absorbs between 420 and 450 nm. In healthy tissue, the peak of fluorescence occurs at 680 nm, but in diseased enamel it occurs at 550 nm, thereby indicating a measurable difference between the healthy and diseased enamels. The two-photon laser imaging technique provides depth discrimination with minimum light loss because of significantly improved resolution afforded by two-photon absorption. Two 850 nm photons can be absorbed by the fluophore, and the resultant fluorescence can be detected with minimum scattering and phototoxicity.

This dental Cr:LiSAF laser provides 100-femtosecond (fs) pulses at the repetition rates of 200 Hz to maintain both the average and peak power levels. However, the laser light focussed on the teeth under dental investigation is in the order of few milliwatts. A red aiming laser beam is also used for obtaining sharp images of the tooth capable of locating the "fluorescent hot-spots" on the tooth surface. This laser-based dental system provides early detection of tooth decay, which could eliminate unnecessary fillings or dental drilling to remove diseased tissue.

8.2.7 Raman Spectroscopic Techniques (RSTs) for Diagnosis and Treatment of Various Diseases

Raman-based spectroscopic diagnostic tools are considered potentially valuable for early detection and diagnosis of life-threatening diseases of the heart, brain, colon, and breast. Clinical studies performed by various medical institutions indicate that Fourier transform Raman spectrometers (FTRSs) using 1064 nm Nd:YAG laser energy reduce fluorescent interference and have successfully demonstrated a technique to distinguish between normal and diseased tissues [4] of the brain, breast, arteries, skin, colon, eyes, and cervix. The RST technique provides reliable diagnostic tools to monitor and analyze various body fluids and abnormalities in organs.

Certain intrinsic shortcomings of this technology have limited the deployment of FFRS sensors for histopathology. The shortcomings include the highly fluorescent nature of the tissue in the UV region of the spectrum and very weak signal because of every one million photons applied to the tissue, only one is a Raman photon. Thus, to obtain Raman spectra with decent SNR, the clinical scientist is required to use exceptionally high laser power densities, thereby leading to rapid biological changes or irreversible damage to the tissue. Despite all these shortcomings, RST provides useful biological information on cancerous tissues or tumors. Peaks in the Raman spectrum correspond to known molecular signatures of a tissue or tumor, thus revealing more structural details about the cell and tissue composition without using dyes, labels, or other contrast-enhancing agents. Since disease symptoms are always preceded by changes in molecular composition [4] of the tissue or body fluids, Raman spectroscopy provides real-time capability for a reliable diagnosis of a wide range of diseases.

8.2.8 Lasers for Cosmetic Surgery

The latest clinical studies indicate that lasers are becoming popular for cosmetic applications including dermatology, hair removal, leg-vein removal, tissue welding, skin resurfacing, and wrinkle removal. Diode lasers for dermatology treatments, diode and ruby lasers for hair removal, dye lasers for leg-vein removal, Er:YAG lasers for skin resurfacing, and Ho:YAG lasers for wrinkle removal are widely used. The current and emerging cosmetic surgical procedures can be performed in the doctor's office with local anesthesia. Leg-vein removal is the most lucrative cosmetic application using dye or diode lasers; it is a compact and cost-effective alternative method.

8.2.9 Optical Tomography for Medical Treatments

Near-IR laser technology is considered the most reliable and effective technique for medical and biological imaging and could be a leading contender in medical-imaging modalities that include magnetic resonance imaging (MRI), computer tomography (CT), and optical tomography (OT). The near-IR laser technology provides improved biological details of specific tissues or organs based on patterns of scattered light at optimum wavelength. Compared to existing medical and biological imaging technologies, OT offers distinct advantages, such as improved reliability, portability, and safety, because it uses no ionizing radiation. Its high reliability and portability features are due to the fact that it does not require large magnets, radiation shields or huge power supplies. In addition, OT technology offers therapeutic treatment potential such as photodynamic therapy (PDT). It is important to mention that in the OT procedure, a laser beam illuminates an object at some point on the surface, light scatters at the bulk material, and an array of sensitive detectors measures the intensity level of the scattered light. The scattered patterns are measured by the detectors, which provide a transmission function of the object for a given source–detector layout. Ultimate solution of this inverse problem for OT imaging requires the identification of the internal optical parameters of the object based on the measured transmission function between the laser source and detector array.

8.2.9.1 Critical Performance Parameters of OT Imaging Systems

Spatial resolution is the most important performance parameter of an OT imaging system and it depends on the number of source–detector pairs in the bulk material [7]. The higher the number of such pairs, the higher will be the spatial resolution. This can be achieved by scanning the laser beam over the object by moving the source across the surface. Another approach to achieve higher spatial resolution is to use prior information about the optical properties of the object from a MRI system. However, the reliability of such an approach is questionable because of continuous physiological changes such as heartbeat and breathing in the biological specimen. OT imaging essentially involves tissue density. It is based on certain assumptions that the medium has weak scattering and absorption coefficients, the distance between two scatters is large compared to wavelength, and the dielectric constant of the tissue changes slowly. Therefore, in optical atmospheric imaging, the scattering of signals is independent and can be treated within the framework of transport or diffusion theory, which usually renders good results.

The key condition of transport theory is the independence of scattering in a multiscattering system [7], where perturbation of the background medium is relatively weak. For optical wavelengths in the human body, the system breaks down because the distance between the scatters in a biological object presents the same order of magnitude as the wavelength of the near-IR spectral region, around 980 nm. In addition, the local deviations of the dielectric constant are large, so scattering cannot be considered as independent. Biological tissues have pronounced microstructures, including biological cells and their organelles, blood vessels, skin, and bone structures, which cannot be treated as noncorrelated. Despite these problems, transport theory remains the promising approach to the problem of optical imaging of biological objects. Consistently good results with real biological tissues are more difficult to obtain because of weak absorption and strong scattering in the near-IR region. This means that more accurate methods need to be investigated before OT will be able to compete with other techniques such as MRI or CT. The light scattering in biological objects is determined by the dielectric constants of the inhomogeneities on different scales (from μms to centimeters with an increase by four orders of magnitude), which will require a tomographic imaging algorithm that takes into account heterogeneities on these scales. This algorithm will compute transmissivity and intensity maps, which are dependent on the number of periodic barriers. Green functions with eigenvalues are used in tomographic imaging to calculate the transmission coefficient of light. Improvements in OT images and spatial resolution are possible through time gating of the signal by estimating time resolution for separate spatial trajectories. Spatial resolution is strictly dependent on time resolution.

8.2.10 Lasers for Mapping of Burned Tissues

Laser scientists are working on a LIDAR (light detection and ranging) system capable of mapping burned tissue with greater precision and reliability. A coherent frequency-modulated continuous-wave (FM-CW) laser radar provides high resolution imaging of burned tissue down to a resolution of 30–50 μms , representing a thickness of only few cells. Clinical scientists hope to achieve a mapping accuracy of 5–10 μms, which will able to determine where the burned tissue ends and healthy tissue begins within a single cell's thickness from (5 to 10 μms).

The laser-based mapping system uses a 100 mW single-mode, distributed-feedback (DFB) laser emitting at 850 nm, a wavelength that is returned strongly by hemoglobin in the blood. The system chirps [8] the 850 nm laser over a 20 GHz frequency range, and the difference in frequency between the incident wave and the reflected wave provides an optical signal proportional to the distance from the target. When a CCD detector array with improved sensitivity is added to the mapping system, the system permits the attending physician to debride the burn, scrapping away the dead flesh with greater accuracy and with minimum discomfort to the patient. Integration of a video camera to the mapping system will provide the physician with a real-time, fast color map of the burned tissue overlaid on a video image of a patient body. A low-power CO_2 laser operating at 10.6 μm can be used to automate debridement, if necessary. This CO_2 with short pulse capability can cut away about 20 μm of burned tissue with great accuracy per pulse and the 850-nm mapping LIDAR system could tell the CO_2 laser when to stop cutting.

The combination of both lasers will able to removal of the dead flesh without taking away any of the healthy tissue and will complete the debridement procedure with minimum time and with least discomfort to the patient. In addition to this civilian application, the military may be the biggest customer for the laser-based mapping system. With current high-power destructive weapons and explosives of modern warfare, burn cases could be expected in much greater numbers than gunshot wounds. Furthermore, this laser-based mapping system would allow doctors to decide whether a small burn requires surgery or could be left to heal on its own, thereby avoiding unnecessary hospitalization.

8.2.11 Laser-Based Endoscopic Technology

Laser-based endoscopic technology plays a key role in the detection of colon cancer and other rectal disorders. Clinical researchers suggest an Erbium-doped YAG laser (Er:YAG) emitting around 3 μm has potential application for a variety of medical and diagnostic procedures including colonology, dermatology, stomatology, and ophthalmology. Laser wavelengths around 3 μms are most attractive because human tissue absorbs most of the optical energy at this wavelength, which makes it possible to ablate both hard and soft tissues. However, the optical fibers currently available for transmission of such wavelength signals suffer from high insertion losses.

An alternate to the 3-μm laser system involves an optically-pumped laser at the distal end of the fiber that converts the light at near-IR wavelength to a hand-held 2.9 μm light source very close to the tissue under treatment. The optical fiber carries the energy from a near-IR Nd:YAG laser to the 2.9 μm hand-held laser. This converter laser unit can be miniaturized to a package with 2 mm in diameter and 20 mm in length, which is longer than a fiber tip and small enough to be inserted into an endoscope. The device can be cooled by water or air, depending on the heat to be dissipated and available type of coolant.

It is possible to employ a Nd:YLF laser operating at 1047 nm to pump ytterbium ions, which then transfer energy to holmium ions. A more efficient method based on direct pumping of the holmium ions involves pumping with a Nd:YAG laser at 1120 nm which directly excites these ions into the upper laser level at 2.9 μms. This pumping scheme offers a miniaturized 200 mJ laser system with a conversion efficiency of about 25%. This laser provides constant energy from room temperature to about 100 °C. operates efficiently without air cooling, and is relatively inexpensive.

8.2.12 Laser Sensors for DNA Analysis

DNA analysis offers a most powerful tool for law enforcement agencies and district attorneys to prosecute criminals. DNA adducts or other fluorescent analytics typically have been embedded into glass and analyzed by low-temperature, fluorescent line-narrowing spectroscopy (FLNS) technology [9]. Selective excitation at the liquid helium temperature of 4.2 K provides high-resolution FLN spectra [9] that can be used in fingerprint identification. Capillary electrophoresis(CE) with FLNS for on-line spectral characterization represents a vital information source for the rapidly evolving field concerned with selective detection methods that provide reliable structural information. A CE-FLNS system provides an additional degree of selectivity in the chemical analysis of structurally similar compounds.

Using nonselective, high-energy laser excitation and automatic translation of the system through laser excitation at the liquid-nitrogen temperature of 77 K, three-dimensional electropherograms are obtained with the relevant characteristic fluorescence spectra determining the fluorescence origin bands known as (0,0) bands, akin to singlet-to-singlet transitions [9]. Once the origin band transitions are known, appropriate laser excitation wavelengths can be selected for the 4.2 K FLNS characterization of the DNA analysis.

8.2.12.1 Laser Excitation Requirements for DNA Analysis

Clinical scientists have found an excimer-laser-pumped dye most suitable for excitation. FLN spectra are generated using a series of laser wavelengths that selectively excite vibronic regions of the first excited singlet state of the compound, each of which reveals a portion of the excited-state vibrational frequencies of the molecule. Fluorescence is then collected at a right angle to the laser excitation beam, dispersed by the monochromator, and detected by a sensitive photodiode array or high-resolution CCD-based camera. High resolution FLN spectra obtained by the FLNS technique with different laser excitation wavelengths can be used as "fingerprints" for spectral identification of a complex compound such as blood or single element such as hair. DNA analysis has potential application in the fields of biological research and forensic sciences, which can identify the relevant evidence relating the scene of crime.

8.2.13 Laser Technique for Vision Correction

Laser surgical procedures have been very successful in 20/20 vision correction and treating other eye diseases over the last two decades. The laser-based photorefrective keratectomy (PRK) technique is most effective in treating nearsightedness. In a PRK procedure, the outer layer of the cornea is polished off and then the laser beam is used to reshape the surface of the cornea. The entire procedure is done at the doctor's Office under local anesthesia and the patient goes home on the same day, experiencing some blurry vision and discomfort for couple of days.

The latest laser-based technique, known an laser in situ keratomileusis (LASIK), allows the surgeon to make a very thin flap in the outer layer of cornea (called the epithelium) and use a "hinge" to lift it up while the inside of the cornea is under laser treatment for few seconds. Then the flap is put back in place on the epithelium. The LASIK technique allows the patient see well the very next day without any discomfort. The results of the LASIK technique are dramatic and patients who have undergone such surgery like it very much. Surgical data and comments available from patients indicate that 70% of the patients have 20/20 vision without glasses or contact lenses and 95% have 20/40 vision.

8.2.13.1 Laser Requirements for Eye Surgical Procedures

Optical coherent tomography (OCT) was originally developed for ophthalmology to provide the 10 μm resolution needed for imaging the retina surface. OTC has been proven very promising for the diagnosis and monitoring eye diseases, including glaucoma or macular edema in diabetic retinopathy, because it can measure the progress of the disease with high reliability.

A number of all-solid-state, femtosecond (fs) pulsed lasers including Nd:YAG, Nd:YLF, and Cr:YAG are available over a wide range of operating wavelengths for eye treatments and surgical procedures. A simple, low-cost, diode-pumped Nd:glass laser with hundreds of femtosecond (10^{-15}) second pulse seems to be best suited for glaucoma treatment because the duration of hundreds of femtoseconds allows the laser energy to get in and out very quickly and the ablation threshold becomes the deterministic parameter. Furthermore, in the hundreds-of-femtosecond pulse laser format, energy can be adjusted to avoid collateral damage, an important goal in all laser medical procedures, particularly in the case of the eye. The shorter the pulse, the broader its optical bandwidth, which is an important factor in some applications such as laser surgical procedures and WDM communication systems. Squeezing a pulse in time concentrates its energy in time and controls laser energy deposition on the area of interest. Short pulses offer high peak power levels for some applications and extreme precautions must be taken to control the energy to avoid any damage to surrounding areas or tissues. Dutch scientists [10] have compressed 13-fs pulses from a Ti:sapphire laser by first extending its bandwidth from 500 to 100 nm, then compressing its duration to less than 5 fs. The scientists state that in theory [10] the lasers can produce bandwidth wide enough to generate 3-fs pulses, but measurement of such pulse width with great precision will be extremely difficult.

8.2.14 Laser-Based Tweezers

Optical trapping (OT) devices, also known as laser tweezers, play a vital role in biophotonics investigation, genetic research, and cell biology research programs. The trapping concept will enhance biophotonic research activities. Clinical researchers have used the laser-based OT technique illustrated in Figure 8.3 for analytical research study of sperm mobility. The sperm, which are transparent to the near-IR laser light at 980 nm, become trapped in the laser's focal volume, as shown in Figure 8.3. When photons are refracted as they pass through a relatively transparent object, they transfer some of their momentum to that object, thereby leading to complete investigation of the vital parameters of the object.

8.2.14.1 Desirable Wavelengths for Optical Trapping

The stream of photons from a laser focused through a microscope objective can impart enough momentum to trap a cell in its beam. Once the cell is caught, it can be moved in three dimensions by moving the laser beam or the microscope objective. The original optical trapping was demonstrated using visible wavelengths. However, the near-IR wavelength is considered best for this application, because the laser light is less damaging to living cells and the inexpensive diode laser-optic at near-IR wavelengths is readily available at minimum cost.

8.2.14.2 Optical Trapping for Biological Applications

Molecular biologists reveal that certain kinds of proteins can generate forces in the piconewton (pN) range. Biologists have used optical trapping to measure the force exerted by the movement of RNA polymerase [10], which transcribes DNA in a cell. RNA polymerase moves as slowly as 4 nm/second, but it moves with a 25-pN force, where 1 pN is equal to 4.448×10^{-12} pound. In comparison, a kinesin protein that binds to microtubules and transport vehicles and particles in a cell, moves at a whipping rate of 800 nm/second, but with a force of only 6 pN, or roughly 26×10^{-12} pounds [10].

The laser-based OT technique offers a unique method capable of measuring biological forces on the order of piconewton range, which is not possible through other means. Optical tweezers can tug on certain types of bacteria, which selectively attach themselves to minerals such as pyrite. These tweezers can be used to study the strength of the bond between the bacteria and the minerals, which can lead to better understanding of how the bacteria function or survive. German biomedical researchers are using laser tweezers to study and develop laser-assisted in-vitro fertilization techniques. The OT devices can be used to examine sperm mobility and to

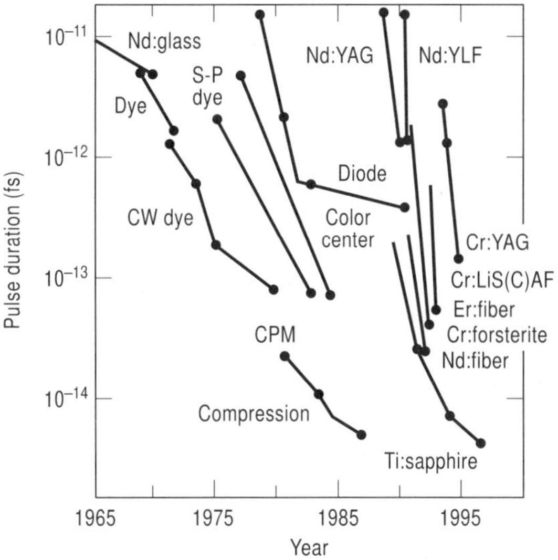

(a)

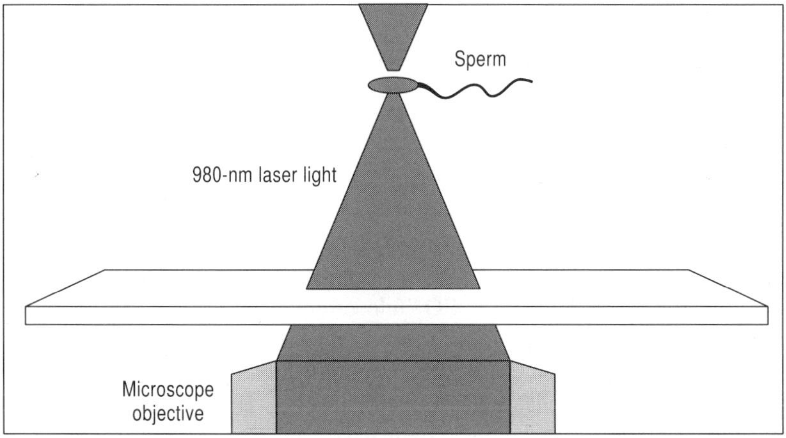

(b)

FIGURE 8.3 (A) Development history of ultrafast lasers and (B) IR laser-based trapping technique.

insert sperm directly into an egg through a hole drilled with an ultrasonic laser drill. Laser exposure of living cells must be carefully controlled to avoid unexpected damage. Clinical studies are needed to determine how much laser light and of which wavelength the living cells can safely tolerate. Excessive photo damage could limit the applications of optical trapping, especially for in vitro studies. Changing wavelengths may not be of great help, because different cellular structures have different spectral sensitivities. For example, exposure to a 760 nm laser may be deadly to cell nuclei. Operating wavelengths yielding excellent results for optical tweezer applications range from 680 to 1064 nm. Optical tweezers can be integrated into a one-beam or two-beam laser system, providing more stable trapping capability; the system can use either the optical beam steering technique or a mechanical method to move the sample. Such a system offers multiple trapping schemes, which can be useful in the investigation of contact between the cells. Commercial optical tweezers are not designed for single-molecule research, which requires higher spatial and temporal resolutions, exceptionally high stability, and precision instrumentation. Clinical researchers would like to see several improvements in key areas, including wavelengths that are minimally harmful to living cells, high-power fiber lasers with tens of watts of output power needed for high trapping forces exceeding 100 pN, and electronic feedback to clamp a cell in place.

8.3 OPTICAL COMMUNICATION SYSTEMS

Switching from RF to optical communication frequencies, one could ultimately accommodate an increase in bandwidth of five to six orders of magnitude. Optical communication in free space offers thousands of communication channels for both voice and data transmission. Much of the technology to meet the ultrawide bandwidth requirement has been developed for military and space communication programs, including the fiber optic telecommunications systems.

8.3.1 Satellite-to-Satellite and Satellite-to-Earth Communication Systems

Space laser experiments performed in late 1998 successfully demonstrated the potential for satellite-to-satellite and satellite-to-earth communication systems. Commercial activity is in progress for satellite-to-satellite optical links. Communication engineering companies are engaged in the design and development of nonlinear optics required for low-earth-orbit (LEO) telecommunications systems, such as the "Internet in the Sky," a project initiated by Teledesic Corp. Kirkland, Washington. Devices and components used in a laser communication terminals are analogous to those used in a RF communication terminal. However, laser communication system offers significant savings in weight, antenna size, and power requirements as illustrated in Table 8.1.

TABLE 8.1 Performance comparison of two communication links between two satellites(each link designed for 1 Gbit/sec data transmission capability)*

Performance parameter	RF link	Optical link
Bandwidth (normalized)	10	10^6
Weight (lb)	500 (min)	100 (max)
Antenna size (inch)	100 (min	15 (max)
Power consumption (W)	350 (min)	100 (max)

*Note: one of the satellites is in LEO, while the other is in geostationary orbit.

The European Space Agency (ESA) launched a remote-sensing (RS) satellite operating in a LEO that used a 60 mW, 800–860 nm semiconductor GaAlAs laser diode source for the transmission of imaging data from SPOT-4 satellite on a 50 Mbits/sec data link. A terminal satellite in a geostationary orbit (GSO) can act as a relay station for transmission of SPOT-4 data to earth. The GSO satellite, due to its much higher orbit, will have a longer time window than the LEO SPOT-4 for transmission of image data to an earth station or terminal. The satellite link uses the GaAlAs laser diode source on SPOT-4, which has direct laser communication capability with a ground station.

Future satellites scheduled for launch into an elliptical earth orbit (EEO) would provide laser communication and tracking functions at 800 nm and 850 nm wavelengths. The EEO satellites will use 150 mW semiconductor GaAlAs laser diodes operating over the 810–850 nm spectral region. This laser wavelength range offers several advantages for satellite-to-earth communication, including good atmospheric penetration, the ability to use smaller rather than longer wavelength technology, and the use of sensitive detectors based on well-matured silicon avalanche photo diode (APD) technology.

A terrestrial fiber optic communication system operating at 1550 nm wavelength offers the most cost-effective performance for satellite-to-satellite laser communication terminals, with data rates ranging from 1 to 100 Gbit/sec. The 1550 nm system offers class-1 safe operation, as opposed to the class-3B level at the 850 nm wavelength. A 155 Mbit/sec data rate, free-space communication system offers an operating range exceeding 5 km with 99% availability.

8.4 SPACE-BASED LIDAR SYSTEMS

Space-based LIDAR systems are best suited to study atmospheric chemistry and physics parameters. A LIDAR system can also measure the mixing ratios of water vapor at various altitudes. Some countries are using LIDARs to investigate the climatic events in the stratosphere and mesosphere (atmospheric region above the stratosphere), including global wind velocities. The term LIDAR typically describes a remote-sensing system having a basic architecture comprises of a modulated laser transmitter, a receiving telescope, and a data acquisition module.

8.4.1 Solid-State Laser Transmitter Technology for LIDAR Systems

Currently, four solid-state laser designs [11] are under development including:

- A diode-pumped Nd:YLF laser with output of 700 mJ/pulse at 1030 nm wavelength
- A KTA optoelectronic pulse amplifier with output of 33 W at 1530 nm wavelength
- A frequency-tripled Ti:sapphire laser source with output of 40 mJ/pulse

- A tandem optical parametric oscillator (OPO) using CdSe and KTA optical crystals capable of yielding emissions ranging from 2–5.3 μms and from 8–11 μm.

Some space scientists are using 600 mJ/pulse, 2-μm, Ho:Tm:YLF lasers and Yb:YAG lasers pumped by InGaAs laser diode bars with impressive optical-to-optical efficiency of better than 30%. Space-based laser components must be protected from radiation hazard levels depending on the altitude of the satellite or spacecraft.

8.5 LASER-BASED SENSORS TO DETECT ATMOSPHERIC POLLUTANTS

Laser-based remote sensors (LBRSs) can measure wind velocities and determine the concentration levels of atmospheric pollutants. In essence, a LBRS allows the measurements of important environmental parameters on both local and global scales. This sensor can also provide early detection of chemical or biological warfare agents and mapping of global wind velocities because of its unique capabilities, such as enhanced sensitivity, wavelength selectivity, and ultrahigh accuracy. This sensor is very similar to a LIDAR system with a basic architecture comprised of a modulated laser transmitter, a fully steerable receiving telescope, and a data acquisition module capable of converting the raw data to a meaningful interpretation format.

8.5.1 Classification of LBRS Systems

The LBRS systems are classified according to the source of optical signal return. In a laser-induced fluorescence (LIF) system described in Figure 8.4, the laser transmitter is tuned to a strong absorption and the return signal is generated by fluorescent decay. This particular system has been used to determine density, temperature, and wind velocities at altitudes between 80 to 105 km. The transmitter beam is steered to several different positions to determine the three components of the wind velocity as well as the atmospheric pollutants. In contrast, both the differential absorption LIDAR (DIAL) shown in Figure 8.4 and the Doppler LIDAR system shown in Figure 8.5 look at the laser light that is scattered elastically by atmospheric aerosols. The DIAL system measures the molecular concentrations by comparing the scattered light at different wavelengths corresponding to local minima and maxima in the absorption spectrum of the atmospheric constituents. In contrast, a Doppler LIDAR system operates in the atmospheric windows with minimum attenuation, using the Doppler shift of the return optical beam relative to the laser transmitter beam [12] to determine the velocity of the scatterer.

8.5.1.1 DIAL System as a Remote Sensing System

DIAL systems operate at two distinct wavelengths, one wavelength is matched to a strong absorption line and the other is unattenuated by atmospheric constituents. These systems are widely used for the measurement of absorbing atmospheric con-

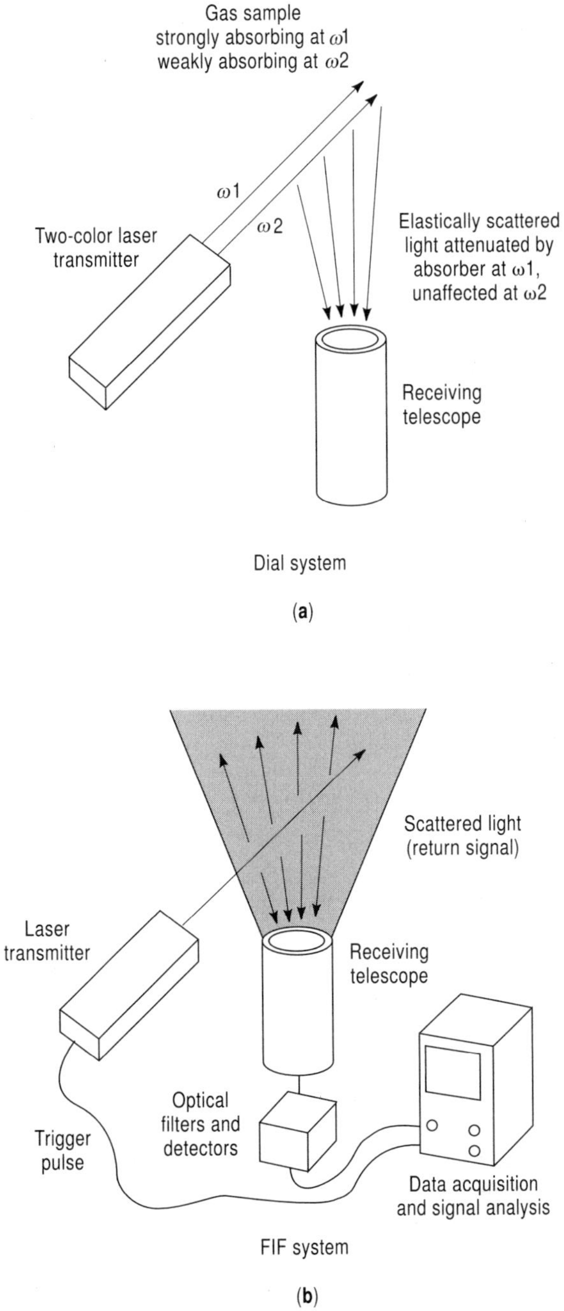

FIGURE 8.4 Design aspects of (A) a dial system using two-color laser transmitter and (B) laser-induced-fluorescence (LIF) system.

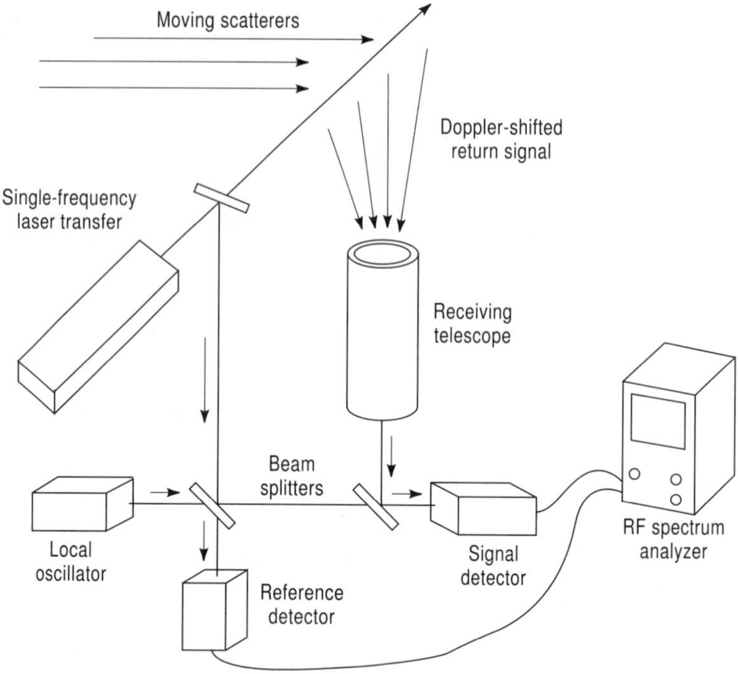

FIGURE 8.5 Critical components used in the design and development of a Doppler LIDAR system.

stituents at various altitudes. Assuming that the transmitter power and the elastic scattering coefficients at two wavelengths are equal, the difference in power collected by the receiving telescope is directly proportional to the density of the absorbing constituents. By comparing the magnitudes of the return optical signal along with laser transmitter signal at two distinct wavelengths, the concentration of the atmospheric gases can be mapped. Since the strongest molecular absorption bands are generally found in the UV or mid-IR regions, a DIAL system is designed to operate in these spectral regions for optimum performance. A DIAL system typically uses a Q-switched laser with output energy in the 10–200 mJ range, which is adequate to provide reasonably good performance level.

8.5.1.2 Performance Capabilities of Doppler "LIDAR" System

As stated earlier, a Doppler "LIDAR" operates in specific atmospheric windows with minimum attenuation. The Doppler LIDAR system shown in Figure 8.5 uses the heterodyning technique to measure the doppler frequency shift, which is equal to the difference between the laser transmitter and the return optical signal beat notes. Accurate measurement of wind velocities is of critical importance, particularly to the aviation industry, because of unexpected presence of clear air turbulence (CAT). A compact Doppler LIDAR aboard an aircraft can detect the presence of

CAT early enough for the pilot to take immediate and appropriate action to avoid it. The early detection of CAT can avoid unexpected discomfort and injury to the passengers and damage to the aircraft. Low-power Doppler LIDAR systems, when placed at appropriate intervals along a runway can provide "wake avoidance" warning to pilots trying to land at busy airports. An orbiting, eye-safe doppler LIDAR system in space provides mapping of the global wind velocity information. Such a system using a DPSS laser transmitter will provide useful aid in monitoring and detecting global weather patterns in addition to the other functions mentioned above. An eye-safe LIDAR sensor is the most suitable candidate for deployment aboard a satellite to provide real-time global weather patterns and early warning of severe weather conditions to various countries that are most likely to be affected by severe weather conditions.

8.5.1.3 LIDAR Sensor for Probing of Outer Space

The receiver in the LIF–LIDAR system shown in Figure 8.4 is capable of looking at fluorescent emissions in the outer space. The laser transmitter beam is electronically scanned through several positions to determine the three components of the wind velocity as well as atmospheric pollutants. The laser transmitter is locked to the peak of a strong absorption for effective excitation. Range information is obtained by modulating the transmitter and gating of the data-acquisition electronics. LIF (Laser-induced fluorescence) sensors operating at UV wavelengths are best suited for the long-range detection of biological and chemical toxins in the battlefield. Because of its high sensitivity to biological warfare agents, a LIF LIDAR sensor can effectively warn ground troops of impending threats in the battlefield. Long-range LIF LIDARs would require high-power tunable lasers, including solid-state Ti:sapphire and the cesium-doped colquirites (Ce:LiSAF and Ce:LiCAF), which have demonstrated significant performance capabilities under battlefield environments in addition to high reliability and safety.

Nonlinear conversion techniques using optical crystals produce tunable lasers directly from a fixed frequency source. For example, a mid-IR optical parametric oscillator (OPO) pumped by a single-frequency Q-switched Nd:YAG laser at 1064 nm can generate wavelengths throughout the near-IR regions. Diode-pumping techniques offer significant improvements in frequency stability, conversion efficiency, and lifetime compared to other techniques. Injection-locking is used to control the output of the LIDAR transmitter. Because of the high cost and complexity of the laser transmitter, long- range LIDAR systems will be exclusively supported by government agencies. However, short-range LIDAR systems, currently under development, will be mostly preferred industrial applications including process monitoring, personnel security and protection, and wind sensing.

8.6 LASER SENSOR FOR SPACE DOCKING VEHICLE

Space scientists have developed a laser rangefinder to provide accurate guidance, navigation, and control needed by unmanned space vehicles to dock with a station in space such as an international space station (ISS). To dock two space vehicles without banging them together, a laser-illuminated video-based guidance system is required. This particular sensor must be capable of acquiring and tracking a filtered corner cube target from a distance of 100 m or less with a docking accuracy of a centimeter. This video-based guidance system is comprised of fiber-coupled diode laser, ruggedized black and white camera, near-IR camera lens, ruggedized computer with video processing cards, VME chassis, optical filters, and other optical processing components. This video-based guidance system has demonstrated higher operating range while tracking within ±2 feet of the shuttle Discovery's hand-held laser rangefinder during the final retrieval phase of SPARTAN. During remote maneuvers in the cargo module, the same sensor tracked the target with an accuracy of ±1.25 cm or ±0.5 in and within ±0.5 degree of the remote system's display.

8.7 INTERSATELLITE LASER COMMUNICATION SYSTEMS

There is a growing trend to deploy intersatellite laser communication (LASCOM) systems for providing wireless voice, video, HDTV, and internet data. The LASCOM technology has fully matured with the development of ultrafast lasers, novel beam-steering capability, high-data-rate receivers, and miniature space-qualified electronics, including connectors and detectors. Laser wavelengths can accommodate megabits to gigabits of pocket data into a narrow concentrated laser beam that can produce a brighter spot with higher gain at the receiving satellite's antenna with minimum input power. Lower power requirements reduce hardware size and weight and cooling requirements, while providing significant savings in payload launch costs. The intersatellite LASCOM will allow transmission of broadband data across a global constellation of LEO satellites with intersatellite spacing not exceeding 250 miles for a constellation comprising of 66 LEO satellites. It is essential that each optical terminal can transfer a thin light beam efficiently to another satellite terminal hundreds or thousands of kilometers away. This will require the two satellite terminals to be in the same or adjacent orbital planes to locate one another, align their optical antennas to within less than one microradian, lock on, then track and maintain the free-space link during the data transfer.

8.7.1 Critical Elements of Laser Communication System

A free-space LASCOM system is comprised of a laser transmitter, drive electronics, regulated power supplies, and optical antennas capable of routing signals from the laser transmitter to free-space, receivers with acquisition and tracking capabilities, pointing mirror, beam-steering mechanism, and optical detectors optimized for the operating wavelengths. A pointing mirror and a slow but accurate beam-steering mechanism is required for acquisition and tracking functions. The detector array and the processing electronics convert the received laser signals into suitable electronic signals. The acquisition and tracking electronics provide control mechanism based on the feedback for appropriate detectors. Finally, the communication and control electronics handle all communications signal processing and control signals to the host vehicle to control the communication terminal.

8.7.2 Communication Link Requirements between Satellites

The principal objective of a free-space laser communication terminal is to establish a reliable, low bit error rate (BER) optical link between cooperating satellites. The BER is the performance indicator of a link and is strictly dependent on the carrier-to-noise ratio (CNR). The BER values are 2.27×10^{-5}, 1.53×10^{-7} and 1.03×10^{-9} for a CNR of 10 dB, 15 dB, and 20 dB, respectively. The higher the CNR ratio the lower will be the BER magnitude. In the case of a LEO satellite constellation comprised of 66 satellites, the intersatellite spacing will be around 360 km. Assuming a free-space loss of 0.125 dB/km at a 860 nm operating wavelength, total free-space loss will be around 45 dB over the 360 km distance. However, for spacing of a few thousands kilometers, the free loss will be extremely high and the link closure will be extremely difficult and will not be able to meet the BER requirements. Distances can vary for geostationary satellite and medium-earth-orbit (MEO) intersatellite links, deep-space probes, and planet-to-planet optical links.

The optical link design must take into account link performance requirements, link degradation, and gases that can affect the signal level. The received signal power can be expressed as

$$P_r = (P_t)\,(G_t)(L_t)(G_r)(L_r)(L_{\text{free}}) \tag{8.1}$$

where P_r is the received power (W), P_t is the transmitter power (W), G_r is the transmit antenna gain, G_r is the receive antenna gain, L_t is the effective loss during the transmit, L_r is the effective loss during the reception, and L_{free} is the free-space loss. Equation (8.1) allows the link design engineer to define the link parameters to achieve the received power requirement and, consequently, the required CNR to satisfy the BER requirement. The slow steering mirror size varies from 10 to 50 cm in diameter, depending on the link architecture selected and performance requirements of the optical link. The smallest element is usually the mm sized fiber interface optics, which directs the optical signal into the μm-sized optical detectors.

8.7.3 Optical Transmitter and Receiver Requirements

The optical transmitter consists of a diode laser source capable of generating an optical signal beam by converting the electronic data-carrying signal into an optical signal level of about 1 W for the LEO networks. This energy is fiber-coupled to an optical antenna. A small pointing mirror adjusts the direction of the signal based on the relative velocities of both satellites and the speed of the light. The signal is then sent into the telescope, which expands the signal into the required transmit beam for transmission across free space.

At the receive antenna of the link, the signal reduces to nanowatt level due to high free-space loss, and is then down-converted to an electrical signal for processing. The optical energy range between the transmit and receive signals is about nine orders of magnitude. The telescope that combines the transmit and receive modules into one unit must have adequate isolation between the incoming and outgoing optical signals.

8.7.4 Acquisition and Tracking Requirements for the Intersatellite Links

Reliable acquisition and tracking functions require precision alignment of the laser beam transmission path. Precision alignment is both critical and challenging. Across the range of thousands of kilometers, a typical optical antenna aboard a LEO satellite is required to point and lock the laser beam into an aperture fractions of a meter in diameter. Point-to-point accuracy of this aperture dimension requires precise knowledge within one or two microradians of the location of both the transmit and receive apertures. The link terminal, after receiving the satellite data, begins to process of acquiring a lock with the target satellite terminal by searching within an initial region of uncertainty of about a 1-degree cone. Once the cooperating satellite has been acquired, control is turned over to the tracking system, which uses precision optics to generate a wide spot for control on the CCD-based acquisition sensor. A tracking beam steering mechanism consisting of a closed-loop feedback control circuit and a high performance mirror driven into two axes is placed directly in the transmit and receive path. Advancements in laser technology and optoelectronic devices will provide the next generation of optical data links in free-space.

8.7.5 Component Performance Requirements for Space Communication Systems

Space environments can have significant impact on the performance level of the devices or subassemblies used in space communication systems. One can avoid extreme temperature fluctuations in space by burying or locating the high-power lasers deep within a satellite or spacecraft. Both high efficiency and reliability require immediate removal of heat generated by the high power lasers. In the case of a compact laser altimeter aboard an orbiting satellite, the operating temperature is around 2 °C. Low-power (10–15 W) electric heaters are used to keep the freezing temperature of space compatible with the required operating temperature. Any pumping-source power decay will cause changes in the laser beam stability, which can destroy the circular symmetry of the laser beam.

For extended missions in deep space or geostationary orbits, lasers generally suffer from optical element radiation darkening. By codoping the laser crystals with chromium, space radiation resistance can be improved. For such mission environments, use of a radiation-resistant Cr^4:YAG passive Q-switch is highly recommended. Some space scientists have selected a slab laser design because of its pulse energy and high peak power of 48 mJ concentrated in 8-ns pulses to provide a compact footprint. The slab laser offers improved reliability, compact packaging, high beam quality, and compensation for thermal lensing along one axis and for beam variations caused by the pump power decay. In summary, a slab laser design can meet the stringent performance requirements during the launch and operation of deep space vehicles.

8.8 FIBER OPTIC LINKS USING WDM AND DENSE-WDM TECHNIQUES

Fiber-optic (FO) links are widely used for transmission of carrier, data, and video signals. Potential FO link configurations, such as central data mixing link, remote data mixing link, and bi-directional data link configurations, are shown in Figure 8.6. These configurations do not use the WDM or dense-WDM techniques. However, implementation of WDM techniques in unidirectional and bidirectional FO links involving two signals is illustrated in Figure 8.7. It is important to mention that the WDM and dense-WDM (DWDM) modulation techniques will meet the growing requirements of Internet access, high-speed data transmission, high-capacity video bandwidth, and high-capacity long-haul transmission lines. Studies performed by the author indicate that dense-WDM-based light- wave transmission systems can provide FO links with data capacities ranging from 40 to 1000 Gbits/second. After the introduction of WDM and DWDM modulation techniques, the demand for high channel densities and data rates is shifting towards improvements in optical fibers, optical amplifiers, and other areas. It is important to mention that it is the optical component technology that determines the performance capabilities of WDM or DWDM links or communication systems. With implementation of WDM and DWDM techniques, FO links are capable of sending 40Gbits/second signals at significantly improved speed and clarity over distances exceeding 600 km without any regeneration, because of significant improvements in erbium-doped fiber amplifier (EDFA) technology [13]. Using ultrawide-bandwidth at 84 nm, a WDM system has a channel transmission capacity of 100 WDM channels of 10 Gbits/sec or overall data rate of 1 THz over a distance of more than 400 km. Studies further indicate that EDFA may not be required for short-distance communication systems operating distances not exceeding 75 km. However, for long-haul transmission systems, one may need both the high-power lasers and the EDFAs to meet the link's operational requirements over the distances exceeding 1000 km or so.

The WDM or DWDM system bandwidth requirement is strictly dependent on the channel spacing and the number of channels involved in the system. Typical WDM channel spacing is about 10 nm in the IR spectral region ranging from 1100 to 1600 nm. However, the attenuation due to Rayleigh scattering and atmospheric absorption is minimum over the 1500 to 1600 nm spectral region. This means that a FO-based WDM or DWDM system must operate over the 1500 to 1600 nm range to achieve optimum performance. Attenuation due to dispersion could be neglected for short-haul transmission lines. However, dispersion in long-haul transmission systems cannot be neglected and requires dispersion-compensation fiber (DCF) between the EDFA stages, which will increase the system cost. Typical dispersion is about 2.8 ps/nm·km over the 1500 to 1600 nm spectral region. Data rates and the number of channels are critical performance requirements in the design of WDM and DWDM systems. Preliminary calculations indicate that a WDM system with

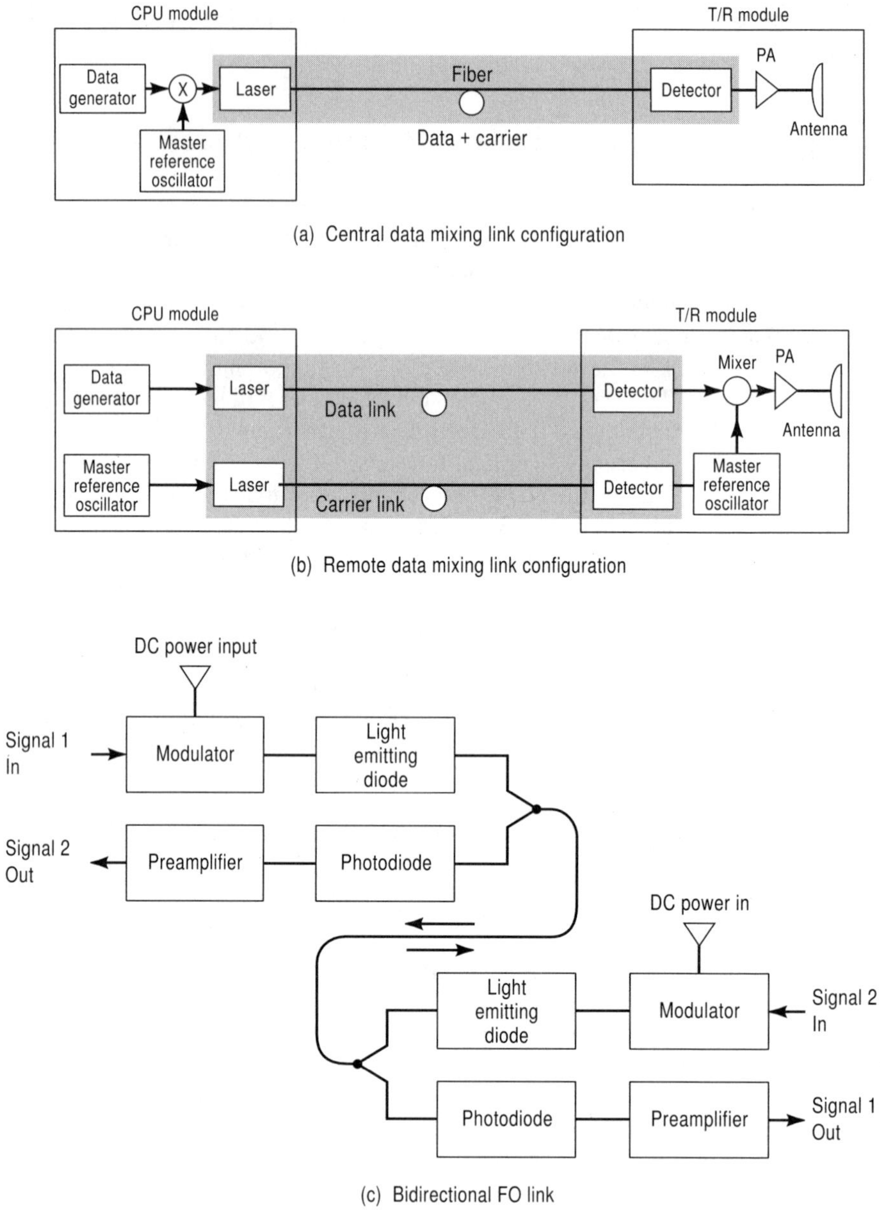

FIGURE 8.6 Various architectural configurations for a fiber optic (FO) link showing unidirectional and bidirectional operations.

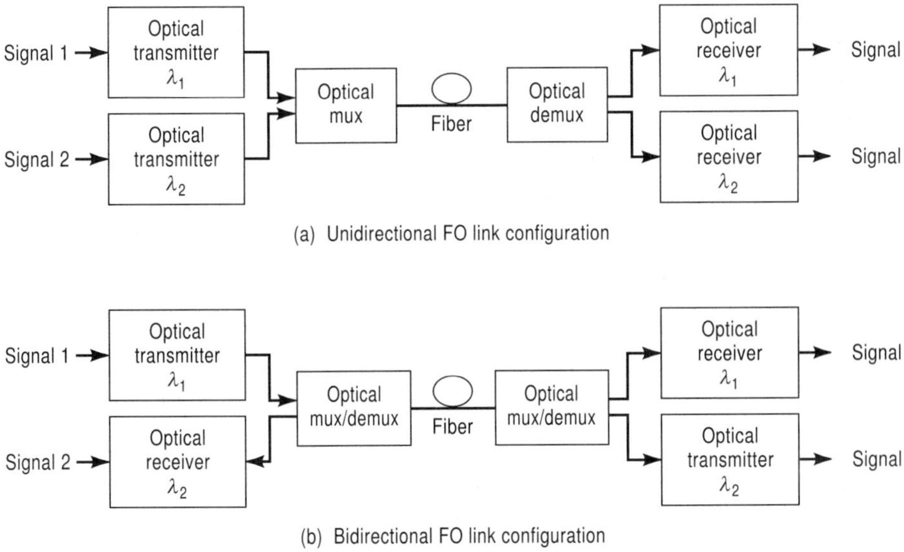

FIGURE 8.7 Implementation of the wavelength-division multiplexing (WDM) technique in unidirectional and bidirectional FO link configurations.

10-nm channel spacing will have an operating bandwidth of 11.8 THz, 6.5 THz, and 6.0 THz, when operated over 1100 to 1150 nm, 1500 to 1550 nm, and 1550 to 1600 nm spectral ranges, respectively. This indicates that wider bandwidth and more channels are possible in the lower spectral region, but the insertion losses are very high. Higher data rate transmission over wide bandwidth and long distances is possible with a DWDM system using state-of-the art optical components such as low-loss, low-dispersion optical fibers, wideband EDFAs, and high-power laser sources.

8.8.1 Optical Fiber Requirements for High Data Rates over Long Distances

As the distances and data rates increase over 1000 km and 10 Gbits/sec, the optical fibers become very sensitive to the dispersion characteristics of the optical fibers. For optical signals to travel with minimum loss over several hundreds of kilometers (300 to 600 km), dispersion compensation is necessary; this can be accomplished through DCFs. Typical dispersion characteristics of a noncompensated fiber, scattering loss and absorption loss for various spectral ranges are summarized in Table 8.2.

It is evident from the dispersion data that dispersion is very severe in the 1500–1600 nm spectral region, which has the lowest losses due to scattering and absorption. Thus, it will be necessary to deploy the DCF technology to reduce the dispersion effects for WDM or DWDM systems. Appropriate lengths of DCF fibers can be introduced between the EDFA stages to minimize or eliminate the dispersion effects in a WDM or DWDM transmission system. However, implementation of DCFs introduces a fair amount of loss in the signal, and thus additional amplifica-

TABLE 8.2 Attenuation and dispersion characteristics of a typical optical fiber for various spectral ranges

Wavelength range, (nm)	Scattering loss, (dB/km)	Absorption loss, (dB/km)	Dispersion, (ps/nm·km)
900–950	1.45	1.85	0.82
1000–1100	0.91–0.65	1.18–0.75	1.45–1.90
1200–1300	0.40–0.32	0.55–0.51	2.25–2.50
1400–1500	0.20–0.16	0.80–0.24	2.72–2.85
1500–1600	0.16–0.15	0.24–0.24	2.85–3.10

tion is required to overcome this loss. A stated earlier, DCF elements can be inserted between the gain stages of an EDFA; nevertheless, such an approach requires additional isolation in the device to retain the performance integrity of the EDFA.

8.8.2 Gain Profile Requirement for Optical Amplifiers

Wide bands of transmission channels require strict control of the gain profile of an optical amplifier over its entire operating band. It is evident from Figure 8.8 that gain of an EDFA is not flat even over the 30 to 40 nm spectral range. When traveling through several EDFAs needed in a 600 km transmission system, in which optical amplifiers are placed say, every 80 km, any variation or difference in gain across the band will start to multiply through each amplifier. This multiplication process will seriously degrade the performance of a WDM or DWDM system. A gain flattening scheme requires more input power as well as a BRAGG grating equalization filter, as shown in Figure 8.8. Both options will add cost and complexity. Studies performed by the author reveal that an optical amplifier that costs twice as much but offers at least twice as many channels may be worth the price when considering the savings of not having to install another fiber, especially for a long-haul transmission system.

Improved WDM or DWDM system performance will require four categories of devices, namely, arrayed waveguide gratings, fiber BRAGG gratings, and dielectric-coated bandpass filters with low insertion loss and a hybrid integrated concept shown in Figure 8.9. The fiber BRAGG gratings ensure high channel isolation, high sidelobe suppression, low wavelength–temperature sensitivity, sharp skirt selectivity, and well-defined passband region when a channel spacing of 0.40 nm is used. Even a channel spacing of 0.80 nm, which corresponds currently to 100 GHz bandwidth, imposes stringent requirements on the components used by a DWDM system.

An alternate approach to achieve higher numbers of channels will requires each channel to have its own laser operating at a specific wavelength. Thus, a DWDM system using 40 different wavelengths at 10 G bits/sec/wavelength will have a data rate capability of 400 G bits/sec. Similarly, a 2.5 G bits/sec system with 80 channels operating at 80 distinct wavelengths and with 200 G bits/sec capability will require 80 different laser transmitters. However, such schemes suffer from very high cost

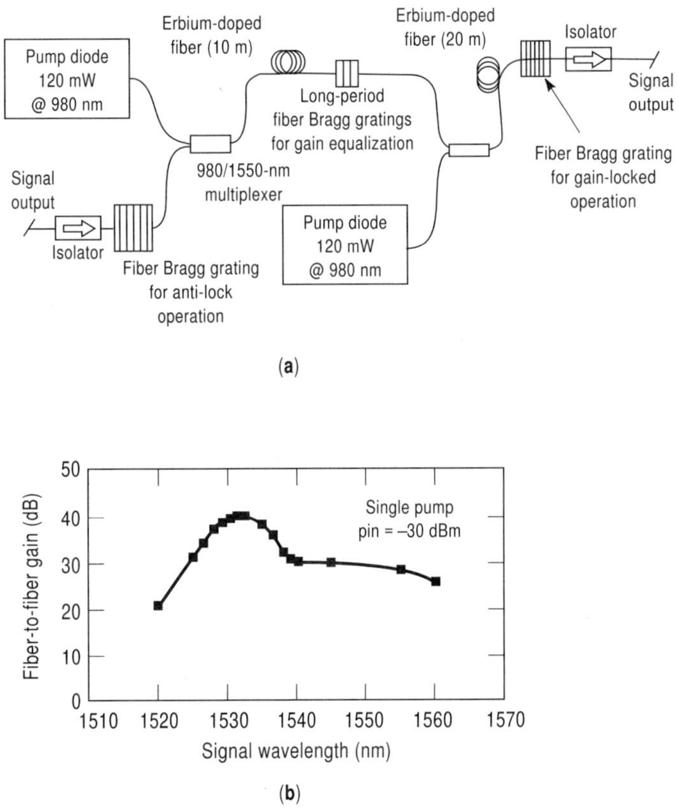

FIGURE 8.8 (A) Gain flattening technique, (B) gain variation versus wavelength, and (C) gain noise figure performance of a EDFA.

and complexity. To overcome such scheme problems, a tunable laser must be considered in the design of a WDM or DWDM system. However such a system configuration would require EDFAs with flat gain and low noise over the 40 to 60 nm bandpass region and tunable high-power pump lasers. Since, the WDM multiplexing scheme shown in Figure 8.9 is more efficient and cost-effective compared to a DWDM system, potential methods must be explored to upgrade existing WDM systems. A DWDM system requires an EDFA incorporating optical gain control and equalization features needed for a multiple-wavelength communication system. Currently, DWDM systems are designed for 16, 32, and 40 channel capacities. DWDM systems with 100 channel capacity are under development [13]. In summary, low noise, high gain, low dispersion, and flat gain in an EDFA over wide band are of critical importance, irrespective whether a system is WDM-based or DWDM based. Future DWDM systems will require gain-locked, gain-flattened, optical amplifiers, regardless of number of channels, signal allocation, or input power.

8.9 SUMMARY

Performance capabilities and critical parameters of laser-based sensors and equipment for applications in medicine, surgery, dentistry, free-space surveillance, optical communications and data transmission, life-science research, industrial and manufacturing process control, wavelength-division multiplexing (WDM) and dense-WDM are summarized, with emphasis on performance, cost, and reliability. Integration of state-of-the art photonic and optoelectronic devices in various optical sensors is discussed. Performance levels of IR sensors and devices used for biotechnology imaging, photodynamic therapy (PDT), noninvasive optical tomography (OT), transmyocardial revascularization (TMR), and multispectral and hyperspectral imaging are summarized. Photonics technology for DNA analysis, space surveillance, and environmental research is thoroughly discussed. Laser-based optical links, communications systems, and telecommunications systems using WDM and dense-WDM techniques are described, with emphasis on performance, safety, and

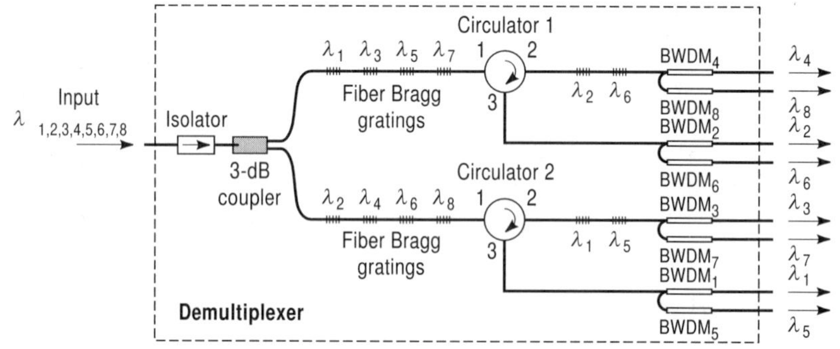

(a) Demultiplexer configuration

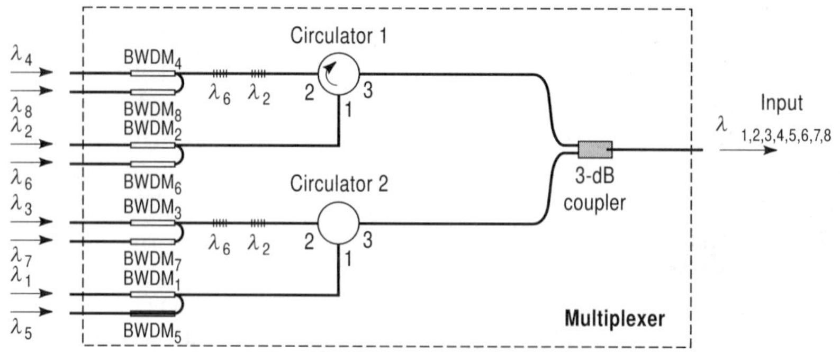

(b) Multiplexer configuration

FIGURE 8.9 (A) Hybrid dense-wavelength-division demultiplexer, (B) dense WDM figuration, and (C) multiplexing scheme to produce very high data rates approaching several gigabits per second.

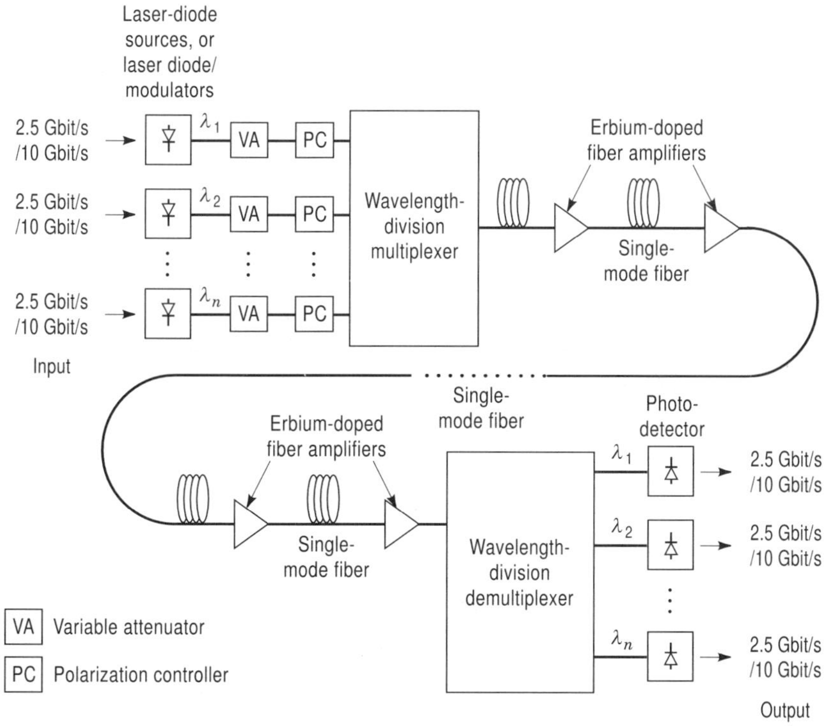

(c) Multiplexing scheme to allow dense-WDM for high data rate

FIGURE 8.9 (continued)

reliability. Critical performance requirements of specific optical components such as erbium-doped fiber amplifiers, DPSS lasers, tunable lasers, BRAGG grating filters, and diode arrays are briefly discussed. Finally, performance improvement of various optical sensors or systems through integration of future technologies and deployment of optical materials are identified.

NUMERICAL EXAMPLE

Compute bandwidth requirements for a WDM system as a function of wavelength and optical frequency using the following expression:
Frequency = (speed of light/wavelength). Computed values are shown below.

Wavelength (nm)	Optical frequency (GHz)	Bandwidth (THz)
1100–1150	272,700–260,900	11.80
1150–1200	260,900–250,00	10.90
1200–1250	250,000–240,000	10.07

Wavelength (nm)	Optical frequency (GHz)	Bandwidth (THz)
1250–1300	240,000–230,800	9.20
1300–1350	230,800–222,200	8.60
1350–1400	222,200–214,300	7.90
1400–1450	214,300–206,900	7.40
1450–1500	206,900–200,000	6.90
1500–1550	200,000–193,500	6.50
1550–1600	193,500–187,500	6.00

Remarks: These calculations are based on a 50 nm bandwidth, which is possible with current state-of-the art, erbium-doped fiber amplifiers used for WDM and DWDM systems. Furthermore, a few years ago the typical channel spacing used in the design of a WDM system was 10 nm. However, the current channel spacing is about 0.8 nm and future spacing of 0.5 nm is expected by the year 2005. Higher bandwidth is possible at lower wavelengths, but minimum loss occurs only over 1500–1600 nm range.

REFERENCES

1. N. McManna. "Camera detects faulty cells." *Photonics Spectra,* p. 123, November 1998.
2. K. Robinson. "Photodynamic therapy offers new medical treatments." *Photonics Spectra,* pp. 219–226, May 1998.
3. L. Lerner. "Hyperspectral imaging opens life-science vistas." *Laser Focus World,* pp. 89–98, October 1998.
4. J. Hetch. "Rare earths create unique long-wavelength lasers." *Laser Focus World,* pp. 135–138, November 1993.
5. K. Kincade. "Raman spectroscopy enhances in vivo diagnosis." *Laser Focus World,* pp. 83–91, July 1998.
6. B. R. Marty. "New research could end dental fillings." *Laser Focus World,* p. 34, February, 1999.
7. I. Khallum. "Inverse problem stymies optical tomography." *Laser Focus World,* pp. 179–188, 1998.
8. N. Savage. "LIDAR system maps burnt tissues." *Laser Focus World,* pp. 30–32, April 1999.
9. D. Appel. "Laser application in DNA analysis." *Laser Focus World,* pp. 46–47, April 1998.
10. A. Baltuska et al. *Optical Letters, 22*(2), 102, 1997.
11. W. B. Gravi. "New push space-based LIDAR." *Photonics Spectra,* pp. 43–44, September 1998.
12. G. J. Dixson. "How remote sensors probe the environments." *Laser Focus World,* pp. 129–136, April 1995.
13. V. Morin. "Flat and fixed gain aplifiiers simplify dense-WDM design." *Laser Focus World,* pp. 127–132, January 1999.

CHAPTER NINE

Application of Photonic and Infrared Technologies for Space and Military Sensors

9.0 INTRODUCTION

This chapter deals with potential applications of photonic and IR technologies to the sensors and devices most suited for space surveillance, tracking and detection of hostile missiles, space-based high-power lasers capable of destroying hostile satellites, and photonic and IR devices for military applications. The sensors and devices described here are partly based on the information available in technical journals and papers presented at international conferences. The IR and electrooptical (EO) sensors described here can be used for ocean surveillance, unmanned aerial vehicles (UAVs), jamming of hostile missiles, battlefield reconnaissance, and target detection, recognition, and identification. Photonic and IR sensors include target acquisition systems,high-resolution IR cameras for UAV applications, space-based antimissile systems, IR countermeasures equipment, IR line scanners, IR search and track (IRST) sensors, laser rangefinders, and multispectral sensors for space environmental research.

9.1 OPTICAL TECHNIQUES FOR 3-D SURVEILLANCE

High-resolution three-dimensional (3-D) images of the ocean is now possible with advances in both high-power pulsed laser and optical reconnaissance sensor technologies. Conventional techniques for remote sensing of the ocean include line-scanning systems and gated-camera imaging systems. A line-scanning sensor uses a narrow beam of CW laser and an optical receiver with narrow field-of-view (FOV). A gated-camera imaging system is comprised of a narrow-pulse, frequency-doubled, Q-switched, mode-locked Nd:YAG laser operating at 532 nm and a gated intensified charge-coupled device (CCD). Although both systems provide high-resolution images within a couple of feet, the images are formed at a fixed range and do not provide 3-D information on the targets under surveillance. Full 3-D imaging capability requires real-time adjustment of the convergence angle or the gate timing for the synchronous scanners and gated cameras.

9.1.1 Alternate Technique for 3-D Surveillance

An alternate technique for the 3-D surveillance involves a pulsed laser and time-resolving receiver that can be used in a synchronous scanning configuration. The new approach uses the streak-tube imaging laser (STIL) and a high-speed computer module. Digitization of the temporal return of each laser pulse permits determination of range parameters. The lateral resolution or the AZ resolution is given by the search rate, which is equal to the ratio of product of swath width and forward platform speed divided by the pulse repetition frequency (PRF) of the laser system [1].

Higher resolution is possible at higher search rates or higher PRF. Based on current data acquisition technology, a scanning LIDAR system is limited to modest search rates and consequently to modest resolutions. Underwater imaging sensors for high-resolution applications such as pipeline inspection can identify objects with horizontal dimensions of 2 to 5 cm at much lower search rates on the order of 0.1 km^2/hr [2]. However, in the STIL surveillance system, the laser beam is imaged onto the slit photocathode of a streak tube, as illustrated in Figure 9.1. The photoelectrons from the photocathode are accelerated, focused, and deflected in time. A sweep signal is applied to steer the laser beam along an axis perpendicular to the fan beam. This allows a range-azimuth image to be formed on each laser pulse. This image is digitally recorded by a CCD-based imaging array. Determination of track dimension requires synchronizing the PRF of the laser with the forward speed of the platform or underwater vehicle. As stated earlier, in a STIL sensor configuration, each laser pulse generates an image across the full fan beam, which can be designed to obtain a large swath width. High search rates can be achieved with modest PRFs that are within the current laser technology and CCD capability. The combination of the fan-beam and push-broom imaging techniques provides for 3-D imaging capability of the ocean without the need for a mechanical scanning mechanism.

9.1.2 Performance Capabilities of STIL System

High-resolution 3-D images have been generated [1] using Nd:YAG laser with 35 ps pulses at 532 nm wavelength with a resolution better than a quarter of an inch in all three directions. Enhanced resolution can be obtained using a 3 ps colliding-pulse, mode-lock dye laser operating at 580 nm. High-resolution imaging in ocean water is limited to the 20 to 30 m range due to attenuation and scattering effects in seawater. A Q-switched, frequency-doubled Nd:YAG has provided high-resolution images through 30 m of water in a deep tank. While the lateral resolution or AZ resolution was on the order of quarter of an inch, the range resolution was in

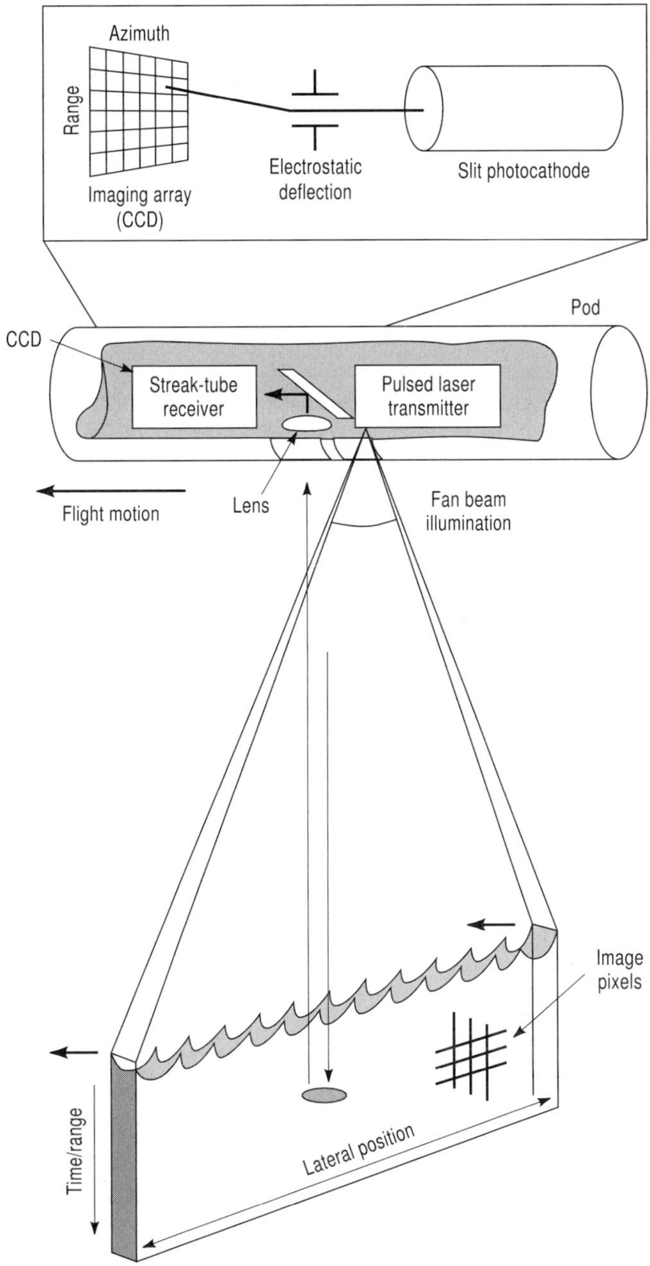

FIGURE 9.1 Pulsed laser sensor for 3-dimensional ocean surveillance.

the vicinity of 4.5 feet due to the 9 ns pulse used by the laser. Dramatic improvement is possible with a Raman-compressed laser pulse of 2 ns. Raman compression technique provides short, high-power pulses for improved range resolution, but in doing so the laser wavelength moves out of the optical blue–green ocean transmission band. A diode-pumped Nd:YAG laser capable of generating narrow pulses on the order of 3 ns at 532 nm wavelength in conjunction with the streak-tube receiver will provide significantly improved range resolution close to 1.5 feet at great depth in ocean water. The STIL approach offers higher search rates with nonscanning, compact hardware. High lateral and range resolutions provided by the STIL technique are not possible with other sensors. This system offers video-quality 3-D imaging at distances four to five times greater than those provided by TV and video cameras.

9.2 INFRARED SEARCH AND TRACK (IRST) SENSOR CAPABILITIES

IRST sensors are capable of detecting and tracking decoys, aircraft, and cruise missiles at long distances with high resolution. When installed on a high-altitude aircraft at appropriate location, this sensor can provide detection ranges close to 100 miles under clear weather conditions. The overall performance of this sensor is dependent on several parameters, such as dwell time, number of detectors, detector sensitivity, photon quantization, and the instantaneous field-of-view (FOV) detector array. Since this is a passive airborne sensor, higher detection and tracking ranges are possible with minimum power consumption, weight, and size, compared to an active sensor.

This sensor, when installed appropriately on a high-altitude aircraft, can provide early warning against hostile fighter aircraft equipped with air-to-air missiles, bombers carrying antisubmarine and cruise missiles, and raid-count information in the presence of RF jamming. Detection ranges in excess of 100 NM are possible with the IRST sensors using a cryogenically cooled, high-sensitivity detector array under clear environments.

9.2.1 Critical Elements and Their Performance Parameters

Detector array, microelectronic components, and processing circuits are the important elements of the IRST sensor. The passive sensor is designed on the basis of anticipated emission wavelengths and IR intensity levels. The design of this sensor is strictly based on IR signature estimates over the 3–5 μm and 8–12 μm spectral ranges, depending on the jet engine, exhaust temperature, thrust level, aircraft engine and operating altitude. The detector array involving cryogenically cooled Hg:Cd:Te detector elements is the most critical element of the sensor. However, optimum cryogenic temperature must be utilized for maximum sensitivity for the specified wavelengths. Studies performed by the author indicate that a two-dimensional, cryogenically cooled detector provides optimum performance for both the staring and scanning systems.

The microelectronic readout circuitry systematically moves photon-generated charges from the FPAs off the processing chip. The FPA technology satisfies the requirement for many larger detector arrays using the integrated circuit approach. The principal objective of this approach is perform all necessary functions prior to signal processing, on the same chip. The signals from all detectors can then be read out to the signal processor by one wire per chip, which is comprised of thousands of detectors. The FPA architecture consists of Hg:Cd:Te detector arrays coupled directly to an array of silicon charge-transfer devices (CTDs). As the IR radiation incident on the detector is converted to an electrical charge, this charge is immediately coupled into a storage device directly beneath the detector. The silicon storage array is a part of the parallel shift register. A series of electrical pulses applied to this shift register transfers each charge packet to the readout register. Another set of pulses applied to the readout register transfers this charge from one row of the array to the preamplifier. These steps are repeated until all rows of the shift register have been read to the signal processor, providing a single-frame image of the IR scene. Incorporating both the detector and CTD arrays into one substrate material will yield optimum performance with minimum cost and complexity.

9.2.2 Critical Performance Requirements of FPA Detector Arrays

The advantages of a FPA include most economical method for high-density packaging of the detectors on the focal plane and allowing signal processing capability on the same focal plane. Both these advantages permit effective implementation of staring and TDI array technologies with minimum cost. Furthermore, these advantages allow tradeoffs between critical system parameters such as aperture size and spectral bandwidth, resulting in optimum design flexibility not otherwise possible. Furthermore, both the system cost and complexity are reduced due to elimination of much of the conventional processing circuit elements used in the existing system.

High-density detector configurations offer greater system sensitivity needed for detection of point sources at long ranges. Improvement in the sensitivity of a scanning array is strictly due to lower noise equivalent irradiance (NEI). A system architecture incorporating several staring FPAs, each having a wide FOV, offers the airborne platform or aircraft the detection capability over full 360 ° spherical coverage against anti-aircraft missiles. Such a threat warning sensor is the first necessary step in the development of countermeasures systems to protect fighters and long-range bombers against air-to-air and air-to-surface missiles. In summary, the passive IRST sensors, when appropriately located on an aircraft, offer 360 ° effective missile detection and tracking capability at high altitudes and long ranges. Low-cost, lightweight, wide-FOV missile threat warning systems are possible using cryogenically cooled, large-area staring IR FPA technology and on-chip processing capability. A large chip with high data rate capability and a staring charge-imaging matrix (CIM) array (128 × 128) fabricated with mid-IR cryogenically cooled Hg:Cd:Te detectors are essential elements of the system to meet the missile threat warning requirements. Deployment of a mosaic array technology in the IRST may provide significantly improved sensor performance.

9.2.3 Detector Types and Materials for Applications in FPAs

Selection of detector type and material is dependent on performance requirements, such as response time, sensitivity or detectivity (D^*), IR spectral range, and reliability. Results of a survey made by the author on various detectors are summarized in Table 9.1.

It is important to mention that higher sensitivity and wider spectral bandwidth are possible in lower cryogenic operations. Hg:Cd:Te detectors offer satisfactory performance at lower cost. However, any substrate impurity will lower the detector yield, sensitivity, and reliability.

9.2.4 IRST Detection Range Computations

The detection range of the IRST sensor is proportional to the square root of the source intensity (J), proportional to the square root of detectivity (D^*), inversely proportional to the fourth power of the detector instantaneous FOV (θ_d), inversely proportional to the fourth power of the dwell time (t_d), proportional to the fourth power of the frame time or observation time (t_f), proportional to the square root of the optics size (D_o), inversely proportional to the square root of the receiver noise bandwidth (Δf), and inversely proportional to the SNR for a given probability of detection and false alarm rate. Analytical studies performed by the author reveal that the detector sensitivity (D^*) and IR source intensity (J) have significant impact on the IRST detection ranges compared to other variables. Parametric values of IRST detection ranges as a function of detectivity and IR source intensity based on the following assumed variables are shown in Table 9.2: optic size (D_o) = 25 cm,signal voltage-to-noise voltage = 5.7 (based on a detection probability of 90% with a false alarm rate of 10^{-6}), numerical aperture (NA) = 0.5, detector instantaneous FOV = $0.1° \times 0.1°$, receiver noise bandwidth = 100 Hz, atmospheric transmission efficiency = 90% (clear atmosphere), and optical efficiency of 80%. These calculations do not take into account the IR background clutter and the atmospheric turbulence effects. Under moderate clutter environments and clear weather conditions, the above ranges might experience reductions of 40 to 50%.

9.2.5 Impact of IR Background Clutter on IRST Performance

Since IRST is a passive sensor, it must make a reasonable number of observations for efficient peak-holding functions. Each observation must be tested against a spe-

TABLE 9.1 State-of-the art detector materials and performance parameters

Detector, material	Response time, (ns)	Detectivity (D^*) (cm $\sqrt{\text{Hz}}$/W)	IR range (μ)	Percent of energy @ 350 K
PbS	100,000–500,000	2×10^{11} @ 300 K	1.80–2.37	0.2
InSb	800	10^{11} @ 77 K	3.0–3.5	5.3
Ge:Cu	50	2×10^{10} @ 5 K	8–25	65
Hg:Cd:Te	100	5×10^{10} @ 77 K	8–14	39

cific threshold level to meet specific false alarm rate and SNR requirements. The target-to-clutter contrast ratio (TCCR), which is dependent on the SNR, can be improved by shaping and optimizing the instantaneous FOV (IFOV) to derive maximum spatial correlation. The contrast of an undetected target brighter than the background can be improved by reducing the IFOV. That is why a wide FOV (typically, $30° \times 20°$) is used for search and narrow FOV (typically, $5° \times 5°$) is selected for tracking a target. The reduction in sensor's FOV is due to the diffraction limits of the optics, which can be avoided by using adaptive or nonlinear optics.

Background IR clutter from clouds and solar radiation can degrade the performance of the airborne IRST sensor. The background clutter reduces the detection range and introduces tracking errors. The tracking errors can be reduced by using low-noise preamplifiers and cryogenically cooled detectors. Spatial filtering of the background clutter can be achieved through two-dimensional autocorrelation functions provided by a complex and powerful data processing scheme. Detection of a cruise missile against the low background clutter level from low clutter photons requires cryogenic cooling of the optics, detector array, and interference filters.

9.2.6 Spectral and Thermal Discrimination Techniques

In an operational situation where spectral discrimination or thermal discrimination is required between the target body temperature and space debris temperature, the spectral band could be divided into two convenient sub-bands, each defined by two separately filtered detector arrays. This can be accomplished by employing a linear or planar detector array filtered over 10 to 12 bands and placing the array in the focal surface normal to the direction of the scan over the target of interest.

This type of spectral band mechanization will permit two alternate decision algorithms capable of providing a two-color discrimination technique. In a situation where the background photon level is high, well-designed, cryogenically cooled optics must be used to achieve a wide FOV for detecting or searching a large number of targets. The false-alarm rate is dependent on the number of detectors used and their response time. If an IRST sensor with 4000 detectors has a response time of 1 m·sec. and the allowable false alarm rate is on the order of one in fourteen hours per detector, the probability of false alarm is better than 10^{-8}. Such a false alarm rate offers a detection probability of 99% with a signal-to-noise voltage ratio of 5.7 or 15 dB, which highly desirable for the detection of low-flying targets including missiles. An IRST sensor scanning over a structured background such as clouds generates a

TABLE 9.2 Theoretical IRST detection ranges (n·m)

IR intensity (W/sr)	$D^* = 10^{10}$ cm $\sqrt{Hz}$/W	$D^* = 10^{11}$ cm $\sqrt{Hz}$/W
1000	202	642
100	64	202
50	45	143
25	32	101
12.5	23	71

response that interferes with target detection and tracking. Such a noise source can be a limiting factor for missile detection and tracking, particularly when the airborne missile is present in the upper atmosphere.

9.2.7 Critical Performance Parameters of the IRST Sensor

The performance level of an IRST sensor is dependent on the performance specifications of several components and devices, such as IR detector array, controlled servo mechanism, target tracking sight control device, control set for Az and EL position selection, cryogenic cooler, and a receiver–converter unit. Cryogenic cooling at 77 K is necessary to achieve optimum system performance. Typical IFOV for search and detection can be 20° × 15° or 30° × 20°. However, the IFOV for tracking is very narrow, ranging from 10° × 5° to 5° × 5°. System stability must be compatible with the tracking accuracy requirements. Typical tracking accuracy of few tens of microradians may be adequate for tracking of an aircraft, provided the platform stability does not exceed 100 or so microradians. The search and tracking ranges are strictly dependent on target source IR intensity, detector IFOV, detector sensitivity, and a host of other key design parameters. Typical detection and search ranges for high-altitude targets could be in excess of 100 NM, provided narrow detector IFOV, effective IR clutter rejection techniques, and detector array with sensitivity of 10^{11} cm $\sqrt{\text{Hz}}$/W or better are used in the system design. However, the tracking ranges will be relatively shorter than detection or search ranges.

9.2.7 False Alarm and Voltage SNR Requirements for Optimum Performance

The threshold current (I_t) is set to achieve a required false alarm rate (FAR) and peak signal-to-noise current ratio (I_s/I_n) to meet specific probability of detection [3] requirements. The average FAR is the average number of false alarms per second for which the output noise current exceeds the threshold current of the detector. The average FAR is written as

$$FAR_{\text{av}} = \frac{0.289}{\tau} e^{-0.5x} \tag{9.1}$$

where x is equal to β^2, β is the threshold-to-noise current ratio (I_t/I_n) and τ is the pulse width or observation time. Computer simulation indicates that the average FAR decreases rapidly with increase in the threshold current. If a matched filter detection method is used, then the above expression can be rewritten as

$$\tau_f FAR = (0.0916)\left(\int e^{-0.5x}\, d\beta\right) \tag{9.2}$$

where τ_f is the postdetection filter time constant. In the case of a sensor with built-in gating circuit, which must be open for the interval of T_i based on maximum detection range, the interval variable can be written as

$$T_i = \frac{2R_{\text{max}}}{c} \tag{9.3}$$

where c is the velocity of the light and R_{max} is the maximum detection range. For a maximum detection range of 10 km, the interval comes to 67 μs. But the average FAR normally occurs as once in a thousand of such intervals; this mean that

$$FAR_{\text{av}} = 1/1000\times 67 \times 10^{-6} = 15/\text{sec} \tag{9.3 A}$$

Assuming a 100 ns pulse width (τ) or observation time, the product of average FAR and observation time comes to

$$\tau FAR_{\text{av}} = 15 \times 100 \times 10^{-9} = 1.5 \times 10^{-6} \tag{9.4}$$

Using this value and assuming a detection probability of 99.9%, the corresponding peak signal-to-noise current ratio (I_s/I_n) comes to 8.1 based on the curves shown in Figure 9.2.

The threshold-to-noise current ratio can be determined from the following equation:

$$\frac{I_t}{I_n} = \sqrt{-2 \ln (3.464)(\tau)(FAR_{\text{av}})} = 4.93 \tag{9.5}$$

At 50% detection probability (P_d), the two signal-to-noise ratios are equal. This means that

$$\frac{I_s}{I_n} = \frac{I_t}{I_t} \tag{9.6}$$

One can directly read the threshold-to-noise current ratio of 4.93from the curves shown in Figure 9.2 using the detection probability of 50% and the numerical value given by equation (9.4). Table 9.3 provides calculated values of detection probability, average FAR, gate interval, and two current ratios as a function of maximum detection range. These computations assume a pulse width or observation time of 100 ns. The calculations indicate that a probability of detection of 90% will require about 2 dB more peak signal-to-noise ratio compared to 50% probability of detection.

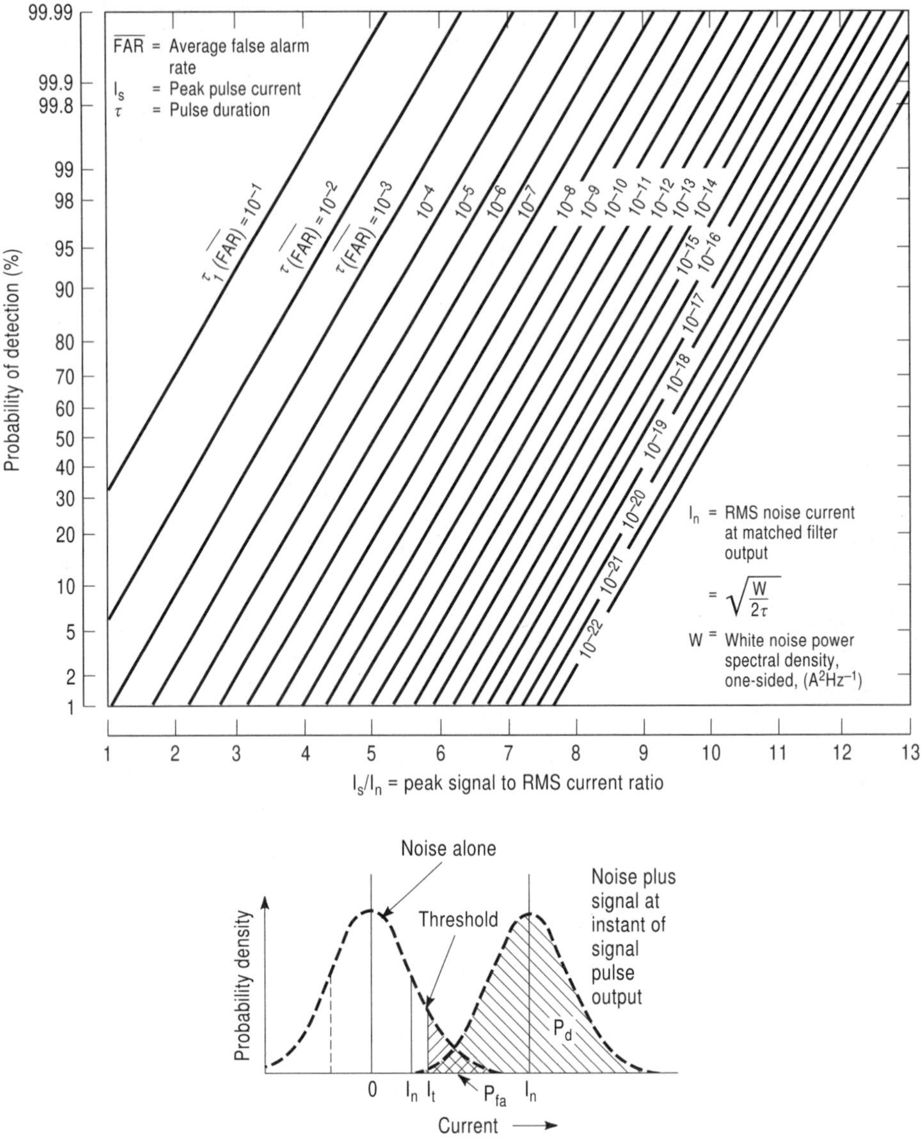

FIGURE 9.2 Probability of detection as a function of peak signal to RMS current ratio, false-alarm rate, and pulse duration.

9.3 FORWARD-LOOKING INFRARED (FLIR) SENSORS

The forward-looking IR (FLIR) is a thermal imaging sensor that provides the pilot a video picture of the lower forward sector ahead of the aircraft needed to locate and identify ground targets under day or night and poor visibility environments. The

TABLE 9.3 Gate opening, FAR, probability of detection, and current ratios requirements for maximum detection ranges

R_{max} (km)	T_i (μs)	FAR/s	T_I FAR × 10^{-6}	$P_d = 90\%$ $I_s/I_n = I$	$P_d = 50\%$ $I = I_t/I_n$
10	67	15	1.5	6.25	4.93
20	134	7.5	0.75	6.30	4.94
30	201	5.0	0.50	6.35	4.95
40	268	3.7	0.37	6.38	4.96
50	335	3.0	0.30	6.40	4.97

FLIR sensor is of critical importance in accomplishing specific military missions in a hostile territory and under all-weather conditions. This sensor provides all-weather, precision navigation and weapon delivery capabilities to a helicopter or fighter aircraft pilot in a hostile and unfamiliar territory. It provides two instantaneous FOVs, one a wide-FOV for search and navigation and other a narrow-FOV for tracking and weapon delivery. In brief, the FLIR provides automatic search, acquisition, tracking, precision navigation, and weapon delivery functions to the pilot. A typical FLIR is comprised of four line replaceable units (LRUs), such as a FLIR optical assembly mounted on a gyro-stabilized platform, electronics module containing all necessary electronics circuits and cryogenically cooled detector array, a power supply unit, and control and processing assembly.

9.3.1 Critical Performance Parameters and Capabilities

The sensor scans the scene of interest in the forward sector and the cryogenically-cooled (77 K) IR detector arrays converts the received signals into the visible light for real-time display to the human observer. High-resolution images are achieved through a high-speed signal processor. The electronic signal processing provides a high degree of contrast most desirable for target classification and identification. The quality of day and night imaging features are shown in Figure 9.3.Appropriate FOV can be selected to achieve optimum performance in a specific mode of operation. Vertical FOV and horizontal FOV requirements as a function of resolution and number of lines are shown in Figure 9.4. As stated earlier, the wide FOV is normally used for navigation and aerial search, while the narrow-FOV is used for target tracking and weapon delivery. Wooden towers, telephone lines, power lines and tall trees that are sometimes difficult to detect by radar can be seen by the FLIR when operating in the wide FOV, which is typically 30° × 20°. A medium FOV is roughly half the size of wide FOV and is useful for target classification and tactical maneuver decisions. A narrow-FOV is typical 5° × 5° and is generally used for acquiring major landmarks at long distances, identification of checkpoints, and precision weapon delivery.

Detection and recognition ranges are strictly functions of target size, weather conditions, and the thermal resolution capability of the sensor. Typical detection ranges using a wide FOV vary from about 4 km under clear to 2.5 km under hazy

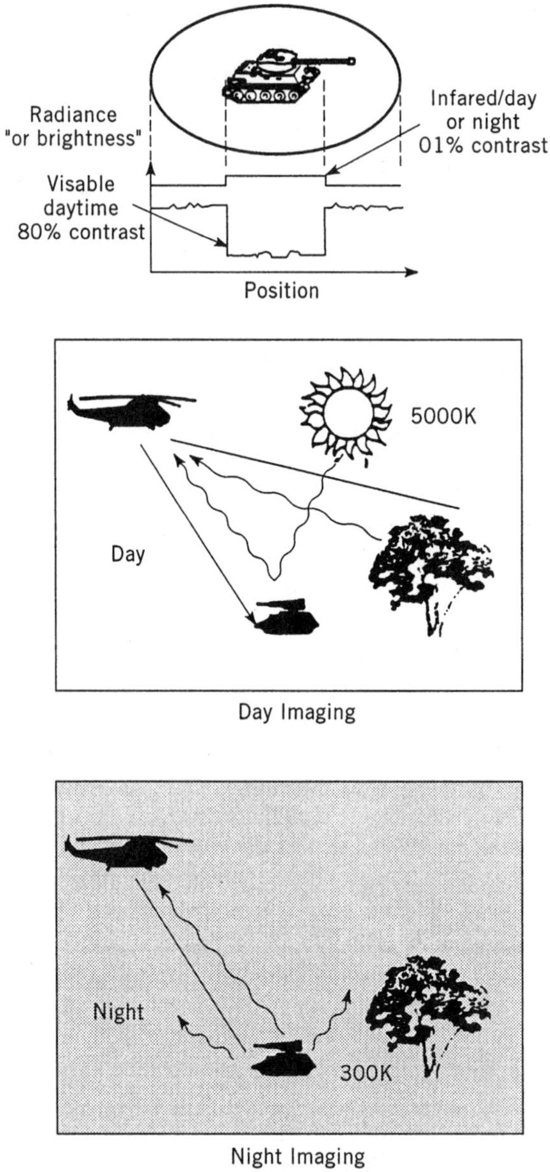

FIGURE 9.3 Quality of day and night imaging available from a FLIR.

conditions for power lines and towers. However, large buildings and hills can be detected as far away as 15 to 20 km under clear to 10 to 14 km under hazy conditions. Typical recognition range varies from 1 to 4 km, depending on the size of the target and weather conditions. In summary, the FLIR is best suited for military applications requiring precision navigation under all-weather conditions; target acquisition,

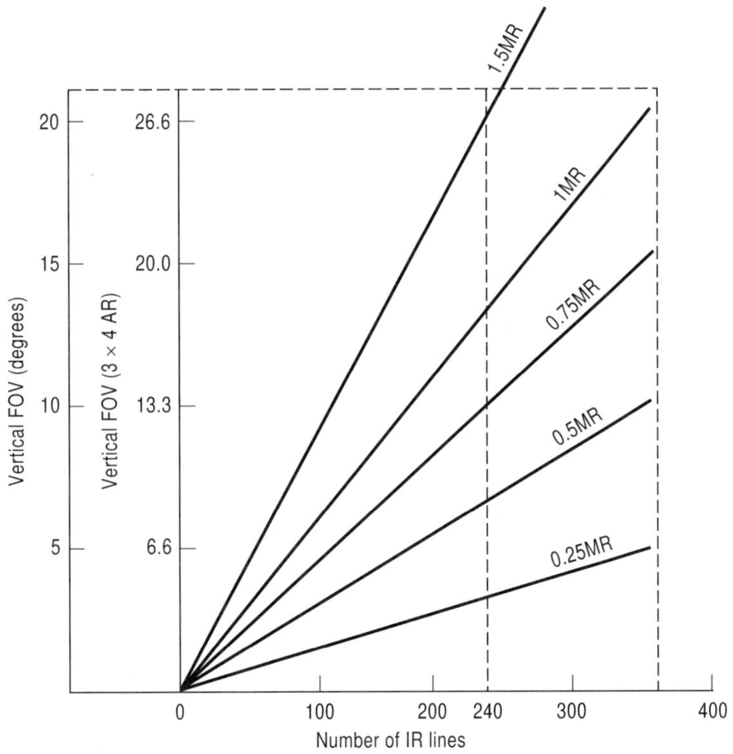

FIGURE 9.4 Horizontal and vertical field-of-view (FOV) requirements as a function of number of IR lines and resolution.

recognition and identification; and precision weapon delivery under hostile battlefield environments. Typical resolution requirements for target detection, recognition, and identification as a function of range are shown in Figure 9.5.

9.4 INFRARED LINE SCANNER (IRLS) SENSORS

IRLS sensors can be used to obtain IR images of the targets of special interest in real-time as well as on high-resolution film in case of specific military applications. This sensor is compromised of high-speed scanning motor, cryogenically cooled detector array, associated optics, and processing electronics. The sensor is generally installed under the belly of the aircraft to record IR images of targets including bridges, airfields, buildings, missile sites, tanks, jeeps, storage tanks, and command and control facilities. Since it provides recording of target images on both sides of the flight direction, this sensor is best suited for reconnaissance applications. The sensor operation is strictly based on the emissivity of the target, which is a function of target surface temperature, and wavelength. Improved resolution and accuracy

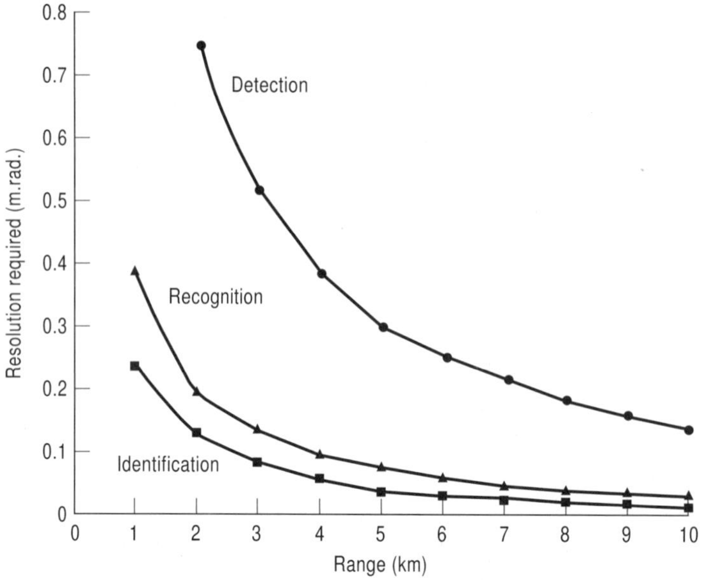

FIGURE 9.5 Typical sensor resolution requirements for target detection, recognition, and identification.

are only possible at cryogenic temperatures. That is why the Hg:Cd:Te detector array is constantly maintained at a 77 K temperature to achieve optimum systems performance. The IRLS provides detection, recognition, and identification of various targets of military importance with resolutions shown in Figure 9.4. Various resolution plots shown in Figure 9.4 indicate that much higher resolution is required for identification of important military target such as parked aircraft, tanks, airfields, and bridges. Most targets can be detected with a resolution ranging from 5 to 25 ft under clear weather environments.

9.4.1 Performance Capabilities and Critical Parameters of IRLS

This sensor is best suited for a day/night, low-level reconnaissance mission with maximum survivability. The sensor provides high-resolution images of targets of military importance, which can be significantly useful for precision weapon delivery missions. Spatial resolution and thermal resolution are the two most critical performance parameters of these sensors. Using state-of-the art technologies, it is possible to achieve a spatial resolution of about 0.25 mrad and a thermal resolution of 0.2 °C, when the sensor is operating in a narrow FOV. The same sensor can provide a resolution of about 1 mrad while operating in a wide FOV mode, which is equivalent to a horizontal resolution of 6.07 ft at a range of 1 n. · mi or about 30 ft at 5NM range. However, the 0.25 mrad resolution of the narrow FOV offers a horizontal resolution of about 1.5 ft at 1 n· mi and 7.5 ft at 5 n· mi range, which is adequate for identification of most targets of military importance.

9.5 SCANNING-LASER RANGEFINDER FOR SPACE APPLICATIONS

A laser rangefinder using state-of-the art laser technology is best suited for acquiring and tracking targets with high accuracy, particularly, in applications involving rendezvous and docking of two spacecraft with ultra-high accuracy and reliability. Such a sensor requires a narrow scanning laser beam of high quality to search and locate a target in space within couple of inches. The laser beam is synchronously scanned with an equally narrow optical receiver FOV without the aid of mechanical gimbals. Because of the stringent size and performance requirements, this scanning laser radar [4] must meet minimum weight, size, and power consumption requirements without compromising the docking accuracy. A high-performance scanning laser offers the required closure rates and angular resolution needed for performing precision docking maneuvers between the satellite or spacecraft and the nearby mother spacecraft. This sensor can be used to aid the rendezvous of space shuttles or newly launched satellites with orbiting space stations so that men, equipment, and supplies can be transferred to the space station.

9.5.1 Laser System Performance Requirements for Acquisition and Tracking

A synchronously scanning laser transmitter–receiver configuration offers electronic beam-steering capability with maximum efficiency but with minimum weight, size, and power consumption. A piezoelectrically driven mirror in the laser transmitter module and electromagnetic (EM) deflection coil in the receiver provide the laser beam scanning electronically. Even small rotation of $\pm 0.25\ ^\circ$ by the mirror can result in relatively large laser beam deflection of $\pm 15\ ^\circ$ at the output of the projector lens assembly, which is used to collimate the output to a desired beamwidth. When a voltage is applied across the piezoelectric crystal, the laser beam is bent proportional to the applied voltage and effectively rotates the attached mirror. Using diffraction-limited laser beam at 900-nm wavelength and two piezoelectrically-driven mirrors with $0.7'' \times 1.0''$ dimensions, a square raster scan pattern can be achieved with approximately 350×350 resolvable scan elements. Higher scan elements offer higher resolution, but at the expense of higher cost and complexity.

Specified current must be applied to the EM defection coils in the optical receiver to generate a narrow instantaneous FOV of the receiver. The photons of the laser light must strike small spot on the high-gain photocathode. When the EM field is varying, the image effectively scans the surface of the photocathode, indicating where the laser spot is located. The ability of the scanning laser to acquire and track moving targets in space is strictly dependent on laser beam geometry, receiver FOV, target range, and SNR.

During the acquisition phase,the maximum target range will be determined by maximum pulse repetition frequency (PRF) and dwell time per scan element. During both the acquisition and tracking, some prior knowledge of target range is needed to use the range gates with high SNRs for high detection probability and low false-alarm rates. The acquisition and tracking rates are dependent on the line-of-sight (LOS) angular rate that the target travels with respect to the boresight axis of the scanning laser radar [5]. The maximum allowable LOS angular velocity determines the worst-case acquisition or best-case acquisition capability. Once the target has been acquired, the maximum allowable LOS velocity of the target with respect to the boresight axis is determined. This angular velocity is directly proportional to the instantaneous FOV (IFIV) of the receiver, but inversely proportional to the product of dwell time per track step and the number of track steps involved. Assuming both the acquisition FOV and track FOV of 30° × 30° each,PRF values of 1000 Hz and 2000 Hz corresponding to an unambiguous range of 80 NM and 40 n· mi respectively,a dwell time of 1 msec, an instantaneous FOV of 0.1° and number steps of 64 for tracking and 375 for acquisition, the computer program provides the acquisition and tracking rates as shown in Table 9.4.

It is important to mention that the target can be tracked anywhere in a 30° × 30° FOV and at two different track steps per cross scan. High track steps per scan requires a slow track rates and vice versa. Computed values of acquisition and track rates indicate that higher rates are possible at higher PRF, but at the cost of reduced operating ranges. Both rates will increase with the increase of instantaneous FOV (IFOV) of the target. An IFOV of 0.1° means an angular resolution of about 1.745 mrad, which represents a docking accuracy of 1.745 ft at a range of 1000 ft or 2.09 inches at a range of 100 ft. A docking accuracy of about 2 inches for a 100 ft range is quite acceptable for docking purposes in space, and is not possible with other sensors.

9.6 SEMICONDUCTOR INJECTION LASER RADAR FOR SPACE

Gallium arsenide (GaAs) semiconductor injection laser radars (SILR) operating over a spectral range of 840 to 900 nm have potential applications for ranging, altimetry, and space rendezvous. Critical elements including modulator, transmitter and receiver, and block diagram of such a laser are shown in Figure 9.6. This SILR has outstanding features, namely, compact size, high internal efficiency, direct modulation with short pulses or high-frequency wave forms, high reliability, and low power consumption and is best suited for space applications. The GaAs diode has a temperature-dependent wavelength shift of about 0.25 nm per °K and will require a

TABLE 9.4 Acquisition and tracking rates for space-based scanning laser

System parameters	PRF = 1000 Hz	PRF = 2000 Hz
Unambiguous range (NM)	80	40
Typical dwell time (msec)	1	0.5
Worst-case acquisition (deg/sec)	0.025	0.049
Best-case acquisition (deg/sec)	0.303	0.606
Worst-case tracking rate (deg/sec)	0.552	1.105
Best-case tracking rate (deg/sec)	1.105	2.209

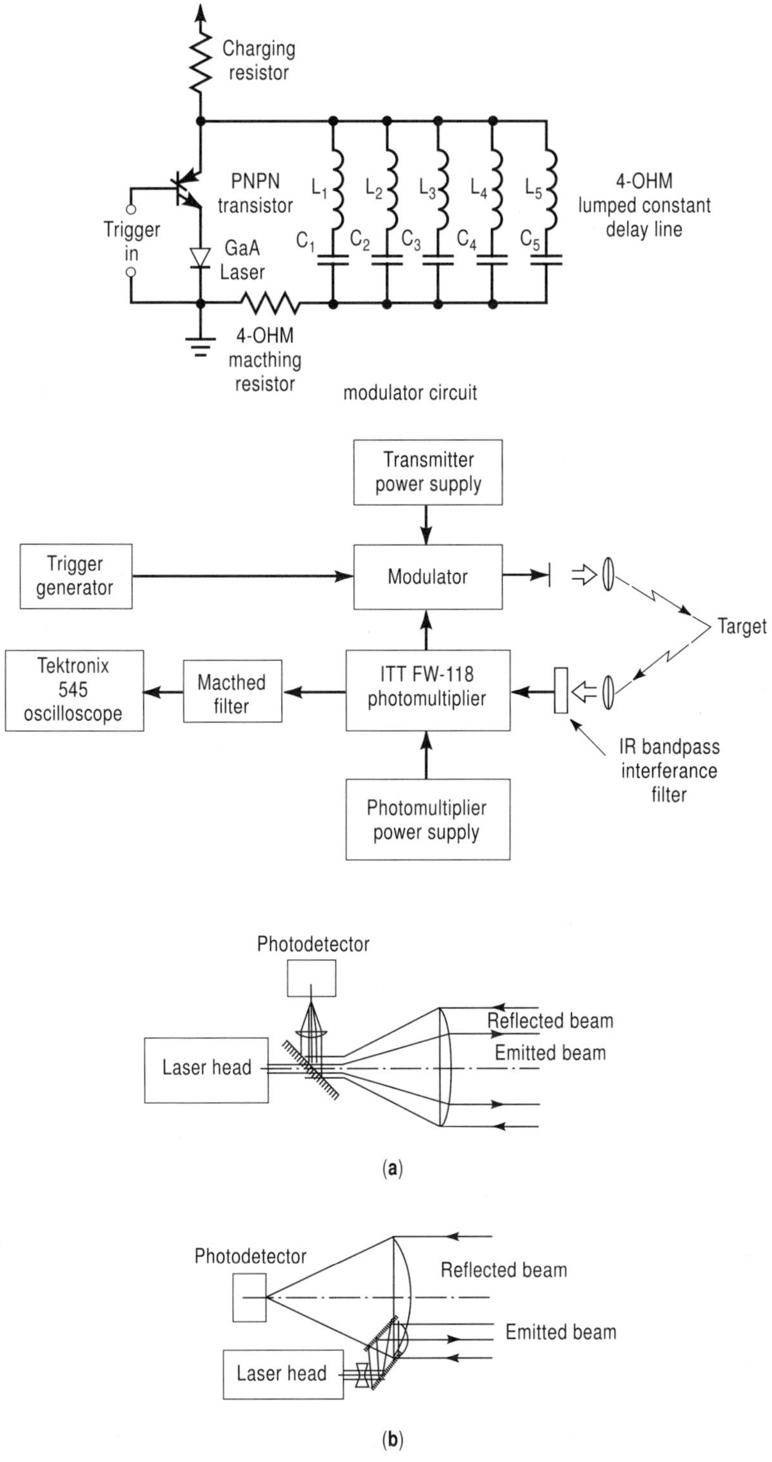

FIGURE 9.6 Critical elements of modular and semiconductor laser.

temperature stabilization network for optimum performance. The range capability of this sensor depends on several systems parameters, operating environments, and whether the radar operation is based on photon-limited operation or background-limited operation. Its performance in space is dependent on the background radiation in space.

9.6.1 Range Capability for Photon-Limited Extended Targets

The range performance of the SILR is signal-photon-limited because of negligible overall noise including shot noise, and thermal noise components contributed by the photomultiplier gain. The detection of targets is dependent on the average number of signal photoelectrons produced by the background radiation and dark current during the observation interval. For a realistic probability of detection, a minimum number of photoelectrons is required, which can vary from 4 to 25, depending on the reflection coefficient of the extended target. Typical values of such parameters are shown in Table 9.5.

9.6.1.1 Photon-Limited Range

The photon-limited detection range is proportional to one-way atmospheric transmission efficiency (0.85), directly proportional to the square root of the reflection coefficient of the target (0.1), quantum efficiency of the photoemissive device (0.003), effective receiving optic area (0.016 mm^2 using a diameter of 5.5 inch), peak laser transmitter power (10,100 and 1000 W), pulse width (10, 50, 100, and 200 ns), inversely proportional to the square root of the number of photoelectrons (10), system loss factor (5 dB), and a constant equal to product of ($\pi h \nu$), which comes to 73.52×10^{-20} using a wavelength of 840 nm for the GaAs diode laser. Computed values of detection range using the assumed values of various parameters shown in parentheses are summarized in Table 9.6. These calculations do not include the integration gain that is possible through integration of pulses. In a photon-limited case, if N is the number of pulses, the signal energy per pulse will be enhanced by a factor of N. But in actual practice, the enhancement will be much less due to pulse overlapping and due to the presence of photoelectron-based dark current. However, in the background noise-limited case, effective signal enhancement will be by a factor close to $\sqrt{N}$, where N is the number of pulses integrated.

TABLE 9.5 Reflection coefficient (Γ) and number of photoelectrons (m_{pe}/pulse)

Target type	Γ	m_{pe}/pulse	Operating conditions
Tree	0.023	6	Nighttime and 21 °C temperature
Telephone pole	0.005	8	Same as above
House with white paint	0.099	25	Same as above
Hanger with sheet metal	1.200	4	Same as above
Retroreflector	1.750	400	Nighttime and 50 °C temperature

TABLE 9.6 Photon-limited detection ranges for GaAs semiconductor laser radar (m)

Pulse width (ns)	10 W	100 W	1000 W
10	125	395	1250
50	280	886	2800
100	396	1254	3960
200	560	1772	5600

9.6.1.2 Detection Range for a Background Noise-Limited Extended Target

When the background radiation noise is large, the detection range of the laser radar is expressed in terms of SNR. The detection for a background-noise -limited extended target is proportional the square root of laser transmitter peak power (10, 100, and 1000 W), inversely proportional to the square root of the product of SNR (12 dB) and system loss factor (3), proportional to fourth root of quantum efficiency (0.003), target reflection coefficient (0.1), receiving optic area (same as in the previous example), pulse width (same as in the previous example), receive background radiation loss (2 dB), and constant equal to 0.614 (using the transmission efficiency of 80%),and inversely proportional to the fourth root of optical filter bandwidth(76 Å), receiver FOV (3.8×10^{-5} sr) and a constant (14.72×10^{-20}). Calculated detection ranges using the assumed values of various parameters shown in parentheses are summarized in Table 9.7.

These calculations do not include the integration of pulses. In a background-noise-limited case, the integration gain is proportional to the square root of the pulses integrated. Thus, a 10-dB enhancement in the signal would require integration of 100 pulses.

9.6.1.3 Detection Ranges for a Corner Reflector-Photon-Limited Case

Much higher detection ranges are possible with corner reflector-photon limited operation, because corner reflectors are the most ideal targets at optical frequencies. When the background radiation and dark current are very low, the detection range in this case is inversely proportional to the fourth power of the product of laser transmitter beamwidth (3.2×10^{-7} sr) and retroreflector beam divergence (2×10^{-10} sr).These two parameters are primarily responsible for significant improvement in the laser detection range. Preliminary calculations indicate that detection ranges of

TABLE 9.7 Detection ranges for a background-noise-limited extended target without integration (m)

Pulse width (ns)	10 W	100 W	1000 W
10	50	159	502
50	112	355	1122
100	159	502	1590
200	225	710	2244

70 km and 125 km are possible with 10 ns and 100 ns pulses, respectively, even with a peak power level of 10 W and with photoelectrons of 10. The detection range in this case is proportional to the fourth power of both the transmitter peak power and pulse width. Optimum laser radar performance requires precision alignment of the transmitter and receiver beams, and a reliable sensor to point the radar to the target and acquire it. In summary, the semiconductor injection laser is best suited for applications where precision acquisition and tracking are of critical importance, such as ranging, altimetry, and space rendezvous.

9.7 EYE-SAFE LASER RANGEFINDER (LRF) SENSORS

Future military sensors capable of providing detection, tracking, recognition, and ranging of military targets with high accuracy require sophisticated electrooptical and photonic technologies. Because of its frequent use for ranging purposes, a LRF must not only provide accurate ranging information, but with utmost safety to the operators. A multifunction LRF offers various functions needed for successful execution of military missions with minimum weight, size, cost, and power consumption. Studies performed by the author recommend the following laser wavelengths to achieve optimum performance for specific mission application:

- 1064 nm for laser decimeter and rangefinder
- 1500 nm for eye-safe laser designator
- 1500 nm for eye-safe laser rangefinder
- 3000–5000 nm for IR jamming applications

9.7.1 Eye-Safe LRF System Requirements

A battlefield or a space-based LRF must meet most stringent performance and environmental requirements, such as light weight, compact size, low power consumption, high reliability, and maintenance-free operation. The LRF system architecture must include practical aspects of modularity, cooling, easy alignment of optical components, and integration of MMIC and solid state technologies to provide high reliability. The modular design concept must satisfy the beam accuracy, firing rates, reliability, safety, mechanical,and environmental requirements. Provisions must be made to replace failed components without alignment. The critical elements of the LRF for military applications include laser rod, flash tube, high voltage transformer for triggering, precision bearings, and output mirror mounted on a stable platform perpendicular to the longitudinal laser axis. The cooling system must be fitted to the laser head as shown in Figure 9.6 for optimum heat transfer efficiency and rapid cooling. In the case of an airborne LRF, a continuous-duty motor is used, whereas in the case of artillery or tank LRF, a quick-starting motor must be used to reduce battery drainage.

The 1064 nm LRF receiver must use high-performance avalanche photo detectors (APDs). Higher receiver sensitivity requires a sensitive APD, preamplifier, temperature stabilization circuit, regulated power supply, and a threshold amplifier. APD devices offer much lower gain compared to photomultiplier tubes (PMTs), which have gains in excess of one million. But higher cost and complexity must be seriously considered before integrating a PMT into the system.

The optics associated with the transmitter and receiver are comprised of an afocal lens to reduce the transmitter beam angle and a focusing lens for the receiver. The timing circuit includes a 30 MHz clock followed by one or more counters and TTL circuits. The power supply must include a storage capacitor, a shaping network for optical pumping, a high-voltage control unit, an auxiliary power supply unit for low-voltage stabilization, and a rotating prism [6].

9.7.1.1 Operating Range Capability of an Eye-Safe LRF

The operating range of the eye-safe LRF is proportional to the square root of receiver-optic effeiciency (80%), diffusion component of the target reflection coefficient (0.85), area of the receiver optics (based on 5 inch diameter), extinction coefficient of the atmosphere (0.5 dB per km for a wavelength of 1064 nm), and power emitted by the laser (10 W), but inversely proportional the minimum detectable power (–43 dBm). LRF operating range, R has been calculated using the assumed values as follows:

$$5 \times 10^{-7} = (1 \times 0.8 \times 0.85/\pi)\,(0.0127 \times 10^{-6}/R^2)\,(100 \times 0.5 \times R/10)$$

or $R = (1.37/0.5) = 2.74$ km with 80% optical efficiency; $R = 3.08$ km with 90% optical efficiency.

This operating range for a solid-state Nd:YAG LRF seems quite satisfactory for battlefield tank or space applications, where accuracy, power consumption, weight and reliability are of critical importance. A typical eye-safe LRF requires the following design parameters:

Laser transmitter power: 10 W
Laser operating modes: CW or Q-switched pulsed
Number of detectors: 12
Receive optic size: 5 inch
Instantaneous FOV: 0.2 to 0.3 mrad
Imaging FOV: 12 mrad (AZ) × 10mrad (EL)
Maximum operating range: 2.75 km (as per calculation)
Estimated weight: 40 lb
Power consumption: 350 W (maximum)

Q-switch pulse operation will require a laser energy of 200 mJ, pulse width of 25 ns, and a PRF of 100 Hz, approximately. So far the detection or operating ranges considered are for the horizontal plane. The slant ranges will be greater than the horizontal ranges and dependent on the visibility, target altitude, and atmospheric transmission coefficient for a specific operating wavelength. Calculated slant ranges as a function of visibility, altitude and atmospheric transmission coefficient are shown in Figure 9.7. It is evident from the curves shown in Figure 9.7 that the atmospheric transmission coefficient increases with the increase in slant range and target altitude. This coefficient is dependent on target altitude and slant range , but independent of elevation angle.

9.8 LASER RANGING SYSTEM FOR PRECISION WEAPON DELIVERY

A laser-based ranging system offers the highest air-to-ground ranging accuracy, which is very important for precision weapon delivery capability to achieve optimum kill probability. Most ranging systems require accuracy of a few meters (5 to 6 m) out to the 20 km range, which is considered adequate for delivery of weapons against most tactical targets. Under ideal conditions, the human eye has an angular error of about 0.5 mrad or 3 feet at a range of 1 NM. However, a laser operator reading error will be further degraded due to pulse rise variations, jitter in the triggering circuit, quantization error, and boresight error. The ranging error is a function of slant range, grazing or depression angle, and angular error of the sensor. The laser sensor ranging error is given as

$$\Delta R = R_s \cos \theta_g \left(\frac{\sin \theta_g}{\sin (\theta_g - \theta_a)} \right) - 1 \qquad (9.7)$$

where R_s is the slant range, θ_g is the depression or grazing angle, and θ_a is the angular error contributed by the laser ranging sensor, which is typically less than 1 mrad for a Nd.YAG laser operating at 1064 nm. Assuming an overall sensor error of 1 mrad, the air-to-ground ranging errors as a function slant range and grazing angle have been calculated and are summarized in Table 9.8. These errors will be reduced to 50% for grazing angles less than 30° if the sensor error is reduced to 0.5 mrad. It is important to mention that an X-band radar with 36-inch antenna-aperture will have an angular error of 2° or about 35 mrad, which is roughly 35 times worse than the laser angular error, leading to much higher ranging errors not acceptable for precision weapon delivery functions. Even a forward-looking airborne radar with a beam sharpening ratio of 20:1 will not be able to match the performance of a laser ranging system with angular error of 1.0 rad. Because of higher air-to-ground ranging accuracy, laser-guided bombs are considered most accurate and effective, leading to high kill probability.

Studies performed by the author on operating wavelengths for ranging applications indicate that a LRF system using CO_2TEA (transverse excitation atmospheric pressure) laser technology operating around 10.6 μm offers optimum performance under battlefield conditions because of its high conversion efficiency and lowest extinction coefficient for smog and fog. Because of high efficiency, high power capability, and large number of IR emission lines, this particular TEA laser is best suited

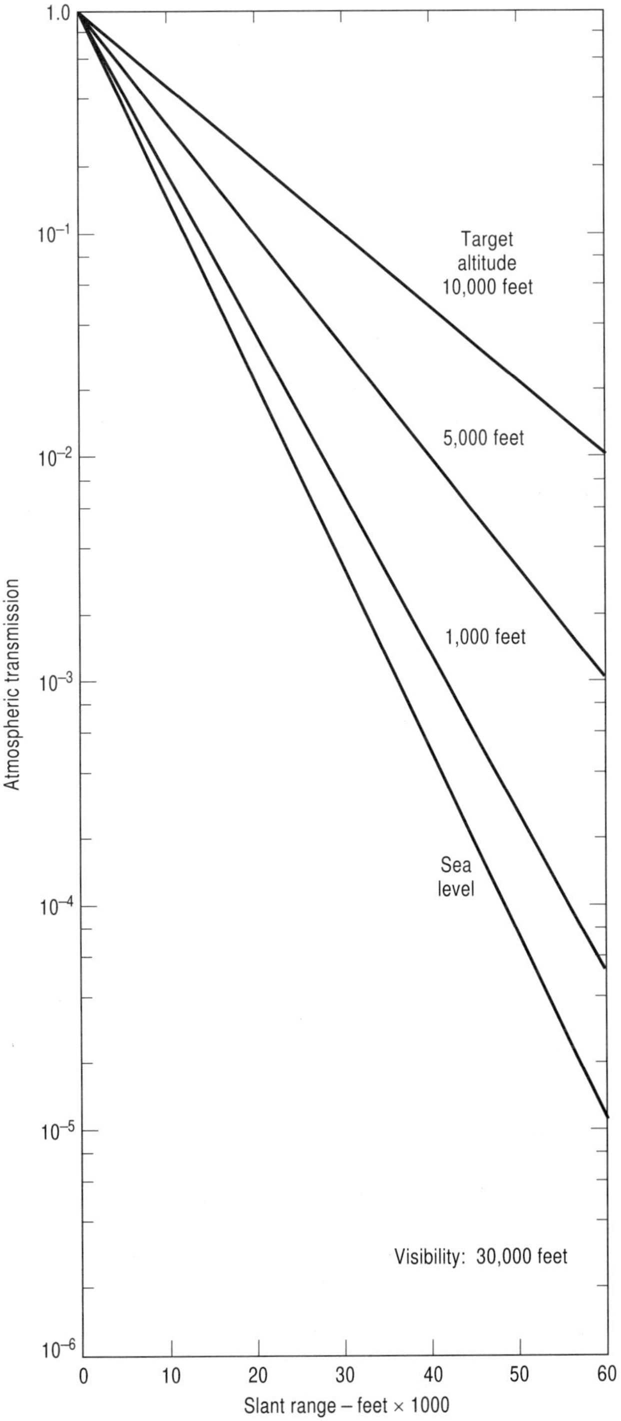

FIGURE 9.7 Calculated slant range as a function of visibility, target altitude, and atmospheric transmission coefficient.

Table 9.8 Air-to-ground ranging errors from a laser ranging sensor (ft)

	Grazing angle (degree)					
Slant range (NM)	10	20	30	40	50	60
1	31	15	10.5	7.2	3.0	1.8
2	62	30	21	14.4	6.0	3.6
5	155	75	52	36	15	9.0
10	310	150	105	72	30	18

for ranging applications. Furthermore, its high power capability (up to 10 GW) makes the TEA laser system most ideal for tracking incoming long-range missiles in midcourse flight and most attractive for training purposes. Typical characteristics of CO_2 TEA laser are:

Operating wavelength: 10.6 μm
Peak power capability: 500 kW (maximum)
Pulse width: 50 ns or less
PRF: 1 to 5 Hz
Beam divergence: < 0.5 mrad
Range resolution: < 25 ft
Gas refill requirement: after every 2 million shots (approximately)

Studies performed by the author reveal that IR signals at this wavelength can penetrate clouds and fog with minimum attenuation. The studies further reveal that IR operations above 12 km altitude will be open for 95% of the time, while operations at 9 km and 6 km will be open for 10% and less than 1% of the time, respectively. According to studies, the operational altitudes can vary from hemisphere to hemisphere because of climatic conditions and seasonal changes. If the IR sensor is located at 6 km and the target at 12 km with a slant range of 80 km, a path loss of about 2 dB can be expected. Despite this loss, the far-IR laser sensor will provide acceptable performance.

9.9 LASER SEEKERS

A laser seeker is a passive sensor capable of detecting laser energy reflected from a target when illuminated by a target designator operated by the forward air controller. This sensor could be mounted under the aircraft belly or in the nose.The laser aids pilots in locating targets for delivering conventional weapons or in attacking with laser-guided bombs known as smart bombs. Helicopter gunships, when equipped with laser illuminators and laser seekers, can provide effective close support, night-time interdiction, armed reconnaissance, precision weapon delivery, and day/night attack capability under fair weather conditions. The laser seekers offer improved accuracy for weapon delivery, reliability, and maintaining logic supports for various kinds of ordinances. Laser spot seekers are available for 1000 lb, 2000 lb, and 3000 lb bombs. Laser-guided weapons yield low circular-error probability (CEP) (not exceeding 20 to 25 ft), according to an article published in *Aviation Week and Space Technology* (page 48, 3 May 1971 issue).

The IR seeker is the most critical element in a laser-guided bomb or IR missile. The seeker assembly (Figure 9.8) is typically comprised of a IR dome capable of meeting stringent optical, mechanical, aerodynamic, and chemical requirements without distorting the IR transmitter or receiver signals; telescope; optical filters to reject solar reflections from clouds and terrestrial objects; reticle; lens assembly; and detector array. The seekers are designed to operate in specific atmospheric windows for optimum performance. The 3.2 to 4.8 μm window is preferred to minimize false targets generated by solar reflections from clouds and terrestrial objects. Generally, cryogenically cooled InSb detectors are used for maximum sensitivity over this spectral range. The seeker detection range is dependent on the visibility and reflectivity of the target, as illustrated in Figure 9.9.

The seeker assembly must be mounted on precision bearings that allow it to rotate freely for optimum performance within the FOV of the seeker. Rotation of the telescope must be done at constant angular speed to determine the location of the telescope assembly relative to the missile centerline and to control the seeker pointing error within the specified limits for optimum performance. The main function of the seeker is to seek the target information and then transfer it to the missile computer to generate guidance, command, and control signals for various elements of missile flight.

It is important to mention that electromechanical gimbals suffer from large size, high cost and poor reliability. The latest IR seeker design configuration using a FOV-multiplexer (FOV-MUX) design offers a rapid capture of scene and generates the necessary guidance signals [7] without relying on the heavy, complex electromechanical gimbals. The FOV-MUX consists of a prism structure, primary and secondary lenses, an image-steering device, conformal entrance widows, diffractive optical elements, and advanced focal plane architectures using micro-ens array technology to improve fill factor. This new seeker technology will significantly improve the weapon delivery accuracy of laser-guided bombs and IR missiles.

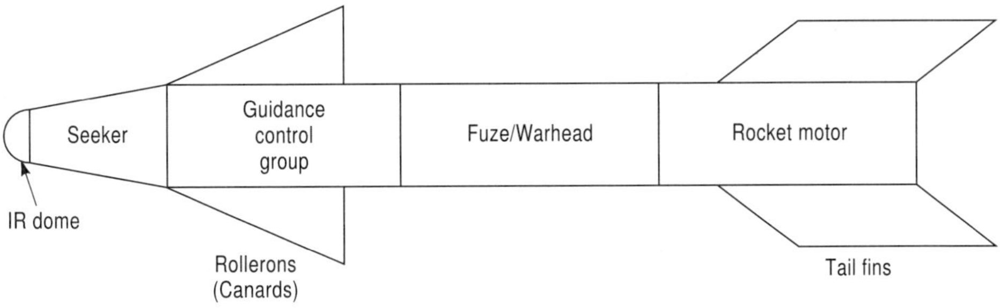

FIGURE 9.8 Block diagram showing the critical elements of an IR guided missile.

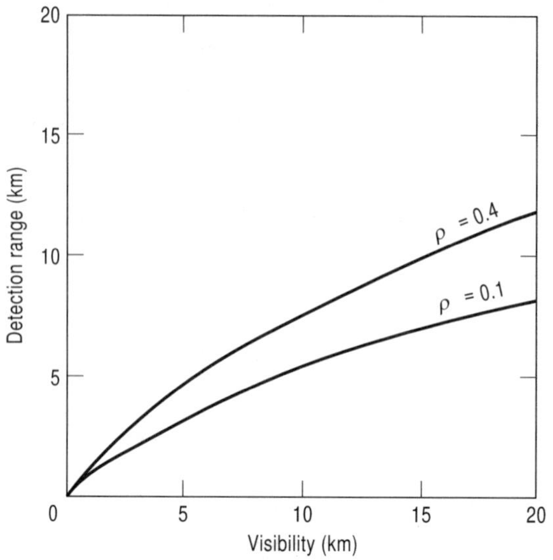

FIGURE 9.9 Detection range capability of passive IR sensor as a function of visibility and target reflectivity.

The latest seeker design [7] employs the state-of-the art staring IR sensor concept incorporating microsystems technologies with advanced focal plane architectures and innovative optical designs. This new seeker architecture simultaneously provides both a large field-of-regard (FOR) and a small instantaneous FOV (IFOV). The FOV multiplexer images several FOVs with high resolution onto a single focal planar array (FPA), while maintaining high optical performance and high data rates. The number of pixels available for the current military FPAs are 256 × 256 or 640 × 256, which permits either a large FOR or small IFOV, but not both at the same time.

9.9.1 Laser-Guided Bombs

Because of extremely low CEP and surgical precision, a laser-guided bomb (LGB) offers the highest kill probability. A laser-guided bomb architecture is comprised of a seeker, computer, fins needed to maneuver during the flight to target, and explosive. The target is illuminated by a laser illuminator known as "flashlight," which guides the bomb to the target under day and night conditions. The CW laser illuminator uses an extremely intense beam of light, which reflects off the target and is visible to the laser seeker receiver located in the nose of the guided bomb. Once the laser seeker in the bomb acquires the reflected laser energy from the target, the received information is passed on its computer that controls the movable fins on the bomb. As the fins change the angle of attack in the air stream, they alter the flight path of the bomb as it falls and guide it to the target. A laser-guided bomb provides a direct hit on the target in a high-altitude drop operation. The laser-guided bomb provides most reliable and effective weapon delivery with impressive accuracy, which is not possible with other bomb guidance technology. LGBs have several advantages. They requires few missions to destroy a specific high-value target with maximum accuracy and high kill probability. Use of LGBs leads to provide fewer losses of planes and pilots.

Pilots can drop LGBs with high accuracy from higher altitudes than unguided bombs, thereby keeping the bomber or strike aircraft out of range of anti-aircraft gun or short-range surface-to-air missiles. LGBs assure the destruction of tactical targets within a given time frame, thereby realizing significant reductions in mission cost and manpower. These bombs greatly reduce civilian casualties because the munitions destroy the target with high accuracy without destroying residential buildings. A LGB provides weapon delivery accuracy of about 6 ft at 1 NM or 30 ft at 5 NM range, which is adequate for short-range targets. High procurement cost is its major drawback.

9.9.1 Laser Designator or Illuminator

Performance of a LGB is strictly dependent on the laser illuminator capability. A laser illuminator is required to guide the bomb to a specified target. The sensitivity of the IR seeker in the bomb assembly depends on the reflections from the target, atmospheric conditions, distance to target, and laser illuminating CW power level, which could vary from 100 W to 500 W, depending on the LGB operational requirements. The reflections from target is dependent on the target size, reflectance, and emissivity. The operating range is dependent on atmospheric conditions, including visibility, cloud, fog, smog, scattering from aerosols, and turbulent conditions. The designator can be operated in CW mode or in high-duty pulsed mode, depending on the cost and space and weight allocation on the host aircraft carrying the illuminator. It is important to mention that laser illuminators with high power ratings suffer from high cost, weight, size, relatively poor reliability, high power consumption, and complex cooling requirement. A laser illuminator operating at 1.064 μm will experience higher extinction coefficient for smoke agents ranging from 3.64/km for oily fog to 2.19/km for smog and fog. However, a C02 laser wavelength of 10.6 μm has the lowest extinction coefficient, ranging from 0.047/km for oily fog to 0.152 for acid fog.Laser designators operating at 10.6 μm have operational advantages such as lower attenuation in fog and haze than Nd:YAG lasers operating at 1.064 μm, better penetration in smog, higher conversion efficiency (ranging from 20–25% compared to 1 to 5% for the Nd:YAG laser) greater eye safety, diffraction-limited beam width, less vulnerability to IR countermeasures, wavelength diversity,and heterodyne detection for greater receiver sensitivity. Despite all these advantages, 10.6 μm laser illuminators suffer from high weight, size, cost, and system complexity.

9.10 INFRARED GUIDED MISSILES (IRGMs)

Electrooptic guided missiles, also known as IRGMs, can either use spot detector technology or image detector technology. Critical elements of a typical IR guided missile are shown in Figure 9.8. The spot detector uses a spot on the target that is generated from an emitting source or laser designator to provide guidance signals. The spot is imaged onto the detector array plane by the seeker optics and the missile scanning mechanism provides the directional information to the control circuit. The spinning pattern in front of the detector array provides directional information to aid in discriminating extended sources (clouds) from the point sources (tactical targets).

Imaging seekers employ an image detecting mechanism such as a TV screen or FLIR to track the target in real-time. The pilot or the weapon operator acquires the target and slaves the seeker to it. Using the electronic gates, the operator can designate the target and fire the missile for a direct hit. The missile uses the guidance signals from the gate area of the scene to track the target and home on it. The electronic gates must be adaptive in size and time to expand as the missile approaches the target in real-time. This particular tracking system is most suitable for slow-moving targets such as heavy battlefield tanks.

EO guided missiles use exotic EO techniques such as optical beam riders, similar to the RF beam rider concept. This technique involves sensitive IR detectors located on the tail of the missile. These detectors sense IR energy when the missile is leaving the laser beam and produces control signals to turn the missile back into the laser beam. IR trackers sometimes use IR or optical flares in the back of the missile to allow operator to track the missile trajectory with respect to the selected target. IR missiles involve costly and complex architectural design, but provide the highest probability of kill because of deployment of the most sophisticated and accurate guidance technology.

9.11 HIGH-ENERGY LASERS TO COUNTER IR MISSILE THREATS

Ballistic missile proliferation in developing or rogue countries poses serious threats to the existence of developed and stable countries. Development of high-energy lasers offers genuine hope to counter the threats posed by tactical theatre missiles such as Russian-made SCUD missiles and Chinese-made M-11 missiles. Once the detection of these missiles is confirmed during their launch and boost phases, high-energy laser-based weapons could provide the most effective deterrent by destroying the missiles while they are in the enemy's own territory. Published articles in leading defense magazines indicate that by the year 2000, more than 20 nations will have short-range and medium-range missiles because of accelerated ballistic missile technology proliferation and technology transfer. Five nuclear powers—USA, Russia, UK, France, and China—are most likely to be engaged in the development of high-energy lasers (HELs) and their adjunct technologies to counter to the threats posed by the proliferation of various categories of missiles. Comprehensive review of HEL technology reveals that long-wavelength chemical lasers, such as hydrogen fluoride (HF) lasers operating around 2.8 μm and deuterium fluoride (DF) lasers operating at 3.8 μm are best suited for such applications. Recent high-power laser research and development activities indicate that a short-wavelength COIL (chemical oxygen iodine laser) laser operating at 1.315 μm appears to be a viable candidate to shoot down incoming hostile missiles in their boost phase or disable or destroy a hostile satellite in orbit. A coherent high-power CO_2 laser design configuration (Figure 9.10) operating at 10.6 μm can be the most suitable ground-based system to knock out a hostile missile or satellite in its boost phase. However, this laser system is not attractive for airborne applications because of excessive weight, large size, and power consumption. An airborne platform with size equal to that of a Boeing-747 will be required to house this particular high-power 10.6 μm laser.

9.11.1 Antisatellite (ASAT) Laser System

An antisatellite laser system capable of disabling hostile orbiting satellites requires a high-energy laser transmitter, complex optics, high-speed computers, high-voltage power supplies, control and monitoring electronics, and other necessary auxiliary devices. A ground-based ASAT weapon system appears to be more practical at present than an airborne system because of heavy weight and high power consumption. An ASAT weapon system with several megawatts of optical power can damage or disable a hostile satellite in orbit under certain operating scenarios. An airborne ASAT weapon system designed to shoot down an incoming missile during the boost phase faces formidable design and operational problems, including aircraft modification, complex cooling scheme, reliability and maintainability, optical alignment under rough flight conditions, and laser beam distortion while transmitting through the atmosphere. However, the latest design of a COIL laser with multi-hundred-kilowatt power level [8] could be viable weapon system candidate to shoot down a hostile missile in its boost phase. An airborne version of this COIL laser can be housed in a large aircraft. However, an airborne COIL system design configuration requires careful evaluation and definition of the critical component requirements, such as optimum mirror size, maximum allowable power consumption, cooling system capacity, and laser transmitter output.

9.11.2 Space-Based Laser Surveillance (SBLS) System

Defense systems managers are evaluating potential space-based IR (SBIR) surveillance systems capable of providing effective discrimination and faster and reliable early warning against hostile missile launchings. For effective detection of worldwide missile launchings, one may have to consider a system comprised of a dozen low-earth-orbit (LEO) Satellites and an airborne SBIR missile warning sensor. An SBIR missile warning system (SBIR-MWS) offers better discrimination between the missiles and decoys and provides early warnings much quicker, particularly when short-range missiles (SRMs) are launched. This warning can be relayed to various locations through the LEO satellites. This system not only provides effective discrimination between missiles and decoys, but also yields hand-off trajectory data to theater antimissile systems (AMSs) to take immediate and effective appropriate action. The LEO satellites will carry multispectral and hyperspectral remote sensing equipment capable of detecting and tracking IR signals coming from the launched missiles.

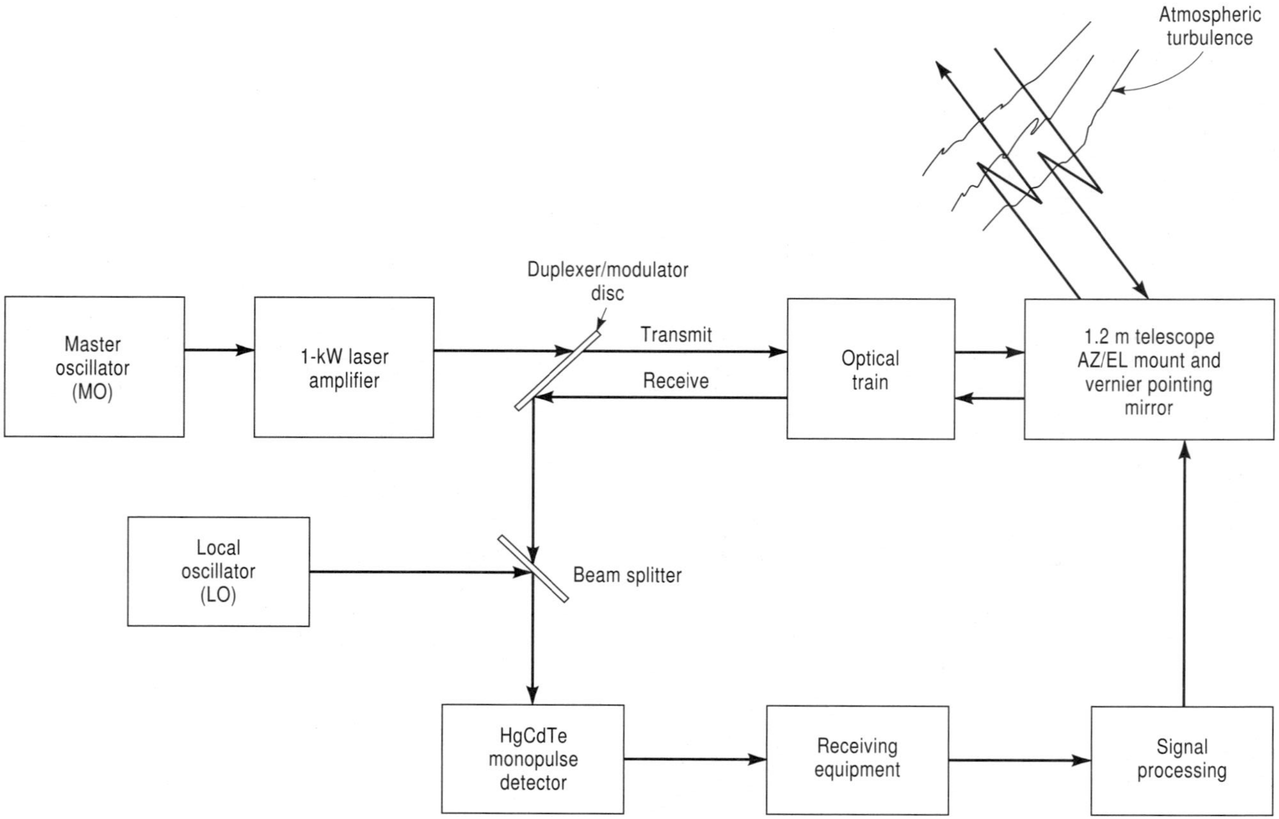

FIGURE 9.10 Block diagram of a coherent tracking laser system operating at 10.6μm. (Courtesy of Lincoln Lab, Cambridge, MA.

9.11.2.1 Laser-Output Energy Requirements for SBIR Surveillance

Defense planners have long recognized the potential applications of high-energy laser systems and photonic-based sensors to achieve defense against the sudden launch of medium-range and long-range missiles by an enemy. Selection of a particular wavelength is dependent on atmospheric propagation characteristics, power consumption limits, logistic aspects, operational requirements, cooling scheme, cost, and complexity. The laser energy output requirement is strictly a function of detection range, energy required to the destroy or melt down the space-borne target, optical system losses, attenuation losses due to scattering, absorptions and diffraction, and other critical system parameters.

The laser peak energy (E_p) requirement to satisfy a given detection range R is given as

$$R^4 = \frac{(E_p)(\lambda^3)(\sigma_t)(D/\lambda)^4(\eta_o)(L_a)^2(0.25 \times 10^{24})}{(\mathrm{S/N})_{\mathrm{RMS}}(\beta_t^2 + \beta q^2)} \tag{9.8}$$

where λ is the wavelength (meter); σ_t is the target RCS (m^2); D is the secondary optical aperture (m); η_o is the overall efficiency, which includes the system efficiency (10%), turbulent mixing efficiency (50%), and quantum efficiency (50%); L_a^2 is the two-way atmospheric loss for the laser system (3 dB for an airborne laser and 30 dB for a ground-based laser); $(\mathrm{S/N})_{\mathrm{RMS}}$ is RMS signal-to-noise voltage ratio (10 dB); β_t is the turbulent beam spread (1); and β_q is beam quality factor (1).

Assuming a detection range of 1275 NM for a high-energy CO_2 laser (10.6 μm), target RCS of 0.01 m^2, RMS S/N ratio of 10 dB, and inserting the assumed parameters shown in parentheses, one gets

$$E_p = (4.05 \times 10^{-24})\left[\frac{(1+1)(1)}{0.012}\right]\left(\frac{10.6 \times 10^{-6}}{1}\right)^4\left[\frac{(10)(31.02 \times 10^{24})}{(0.01)(10.6 \times 10^{-6})^3}\right]$$

$$= 213 \text{ J or W·s}$$

It is important to mention that the RMS S/N voltage ratio of 10 dB offers very high probability of detection (99.9%) with extremely low -alarm rate, which is absolutely necessary to meet a specific kill propbability requirement. The laser energy requirement is proportional to the operating wavelength, target size, S/N voltage ratio, and fourth power of detection range, but inversely proportional to the fourth power of optic size. Parametric trade-off studies must be performed to select the design parameters to achieve the most cost-effective design of a ground-based or airborne system. Lower operating wavelength and large optics require less laser energy, thereby yielding the most cost-effective design. Calculated values of energy levels for airborne and ground-based lasers as a function of various critical parameters are summarized in Table 9.9.

These calculations are performed for a 10.6 μm CO_2 laser using 3-dB loss for the airborne design and 25 dB for the ground-based system, S/N ratio of 10 dB, and optic size of 1 m. For a COIL airborne laser system operating at 1.315 μm and with a

TABLE 9.9 Energy level requirements for airborne and ground-based lasers (J) operating at 10.6μm

	Radar cross-section of target (m^2)					
	Airborne laser system			Ground-based laser system		
Range (NM)	0.001	0.010	0.100	0.001	0.010	0.100
1000	810	81	8.1	10,080	1008	100.8
1250	1970	197	19.7	24,610	2461	246.1
1500	4080	408	40.8	51,030	5103	510.3
2000	12,960	1296	129.6	161,280	16,128	1612.8

radar cross-section (RCS) of 0.001 m^2, the energy requirement comes to 99.4 J , 241.7 J, 500.6 J, and 1590 J, for a detection range of 1000 NM, 1250 NM, 1500 NM and 2000 NM, respectively, provided other assumed parameters remain the same (Table 9.9). This clearly indicates that laser output energy requirements will be significantly lower at shorter wavelengths, thereby providing the most cost-effective design for a laser-based surveillance system to detect long-range hostile missiles while they are in the launch and boost phase. It is important to mention that detection of an intercontinental ballistic missile (ICBM) with a 6000 NM range even at 3000 NM by a high-energy laser surveillance system will yield a warning of better than 15 minutes to take appropriate action to intercept or destroy the incoming missile.

Studies performed on potential laser wavelengths by the author indicate that the 10.6 μm wavelength has better penetration in metals, whereas the DF laser wavelength of 3.8 μm is better in maritime atmosphere. As stated earlier, that shorter-wavelength laser such as COIL lasers would require lower energy level and offer a cost-effective design. The computed values of energy level assume that the laser beam remains fairly stationary and heats the same column of air. However, under turbulent conditions, thermal blooming or defocusing of the laser beam will degrade laser detection performance. Furthermore, as one gets closer to the coastline or seashore, it will be harder to detect the incoming missile because of significant attenuation in the vicinity of water, which will further reduce the detection range capability and reaction time to spot or intercept the hostile missile.

9.11.2.2 Factors Impacting Laser Detection Range Capability

The factors, which affect the laser performance level, are strictly dependent on whether the laser system is a ground-based system or an airborne system. The detection capability and the overall mission effectiveness of space-based laser systems designed to defend against long-range ballistic missiles are not only dependent on the laser output energy level, but also on several other factors, such as atmospheric propagation characteristics, laser frequency shift due to pressure variation, nonlinear thermal blooming, laser frequency jitter, mechanical mirror vibration, medium refractive index variations, multimode beating, and transmission line stability.

These factors can degrade the quality of the laser beam and its intensity. Critical atmospheric effects such as scattering due to molecular and aerosol distribution, random wander, spreading, distortion of the laser beam due to atmospheric turbulence, nonlinear thermal blooming due to laser absorbtion of the laser beam power, and strong attenuation due to plasma would have an adverse impact on the laser-based missile surveillance system.

Studies performed by the author reveal that the mid-IR wavelengths are favored for turbulence and aerosol scattering conditions. At longer wavelength such as 10.6 μm, thermal blooming is dominant due to intense molecular absorption, while at the shorter wavelengths such a 3.8 μm for a DF laser or 2.8 μm for a HF laser, turbulence-induced beam spreading and aerosol absorption and scattering effects are more critical. The studies further indicate that the RMS beam radius is the sum of the squares of radii changes due to diffraction, jitter, and turbulence. The RMS beam radius is the $(1/e)$ radius of the mean irradiance profile in the absence of thermal bloom. The peak irradiance, which represents the laser peak limit, is directly proportional to the laser transmitter power, inversely proportional to the square of beam radius, and inversely proportional to average atmospheric loss contributed by the molecular and aerosol distributions. Thermal blooming [10] due to heating of an atmospheric layer contributes to beam spreading (proportional to wavelength) and distortion and must be avoided if optimum system performance is desired The turbulence-induced beam spreading (proportional to $1/\lambda^{0.2}$) is strictly dependent on operating range, wavelength, the lateral coherence length of a spherical wave, the level of turbulence, and refractive index of structure constant.

9.11.2.3 CW Thermal Blooming Effects

Continuous wave (CW) thermal blooming refers to the self-induced spreading, distortion, and bending of the laser beam due to the absorption of a small fraction of the laser power by the molecular and aerosol constituents in the atmosphere. The heating effect distorts the laser beam and can limit the maximum irradiance that can be propagated to a target regardless of the transmitter power output. Studies performed on thermal distortion indicate that the shift of the peak irradiance into the flow direction is about 10 times the undistorted beam radius, even though a reduction of the peak irradiance level due to thermal distortion is only 2% of the undistorted value.

9.11.3 High-Power, High-Energy CO_2 Laser Sources for Missile Defense Systems

High-power, high-energy (HPHE) CO_2 lasers have a potential applications in missile defense systems (MDSs), regardless whether the system is an airborne system or ground-based system. Longitudinally excited CO_2 lasers using high-voltage pulses offer many orders of improvement in output power and energy levels. Transverse excited at atmospheric (TEA) conditions laser transmitters are capable of producing several megawatts of pulsed output levels with conversion efficiencies between 15 to 25%. The TEA lasers are relatively simple and compact and are best suited for airborne missile surveillance systems for detection of hostile theater missiles. However, the HPHE-CO_2 lasers with output levels in the gigawatt range will be most attractive for ground-based missile surveillance systems for the detection of long-range missiles while they are in launch and boost phases. The latest laser technology indicates that the COIL laser with reasonable power output level could be the most suitable airborne missile surveillance system. However, more research and development activities must be directed to improve its output and energy levels close to those offered by the CO_2 lasers.

It is important to mention that a HPHE-CO_2 laser requires arc-free excitation of large volumes of gas mixtures with beam cross-sections sufficiently large so that the radiation power density within the gas medium remains below the safe optical breakdown threshold. A transverse electron discharge (TED) scheme permits the excitation of large volumes of gas mixtures of C O_2, N_2 and He gases with beam cross-sections up to 60 cm^2 in area. Reproducible discharge conditions in 30% CO_2 gas mixtures are possible with input energies exceeding 300 J. When a number of discharge modules are employed in series in a simple oscillator configuration, laser output energy level exceeding 300 J with peak power levels of several gigawatts are obtained, which can meet the ground-based system power requirements needed to destroy the long-range missile in cruise phase or in boost phase. Under the optimum gas mixture of the above-mentioned three gases (30%, 20%, and 50%, respectively), a pulsed CO_2 laser system using seven discharge modules, electrode spacing of 7.5 cm, optical resonator length of 6.8 m, active cross-sectional area of 58 cm^2, focussing mirror with focal length of 3 m, and charging voltage of 75 kV, demonstrated an output energy exceeding 300 J and maximum power greater than 3 GW with conversion efficiency of about 8% and brightness exceeding 4×10^{13} $W/cm^2/sr$ [11]. The same laser with operating voltage of 55 kV, cross-sectional area of 26 cm^2, electrode separation of 5 cm, and resonator length of 5.8 m demonstrated an output energy in excess of 220 J and maximum output power greater than 2 GW with conversion efficiency better than 10%.These experimental data [11] indicate that higher conversion efficiency is only possible at reduced energy output and power output levels. However, by using a very high voltage trigger source, several discharge modules in a simple oscillator configuration, and large beam apertures, one can achieve both the energy level and power output capable of meeting the laser-based missile surveillance system requirements.

9.11.4 Detection and Tracking Requirements for a Missile Surveillance System

Detection and tracking capabilities of the laser-based missile surveillance system are strictly dependent on the type of detectors used and their sensitivities, the detector's instantaneous FOV (IFOV), source temperatures, atmospheric environments, and IR source intensity levels during various phases of the missile's flight, such as launch, boost, cruise, and terminal phases. Since both the source temperature and IR intensity levels are extremely high during the missile launch phase, detection of missiles is possible with a cryogenically cooled FPA detector array with detector

IFOV better than 0.5 × 0.5°. However, detection and tracking of the missiles in the reentry phase can be accomplished with detector IFOV of about 5° × 5°. If the source temperatures and intensity levels are significantly low during the reentry and terminal phases, reliable tracking may require a detector IFOV close to 1° × 1°, based on preliminary calculations with various assumed parameters. This means that smaller detector IFOV is required to detect or track low-IR observable missiles. Critical parameters such as dwell time, frame time, cryogenic operating temperature, and number of detectors in the FPA determine the detector's instantaneous FOV.

9.12 LASER-BASED SYSTEM OFFERS DEFENSE AGAINST MACH-SPEED MISSILES

The latest research activities have led to the development of high-energy free-electron laser (FEL) weapon systems capable of providing defense against Mach-speed missiles such as SCUD missiles. This type of laser weapon system operating at 4.9 μm and with CW output power of 500 W can be very effective [12] as a ship defense weapon system. A high-energy FEL weapon system offers several advantages in terms of power, size, cost, and tuning capability. A FEL laser with large mirrors having a small radius of curvature will spread the beam at the surface of the mirror, thereby enabling the mirrors to withstand high energy levels. A FEL requires a small-footprint, high-power source, which is ideal to provide an effective defense against short range, Mach-speed missiles. An optical tracker can locate the SCUD missile in flight and direct the high-energy beam of FEL or DF lasers onto its warhead, detonating it seconds later after launch. Increasing the Rayleigh range [10] between the mirrors will allow diffraction to distribute the beam energy over a wider mirror, where the high energy is focused. This type of laser system offers a low-cost, fairly effective, mobile, short-range antimissile defense system.

Research activities are being focused on the design of DPSS lasers operating at 1.06 μm capable of delivering 100 kW of CW output level, which can be just adequate for short-range, mobile, antimissile defense system application. The thrust of this laser technology hinges on a technique called heat capacity lasing. This DPSS laser technology offers higher reliability, excellent maneuvering capability, circumvented logistical issues involving transport, fuel reserve, special storage, and a cost-effective weapon system from which thousands of shots can be fired with minimum cost.

9.13 PHOTONIC TECHNOLOGY FOR BATTLEFIELD APPLICATIONS

Military planners are exploring photonic technology in the design and development of devices and sensors for battlefield applications. Off-the-shelf components (OTSC) are being investigated to reduce the cost of the weapon system. Early developments of photonic and optoelectronic devices such as CCDs, microlasers, IR imaging devices, tunable vertical cavity surface emitting lasers (VCSELs), laser-guided bombs, EDFAs, CCD-based video cameras, high-speed IR digital cameras, and optoelectronic devices using cutting-edge technologies such as integrated microelectromechanical (MEM) mirrors on the chip and fetmosecond-laser technology have paved the way for their applications in military sensors and equipment. The first-generation of VSCEL devices emitting at 850 nm and 980 nm offer tuning range capability of 30 nm and 40 nm, respectively. The second-generation devices operating at 1550 nm offer a tuning range exceeding 60 nm. Third-generation VSCEL devices are expected to extend the tuning range to 120 nm for a 1550 nm laser with output power levels exceeding 5 mW.

Photonic and electrooptic device technologies are widely used in IR beacons, roof-mounted optical sights, night-vision binoculars, laser rangefinders, night observation devices for battlefield troops, laser-guided antitank missiles, pocket image intensifiers, fiber optic gyros, and laser warning devices. The latest photonic and electrooptic technologies can significantly improve the performance level of forward-looking IR scanning sensors to alert helicopter crews of enemy targets, helmet-mounted display with night-vision sensors, pulsed laser diode aiming light sensors capable of providing pilots with the topography of the land ahead to avoid any obstacles and laser diode targeting machine guns to provide target acquisition and designation capabilities.

Several electrooptic and photonic devices developed under Strategic Defense Initiative (SDI) programs have significantly improved the combat capabilities of the armed forces. These devices were used to develop rotating turret assemblies complete with beam expanders and 1.5 meter telescopes for motion compensation, computer-based fire-control systems capable of providing all functions of target engagement from acquisition to kill determination, "beam" walk mirrors guided by HeNe lasers and sensor arrays to maintain the precision alignment of megawatt-class 1.315 μm COIL laser beams and to compensate for jitter caused by aircraft flexing while in flight. Such a COIL laser has an operational range capability exceeding 200 NM, depending on atmospheric conditions and platform altitude. An airborne COIL laser system can provide effective defense against theater missiles with minimum complexity. However, this laser system requires sophisticated and costly cooling schemes to cool the high-energy laser transmitter and affiliated pumps.

9.14 IR TELESCOPE FOR SPACE APPLICATIONS

Large amounts of energy in the universe are irradiated over the IR region from cool objects such as stars, planets, and ionized gases as solar radiation. IR telescopes play a key role in astronomical and atmospheric research studies over the IR spectrum ranging from 1 to 1000 μm. Measurements of IR radiation from ground stations are extremely difficult because of atmospheric absorption and scattering effects. Space-based IR telescopes provide reliable and accurate astronomical and atmospheric research data based on observations in the higher atmospheric regions of the earth. Optimum performance of such telescopes requires IR detector operations close to a cryogenic temperature of 2 K and reduced thermal emissions of the telescope itself. Cryogenic cooling and focal plane instrumentation are absolutely necessary for acquisition of high-resolution data across the wide IR spectral region.

9.14.1 Critical Performance Requirements of IR Telescopes

The most critical elements of the IR telescope include the toroidal coil, cryostat unit, telescopic tubing, and thermal and radiation shields. These shields are attached to the outer shell of the telescope by fiberglass support structures of high mechanical strength and low thermal conductivity to provide effective thermal isolation. The temperature gradient varies from 2 K for the detectors and 10 K for the mirrors to about 20 K for the telescope aperture. The heat leak strictly depends on radiation shield temperature and varies from 1740 mW at T_R of 100 K to 14.1 mw at T_R of 30 K, where T_R is the radiation temperature of the shield. Both the thermal shield and radiation shields are cooled by helium gas through an efficient heat exchanger to maintain optimum performance of the telescope.

IR detectors require cooling temperatures as low as 0.3 K to achieve high sensitivity to obtain reliable data on upper sky and terrestrial parameters without any atmospheric disturbances. The detector array is designed to provide maximum sensitivity over the 8–120 μm spectrum. The detector array is comprised of hundreds of sensitive, cryogenically cooled detectors placed at the focal plane of the telescope. The upper end of the cryostat acts as a cold radiation baffle that prevents earth and solar radiation from entering the telescope. An uncooled solar shield is attached to the upper end to allow gathering of sensitive astronomical data up to a 40° angular distance to the earth's horizon and about 90° angular distance to the sun. Under these operating conditions, the detector array allows reliable detection of radiation flux levels as low as 10^{-15} .

9.14.2 Telescope Performance Requirements for SPACELAB

IR telescopes have potential applications in Spacelab or shuttle missions. Because of the large mirror, compact shields, instrument point system, cryostat with coolant, and other accessories, the overall weight of the IR telescope for Spacelab could be several hundred pounds. Furthermore, frequent optical alignment may be required to maintain the required directional accuracy of ±10 arc second. The excessive weight of the IR telescope is strictly due to storage of considerable amounts of helium II coolant needed for extended mission periods exceeding 20 to 30 days. The helium introduced into the cryostat prior to launch must be continuously pumped during the telescope operation. The mass flow of the evaporated helium must be precisely controlled to meet temperature requirements and to absorb varying heat loads during different experimental phases of the space mission. An annular helium tank can be attached to the outer shell of the cryostat unit with a support structure of high thermal efficiency and mechanical integrity. The attached tank capacity depends on the duration of the space mission requirements. High performance insulating materials must be used in all cryostat designs for space applications to ensure continuous optimum performance over the extended mission periods. High-quality superinsulation is necessary to prevent helium evaporation leakage and to maintain required temperature of the system during the entire mission period. It is important to mention that temperature requirements for scientific payloads varies from 10 K for the primary mirror to 20 K for front end of the cold baffle plate. However, the heat load for various IR telescope components varies from about 200 mW for the scientific payloads to 50 mW for the chopper to 70 mW for the helium tank support structure, filling tube, and electronic wires and circuits.

9.15 GROUND-BASED SURVEILLANCE AND MISSILE WARNING SYSTEM

A ground-based surveillance and missile warning (SMW) system [13] is absolutely essential to neutralize the growing threat posed by the continued proliferation of tactical and theater missiles, cruise missiles, and unmanned aerial vehicles (UAVs). These targets are becoming faster, more maneuverable, and stealthy. Due to improved propulsion technology and more accurate guidance system, missiles can approach from far distances on non-line-of-sight (LOS) trajectories. With supersonic speeds, only limited time (less than 30 minutes) is left from their first detection to threat confirmation for weapon release. To counter the threats posed by various supersonic missiles, a SMW system is required, which performs threat detection, target classification, and target designation functions within the very short time of target approach. A ground-based SMW system will be able to provide covert surveillance and reconnaissance, fully automatic warning and target acquisition, and detection with high probability and low false-alarm rate. This tactical reconnaissance and SMW system must use passive sensitive IR search and verification sensors, as shown in Figure 9.11. This figure illustrates a practical concept of multispectral reconnaissance and missile warning system comprising of IR passive search and verification sensors with FOV of $1° \times 1°$ and instantaneous FOV of $0.1° \times 0.1°$ capabilities using state-of-the art system components. The system [13] configuration shown in Figure 9.11 offers passive detection of high-speed missiles at distances exceeding 20 km, total coverage of the hemisphere in a continuous search mode, sufficient number of samples for target acquisition with frame rate of 2 Hz, real-time signal processing capability to perform target classification in the presence of severe clutter, two-dimensional images with IR and UV signatures, three-dimensional tracks to suppress false targets and discrete clutter from a target, and a false, alarm rate of 1 in 24 hours during the surveillance and missile warning operating phases of the system.

9.15.1 Performance Requirements for a Surveillance and Missile Warning (SMWS) System

An effective SMWS must provide target detection at estimated ranges exceeding 20 km [13] with high detection probability (90% or higher) and low false-alarm rate of

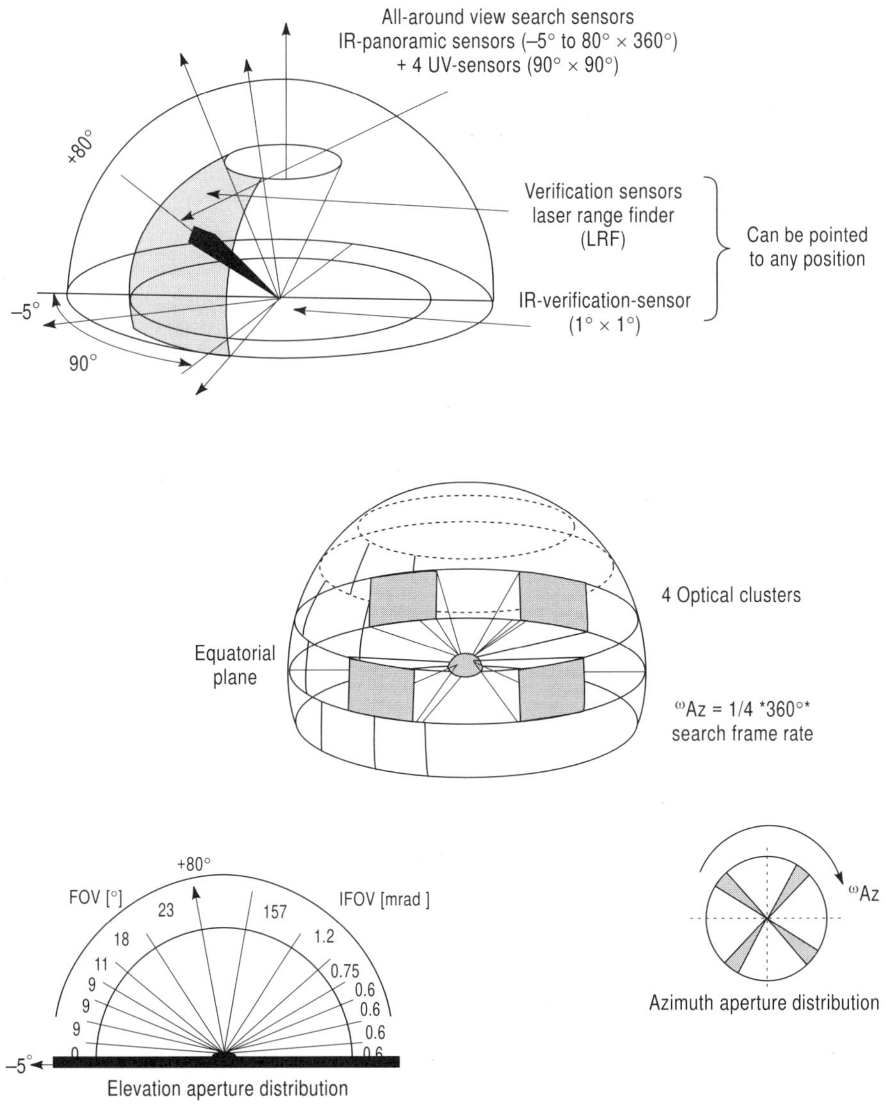

FIGURE 9.11 Conceptual design of a multispectral reconnaissance, surveillance, and missile warning system (MWS) using advanced technologies.

1 in 24 hours The SMW system, when located on a hill could, provide excellent look-down capability over 360° AZ plane to detect low-altitude terrain-hugging missiles with no impact from solar reflection or sea glint. This system provides the excellent spatial resolution during search mode needed to achieve a high signal-to-clutter (S/C) ratio. The total number of pixels per field-of-regard (FOR) rises dra-

matically with increasing resolution. A resolution of 0.6 mrad would require about 17 megapixels to cover the 2π sr of a hemisphere. If this pixel rate is updated to twice per second, the real-time computing capacity is available for on-line-image processing and target classification. Using a cryogenically cooled staring FPA comprised of Hg:Cd:Te detectors and time-delay integration (TDI) technology, sampling rates of 2–5 samples per second, integration times of a few μs per element, and frame rates of 200 to 300 Hz [13] are possible. The SMW system shown in Figure 9.11 uses four identical optical clusters rotated in the AZ plane at a speed of 180°/sec or 30 rpm. Each optical cluster consists of seven individual optical channels with defined FOVs ranging from 9° to 23° to match the image resolution requirements compatible with changing background clutter level. When radiation collected by the optical objective is imaged onto a 256 × 256 element FPA, the IFOV varies from 0.6 mrad to 1.6 mrad (Figure 9.11). If image distortion is less than one pixel, an integration time of 160 μs ($1/256 \times 256 = 160 \times 10^{-6}$ s) is possible by sampling of the hemisphere with a 2 Hz search frame time. The search sensor design is based on switching element arrays compromised of micromirror arrays, and microlens arrays as shown in Figure 9.12. Arrays of switchable elements are incorporated in every optical path of the search sensor clusters.

With a micromirror array, radiation from the outside scene can be blocked off from all channels with micromirrors in the off position. With a microlens array approach, the entrance of the primary lens is partitioned into several subapertures [13]. Integration of multiplexing techniques involving both the micromirror and microlens arrays into the SMW system configuration offers reliable target search capability.

The laser rangefinder operating at the eye-safe wavelength of 1.54 μm, which is a subsystem of the SMW system, provides an instantaneous FOV (IFOV) of 0.1 mrad with a narrow FOV of 1° × 1°. This laser sensor offers a range capability of 20 km at a PRF of 12.5 Hz [13]. Micro-mirrors with dimensions of 100 × 100 μm exhibit excellent performance in terms of switching speed and stability at accelerating loads up to thousands of g's. Complex algorithms used by the signal-processing unit of the SMW system produces about 300 image frames twice per second with a 2 Hz frame rate. Since noise and clutter are dominant in every image, front-end filtering will produce between 100 to 1000 initial events per frame, as illustrated in Figure 9.12. This adds up to 60,000 to 600,000 events per second in a 2-second search mode that require further processing. The track preclassifying algorithms will reduce the initial alarms of 1000/second to 10/second within a few seconds after initial detection of the target. Three-dimensional tracks are generated by laser rangefinder and high-resolution IR target signatures, which will further reduce the false alarms to less than one during 24 hours.

The ground-based SMW system uses passive IR/UV sensors and a laser rangefinder. Combination of these sensors using a distributed aperture optical configuration provides effective search, detection, and tracking capabilities over a hemispherical sector with high search frame rates and long dwell times. Laser ranging permits 3-D tracking with ultra-low false alarm rates and high target-to-clutter ratios.

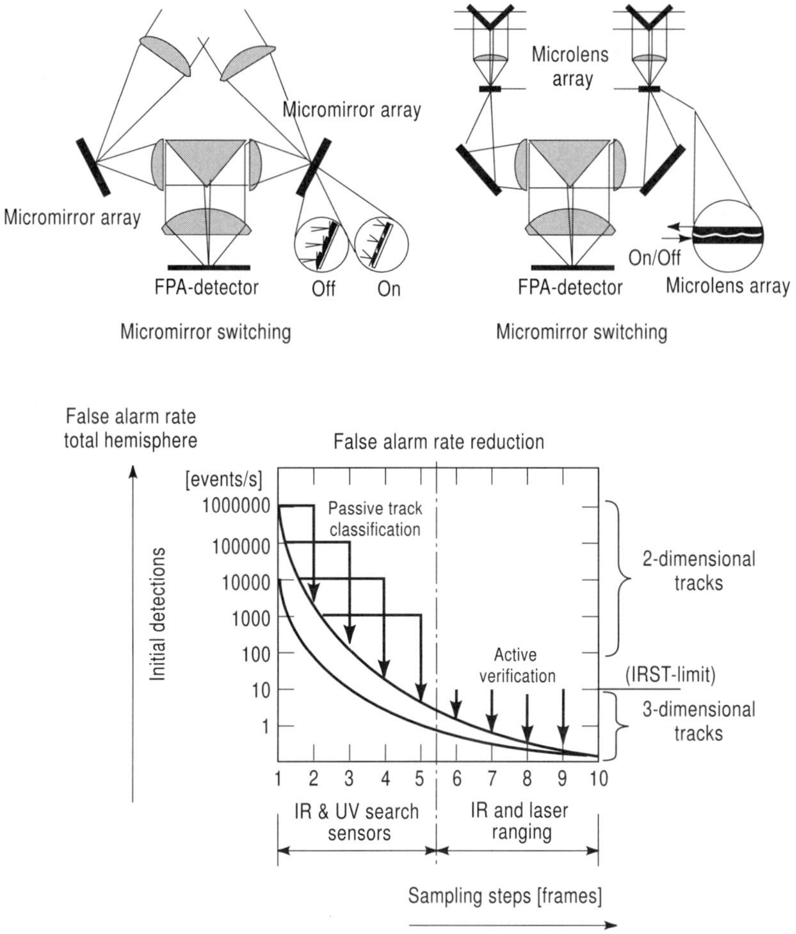

FIGURE 9.12 Performance improvement in IR\UV search sensor due to integration of microlens and micromirror technologies.

9.16 SUMMARY

Potential applications of IR and photonic technologies in military sensors and systems are discussed with emphasis on performance, cost, mission success, and reliability. State-of-the art performance capabilities and limitations of IR and electrooptic devices for applications in military systems are summarized. Integration of photonic and optoelectronic techniques in IR systems is identified. Critical performance parameters and unique design aspects of IR sensors are elaborated, including IR line scanners; IRSTs; eye-safe laser rangefinders; LIDARs for 3-D ocean surveillance; FLIRs for navigation, detection, and tracking of military targets; lasers for docking in space; laser seekers; laser-guided missiles, laser designators, high-energy laser for space surveillance, photonic sensors for battlefield applications; IR telescopes; and ground-based surveillance and missile warning systems to provide defense against tactical, theater, and cruise missiles. Mathematical expressions are provided to compute the detection range of high-energy laser weapon systems. Particular emphasis is placed on critical system performance parameters such as IR background clutter from solar reflections and clouds, thermal stability, spectral discrimination, false-alarm rates, detector instantaneous FOV, and FPA detector sensitivity. Readers are advised to refer to appropriate technical books if detailed information on operating principles and design aspects of specific sensors or devices is desired. Conceptual design concepts of some IR sensors incorporating photonic and electrooptic devices are based on the current state-of-the art device technology and the experience of the author in these areas. Estimated performance projections of some space-based and ground-based surveillance sensors are based on the information available in International conferences on IR and MM wave technologies. Computed values of detection, search, and tracking ranges for laser ranging and high-energy laser surveillance system are provided for readers to have a good understanding of their performance capabilities. Parametric studies performed by the author on laser wavelengths indicate that the energy requirements for a COIL airborne laser operating at 1.315 μm is roughly eight times less than those for a CO_2 laser operating at 10.6 μm under identical operating conditions.

NUMERICAL EXAMPLE

Compute the peak energy (E_p) requirements for a COIL laser-based airborne system operating at 1.315 μm using equation (9.7) and the previously assumed parameters to detect a target as a function of following target RCS and detection ranges:

Target RCS: 0.001, 0.010, 0.100, and 1.000 m^2
Detection ranges: 1000, 1250, 1500, and 2000 NM

Computed values of peak energy requirements for a COIL Laser Radar (J)

	Target RCS (m^2)			
Range (NM)	0.001	0.010	0.100	1.000
1000	100.5	10.05	1.005	0.100
1250	244.4	24.44	2.444	0.244
1500	510	51	5.1	0.51
2000	1608	160.8	16.08	1.61

REFERENCES

1. J. W. McLean et al. "Streak-tube LIDAR allows 3-D ocean surveillance." *Laser Focus World,* pp. 171–176, January 1998.

2. A. R. Jha. "IR sensor Detection range computations." Technical Report, Jha Technical Consulting Services, Cerritos, CA, 1989.
3. *Electro-optic Handbook.* RCA, Commercial Division, Harrison, NJ, pp. 110–112, 1974.
4. B. S. Goldstein et al. "GaAs injection laser radar." *Proc. of IEEE, 55,* 2, 181–18, 1967.
5. *Applied Optics, 11,* 2, 290–293, 1972.
6. P. Hermeli. "Design of a rangefinder for military applications." *Applied Optics, 11,* 2, 273–275, 1977.
7. H. D. Thollet, et al. "New IR seeker technology." *Proc. of SPIE, 3436,* 484–493, 1997.
8. B. Hines. "An Airborne High Power COIL Radar with multihundred-kW power level." *Laser Focus World,* p. 98, August 1998.
9. Contributing Editor. "Spaced-based IR systems." *Aviation Week and Space Technology,* pp. 56–58, April 1998.
10. P. G. Gebhardrt. "High power laser propagation *Applied Optics, 15,* 6, pp. 1479–1483, 1970.
11. M. C. Richardson et al. "A 300-J multigigawatt CO_2 laser." *IEEE Journal of Quantum Electronics, QE-9,* 2, 236–245, 1972.
12. D. McCarthy, "A few good beams: A light-speed defense against Mach-speed missiles." *Photonics Spectra,* pp. 102–103, April 1999.
13. W. J. Bernard. "Technology demonstrator for ground-based surveillance and missile warning." *Proc. of SPIE on Infrared Technology and Applications XXIV,* Vol. 3436–3456, pp. 494–502, 19–24 July 1998, San Diego, CA.

CHAPTER TEN

IR Signature Analysis and Countermeasure Techniques

10.0 INTRODUCTION

This chapter will focus on methods to estimate IR signatures of various man-made IR sources and IR countermeasure (IRCM) techniques to defeat the increasing threats posed by airborne IR and mobile surface-to-air (SAM) IR missiles. Computer simulation methods are summarized to estimate the IR signature of commercial and military jet engines and complex aircraft surfaces. Mathematical expressions are provided to compute the IR signatures of specific sources as a function of emission wavelength, radiant intensity, and observation range. Estimates of spectral radiance levels from various IR sources are provided. Contours of exhaust temperature for tubofan and turbojet engines used in commercial transport aircraft are furnished to provide readers with radiation intensity levels as a function of plume temperature and radial and axial distances from the face of the engine tailpipe. These contours indicate the location of the various regions of maximum IR radiation levels that could be targeted by enemy IR missiles.

Advanced IR countermeasure techniques will be discussed. State-of-the art IR sensors capable of picking up the IR signatures of SAMs and tactical fighters will be identified. Performance requirements for IR noise jamming equipment capable defeating the threats posed by SAMs and other tactical IR missiles will be described. The latest IR missiles with "robust all-aspect" approach capabilities are not susceptible to some existing IR countermeasure techniques. Directional IR countermeasure (DIRCM) techniques are required to neutralize the threats posed by such sophisticated IR missiles. Such DIRCM systems must be designed to detect, acquire, and track these missiles while they are in the launch or boost phase. Passive IR missile warning systems are described, with emphasis on high detection and tracking capabilities with minimum false-alarm rates. Computer simulation plots of IR signatures of a jet engine based on assumed key parameters such as thrust, exhaust temperature, engine speed, altitude, and aspect angle are provided. Performance levels of IR flares and conventional jammers are discussed. Most of the material provided in this chapter is based on the information obtained from technical journals and magazines such as *Aviation Week and Space Technology, Journal of*

Electronic Defense, Electronic Warfare, Microwave Journal, and product sheets from various suppliers.

10.1 IR Radiation Sources and Their IR Signature Levels

Infrared energy is an electromagnetic (EM) radiation that can be seen in the IR spectrum ranging from 0.75 μm to 1000 μm. Radiation in this region is referred to as thermal radiation, because heating a body or object generates it. The heated object radiates IR energy in all directions, but with maximum intensity only in a specific direction or angular region. Sources of IR radiation include commercial and military jet engines, rocket engines, missile exhaust nozzles, trucks, jeeps, battlefield tanks, ships, landing lights, armored personnel carriers, oil heat exchangers, and hot aircraft surfaces near the engines. The radiation levels from jet engine exhaust pipes and missile propulsion systems are of the most critical importance to the pilot of a tactical fighter or heavy bomber. The peak of the IR signature is directly behind the engine and slightly below the aircraft centerline. Radiation intensity patterns from a high-performance jet aircraft as a function of AZ and EL angles and longitudinal distance from the tailpipe are shown in Figure 10.1. It is evident from these patterns that the radiation intensity is maximum in the axial direction of the AZ plane. However, the maximum intensity seems to be about 10° below the axial axis or the longitudinal axis in the EL plane. Contours of exhaust temperature for turbojet and turbofan engines used by the commercial transports are illustrated in Figure 10.2. It is clear from these contours that the exhaust temperature and consequently, the radiation level drops rapidly as one moves away from the tailpipe face in the rear of the engine. One will notice from these curves that the radiation intensity patterns from a military jet are narrower in the AZ plane than those for commercial jet engines. However, the EL signature patterns appear to be similar for multijet military aircraft and commercial transport aircraft.

The contours shown in Figures 10.1 and 10.2 reveal that the peak radiation intensity occurs at the face of the tailpipe, regardless of whether it is a military or commercial jet engine. These contours further reveal that maximum radiance occurs within a radial distance of about ±2 ft from the engine centerline, whereas the weak radiation levels continue to occur up to a distance ranging from 100 to 150 ft from the face of the tailpipe, depending on the engine thrust and speed. Exhaust temperature contours for a military jet aircraft with afterburner operation (thrust level of 23,500) and without afterburner operation (thrust level of 15,800) are shown in Figure 10.3. It is interesting to note that the 66 °C temperature level with afterburner operation can be seen beyond 400 ft from the face of the tailpipe, whereas the same temperature level without afterburner operation stops at 100 ft from the face of tailpipe. Based on these contours, one can conclude that maximum IR signature from a commercial jet aircraft occurs at takeoff, whereas the maximum IR signature occurs at takeoff as well as with afterburner operation in the case of high-performance military jets. These high IR radiation levels can be detected by IR missile seekers at long distances ranging from 20 to 40 km, depending on the IR receiver sensitivity. It is important to mention that the radiated intensity level from a turbo-

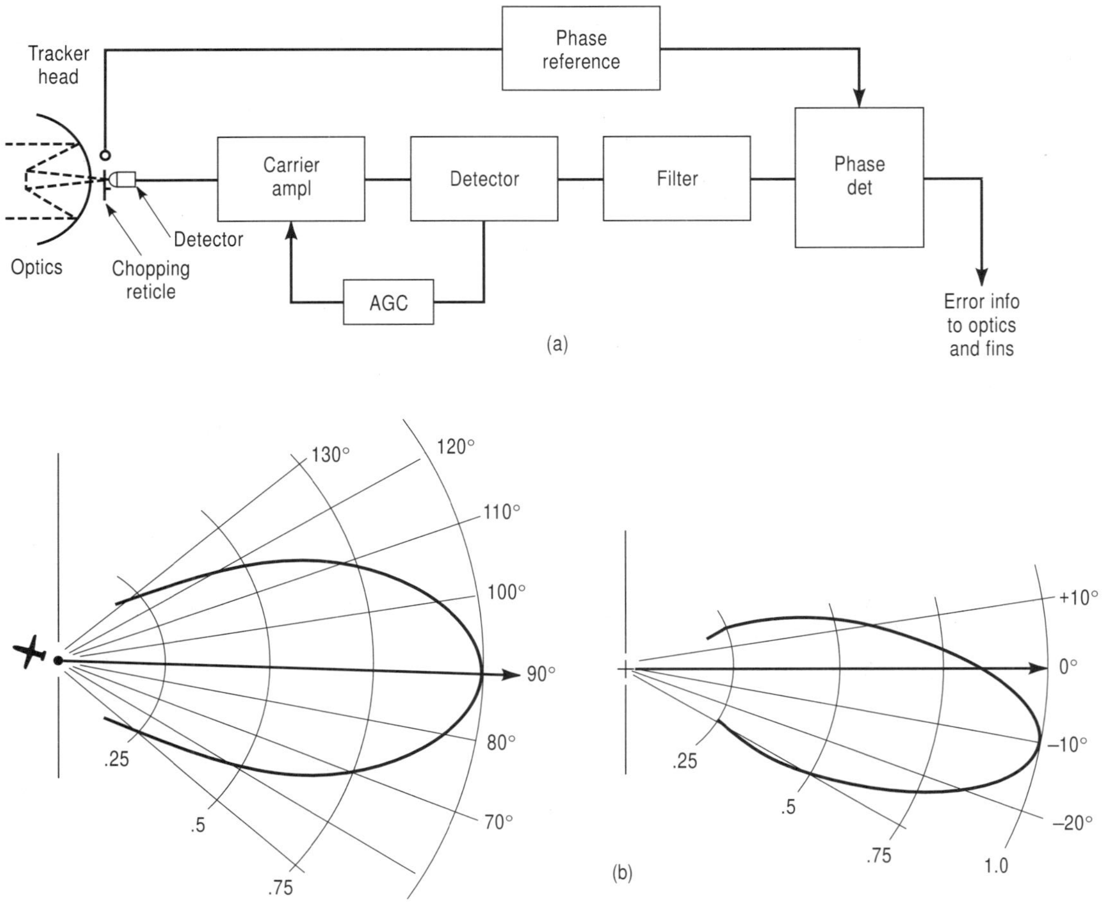

FIGURE 10.1 (A) Block diagram of a IR missile and (B) normalized IR signatures of a tactical aircraft.

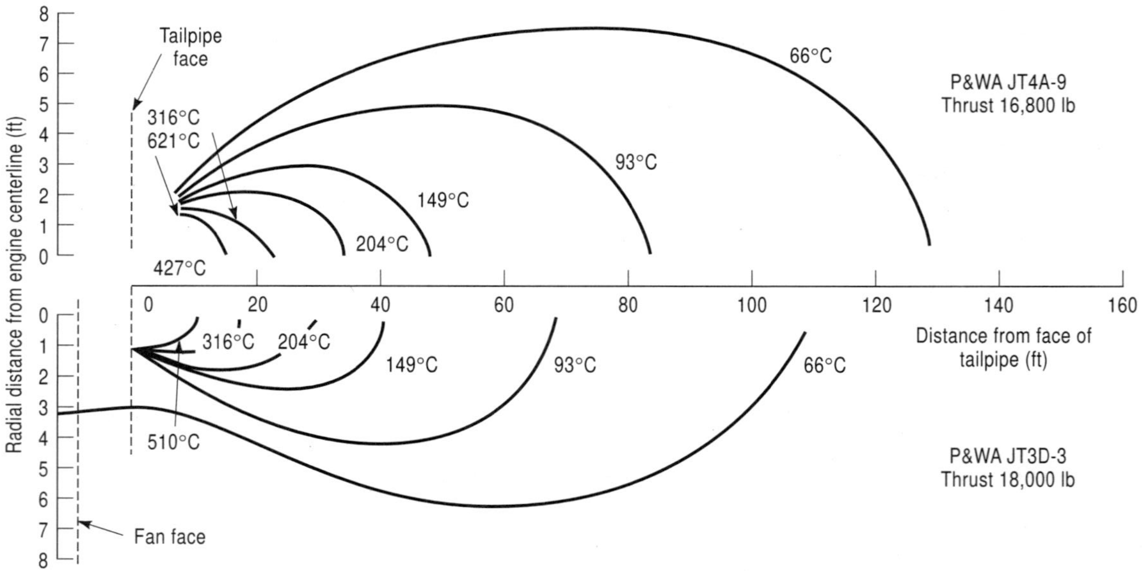

FIGURE 10.2 Contours of exhaust temperature for turbojet and turbofan engines used by commercial transports.

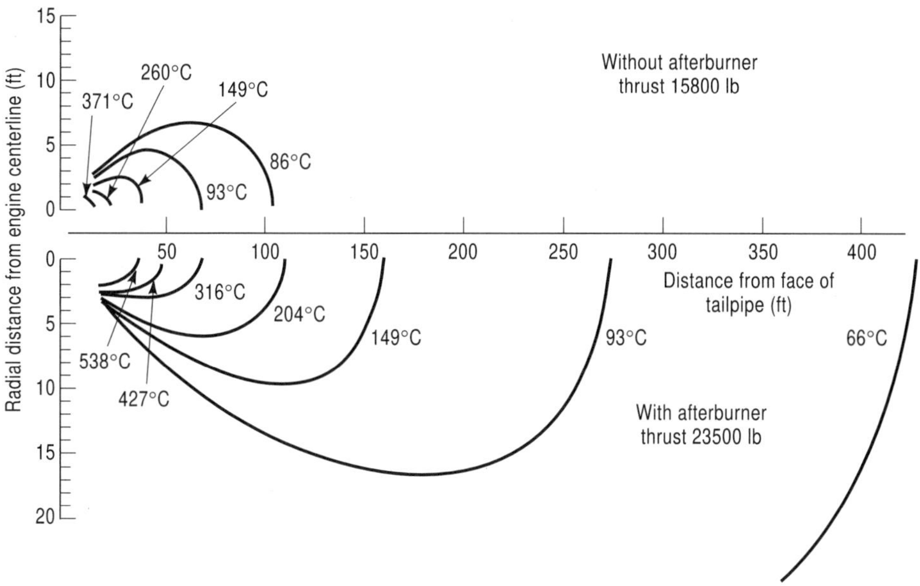

FIGURE 10.3 Sea-level exhaust temperature contours for a turbojet engine with (bottom) and without (top) afterburner.

fan engine is about 75% of that for a turbojet engine, which corresponds to a detection range of an IR missile seeker equal to 87%, based on the IR signature of a turbofan engine compared to the 100% detection range based on the IR signature of a turbojet engine. The IR seeker detection range is proportional to the square root of the radiant intensity of the source.

10.2 COMPUTATION OF IR RADIATION LEVELS FROM VARIOUS JET ENGINES

The radiation level or IR energy from a jet engine is dependent on several factors such as type of engine, thrust rating of the engine, afterburner operating mode, speed, operating altitude, atmospheric conditions, exhaust temperature, tailpipe cross-sectional area, and exit nozzle exit velocity. The total radiance (R) level from a jet engine tailpipe can be expressed as

$$R = \frac{(\varepsilon)(\sigma)(T^4)}{\pi} \tag{10.1}$$

where R is the radiance (W/cm^2/sr), ε is emissivity of the of the tailpipe material (which is a function of temperature and wavelength), T is the exhaust gas temperature (K), and σ is a constant with a value of 5.67×10^{-12}) W/cm^2/K^4.

The radiant intensity (J) is written as

$$J = (R)(A) \text{ W/sr} \tag{10.2}$$

where A is tailpipe cross-sectional area in cm^2. The radiant emittance (W) is defined as

$$W = (R)(\pi) = [(\varepsilon)(\sigma)(T^4)] \text{ W/cm}^2 \tag{10.3}$$

The radiant emittance is a function of both the wavelength and temperature of the source or body that is emitting IR radiation. Assuming an emissivity of 0.9 and exhaust gas temperature of 900 K, one gets

$$W = (0.9)(5.67 \times 10^{-12})(900^4) = 3.348 \text{ W/cm}^2$$

Thus, one can get the radiance, $R = (W/\pi) = (3.348/\pi) = 1.066$ W/cm^2/sr.

Assuming, jet engine tailpipe area of 3600 cm^2, one gets the radiation intensity, $J = 1.066 \times 3600 = 3839$ W/sr. An object at a temperature of 900 K transmits only 16.5% of the total IR radiation in the 2–3 μm spectral range. This means the radiant intensity (J) over the 2–3 mμ region will be roughly 633 W/sr (3839 × 0.165). It is important to mention that the overall radiant intensity can be obtained by multiplying this value by the number of engines on the aircraft. In a fighter aircraft, the typical number of engine varies from 1 to 2, whereas in a commercial transport the number varies from 2 to 4.

The percentage of total radiant emittance concentrated over a specified spectral bandwidth ranging from λ_2 to λ_1 can be expressed as

$$\Delta W(\lambda) = \frac{W(\lambda)}{\sigma T^4} \tag{10.4}$$

where

$$W(\lambda) = \int_{\lambda_1}^{\lambda_2} \left[\frac{(2\pi)(c^2)(h)(\lambda^{-5})}{(e^{-x} - 1)} \right] d\lambda \tag{10.5}$$

where $x = (1.439/\lambda T)$, c is the velocity of light, and h is a constant equal to 6.626×10^{-34} W/sec^2. Inserting various constants, assuming the tailpipe temperature of 1000 K, and integrating over a spectral range of 2–3 μm, equation (10.5) can be rewritten as

$$W(\lambda) = \int_{\lambda_1}^{\lambda_2} \left[\frac{(378.8 \times 10^{-14})\lambda^{-5}}{(e^{-x} - 1)} \right] d\lambda = 0.924 \text{ W/cm}^2 \tag{10.6}$$

and

$$\sigma T^4 = 5.67 \text{ W/cm}^2 \tag{10.7}$$

thus

$$\Delta W(\lambda) = \frac{0.924}{5.67} = 16.3\% \tag{10.8}$$

Total irradiance or emitted power (P_e) is the product of the radiant emittance (W) and the area of the emitting body (A). It is important to mention that the emitted power decreases as an inverse square of the distance or range from the emitter. Assuming an emissivity of 0.8, emitting area of 1 m^2, and a range of 500 NM, the emitted power comes to 23 W, 60 W, and 165 W at tailpipe temperatures of 600 K, 800 K, and 1000 K, respectively. However, at 200 NM range, the emitted power levels are 150 W, 415 W, and 1000 W at tailpipe temperatures of 600 K, 800 K, and 1000 K, respectively, assuming the same area and emissivity.

Computed values of percentage of total radiant emittance over a specified spectral region as a function of temperature are summarized in Table 10.1.

10.2.1 Aircraft or Missile Skin Temperature as a Function of Speed

Due to aerodynamic heating, the skin temperature of an aircraft or a missile will undergo radical changes. The skin temperature is dependent on the aerodynamic shape, speed of the aircraft, altitude, and atmospheric conditions. As the speed increases, the surface or skin temperature due to laminar flow over a body surface flying above 37,000 ft altitude increases according to the following expression:

$$T_{\text{skin}}\,(\text{K}) = 216.7\,(1 + 0.16\,M) \tag{10.9}$$

where M is the speed of the missile or aircraft and the temperature is expressed in K. The skin temperature and the wavelength of peak radiation intensity as a function of speed are given in Table 10.2.

These calculations are based on certain assumptions and, therefore, they are estimated values only. The exact speed ranges for various missiles are not known and the values quoted simply indicate the trends. The wavelengths shown in Table 10.2 represent the values corresponding to peak spectral energy. The product of wave-

TABLE 10.1 Percentage of radiant emittance over a specified spectral bandwidth at various temperatures

Temperature (K)	$\lambda_2 - \lambda_1$ (μm)	% of total radiant emittance
500	6–4	4.1
500	6–5	11.3
1000	3.00–2.00	20.4
1000	3.00–2.25	16.3
1000	3.00–2.50	11.2
2000	1.50–1.00	20.6
2000	1.50–1.20	13.3
2000	1.50–1.25	11.2
3000	1.00–0.50	26.1
3000	1.00–0.75	16.4
3000	1.00–0.85	10.1

TABLE 10.2 Skin temperature and peak radiation wavelength (λ_{peak})

Speed (M)	T_{skin} (K)	λ_{peak} (μm)	Remarks on platform
1	252	11.50	Regular jet
2	288	10.06	Fighter aircraft
3	323	8.97	High-performance jet
4	359	8.07	Short-range missile
5	394	7.35	Medium-range missile
10	572	5.07	Theater missiles
15	750	3.86	ICBM

length for peak radiation and temperature is about 2898 μm K, whereas for maximum number of photons, the product is equal to 3670 μm K.

10.2.1.1 *Critical Parameters Affecting the Radiant Intensity*

Studies performed by the author on rocket motors and supersonic jet engines indicate that the radiant intensity (J) is dependent on several motor or engine parameters, including the thrust rating to specific range requirement, type of propellant used, thermal characteristics of the sources, missile or aircraft speed, operating altitude, aspect angle, atmospheric characteristics, and flight envelopes of the engine type such as ramjet, scramjet, turbojet, or ATR (air turbo-rocket). Typical flight envelopes of various jet engines as a function of speed and altitude are illustrated in Figure 10.4. The studies further indicate that maximum radiant intensity of military jet engines occurs over the 2.5–3.2 μm spectral range, approximately, whereas the peak radiation intensity for flares or jet engines operating in afterburner mode occurs roughly over the 1.2–1.8 μm range, as shown in Figure 10.5.

10.2.2 Impact of Aircraft Maneuver on IR Signature

It is important to mention that the IR signature of an aircraft under static and sea level conditions is quite different from that under dynamic conditions as a function of speed and altitude. When the aircraft is flying under yaw, pitch, or roll conditions, its IR signature will be radically different. Computer simulations performed based on certain assumptions indicate that the maximum radiation level from a jet aircraft occurs at its tail aspect or at a 180° aspect angle, whereas the minimum IR radiation intensity occurs at the nose aspect or 0° aspect angle, as shown in Figure 10.6, with and without atmospheric effects. One can see that the peak intensity of about 1000 W/sr at the tail aspect is reduced to 450 W/sr at the nose aspect, and in the forward flight sector the radiation levels are close to negligible. Impact of roll and pitch angles on the radiant intensity is evident from the curves shown in Figure 10.7

The radiant intensity experiences significant reduction due to atmospheric attenuation, regardless of aircraft maneuver angles. The reduction in IR intensity level is relatively moderate for pitch angles ranging from + 10° to –20° at altitudes exceeding 5000 ft and slant ranges less than 5 km. Computer simulation reveals that maximum intensity occurs over the 2.8–3.2 μm spectral region with tailpipe temperatures

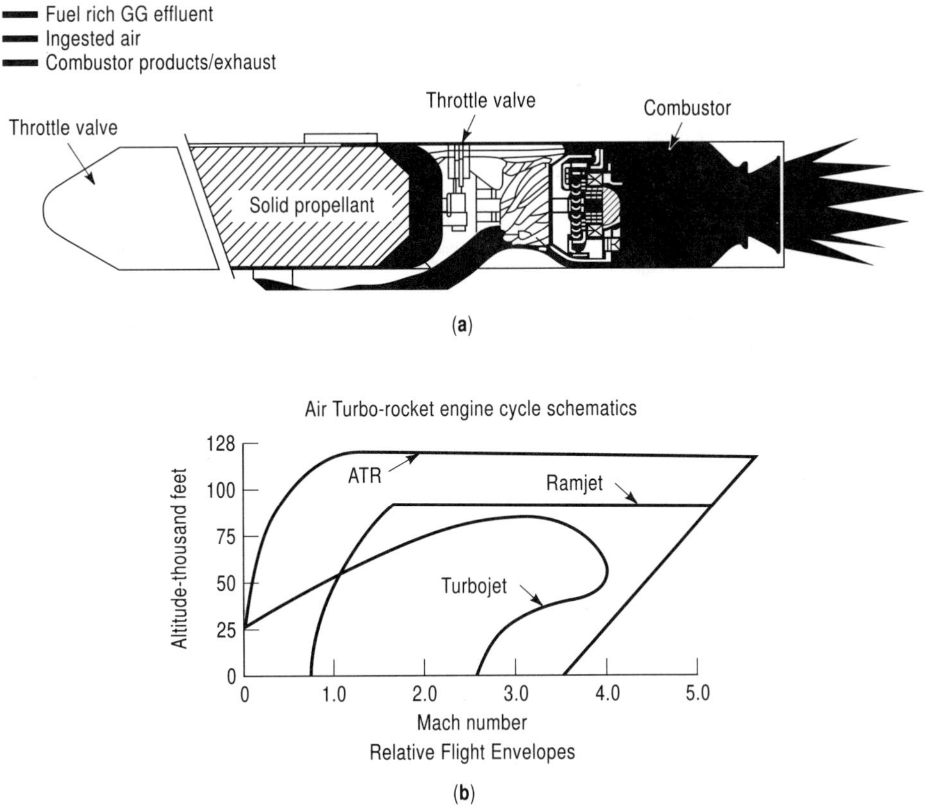

FIGURE 10.4 (A) Block diagram of a hybrid air turbo-rocket power plant used by an airborne missile and (B) flight envelopes for various jet engines as a function of speed and altitude.

ranging from 900 to 1050 K. However, when the aircraft is operating under afterburner mode, the maximum radiation intensity will be about 30–40% higher, due to high temperature and thrust level over the 1.3–1.8 μm spectral region, as illustrated in Figure 10.5. When operated in an afterburner mode a high-performance fighter/strike aircraft must complete its mission in minimum time to avoid lock-on by enemy's IR missiles, but with maximum safety in hostile territory.

10.2.3 Impact of Emissivity on IR Signature

Mathematical expressions previously used to determine both the radiant emittance (W) and radiant intensity (J) involve the emissivity of the materials used by the exit nozzle or the tailpipe. Both these quantities increase with the increase of emissivity, which is a function of wavelength and temperature. Only a few materials are available that capable of retaining high emissivity, thermal performance, and mechanical integrity under extremely high temperature and thrust conditions. Emissivity is a

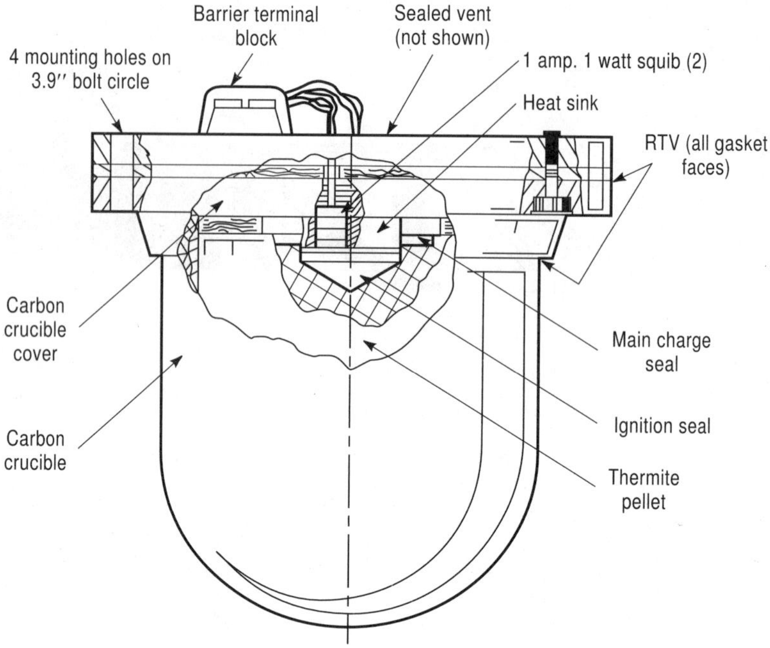

(**a**) Critical elements of an IR flare

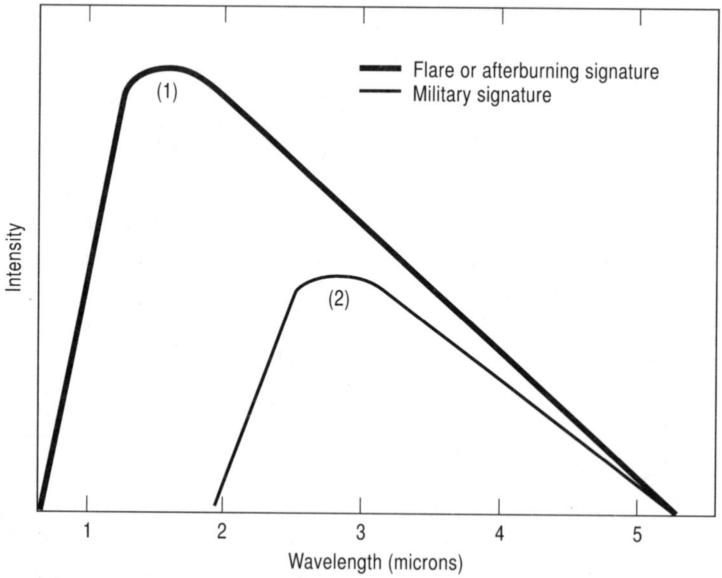

(**b**) Intensity of a (1) Flare and (2) Jet engine

FIGURE 10.5 (A) Critical elements of flare and (B) intensity from flare (1) and military jet (2).

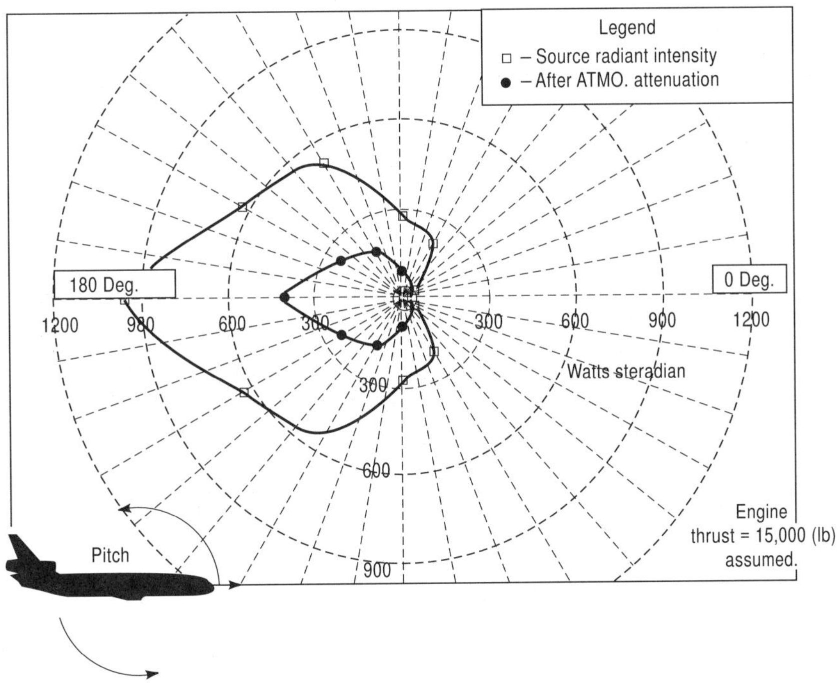

FIGURE 10.6 Simulated IR signature or radiation intensity (W/sr) from a jet engine with and without atmospheric attenuation at various aspect angles.

complex function of wavelength (λ) and surface temperature (T) and is denoted by $\varepsilon(\lambda,T)$. From the emissivity data summarized in Table 10.3, one will observe that polished surfaces will have lower emissivity values compared to unpolished surfaces under similar operating conditions. Low-temperature emissivity values of selected metals with high mechanical integrity are shown in Table 10.3.

10.3 COMPUTER PROGRAMS TO PREDICT IR SIGNATURES

Various computer programs are available to predict the IR signatures of emitting sources. The most practical and earliest computer program, known as SCORPIO-N, was developed by the engine division of General Electric Company. This program [1] offers reliable IR signature estimates of the radiation emitted by turbojet, turbofan, and other propeller-based aircraft. Currently, most advanced programs such as SCORPIO-IIIA, SPRITE, and upgraded SCORPIO-N are available to predict the IR signatures from various aircraft engines. The SCORPIO-N program can also compute the lock-on ranges for specific IR missiles to access the vulnerability of the aircraft, regardless of the flight conditions and AZ and EL aspect angles. This program is capable of computing the IR radiation levels over specified spectral regions

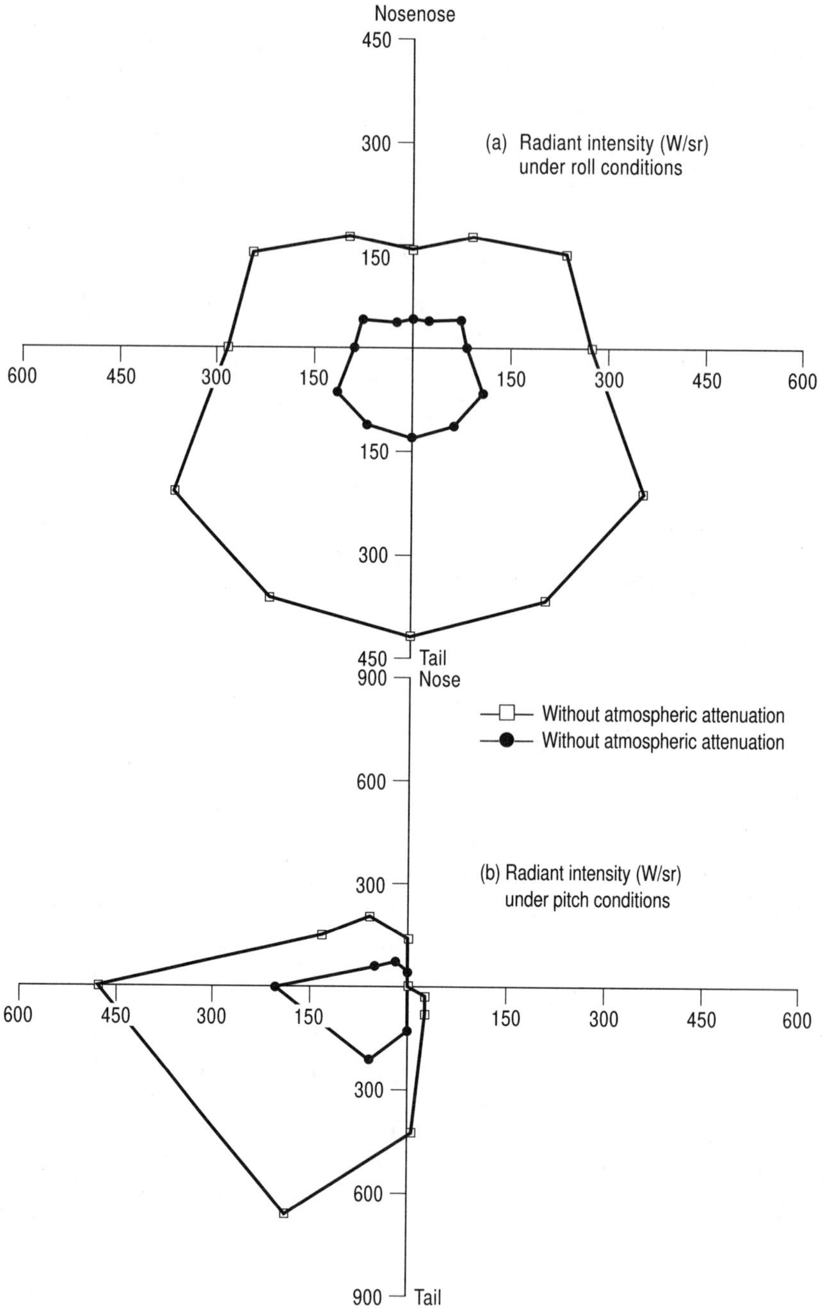

FIGURE 10.7 Typical radiant intensity from a jet engine under aircraft roll and pitch conditions with and without atmospheric attenuation.

TABLE 10.3 Emissivity of selected high-temperature materials as a function of temperature and operating wavelength

Material	Temperature (K)	Wavelength (μm)	$\varepsilon(\lambda,T)$
Carbon	293	9.89	0.95
Graphite	293	9.89	0.98
Magnesium (oxidized)	473	6.13	0.37
Magnesium (polished)	293	9.89	0.07
Nickel (oxidized)	473	6.13	0.37
Nickel (polished)	293	9.89	0.05
Nickel (unpolished)	293	9.89	0.11
Steel (polished)	373	7.77	0.07
Steel (oxidized)	473	6.13	0.79
Stainless steel (buffed)	293	9.89	0.16
Stainless steel (oxidized)	333	8.70	0.85

ranging from 1 to 20 μm, provided appropriate values of the variable and constants involved are specified [1].

It important to point out that advanced IR missiles in the hands of adversaries could be more sensitive and deadly. In order to avoid detection by enemy IR seekers, the engine designer must design engine configurations that present a minimum IR signature. Diverting the hot exhaust gases in a different direction, instead in the rear of the aircraft, may reduce IR signature seen by the enemy's IR seeker. The engine designer must know the IR emissions from the engine's tailpipe and deploy techniques to suppress these emissions. Available tools must be used to reduce the vulnerability of a fighter or surveillance aircraft to IR guided missiles.

10.3.1 Evaluation Methods of IR Emissions from an Aircraft

Evaluation of instantaneous vulnerability of a jet-powered aircraft to a high-performance IR missile must address the following issues:

- Emissions from the hot internal parts of the jet engine such as inlet fan, compressor, gas turbine, exhaust nozzles, and tailpipe
- Distribution of the heat flow parameters and directions of hot gases for minimum IR signature levels
- Evaluation of gaseous emissions and their transmissions in the atmosphere
- Airframe IR signature contributions, including emissions, reflections, and installation factors in the vicinity of tailpipe
- IR missile characteristics to determine the range at which the incoming missile can discriminate a target from the background clutter

10.3.1.1 Emissions from the Hot Internal Parts of the Engine

The internal parts that contribute to IR emissions include inlet fan, multistage compressor, multistage gas turbine, exit nozzle, and tailpipe. The spectral IR emission level or the spectral radiant exitance (M_λ) from a solid surface at uniform temperature T can be obtained using the Planck's law, which is defined as

$$M_\lambda = \frac{(A)[\varepsilon(\lambda)](C_1/\lambda^5)}{e^x - 1} \tag{10.10}$$

where A is the surface area, $\varepsilon(\lambda)$ is the wavelength-dependent emissivity, T is the surface temperature (K), C_1 is a constant equal to 3.742×10^{-16} W·m^2, and parameter x is equal to $0.01439/\lambda T$), when λ is expressed in meters.

The total IR radiation arising from a particular surface consists not only of the direct emissions, but also of the reflections of emissions from the engine component surfaces. This means that the radiant emissions from the interior engine surfaces are dependent on internal surface temperatures, the emissivities and the reflectivities of various internal parts, the projected areas, and the interchange factors between each pair of surfaces.

10.3.1.2 Temperature Estimation for Various Internal Parts

Temperature of various hot engine parts can be computed using standard heat transfer two-dimensional equations. Since the radiation is dependent on temperature exponentially, each temperature difference of 50 °F or so is represented by a different surface. In the case of a typical jet engine application, at least 20 to 25 different surface temperatures must be used to compute the radiation level from a jet engine based on temperature distribution function.

10.3.1.3 Impact of Geometry and Surface Reflection on IR Radiation Level

The geometry can be represented by the projected areas of given temperature (T) and emissivity (ε), which are located and orientated with respect to the observer. The view factor of each surface to every other surface must be included into the program along with the projected areas and emissivities in order to include the diffuse reflections in the expression for total radiation level.

The reflection of radiant intensity (J) at a surface can be a function of either specular or diffused absorption, specular or diffused reflection, angle of incidence, wavelength and surface temperature. Both the reflections and emissions will vary with the distributions of the flow fields. The analytical program solves a nonuniform axially symmetric flow field with external flow. The analysis must include axial momentum and energy and diffusion conservation equations to obtain acceptable accuracy of IR radiation level estimation. Higher order of accuracy is possible if the analysis also includes the conservation of turbulent kinetic energy and conservation of tangential momentum. The flow field computations must include the shock-free supersonic flow condition by assuming sudden expansion or contraction from a high nozzle exit pressure to a different atmospheric pressure depending on the operating altitude.

10.3.1.4 Impact of Emissions on IR Signature from the Airframe

The airframe itself emits radiation at lower temperatures and can be a major contributor for the target signal because of the large surfaces, if viewed from the forward sector or nose aspect. In addition to the direct emissions from the engines, the airframe can affect the IR radiation by blocking the view of the hot plume and/or engine at certain aspect angles under dynamic flight conditions. In the lower wavelength regions, reflections from the sun known as solar glint will dominate all other contributions. The reflections of the plume off the frame could be significant because of large surface areas of the aircraft. The reflections of the plume must be considered along with the direct emissions from the hot engine parts and blockage of the plume.

10.4 DETERMINATION OF OVERALL RADIATION LEVEL AND LOCK-ON RANGE USING THE SCORPIO-N SOFTWARE PROGRAM

The overall radiation level depends on the radiation from the hottest engine section, the blockage of the plume by the airframe, the obscurance of the airframe and hot engine parts by the plume, aspect angles, number of engines used by the airframe, locations of the engines, mixing of the two or four plumes in the case of multi-engine configuration, effects of external flows when the engine axis is not parallel to the longitudinal axis of the aircraft, and the nonaxial hot gas flows resulting from noncylindrical surfaces. The SCORPIO-N software program [1] developed by General Electric Company around 1978, offers reliable prediction of the overall IR radiation level from a jet-powered aircraft by taking into account all the factors and issues discussed previously. As stated earlier, this program can be used to determine the IR missile lock-on range. When an IR missile receives target radiation, the missile computer processes the received signal and then discriminates it from the background induced clutter. It is necessary for the missile electronics to reject the background clutter using clutter rejection filters to maintain lock-onto the target. It will be difficult to acquire a lock-on capability on a maneuverable target, because of lack of instant information on the relative velocity and direction of missile and target. SCORPIO-N offers an approach to the missile–target interrelation as a first step in the problem solution. This program assumes that all relevant missile parameters, such as its speed, range and altitude, are known and then computes a lock-on range for minimum and maximum S/N ratios. The program computes the effective radiation level based on a minimum of two distances for each aspect angle at which the lock-on is required. The lock-on range for a particular class of missile must be determined for each aspect angle and engagement geometry. The lock-on range is dependent on net target signal-to-noise ratio, target area, total target spectral radiant level, missile-to-target range, noise-equivalent input (NEI) of the missile, and normalized missile spectral sensitivity, which is a function of missile plume temperature and associated wavelength.

This program uses molecular band models to describe the spectral distribution of the average intensity and width of the spectral lines within the bands. These models based on quantum theory are modified to match homogeneous empirical data for wide range of predetermined pressures, temperatures, and wavelengths occurring in various spectral regions. The software program takes into account all relevant geometrical factors defining the exhaust system exit, location and orientation of a second plume from a second engine, and location of the observer. In the case of a two-engine fighter aircraft, the second plume must be located symmetrically about the airframe centerline, which is known as the crosswind direction. In the case of a four-engine surveillance aircraft, the second plume is generally located on the same side of the fuselage in the horizontal plane. In the case of four-engine configuration, two staggered parallel exhaust systems are located on each side of the fuselage section. However, in the case of a two-engine aircraft, each exhaust system is located on the rear end of the aircraft but equidistant from the centerline of the aircraft. In case of a multiengine aircraft, IR radiation levels will be significantly higher and thus can be easy prey to an IR missile attack. Regardless of number of engines used by an aircraft, the angle between the exhaust exit plane and the cross-flow direction will affect the IR radiation intensity as seen by the missile seeker.

As stated earlier, the airframe is comprised of various structural sections, such as nose, fuselage, wings, stabilizer, rudder, and other structural elements. These sections will have different temperatures, surface conditions, areas, and materials. The surface conditions and the materials and their emissivities will be different for these structural sections. Furthermore, the emissivity is a function of temperature, surface condition, and wavelength. For a given material and temperature, the emissivity is slightly dependent on the aspect angle and could vary from 0.9 at zero-aspect angle to 0.8 at 90° aspect angle. The program must take into account the velocities and temperatures of hot gas leaving the tailpipe into atmosphere. Under these variable conditions, it will be extremely difficult to obtain the exact estimation of the IR signatures of the engine.

Based on the results obtained from this program, both the radiant intensity J (W/sr) and radiance R (W/cm^2/sr) for a jet-engine tailpipe are roughly 25 times that of plume under cruising flight conditions, which further increase by a factor of 20 to 30 under afterburner operations. This indicates that a high-performance fighter/strike aircraft could be more vulnerable to an IR missile attack when operating in an afterburner mode.

10.5 RADIATION LEVELS EXPECTED FROM SHORT-RANGE, MEDIUM-RANGE AND INTERCONTINENTAL BALLISTIC MISSILES (ICBMs)

Propulsion system requirements for a missile are strictly dependent on type of missile, operating range, payload, and initial thrust. Liquid-fueled ramjet and scramjet propulsion systems are considered most ideal for the next generation of beyond-visual-range (BVR) air-to-air (A/A) missiles to achieve enhanced range capability and to provide sustained power for increased endgame agility. Ramjet propulsion technology is considered a key ingredient for the current antiradar missile (ARM), but future ARM systems are most likely to employ a solid-fuel, variable flow-directed ramjet technology capable of offering significant performance improvement in terms of range and time to target. Still-higher missile performance is possible with a ram rocket motor using four inlets in the center of the missile body and a high boron content in the sustainer propellant for high specific impulse energy with low volume [2].

It is important to mention that the missile plume radiation level is dependent on the initial thrust, rocket exit temperature, and the distance between the missile and an observer or IR sensor. Heat transfer calculations are based on kinetic theory, and the reentry cooling of a missile is based on dissociation of oxygen and nitrogen in the ionosphere, which extends roughly from 50 to 200 miles in altitude. When the missile travels in the exosphere, which extends from 200 to 700 miles in altitude, the gas molecules have no impact on its travel and the missile surface temperature is strictly governed by the radiation to and from outer space. The temperature of a missile in space depends on its surface configuration and solar radiation intensity and also whether or not the missile is rotated. The equilibrium temperatures of the missile surfaces in outer space that are exposed to solar radiation depend on the heat balance between the surface absorption and solar radiation. The IR radiation is a function of the equilibrium temperature and surface emissivity.

As the missile warhead enters the lower atmosphere in the reentry phase, intense aerodynamic heat is generated, particularly when the missile reaches an altitude of about 50 miles. For reentry speeds of 20 Mach or higher, extremely high temperatures pass over the missile surface, leading to both excessive heat transfer rates and high IR radiation levels, perhaps close to 1000 W/sr depending on the missile nose cone design (pointed nose or blunt nose design) and its radar cross-section (RCS).

10.5.1 IR Emissions from Short-Range Ballistic Missiles (SRBMs)

Short-range, SCUD-type missiles generally use air-breathing power plants that combine both rocket and turbofan technologies. The air-turbo rocket (ATR) cycles provide several advantages for air-breathing propulsion systems (ABS) such as low-cost construction, high thrust-to-weight ratio (ranging from 20 to 40), and long-term storability. The long-term storability is of critical importance for tactical and theater missiles to reduce costly logic problems. Using a liquid-oxygen/RP-1 fuel mixture, ATR applications to other vehicles including launch boosters are possible. An ATM-based propulsion system using solid-fuel technology [3] can reach altitudes exceeding 100,000 feet or 20 miles with speeds up to 20 Mach and thrust per frontal area as high as 10,000 lb/ft^2, which is adequate for SCUD-type missiles with operating range from approximately 300 to 600 miles (Figure 10.8). The estimated IR radiation intensity levels from these missiles using solid-fuel-based ATR propulsion systems could vary between 3000 and 5000 W/sr. According to the published literature [3], most SRBMs have operational ranges from 200 to 950 km (125 to 594 miles) with maximum payload capability not exceeding 1000 kg or 2208 lb. An inverse relationship exists between the missile range and the weight of the warhead [3] that it can carry. A missile capable of hurling a 1000 kg warhead to a range of about 400 km can also send a 700 kg warhead to about 500 km or a 200 kg warhead to a range exceeding 800 km. In the case of a nuclear warhead, the weight of such a warhead sets the upper limit on the operational range of the missile. In the case of SCUD-type missiles, the range to the target sets the upper limit on the amount of

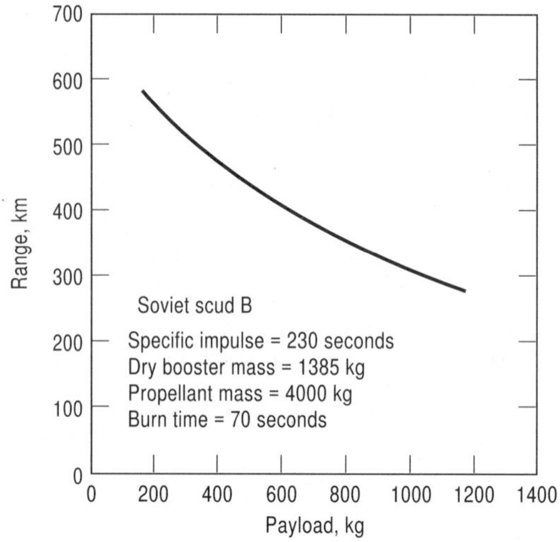

FIGURE 10.8 Estimated range performance of a short-range SCUD-8 ballistic missle as a function of payload.

the high explosives that can be safely carried to the designated target. The technical article [3] further indicates that the operating range of a SRBM can be increased by lengthening the missile and increasing its propellant load by about 25%.

The thrust requirements for a rocket engine is dependent on the missile to target range, payload weight, and warhead weight. Higher thrust levels are possible by adding more stages or strapping two or four rocket engines on the first stage, which is more practical and cost-effective. A SRBM developed by North Korea [3] with a diameter of 1.3 m, length of 16 m, and propellant load of 16 tons uses four rockets engines and produces a thrust exceeding 116,000 lb, which is sufficient to send a 1000 lb payload to a distance exceeding 1000 km. Missile operational range capability can be further enhanced by using improved high-temperature materials and aerodynamic design. The IR signature at ignition depends on the rocket energy, thrust, plume temperature, and atmospheric conditions in the vicinity of the launch pad. Spectral radiant emittance ($W/cm^2 \cdot \mu m$) is strictly dependent on the exhaust temperature as observed during the launch phase. Computed values of spectral radiant emittance using equation (10.10) as a function of exhaust temperature are summarized in Table 10.4, from which IR signatures of the missiles during the launch can be predicted with reasonably good accuracy.

10.5.2 Range and Thrust Requirements for MRBMs and ICBMs

High-power boosters are required to launch MRBM and ICBM missiles towards designated targets. Regardless of missile type, critical missile performance requirements are target to range, amount of fuel to satisfy mission requirement, and warhead weight. Typical range capability for a MRBM varies from 1500 to 4000 km,

TABLE 10.4 Maximum spectral radiant emittance (*E*) for a black body as a function of temperature and radiation wavelength

Temperature, T(K)	Spectral radiant emittance, E (W/cm^2-μm)	Wavelength, λ (μm)	Product T_λ (μ K)
6000	11,000	0.48	2880
4000	2,000	0.73	2920
3000	200	0.96	2880
2000	140	1.45	2900
1000	1.00	2.91	2910
500	0.45	5.80	2900
300	0.005	9.70	2910
200	0.0005	14.50	2900

Remarks: According to Wien's displacement law, the maximum spectral radiant emittance occurs when the product T_λ is equal to 2896 μm K.

whereas for an ICBM it ranges from 5000 to 10,000 km, approximately. Exact thrust level requirement as a function of range and payload for these two type missiles are beyond the scope of this book. However, rough estimates of thrust vary from 250,000 to 500,000 lb for a MRBM, whereas for an ICBM the estimate could vary anywhere from 1 to 2 million lb, approximately. After the missile leaves the launch pad and reaches its maximum height between the postboost phase and midcourse phase, its IR signature level continues to drop. One will find the maximum radiant intensity during the launch phase and minimum in the terminal or final phase. Typical flight times of these missiles in various phases are shown in Figure 10.9. Detection of these missiles will be extremely difficult during the launch and terminal phases because of the short flight time of about 1 minute and less than 1 minute, respectively. Detection of these missiles is possible during their midcourse flights because the flight times are generally more than 20 minutes. This means that a missile defense surveillance system detects the missile during its midcourse flight time with high detection probability and low false-alarm rate. Such a missile defense surveillance system can detect a missile during its midcourse flight time with high detection probability and low false-alarm rates. A North Korean two-stage MRBM demonstrated a range of 1800 km with a 1 ton payload and even longer range of 2500 km with a payload of 500 lb. The North Koreans are working [3] on long-range missiles with range capability from 4000 to 6000 km with a 1 ton warhead. Chinese CSS-2 IRBM missiles [3] supplied to Saudi Arabia in 1987 have a range of 3000 km. The operating range of a missile can be increased if the body of the missile is made from high-strength aluminum alloy instead of steel. Nevertheless, one must note that tensile strength at elevated temperatures and thermal performance of steel are much better than those of aluminum. The room-temperature tensile strength of even ordinary steel is between 120,000 to 170,000 lb/in^2, whereas it is hardly 29,000 lb/in^2 for an aluminum alloy.

The warhead capability varies from 1 to 3 tons for a MRBM, whereas for an ICBM it varies from 1 to 10 tons, depending on the types of explosives deployed

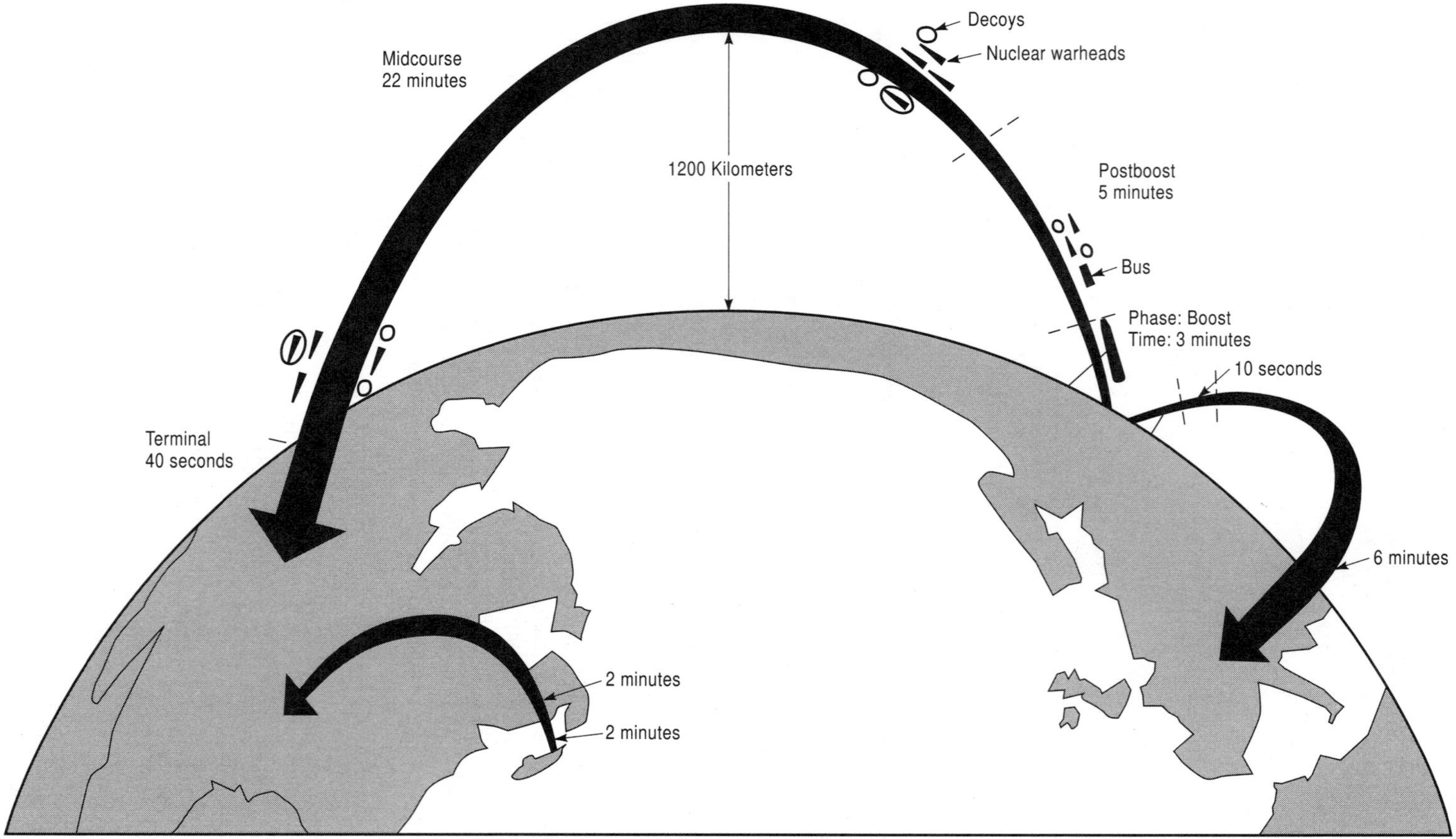

FIGURE 10.9 Boost phase, midcourse phase, and terminal phase trajectories of an intercontinental ballistic missile (ICBM) with their estimated flight times.

and range involved. Smaller multiple warheads can be deployed in an ICBM [4] with longer-range capabilities. Estimation of missile accuracy is most complex and the accuracy, which is known as the circular error probability (CEP), is strictly dependent on the missile's guidance system accuracy. The missile guidance system guides the missile during its powered flight profile without external disturbances during the boost, midcourse, and terminal phases, as shown in Figure 10.9. The missile accuracy further depends on the severity of buffeting and uncontrolled aerodynamic lift due to cross winds during the reentry phase. Considering all adverse factors, rough estimates of CEP vary from 2 to 4 km for a SRBM, 4 to 6 km for a MRBM, and 6 to 8 km for an ICBM [3]. Better accuracy can be achieved with advanced inertial guidance systems. Higher accuracy is possible with a GPS receiver-based system.

10.5.2.1 Factors Affecting the IR Radiation from MRBM and ICBM systems

The IR radiation levels from these missiles are strictly dependent on the thrust level, plume intensity, and atmospheric conditions at a given altitude and range. Thrust requirements for an ICBM rocket vary from 1 to 4 million lb, approximately, depending on the range, fuel storage, and weight due to warhead and additional stage rockets. As stated earlier, the rocket thrust requirements for these missiles vary from 200,000 lb to 4 million lb, approximately, depending on the range, type of missile, number of rocket stages, propellant weight, type of fuel, fuel storage capacity, and weapon payload. In the case of a multistage rocket propulsion system, the first stage must have the highest thrust level because of various weight elements involved, including missile body, weight of additional fuel storage, weight of additional stages, and payload weight. This means the IR radiant intensity from the exhaust of the first stage rocket will be the maximum and could be in excess of 10,000 W/sr. In case of an ICBM, the first stage generally includes four to five high-thrust rocket engines strapped together to provide the thrust required. The second stage generally consists of two side-by-side rocket engines with medium thrust ratings, and the third stage is comprised of a single engine of high thrust rating capable of carrying the warhead to the designated target. The combined thrust provided by the first, second and third stage must be sufficient to launch the missile, fly it, and guide it to the target of interest. It is important to mention that the IR radiation intensity will be maximum from the first stage rocket and minimum from the third-stage booster rocket.

IR radiation intensity is strictly dependent on the exhaust temperature of the missile and the atmospheric conditions existing during the launch, boost, midcourse, and terminal phases. Several factors such as uncontrollable aerodynamic drag and turbulence-related atmospheric environments will affect the radiation intensity during the various phases of the missile's flight. In the case of terminal and boost phases, atmospheric characteristics will have significant impact on the radiation intensity level from rocket exhaust gases. Based on published technical data on rocket technology, it can be stated that the radiation intensity from a plume is about 10% less than from the hot gases of rocket engine exhaust. Both the plume and rocket engine temperatures are dependent on atmospheric conditions and the radiation intensity of these sources is contingent on the time of detection by the IR surveillance sensor. If the source temperature is known, then the spectral radiant emittance (W/cm^2-μm) can be estimated from the data shown in Table 10.4.

10.5.2.2 Detection range estimates for missiles

Detection range capability of a missile is dependent on the rocket thrust level, plume temperature, atmospheric conditions, IR source intensity level under a specific missile phase and IR surveillance performance parameters. In general, the IR surveillance receiver detection range is directly proportional to the square root of source intensity (J), detectivity parameter (D^*), and optic size, and inversely proportional to the square root of the peak-signal-to noise voltage ratio, and to the fourth root of the product of detector IFOV and IR receiver noise bandwidth. In brief, significant improvement in detection range is possible with higher source intensity, larger optic size, higher detectivity, and narrower detector IFOV. Improvement due to other IR receiver parameters is relatively low.

Since the IR intensity level will be different during the launch, midcourse, and terminal phases (Figure 10.9), the detection range will be different. Preliminary detection range calculation have been performed assuming an IR source intensity of 10^6 W/sr during the launch phase, optic size of 1 m, numerical lens aperture of 0.25, detectivity (D^*) of 10^{10} cm $\sqrt{\text{Hz}}$/W, detector instantaneous field-of-view (IFOV) of $0.1° \times 0.1°$, peak signal-to-noise current ratio of 5:1 (14 dB) to ensure a detection probability of 99.9% with ultra-low false-alarm rate, and other relevant parameters of a missile surveillance receiver. The computed results shown in Table 10.5 indicate a detection range of about 4870 NM or 9014 km for an ICBM based on the above assumed parameters. Calculations are repeated for the detection range of the missile during the midcourse phase using the same parameters, except the source intensity is 10^5 W/sr. Under these assumptions, the midcourse detection range of the ICBM comes to about 1540 NM or 2850 km (Table 10.5). Calculations performed for the terminal phase, assuming the source intensity of 10^4 W/sr, indicate a terminal phase detection range of 487 NM or 901 km (Table 10.5). The detection range is proportional to the square root of the source radiation intensity (J). If the actual source intensity levels will be higher or lower than the assumed values during various missile phases, the detection range will be relatively higher or lower based on the square root relationship between the detection range and source intensity. High-

TABLE 10.5 Missile detection range as a function of various parameters during various missile phases (NM)

Missile phase	Source intensity, J (W/sr)	Detectivity (D^*) cm $\sqrt{\text{Hz}}$/W	
		10^{10}	10^{11}
Launch	10^6	4870	4870
Midcourse	10^5	1540	1540
Terminal	10^4	487	487
Instantaneous FOV (deg)		0.1×0.1	1×1

er detection ranges can be achieved using improved detector sensitivity or detectivity, which is possible under cryogenic operations.

As stated earlier, higher source radiation intensity is possible with higher plume or exhaust temperature and large rocket exhaust area (Figure 10.10). Assuming a rocket exhaust diameter of 1 m, exhaust plume temperature of 3000 K in the launch phase, and emissivity of 0.9, one gets the radiation intensity (J) of about 1.03×10^6 W/sr ($J = \varepsilon\sigma T^4 A/\pi$, where ε is the emissivity, A is the area, T is the temperature, and $\sigma = 5.67 \times 10^{-12}$ W/cm²/K⁴). The assumed value of 10^6 W/sr for the radiation source intensity during the missile launch phase reveals a rocket exhaust temperature of about 3000 K. If the temperature (T) is higher or lower, the source intensity J will follow the proportionality law of T^4, as shown in the parentheses above. Computed estimates of detection range as a function of various parameters are shown in Table 10.5.

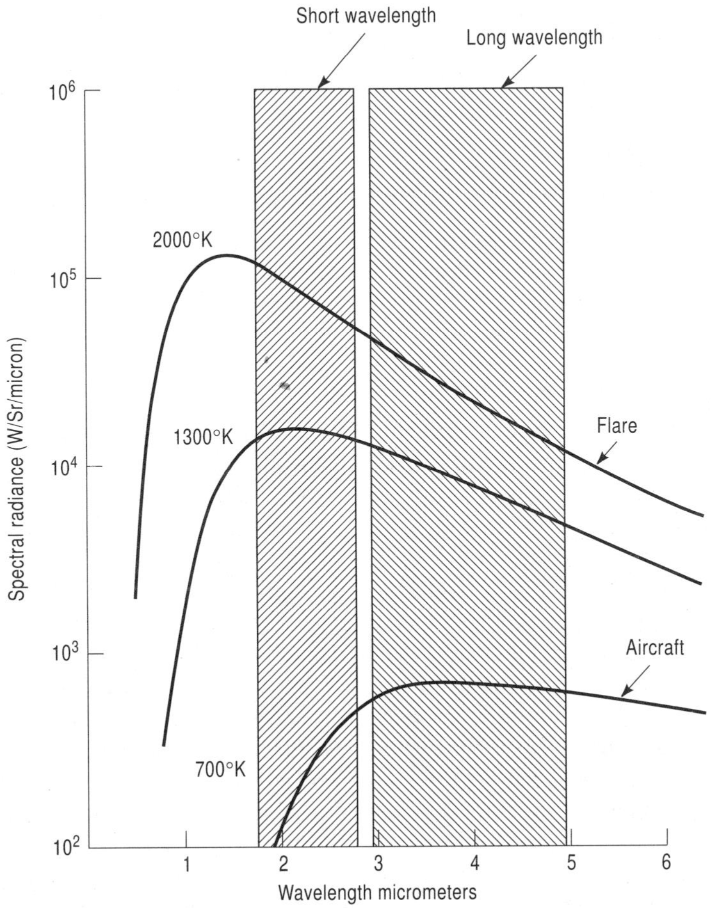

FIGURE 10.10 Typical radiance levels from various IR sources as a function of wavelength and source temperature.

Preliminary calculations indicate that a cryogenically cooled PbS detector at 77 K and InAs detector at 4.2 K with a detectivity (D^*) of 10^{11} cm $\sqrt{\text{Hz}}$/W over the 0.2 to 3.2 μm spectral range can provide a detection range of 15,000 NM with a 0.1° × 0.1° detector's IFOV and 4870 NM with a 1° × 1° detector IFOV. All these calculations assume atmospheric transmission efficiency of 90%, optical efficiency of 80%, numerical aperture of 0.25, a voltage S/N ratio of 14 dB, and surveillance receiver noise bandwidth of 100 Hz.

10.6 INFRARED COUNTERMEASURE (IRCM) TECHNIQUES

Threats posed by hostile IR missiles, cruise missiles, short-range missiles, and manned fighter/bomber aircraft must be neutralized to protect costly tactical assets including the defense and offensive weapon systems, command and control centers, runways, and bridges. To provide adequate protection, it will be necessary to locate and detect IR threat sources and deploy appropriate techniques to neutralize them effectively. IRCM techniques include IR decoys or flares, IR deception, IR noise jamming and other concepts that may be developed in future.

10.6.1 Passive IRCM Techniques Involving Flares and Decoys

Passive IRCM techniques involving flares or decoys are relatively cost-effective if the precise type and nature of IR threats are known. These techniques do not require large power consumption, thereby providing significant savings in fuel and other accessories. The flares act like decoy IR targets, forcing the incoming missiles away from the strike/fighter or reconnaissance aircraft. The flares can be designed to be most effective against a specific IR threat. The conventional flare shown in Figure 10.11 could be most effective against a fighter aircraft operating in an afterburner mode because its spectral radiance level is much greater than that of the aircraft under consideration. The plume temperature could be greater than 1400 K at aircraft speeds exceeding 2 Mach, depending on the atmospheric conditions and altitude of the target aircraft.

The flare is designed to operate as a black body radiator (BBR) that provides a high-intensity radiation source of specific spectral radiance level and radiant intensity. The BBR shown in Figure 10.5 features a carbon receptacle into which a main charge is placed. An initiating firing device is attached and sealed to the carbon receptacle to provide a moisture-resistant enclosure. When properly initiated, the pyrotechnic charge heats the carbon receptacle containing the thermite pellet that generates an intense IR signal for a minimum period of 2 min at a steady radiation intensity exceeding about 90 W/sr. The pyrotechnic reaction provides an IR signal free of smoke or flame interference. Its typical characteristics include a reaction time less than 10 seconds to initiate an IR output signal of intensity greater than 75 W/sr in 30 sec in three to five selected spectral bands compatible with IR threats, and minimum peak radiation intensity of 90 W/sr. The flare typically weighs around

3.5 lb and several flares can be stored in a cylindrical housing with specific firing sequence.

To provide an effective IRCM capability, a large number of flares are required in the proximity of the target to be protected against the IR missiles. Low-cost conventional flares (Figure 10.11) offer reasonably good performance against most high-performance fighter/strike aircraft. Since, such protection is limited only to a couple of minutes, the tactical aircraft will not receive protection throughout the remaining period of the mission. An active IRCM technique or IRCM dispenser system capable of ejecting multiple flares simultaneously is required if continuous protection from a fighter/strike aircraft or bomber is the principal objective. The latter technique offers a tailored response best suited to a partial tactical situation.

Deployment of the flare technique has some limitations. Flares require confirmation of missile launch either visually or with an electronic missile launch detection system. Furthermore, the flares offer protection only for a short-duration operation. Excessive false alarms and lengthy mission duration leave no protection against air-to-surface IR missiles or strike aircraft.

10.6.2 IR Signature Reduction Techniques

IR signature reduction techniques are passive in nature, but involve sophisticated design of the exhaust section of the aircraft engine to suppress the IR radiation level from the hottest part of the engine as well as the tailpipe. The exhaust section containing the tailpipe must be designed in such a way that directs the hot exhaust gases in several directions, thereby significantly reducing the IR intensity level in a specific direction. The exhaust section design can incorporate a structural feature that bleeds a portion of the hot gases for reheating the input airflow to the compressor, thereby realizing further reduction in the IR radiation intensity. Diverting the

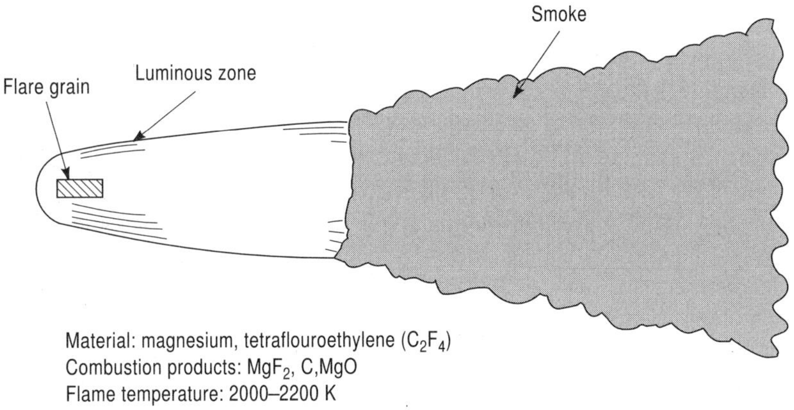

FIGURE 10.11 Specific details of a flare used for IR countermeasures to jam an enemy IR missile seeker receiver.

direction of the hot gases could affect the jet engine efficiency to some extent, which is not fully known. This technique could be equally effective in reducing the IR signatures of rocket engines, cruise missiles, air-to-air radar guided missiles, and air-to-surface missiles. Application of this particular technology may not be attractive for SRBMs, MRBMs, or ICBMs, because it can have an adverse impact on the rocket thrust level during the missile launch phase.

Injection of appropriate chemical agents into the exhaust gas flow leading to the exhaust nozzle could suppress the IR signature to some extent. This technique could alter the actual spectral radiance level. However, this technique is very complex and could affect the integrity of the exit nozzle. The effectiveness of this particular technique is not yet fully known.

10.6.3 Active IRCM Techniques

As stated earlier, active IRCM techniques are effective in providing continuous protection to a tactical aircraft, but they are complex and expensive. For a typical airborne IR jammer, radiating area is most critical. This means that smaller but high-temperature IR sources with variable spectral radiance capability must be used to neutralize the IR threats posed by the IR missiles including surface-to-air (SAM) missiles and air-to-air heat-seeking missiles or tactical aircraft. Today's IR jammers [4] use two different source/modulation concepts: the mechanically modulated constant temperature source and electronically modulated alkali metal vapor lamp. The first type generally uses silicon carbide or doped graphite sources typically operating between 1725 and 2000 K [4]. The electronically modulated IRCM systems typically employ cesium- or rubidium-filled lamps capable of achieving temperatures ranging from 3600 to 4000 K per pulse. These lamps produce maximum IR energy per unit area and are most suited for high-performance tactical strike aircraft or heavy bomber applications. In the case of cavity-based jammers, fuel is injected into a cavity and ignited and the source is heated to produce IR energy, which is directed against the incoming missile to introduce errors in the missile track/guidance system.

Tactical strike and reconnaissance aircraft are vulnerable to unwarned IR missile attack from covert installations. High-power active IR jammers provide the pilots the needed protection during their critical tactical missions. The guidance unit of the IR homing missile acquires and tracks the IR energy emitted from the hot engine parts radiated in both the AZ and EL planes, as illustrated in Figure 10.1. The principal function of the active IR jamming system is to confuse the tracking and guidance unit of the missile, thereby causing it to miss the intended target by some reasonable distance. The active IR jamming system carried by a tactical aircraft transmits the correctly modulated IR signal that negates the tracking capability of the hostile missile.

Since the jammer IR source is located not too far from the aircraft engine or engines, the IRCM's effectiveness is range-dependent, as both source intensities decrease as inverse functions of the slant range to the missile. This means that if the active IRCM source is four times more intense than the aircraft engine exhaust and it has the same transmitted patterns, it will be four times more intense at all points in space. In the case of a tactical situation, the IR jammer is expected to be in continuous operation and, thus, the tactical aircraft carrying the jammer will have continuous protection [5] throughout its mission. It is important to mention that the designer of an IR jammer must pay special attention to maintenance, reliability, and part-replacement aspects, which are considered most critical under battlefield environments.

A block diagram of a low-cost, pod mounted, ram air turbine generator (RATG) active jamming system [6] is shown in Figure 10.12. This system can offer effective IRCM capability against threats posed by SAMs, air-to-air missiles, strike/fighter aircraft, and helicopters. The system illustrated in Figure 10.12 shows the system extended on its slide rails to the designated maintenance position. The RATG [5] design configuration makes the system completely independent of the host aircraft

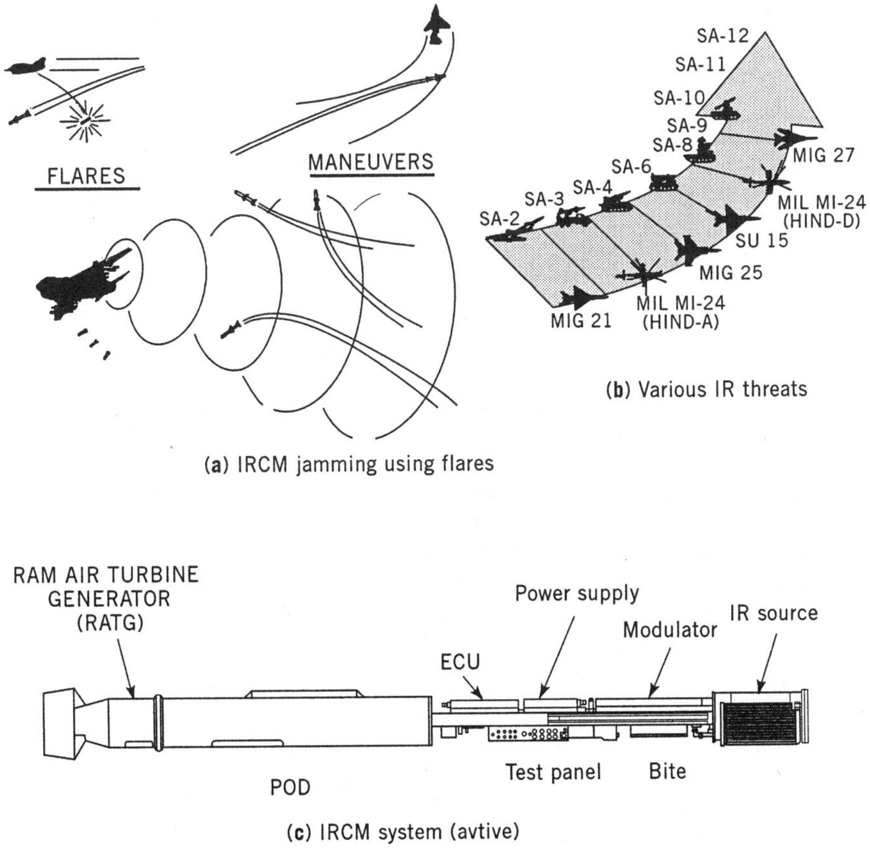

FIGURE 10.12 Passive (A) and active (C) IRCM techniques to counter the threats (B) posed by various tactical missiles and fighter aircraft.

electrical power system. A liquid-to-air heat exchanger [5] with fins is mounted adjacent to the IR source, providing cooling for the modules including power supply, modulator, electronic control unit (ECU), and BITE circuit, all mounted on the cold plate, except for the IR source. The ECU provides for preprogrammed start-up of the IR lamp power for standby operation and for a low-frequency generator to supply appropriate trigger signals to the modulator. The intensity of the IR jamming power over the required spectral band can be maintained to neutralize the threats posed by IR missiles or tactical aircraft.

10.7 ELECTROOPTICAL THREAT WARNING (EOTW) SYSTEM

Incoherent and coherent electrooptical (EO) active jammers have been developed to foil the accuracy of the radar, laser rangefinder, and laser-guided weapons [7]. Before any jamming can be applied, the EOTW system is required to detect and locate the threat in question. A laser warning receiver (LWR) can pinpoint the location of a hostile laser radar or laser-illuminated missile by detecting the sidelobe energy diffracted off axis into the far field pattern as well as from the atmosphere. A LWR requires several hundred detectors to meet the wide field-of-view (WFOV), spectral bandwidth, angular resolutions, and false-alarm rate requirements. A LWR is capable of detecting, discriminating, and locating of multiple threats in a mixed weapon environment.

A laser homing and warning system (LAHAWS), shown in Figure 10.13, uses CCD-based imaging arrays instead of a large array of avalanche or PIN photodiodes to provide the necessary angular resolution over the desired FOV with minimum cost. The LAHAWS uses a pair of PIN photodiodes only to detect the presence of a radiating laser source; the location of the laser threat is accomplished by a 100 × 100 element CCD array [7], which provides an output in the x×y coordinate system for the CRT display. When a laser source is detected, an audible and visual warning is given to the operator. This system provides a target location accuracy of about 1% using the 100 × 100 element CCD array. Higher accuracy is possible with a 256 × 256 element CCD array, but at the expense of additional cost and complexity, since frame-on-frame subtraction is performed with digital memory. Since the CCD array generates a raster-type analog output, it can be converted into a digital code using an A/D converter (ADC). The number of bits in the ADC is determined by the ratio of the CCD saturation level to the system noise level, the dynamic range of the CCD array, and the false-alarm rate.

The LWS accuracy can be improved using a higher-element array comprised of 244 × 190 or 488 × 380 elements. Wide dynamic range is possible with an ADC on-chip amplifier that permits low-light-level (LLL) operation at room temperature (300 K). The technology that is used by the CCD array consists of a combination of an integrated circuit (IC), silicon gate technology, and buried channel concept. Large array range, high location accuracy, high design flexibility, and inherent high storage capacity of the CCD are the major advantages of the LWR system. The

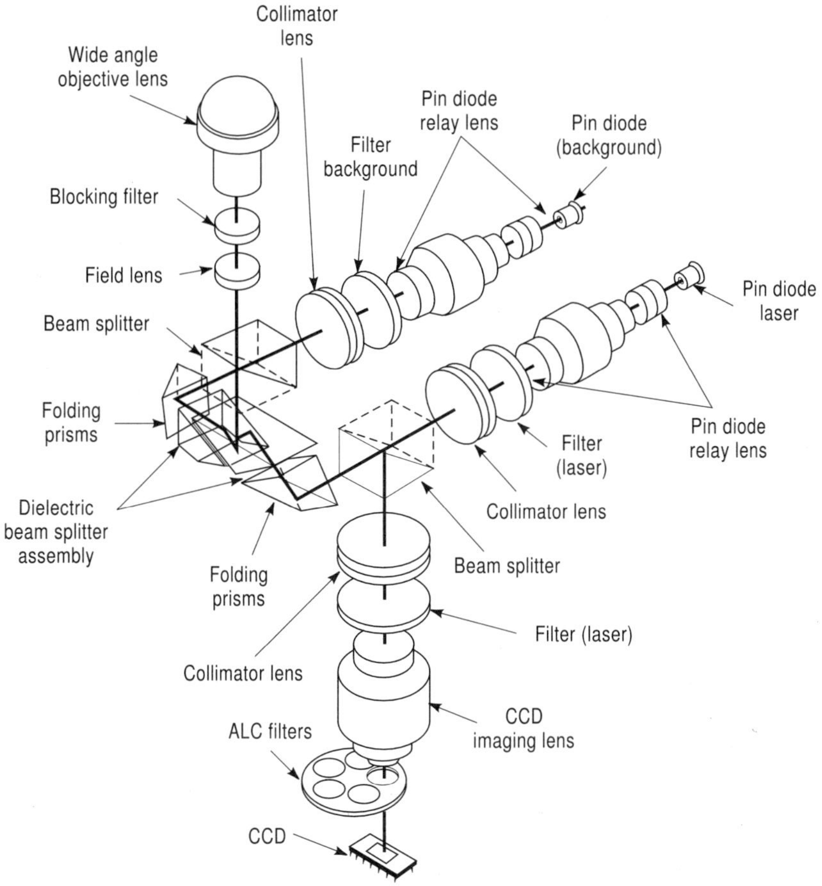

FIGURE 10.13 Laser homing and warning system layout (LAHAWS) showing critical elements of the system.

LWR provides quadrant warning and discrimination functions against laser rangefinders, beamriders, and designators. The sensor provides detection and identification of threats of specific wavelength with improved sensitivity and directional accuracy better than ±1°. Furthermore, the sensor uses a unique design configuration that provides IR clutter background rejection, high probability of detection, and low false-alarm rates to ensure the survivability of a tactical aircraft. Combinations of spatial, spectral and temporal filtering methods of discrimination are necessary to overcome the IR clutter background and to achieve respectably low false-alarm rates. The spectrum correlation technique used by the sensor offers identification of exact characteristics of the incoming radiation and dramatically enhances the background immunity of the IR warning receiver under severe clutter environments.

10.8 LASER-BASED DIRECTIONAL IR COUNTERMEASURE (DIRCM) SYSTEM

A laser-based DIRCM system [8] provides missile warning, detection of missile threats, and activation of countermeasures to defeat impending missile threats including IR-guided SAMs and advanced IR missiles with "all-aspect approach" capability. There are two fundamentally different design concepts for the DIRCM system. One is called an "open-loop" system; it attempts to disrupt or confuse the IR missile seeker by directing the IR energy beam at it. The IR energy can be generated by low-cost, noncoherent sources such as flash lamps or coherent lasers. The other technique, known as "closed-loop," requires a sophisticated laser system called "SMART DIRCM" [9]; it first bounces laser energy off the missile seeker and then analyzes the IR return signal to determine the type of IR missile. After identifying the IR missile type, the DIRCM selects [8] a laser modulation that is most effective in neutralizing the threat posed by a particular missile. The "closed-loop" system design is more flexible and is better capable of countering the threats posed by new and more sophisticated IR seeker designs; nevertheless, it is more complex and heavy and requires a powerful laser. A solid-state laser with output power level of 10–20 W can be adequate to obtain the return echo from a seeker.

A conventional IRCM system simply cannot radiate enough energy in all directions to provide all-aspect jamming of modern IR threats. However, a DIRCM system is most effective in detecting, acquiring, and tracking IR missile launches and then accurately focussing intense IR energy in the spectral region of the missile. This system is designed to provide fast, accurate location of an approaching missile. The missile warning subsystem of the DIRCM system detects the missile launches and performs a threat assessment during the missile flight duration. The DIRCM system [8] consists of a high-power laser, arc-lamp-based modulator, high-intensity IR jammer with 5-degree beam, four-axis pointing and fine-tracking IR (FTIR) sensor, and passive missile warning system (MWS). The MWS sensor permits detection of the missile plume, tracking of multiple energy IR sources, defeating algorithms, classification of missiles under IR clutter environment, fine angle-of-arrival (AOA) information, and rapid and accurate target hand-off input to the IRCM pointing and tracking subsystem. The fine AOA capability provides detection ranges almost twice those of existing passive systems, regardless of the weather, altitude, and IR clutter environments.

The MWS offers extremely high resolution at ranges exceeding 10 km [13] and permits tracking of IR missiles through all modes of operations including its post-burnt out phase. During the track operation, the system electronically processes the incoming missile image based on its IR signature. The electronic image is then used to lock the IR jammer transmitter on the incoming IR missile. The high intensity jamming energy will destroy the missile guidance system, thereby terminating the missile threat.

The DIRCM system is not only capable of defeating modern IR SAM threats, but also offers a provision for rapid upgrading to defeat future sophisticated SAM threats. With a built-in laser pathway and modular system design, the DIRCM can be equipped with high-power laser source capable of neutralizing the threats posed by laser-guided missiles.

10.8.1 Missile Warning Receiver (MWR)

The missile warning receiver (MWR, also known as MWS), is the most critical subsystem of the DIRCM system. As stated earlier, the MWS detects missile plumes under various phases of missile flight, provides fine AOA information regardless of IR clutter environments, and classifies missile types within the shortest possible time frame. Computed detection ranges by a MWS as a function of source intensity and temperature are shown in Figure 10.14, using the assumed parameters specified therein. MWSs can use either IR sensors or UV sensors [10]. An IR-based MWS can detect the IR emissions from a missile's rocket plume at a greater range than a UV-based MWS. Some current IR sensors suffer from high false-alarm rates, particularly at low altitudes, due to battlefield heat sources. UV-based sensors are less vulnerable to IR clutter at low-altitude operation; hence, they are best suited for protection against shoulder-fired IR missiles such as Stingers. MWSs using UV-type sensors will be most effective in protecting helicopters and low-flying military transports. Current UV sensors do not use cryogenic cooling, which could make them more susceptible to false alarms when used for detection of high-speed targets. On the other hand, IR sensors require cryogenic cooling to achieve the high sensitivity needed for detection of missiles at long distances and such sensors are

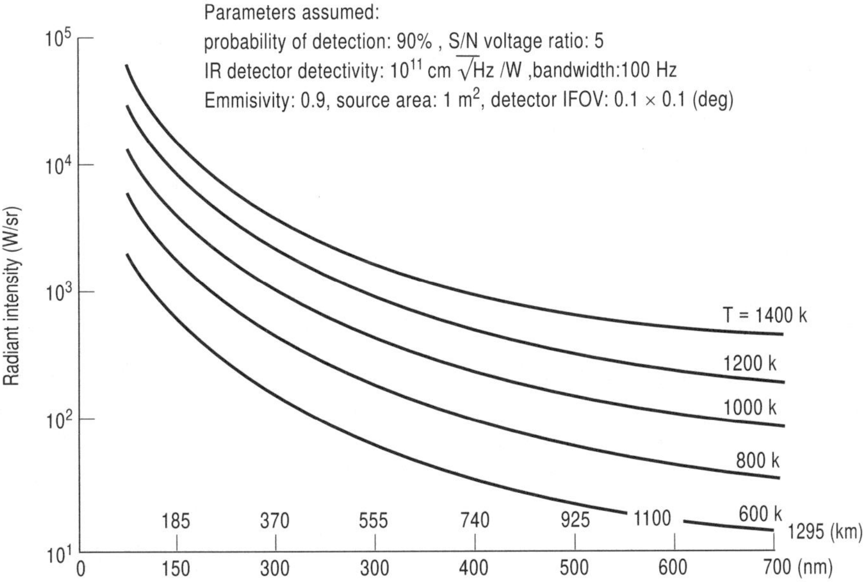

FIGURE 10.14 Calculated detection range capability of a missile warning system as a function of IR source temperature and radiant intensity.

best suited for detection of targets such as high-performance, high-speed strike aircraft. The false-alarm problem for IR sensors has been eased significantly by using a "two-color" (multispectral sensor) IR sensor that measures IR emission from the rocket plume at two different wavelengths with appropriate separation in the mid-IR spectral region (3 to 5 μm) corresponding to approximate rocket plume temperatures of 966 K and 580 K, respectively. With the latest technology, the MWR can be housed in a single 20–30 pound assembly that includes all electronics, MWS computer capable of providing all necessary computations, a two-color rotating scanning mechanism to provide full spectral sensing, and associated accessories. Higher scan rates are desirable to provide more rapid response needed against high-speed, short-range IR missiles operating over the 3 to 5 μm spectral region of the infrared zone.

It is interesting to mention than an infrared search and track (IRST) sensor is capable of providing MWR functions with the same accuracy and reliability. The passive IRST sensor can detect IR emissions from the rocket engine plume of an attacking missile. The IR focal planar array (FPA) detector used by the IRST sensor provides continuous high-resolution spherical coverage. Such a sensor is best suited to provide needed early warning, defensive search and track functions, target cueing, bomb damage assessment, and navigation under IR threat environments. FPA detectors are available with 256 × 256 elements to provide satisfactory resolution capability. An IRST sensor equipped with a cryogenically cooled, 512 × 512 element FPA detector will effectively improve the resolution by a factor of two. Multiple IRST sensors that can operate in two different IR bands, namely, 3–5 μm (mid-IR) and 8–12 μm (far-micron), will provide effective IR threat warning coming from the full 360° azimuth sector and reasonable elevation sector. The mid-IR FPAs are best suited for detection of rocket plumes, whereas the far-IR FPAs are best for tracking a missile after rocket burnout. It is important to mention that IR-based FPAs provide the sensitivity and ability to see through light rain and fog. Based on these statements, it can be concluded that both the IRST and the missile warning receiver using cryogenically cooled FPAs will provide significantly improved missile warning performance, irrespective of weather, aspect angle, and IR clutter environments.

10.9 IR COUNTER-COUNTERMEASURE (IRCCM) TECHNIQUES

Sophisticated IR-guided missiles are designed with built-in capability to detect real IR-emitting targets while rejecting flares. IR missile seekers use IR energy emitted by the target as a tracking source. All objects emit IR energy, depending on the body temperature, emissivity, and surface conditions. The hotter the body, the higher the emitted energy level. As the temperature rises, the wavelength of the peak emission becomes shorter. A tactical aircraft will have a peak energy emission in the 1.5 μm range, whereas the aircraft operating under military power scenarios will have peak emissions in the 3 μm region. Sharpness of the peak energy response and fluctuation in the intensity level would make it possible to differentiate between the real IR target and the flare. But this method requires a comprehensive knowledge of flares and IR target signatures.

10.9.1 IRCCM Capability Offered by Various Design Aspects of Missile Seekers

The earlier IR missile seeker types [11] included spin-scan, rear-aspect, and hot-spot tracker mechanisms to "bang bang" the control surfaces. These missile seekers evolved into conical-scan trackers with all-aspect tracking capability and proportional navigation guidance. Imaging missile seekers are being used by several adversaries because of their minimum cost and low-cost technology insertion.

Practically all spin-scan seekers use amplitude modulation tracking. In spin-scan seekers, the IR emitted energy from the target is collected by the seeker optics, focused through a rotating reticle, detected by an IR detector array, and processed to derive the amplitude and phase of the target signal. The IR clutter rejection network is included in the processing unit to eliminate background clutter from clouds, ground, and aerosols. The error signal is passed on to the guidance control unit, where it is used to control the missile guidance fin mechanisms.

Practically all air-to-air (A/A) missiles use proportional navigation guidance laws. In a proportional navigation system, the missile leads the target by flying to an intercept point projected in front of the target, which provides the missile the shortest flight path to the intercept point with minimum maneuvering. To fly such a flight path, the missile seeker's line-of-sight (LOS) rate with respect to the target must be zero. This keeps the seeker at a constant angle relative to the incoming missile fuselage. Target maneuvering and missile velocity will change the seeker look angle required to fly the proportional navigation path. The missile guidance unit commands the missile to fly a path driving the tracking error signal to zero. When a zero error signal is achieved, the missile will be on a proportional navigation path.

Most current IR missile seekers seem to use cryogenically cooled InSb detectors or their equivalent to detect the longer-wavelength radiation. Cryogenic cooling provides improved detection range and offers all-aspect tracking capability. In case of center-spun, spin-scan-based missile seekers, the seekers are relatively insensitive when the target is in the center of the seeker scan. An error signal is generated only when the target falls away from the center of the reticle, causing the missile to enter a path leading to the target.

Conical-scan missile seekers solve some of the problems associated with spin-scan seekers, notably the lack of error response when the target is near the center of the seeker FOV. In a conical-scan seeker, the reticle is fixed and does not spin. Instead, the secondary mirror is tilted and spins, causing the target image to scan in a circular path around the outer edge of the reticle. Conical-seeker optics generate the greatest amount of FM modulation for a given tracking error, thereby providing the most sensitive and tight tracking loop desirable for optimum performance.

Imaging missile seekers do not use a reticle. Instead, they use an array of detector elements that detect IR energy from the scene and generate a spatial map of that scene. An image can be formed with a two-dimensional array of detectors that stares constantly at the scene.

The latest versions of some of these missile seeker configurations offer reasonable built-in IRCCCM capability, but the effectiveness of the IRCCM varies from one seeker to another. Because of the sensitivity issue, discussion on the degree of IRCCM effectiveness for these IR missile seekers is beyond the scope of this book.

10.9.2 Potential IRCCM Techniques

IR missiles need sophisticated circuit technology and exotic signal processing capability to detect the presence of flares and reject them as false targets while allowing the seeker to continue tracking the real target. The IRCCM techniques for possible integration into a missile seeker must include two critical components: The first is "flare switching," which detects the presence of a flare in the seeker FOV and activates the second component, called the "response" circuit, which asks the missile seeker to take immediate action to reject the radiation from a flare. To be more effective, a missile must detect the flare before initiating any IRCCM action. When a missile seeker detects a flare, the seeker will switch on the IRCCM "response" circuit to reject the flare.

Switching techniques to reject the flares include rise time, two-color approach, and kinetic and spatial discrimination. An IR missile using rise time switching monitors the energy level of the target it is tracking. A sharp rise in the received IR energy expected within a specific time indicates a flare presence in the seeker FOV, because of significantly higher IR energy (about two times) than the target. A seeker with a temporal discriminator would switch on the IRCCM if the seeker detected an IR energy increase above a present threshold (i.e., flare-to-target energy level greater than 2.5:1) within a preset time limit. A threshold of a 2.5:1 energy increase within 40 to 50 msec is generally sufficient to indicate the flare detection in all aspects, while ignoring the relatively slow energy rise time from an afterburner ignition, which normally takes a minimum of 1 to 2 seconds. The IRCCM must be switched off when the received IR energy level drops to a preset value indicating the absence of the flare in the seeker FOV.

A missile seeker equipped with a two-color switch samples the energy level in two or more different spectral bands. A sudden jump in band A energy compared to band B energy indicates the presence of a flare in the seeker FOV. The seeker then uses two different detectors with optimum sensitivity in the bands (for example, PbS detector for the short band and InSb for the long band) or uses a detector integrated with two bandpass optical filters on the reticle spokes.

It is important to mention that flares separate very quickly from the dispensing aircraft due to their higher aerodynamic drag. In a beam-aspect engagement, the missile seeker will have a large sudden change in the LOS rate due to rapid deacceleration of the flare. The kinematic switch senses the rapid change in real motion between the missile and targets, indicating the IRCCM response. Kinematic discrimination tends to have difficulty in head-on and stern engagement geometries.

Advanced IR missiles with a spatial discrimination switch use the physical separation of the flare from the target to discriminate between the two. As the flares separate from the rear of the aircraft, the missile seeker will see the target on the forward side of the FOV and the flare on the rear side of the FOV. Once the two hot objects are observed on opposite sides of the FOV, the IRCCM mode switch is "on" position. Most IR seekers have a FOV that is less than 2.5°. At long ranges, the flare will remain in the seeker FOV, whereas for short-range engagements, the flare will be only in the FOV for a very short time. Furthermore, as long as the flare is in the seeker FOV, the missile is not tracking the target.

Electronic FOV gating techniques can be used to discriminate between the real IR target and flare. At some time after the flare is dispensed, the target and flare will no longer be in the same lobe of the seeker scan pattern. The time phase blanking concept can be used with seekers that have multiple detector elements. The detector will put out a pulse when the target is detected in its IFOV. The time between pulses for different detectors will remain the same for a single target centered on the seeker FOV. Since the seeker and the target do not occupy the same point in the seeker FOV, the output of the detector due to the flare pulse will not arrive at the time the missile expects to see a pulse. The missile seeker only acquires the pulses that arrive at the expected or predetermined time and rejects all others, thereby rejecting the flare in the seeker FOV.

Imaging seekers are inherently highly resistant to flare countermeasures. Flares will not easily decoy them, because the point source simply does not look like an expanded aircraft with an exhaust plume. Furthermore, implementation of software processing with discriminating algorithms within the seeker would discriminate between the real IR target and the flare decoy.

10.10 SPACE-BASED IR (SBIR) SURVEILLANCE SENSORS

SBIR surveillance sensors will provide detection, tracking, and surveillance of space-orbiting objects, hostile satellites, and weapons of mass destruction. A distributed LEO satellite constellation consisting of SBIR surveillance sensors will provide precision midcourse tracking of hostile space-based weapons that is critical to effective ballistic missile defense and will also provide advanced warning to take appropriate and swift action to destroy or disable the missile while in flight. The baseline system can contain constellations of a few LEO satellites, each equipped with high-performance IRST and visible optical sensors to provide continuous detection and tracking of various space targets including IR missiles and other satellite-based nuclear weapons. This SBIR surveillance system might be comprised of an IRST sensor and missile warning sensor equipped with advanced processor including discriminating algorithms and cryogenically cooled FPA detector array with detectivity better than 10^{11} cm $\sqrt{\text{Hz}}$/W.

10.11 SPACE-BASED ANTISATELLITE (SBAS) SYSTEM

A space-based antisatellite (SBAS) system has potential application to destroy incoming high-altitude missiles or satellites with hostile intent. A megawatt-class space-based chemical oxygen iodine laser (COIL) system is best suited for such application [12]. Studies performed by the author indicate that a megawatt-class deuterium fluoride (DF) chemical laser operating in the 3.6 to 4.2 μm range can be an excellent choice for an antisatellite weapon system. One megawatt of CW power can be adequate to destroy a refrigerator-sized satellite orbiting at 260 miles above the earth. The space loss is proportional to operating wavelength. Preliminary calculations reveal that a two-way loss for a CO_2 laser (10.6 μm) is about 30 dB, for a DF laser (3.6 mμ) about 5 dB, and for a COIL laser (1.315 μm) around 4.2 dB. Therefore, even putting a 100 kW, mid-IR, chemical laser operating at 3.6 μm aboard a satellite or space station can destroy a nuclear-armed hostile missile or satellite. Comprehensive examination of space losses at various wavelengths indicate that a chemical laser operating in the mid-IR spectrum can avoid high atmospheric attenuation and distortion of the laser beam at LEO satellite altitudes. The space loss at 350 km altitude is about 0.0002 dB/km due to absorption and scattering and, thus, lasers with output power levels of 100 kW will be sufficient for such applications.

The SBLS system requires compact optics, high-efficiency cooling scheme, and large amount of input power, even for a 100 kW laser. Availability of electrical power even to supply a 100 kW laser operating in space presents challenging problems. Most of the power in space is only available from solar panels with conversion efficiencies ranging between 20 to 25% for state-of-the art photovoltaic cells. Currently, a spacecraft equipped with a large number of solar panels is capable of providing 15 to 20 kW of electrical power, which is just sufficient to meet the power requirements for sensors and auxiliaries in the spacecraft. Rapid advancement in the state-of-the art solar cell technology is necessary to achieve 20 to 50 kW of power from solar panels attached to a large spacecraft, which may be just sufficient to drive a 10 kW (CW) laser transmitter. Because of the nonavailability of electrical power in space, it will be extremely difficult, if not impossible, to equip the space station or a satellite with a megawatt-class CW chemical laser in near future. However, a mid-IR chemical laser with peak power of 10 kW operating at 2% duty factor can be used in a commercial imaging sensor or a reconnaissance satellite capable of providing high-quality images with 1 meter resolution.

10.12 SUMMARY

Mathematical expressions are provided for computation of IR signatures of various man-made sources. Potential IR sources and sensors for military applications are identified. IR emission levels from various sources as a function of temperature, emissivity, and surface condition are calculated. Infrared intensity levels from missile rockets as a function of temperature and thrust are computed using the MathCad computer program. Radiative power, radiation intensity, and radiant emittance as a function of exhaust temperature and exit nozzle area are calculated for commercial and military jet engines, assuming relevant parameters. Percentage of total radiant emittance over specific IR bands are computed at various temperatures ranging from 500 to 3000 K. Skin temperatures for aircraft and missiles as a function of speed are calculated for jet engines and cruise missile rocket engines. Flight envelopes for ramjets, scramjets, turbojets, and air-turbo rockets (ATRs) are identified based on thrust and speed. Computer simulation techniques using analog method to compute the IR signature of a jet engine as a function of temperature, thrust, and aspect angle are described. IR signature contributions from engine cavity, compressor and turbine stages, airframe, and tailpipe sections are briefly discussed. Missile lock-on ranges are computed, assuming appropriate values of critical parameters such as target spectral radiant intensity, background radiance level, missile spectral sensitivity, and noise equivalent input (NEI). Detection ranges for SRBM, IRBM, and ICBM missiles are calculated as a function of detector IFOV, optics size, source intensity, detector sensitivity, and relevant efficiency parameters. Passive and active IR countermeasure techniques and systems are discussed with emphasis on IR background clutter rejection techniques and false-alarm reduction methods. Laser-based DIRCM systems capable of providing missile warning, detection, and tracking of long-range IR missiles and activation of active countermeasures techniques to defeat the threats posed by IR guided A/A missiles, SAMs, and advanced IR missiles with all-aspect capability are fully described. Capabilities and critical performance parameters of missile warning systems are discussed, with emphasis on false-alarm rates and probability of detection. Performance capabilities of IR seekers are briefly summarized with emphasis on seeker FOV and IR discrimination techniques. Flare rejection and discrimination techniques in the design of missile seekers are discussed, with particular emphasis on false-alarm rates and the two-color discrimination concept. Performance capabilities and limitations of space-based IR surveillance systems are summarized. Operational capabilities of ground-based laser surveillance systems are identified, with emphasis on their critical system elements. Conceptual design aspects and technical risks involved in the design of a space-based antisatellite laser system are described, with emphasis on limited electrical power available from solar panels.

NUMERICAL EXAMPLE

Compute the radiation intensity as a function of temperature and surface emissivity with the following assumptions:

Temperature (T): 600, 800, 1000, 1200, 1400, and 1600 K
Emissivity (ε): 0.8 and 0.9
Area (A): 0.1 and 1.0 m^2
Parameter (σ): (5.67×10^{-12}) W/cm^2 K^4
Radiation intensity (J) = $[(\varepsilon)(\sigma)(T^4)(A)]/(\pi)$ W/sr

Inserting the assumed variables and constants in the above, one gets the computed values of radiation intensity J (W/sr) as shown in the table below.

Radiation intensity as a function of temperature, emissivity, and surface areas of the emitting source (W/sr)

	$\varepsilon = 0.8$		$\varepsilon = 0.9$	
Temperature (K)	$A = 0.1$	$A = 1$	$A = 0.1$	$A = 1$
600	150	1496	168	1683
800	473	4729	532	5320
1000	1154	11,543	1299	12,986
1200	2394	23,936	2693	26,928
1400	4434	44,344	4989	49,887
1600	7565	75,648	8510	85,104

REFERENCES

1. SCORPIO-N Program by General Electric Company: To predict the IR signatures from aircraft engines and surfaces. 1978–1979. Technical Report, G.E. Aircraft Engine Division.
2. M. A. Dornheim et al. "Missile makes eyeing ramjets." *Aviation Week and Space Technology,* p. 96, 7 September 1998.
3. R. Brahain, Spectra Editor. "The new global threat." *IEEE Spectrum,* pp. 21–32, March 1999.
4. J. Career. "Active IRCM provides continuous protection for today's aircraft." *Infrared Countermeasures Surveillance,* pp. 343–347, 1982.
5. R. Farmer. "AN/ALQ-123 IRCM pioneer." Electro-optical Systems, Xerox Corp., Pasadena, CA, pp. 45– 49.
6. Contributing Editor. "IRCM jamming techniques." *EW Journal,* pp. 33–45, December 1975.
7. H. Sadowski. "CCD: A candidate technology in electrooptical countermeasures." *EW Journal,* pp. 65–74, January/February 1976.
8. Contributing Editor. "USAF plans "SMART" laser-based DIRCM." *Aviation Week and Space Technology,* pp. 40–41, 10 October 1994.
9. Contributing Editor. "NEMESIS: Imaging technology provides countermeasures to defeat the IR missile threat." *Defense and Security Electronics,* pp. 28–31, December 1995.
10. P. J. Klass. "British MWS tests to assess IR, UV sensors competition." *Aviation Week and Space Technology,* pp. 78–81, 30 March 1998.
11. Maj. C. M. Deyerle. "Advanced infrafed missile counter-countermeasures." *Journal of Electronic Defense,* pp. 47–50, January 1994.
12. D. Appell. "Military may test powerful laser in space." *Laser Focus World,* pp. 20–26, November 1997.

CHAPTER ELEVEN

Future Applications of IR and Photonic Technologies and Requirements for Auxiliary Equipment

11.0 INTRODUCTION

This chapter first describes the future applications of IR and photonic technologies in various fields. Later on, performance requirements of critical auxiliary devices and components such as control circuit, power supply, and cryocooler used by the IR systems are summarized. Future applications of IR, photonic, and optoelectronic devices and sensors are discussed, with emphasis on performance and cost-effectiveness. IR and photonic systems using these devices include communications equipment, nuclear power reactors, unmanned aerial vehicles (UAVs), remotely piloted vehicles (RPVs), mini-RPVs for tactical missions, photonic-based ring lasers, compact smart weapons, precision wind-profiling systems, laser-based micromachining tools, encryption schemes for secure communications systems, laser-based sensors for diagnosis of plumbing problems, ellipsometers, photonic-based food inspection sensors, laser-based underwater mine detection systems, and invasive diagnostic and therapeutic systems.

The last portion of the chapter is dedicated to define the performance requirements of auxiliary components and circuits such as power supplies, cryocoolers, temperature control circuits, thermoelectric coolers, and laser drivers. In addition, potential optical software programs are briefly discussed, with emphasis on algorithm generation, standardization of optical elements, and optical design and analysis.

11.1 LASER SCANNERS FOR UNDERWATER MINE DETECTION

Generally, sonar-imaging systems are used to search and detect underwater areas that might contain airline crash debris or undersea mines. In the 1999 Swissair crash just off the Nova Scotia coast, a sonar imaging sensor successfully located large areas of dense wreckage, but did not provide enough resolution to determine immediate priorities for divers to identify clues as to the cause of the accident involving the MD-11 aircraft. The investigators then turned to a high-resolution laser-line scanning imager that was developed specially for the detection of underwater targets in very shallow water. Three different 21 inch diameter imaging systems were deployed, namely, a low-resolution forward-scanning source, a high-resolution side-scanning sonar, and a high-resolution laser-line scan imaging sensor with a resolution of 0.25″ × 0.25″. It was the high-resolution laser-line scan imager that aided in the recovery of the Swissair MD-11 wreckage by retrieving images of the aircraft debris from the murky ocean floor.

In the case of mine countermeasures, a sonar is used first, and then a laser-line scan-imaging sensor is deployed, if and when warranted by the target of interest. A laser-line scan system (LLSS) uses a frequency-doubled Nd:YAG laser operating at 532 nm to illuminate a spot on the ocean floor. The wavelength of 532 nm is considered most attractive, because the optical signals between 450 and 550 nm transit through the ocean water with minimum attenuation. The laser spot size determines the image resolution, which strictly depends on optics size. However, the detection range is dependent on the gain of the photomultiplier tube (PMT).

The imaging system consists of four compact rotating mirrors capable of scanning the ocean floor and a PMT with low noise and high gain needed to detect the lowest amount of reflected light. As the scanning sensor housing is towed through the ocean water, the forward movement of the imaging sensor generates a two-dimensional (2-D) image using one thin slice at a time. The 2-D image is created the same way as in a laser-based computer printer with high-resolution capability. The entire sensor can be installed into a cylinder with a diameter of about 21 inches and a length of 30 inches. This sensor can be installed in different types of towed or remotely operated naval craft.

11.2 PHOTONIC-BASED ANTHROPOLOGICAL (PBA) SENSORS

PBA sensors are widely used to obtain anthropological information on molecular components and structures. Anthropologists have used gas chromatography–mass (GCM) spectrometry systems, scanning electron microscopes (SEMs), Raman spectrometers, and Fourier transform IR (FTIR) microscopes for noninvasive analysis of samples and small objects. The high cost of Raman spectroscopes (around $100,000) generally compel investigators to use the FTIR microscopes because of their low cost. However, the Raman microscope offers a resolution of 1 μm or 0.0001 cm and can be focused on single grating to determine whether a color is produced by one pigment or by a blend. Furthermore, this sensor can provide evidence of degradation when the white lead pigment is turned to black lead sulfide. Studies performed by the author on spectrometers indicate that the Raman spectroscope offers the highest resolution and this sensor must be preferred over other sensors if performance, not cost is the principal objective.

On then other hand, a low-cost photonic-based IR camera costing around \$25,000 and incorporating microbolometer and ferroelectric techniques is the next best sensor for such applications. These IR cameras eliminate the use of costly and bulky design aspects of cryogenic cooling. Even the state-of-the art, uncooled IR cameras cannot match the performance of the Raman microscope. Photonics often deliver nondestructive and in situ compositional analysis of fragile or otherwise nontransportable historical objects. However, conservation scientists who use Raman microscopic techniques on art works reveal that Raman microscopy offers great detail of the pigment grains, within a single micron cross-section, to see if they are the same or several distinct colors. This capability of the Raman spectroscopic technique makes it an order of magnitude or two better than infrared technology, which focusses on a wider spot (about 30 μm across) and delivers a broader signal mix and less spatial resolution.

Some conservation scientists feel that the tight resolution (1 μm) of the Raman microscope is extremely necessary in determining whether a color is produced by a single pigment or a combination of pigments or even the size of the grains. No other sensor can match or outperform the capabilities of the Raman microscope. These scientists further feel that a tight resolution is crucial to confirm the use of white lead as a pigment in a 13th century Byzantine illuminated manuscript [1]. The potential danger of Raman microscopy's fine-point laser accuracy is that it can thermally damage the specimen if excessive laser energy is abruptly applied. To prevent damage to the specimen, a neutral density filter must be used to decrease the laser power at the surface to less than 0.5 mW. Higher laser energy can be applied, depending on the thickness and type of material under examination. The Raman microscopic technique has been used by various scientists to analyze stained glass, ancient burial linens, and high-quality oil paintings with Nd:YAG lasers operating at 1064 nm and with power levels not exceeding 40 mW. Impressive results have been obtained at this power level without any damage to the specimens.

Some conservation scientists have used argon (485 and 514 nm) and krypton (647 nm) lasers with power levels not exceeding 5 mW for such applications. However, their resolution capabilities did not even come close to 50% of that of Nd:YAG lasers. Scientist who cannot afford a Raman microscope generally prefer a GCM spectrometry system or SEM. Raman microscopic (RM) techniques can detect inorganic, mineral-based pigments, which is not possible with other techniques. The RM method represents a lateral advance in archaeometric technology. Low-tech, low-cost methods such as UV lamps can hint immediately at the work of unscrupulous dealers who have made restorations and then tried to conceal their revisions. These dealers apply varnishes mixed with UV absorbers or phosphors, causing a varnished surface to appear uniformly dark or bright for a short time.

Some scientists have used vidcon IR cameras costing around \$8,000 to \$10,000, but they are less effective than current instruments and require more controlled environments to obtain an image. X-ray imaging systems can achieve similar results but suffer from excessive size and weight. Furthermore, in their registration of density and molecular weight, X-ray images will show many light-colored, high-molecular-weight pigments as dark, whereas areas covered with darker carbon-based pigments appear white.

11.3 PHOTONIC-BASED SENSORS FOR FOOD INSPECTION

Photonic-based sensors (PBSs) are widely used for rapid analysis and precision monitoring in the food processing industry. Food analysts are responsible for identifying the presence or absence of deadly bacteria in a food mixture or in a cooked item to avoid any problems before they crop up. Unreliable food analysis or inspection could cause food poisoning, leading to costly lawsuits by consumers. PBSs are best suited for reliable, efficient, and foolproof food analysis and monitoring [2].

Meat and grain processors, cake mix producers, brewers and ice cream makers all use Raman microscopes, spectrophotometers, FTIR spectrometers, and a whole array of other photonic instruments. However, a particular sensor is only best suited for a specific food analysis. Some sensors may be too costly and sophisticated for certain food processing applications. Various IR and photonic sensors will be briefly discussed, with emphasis on decontamination level, exposure time, energy density requirement, and the sensor's effectiveness.

11.3.1 PBSs for the Beer and Juice Industries

The Near-IR spectroscopic technique is best suited to study beer brewing and juice blending processes and to monitor important characteristics including alcohol content, color, turbidity, and food dynamics. Brewers generally use this technique to evaluate the quality of ingredients such as malt, cereal, and hops. Implementation of an acoustooptic tunable filter to a near-IR spectrometer will allow simultaneous measurements of sugar and acidity in juices. Furthermore, a 50 W halogen optical source operating over 1500 to 1900 nm spectral region is considered most effective for juice blending processes.

11.3.2 IR Sensors for Oil Analysis

Various spectrometers are used by manufacturers to analyze olive oils and seed oils. Generally, nuclear magnetic resonant (NMR), UV spectroscopy, and atomic absorption spectroscopy techniques are used to evaluate edible oils and their constituents. Analysis data obtained using NMR spectroscopy, IR spectroscopy, and Raman spectroscopic instruments indicate that the spectroscopic method is much faster than the traditional wet chemistry method of analysis, but with comparable results. The FTIR and Raman spectroscopic methods take only a few minutes, whereas the chemical methods take hours.

Raman spectroscopic and FTIR techniques are extremely useful where identification of geographical origin and detection of adulteration of virgin olive oils are of prime importance. With FTIR spectroscopy, food analysts are able to identify geographical origin easily, whereas the Fourier Transform Raman spectroscopy (FTRS) method is best suited for discrimination between authentic oils and adulterated mixtures. Analysis data indicate that that FTRS technique is rapid, reproducible, and quite reliable for the analysis of oils compared to traditional time-consuming, wet chemistry methods.

11.3.3 IR Sensors for Food Processing

Optical light microscopes or near-IR spectrometers are most ideal for the analysis of processed foods. Near-IR spectrometers get some results within few seconds, provided the machine is well calibrated. However, it can provide a definitive, foolproof analysis if one can wait for 30 seconds or more. Raman spectroscopy is more sophisticated than necessary, but provides more reliable and accurate results. Near-IR systems have been in service for decades in flour mills to measure protein and moisture content in minimum time. These sensors are used to make quantitative measurements of such foods as sugar, shortening, and flour. However, these sensors require significant amounts of work to calibrate the equipment, including testing dozens of samples to obtain an optimum operating wavelength for the food sample under test. But once the equipment is calibrated, the near-IR spectroscope works very quickly with reasonably good accuracy. Because speed is the most important performance parameter in the food processing industry, near-IR sensors are used for a wide range of foods such as cookies, cereal, chips, beer, sodas, and dog foods. Pet food suppliers are the major users of high-speed near-IR spectrometers because of use of scrap ingredients in their products.

IR food processing sensors provide accurate analysis of ingredients to ensure the right protein level. Food analyzers use IR energy in the 1100 to 2500 nm spectral region and food samples to be tested need not be specially prepared, as required with IR spectroscopy technology. That is why it is well suited for low-cost, conveyer-belt analysis. If the food analysts require more precise and quantitative measurements, they can turn to other IR sensors such as the FTIR spectroscope, which can get molecular information for positive identification of materials such as sugar or shortening. Some applications require special processes, depending on the quality and the quantity of the food involved. In the case of ice cream, video and digital cameras can be hooked to optical light microscopes to study images of ice cream layers with emphasis on the size and the distribution of the ice crystals. If food processors find a black streak in a food product, the analyst might first look at it with a microscope to determine the type of streak. If it is grease, one would look at it using the FTIR technique. If it is a metal, one could detect it with a scanning electron microscope.

11.3.3.1 Photonic Sensors for Food Safety

Imperfections in food samples due to inadequate baking or presence of strange colors will draw immediate attention if the imperfection indicates the presence of health-threatening bacteria. To provide an ironclad guarantee against food poisoning or presence of bacteria, immediate testing for pathogens and pesticides in food samples must be conducted. According to the food processing industry, detection of pathogens is necessary because they pose a greater health risk than pesticides. Photonic devices are available to detect various forms of bacteria, pesticide residues, and other forms of contamination.

The meat processing industry, in particular, faces serious concerns about harmful microbes and other contaminants. Scientists are developing state-of-the art photonic devices capable of detecting pathogens such as *E. coli* in meat within a matter of few hours, compared to several days or weeks required by current methods. Laser-based biosensors are best suited to detect *Salmonella* in meat products much more quickly than conventional methods can. Changes in the optical properties of meat can signal the presence of bacteria and the test results can be available in a matter of hours rather than days. Some analysts use UV imaging fluorometers and multivariate analysis algorithms to detect and identify protein, DNA, and bacteria. A hand-held detector uses laser-induced fluorescence spectroscopy to illuminate fecal contaminants, which is a major source of bacteria in meat.

Scientists are currently developing IR and near-IR lasers to kill dangerous bacteria such as *E. coli* in food and water. Sterilization can be achieved with a 200 W Nd:YAG laser operating at 1060 nm, but it requires a much longer exposure time and an energy density more than 500 times higher than that of a CO_2 laser operating at 10.6 μm. Sterilization of bacteria is also possible with a frequency-tripled Nd:YAG laser operating at 355 nm (1065/3 = 355). The most effective sterilization of bacteria is achieved with a 600 W CO_2 laser. Research scientists indicate that the Nd:YAG laser is best suited for small-scale decontamination of water, whereas a CO_2 laser is most ideal for decontamination of surfaces. Speed is the most critical factor as far as killing of *E. coli* bacteria is concerned. Heating of water to 50 °C for about 20 minutes will drop the bacteria viability by 90%, whereas an exposure to a 100 WNd:YAG laser for just 25 seconds will drop the viability by 99.9%.

11.4 LED-SOURCE FOR SEWER INSPECTION

Multichip LED (MC-LED) light sources operating in the lower IR spectral region (450 to 750 nm range) are available to develop new tools and capabilities badly needed by the plumbing industry and sanitation departments. Remotely controlled TV systems using MC-LED light sources have been designed for applications in sewers for underwater inspection of the pipes and drainage. Plumbers now can use these low-cost MC-LED sources to discover and clearly view in greater detail foreign objects in sewer or water piping systems. Older systems using single-chip LED (SC-LED) sources suffer from several drawbacks such as large size, high cost, and poor performance. The MC-LED source with its smaller camera permits rapid maneuvering in very small pipes and corner regions, which is not possible with other methods. The camera is connected to a flexible and rugged plumbing "snake" allowing it to negotiate into tight 90° bends and go deep into the pipes, which is not possible with existing systems. The new system, comprised of MC-LED source and minicamera, permits a clear view of cracks and blockage problems throughout the water system. This particular LED-based system allows plumbers to diagnose and video tape the trouble. The MC-LED source uses 48 LED chips distributing the light evenly across the scene for clear observation and detection of the obstruction or cracks in water pipes.

11.5 PHOTONIC-BASED SENSORS FOR SEMICONDUCTOR PROCESS CONTROL

With the current high production costs, the semiconductor industry requires effective and efficient inspection quality control practices. Reliable process control is essential to reduce product costs. Any number of defects can reduce wafer-producing rates during the more than 500 processing steps in the semiconductor industry. Processing techniques include crystal growth, ion implantation, metal deposition, and film deposition and polishing. By the time a wafer goes through lithography and etching and reaches the end of the product line, it may cost as much as $300. Critical processing steps involved in semiconductor manufacturing are shown in Figure 11.1.

A laser-scattering band-defect detection system, shown in Figure 11.1, measures the intensity of the dark-field signal at one or more angular locations with respect to the wafer by scanning the laser spot across the wafer. The wafer is translated under the laser-scanned beam, and a map of all the signal levels is accumulated. This process offers signal-to-noise ratio instead of image resolution. A hybrid approach is comprised of the above detection system and a linear CCD sensor [3] operating in time-domain integration mode to acquire an image of the scanned laser spot. The hybrid approach combines the sensitivity of the CCM imaging sensor with the speed of the conventional dark-field detection. Laser power stability, operating wavelength, spot size within tens of μm, improved detection efficiency greater than 50%, low electronic-noise floor better than –60 dB (less than 0.1%) of the detected signal, and dynamic range more than 60 dB are the most impressive performance parameters of the hybrid detection system.

To find a few submicron-size defects on a semiconductor wafer, a billion fields of view or scanned spots must be checked out. Effective defect reduction techniques involve a high-throughput laser scattering tool or a slower optical imaging tool to build a coordinate map of the defects in the wafers. A few of the these recorded defects must be reviewed under a high-resolution optical microscope or scanning electron microscope to classify the defect types. Analysis techniques such as secondary ion mass spectrometry, transmission electron microscopy, energy-dispersive X-ray, and Auger spectroscopy are available for further identifications of the composition of the defect. All the information obtained to this point is used to perform a forensic analysis to identify the source of the defect.

Metrology activities in semiconductor processes focus on microscopy, overlay registration, critical dimensions, film thickness, and profile measurements. Critical dimensions including line widths, contact diameters, and sidewall angles must be

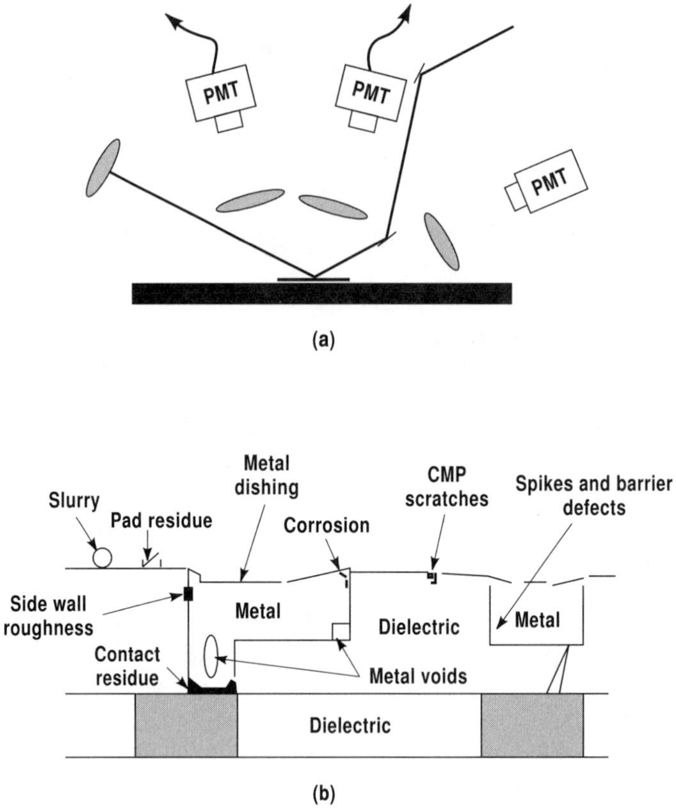

FIGURE 11.1 Laser-based (A) defect detection system; (B) processing steps in semiconductor wafer manufacturing.

achieved to meet the specified tolerances. Photonic-based equipment, electron-beam machines, and X-ray and ion-beam imaging sensors are used for metrology and failure analysis because of the striking features. Metrology detection is possible with high-resolution microscopy using bright-field detectors [3]. Optical microscopes can reliably resolve features greater than 300 nm, but for smaller dimensions, a scanning electron microscope is best suited. To detect defects on patterned wafers, state-of-the art bright-field imaging-based systems using pattern filtering and a host of sophisticated image-processing algorithms are best suited to detect defects as small as 120 nm.

Thin-film characterization, which includes determination of film thickness, refractive index, and extinction coefficient over a wide range of wavelengths ranging from 250 to 800 nm to support several processing steps, can be accomplished through spectroscopic ellipsometry (Figure 11.2), interferometery, and other systems. The spectroscopic ellipsometry technique, shown in Figure 11.2, involves a light source, polarizer, detectors, and grating, and is most suitable for measuring

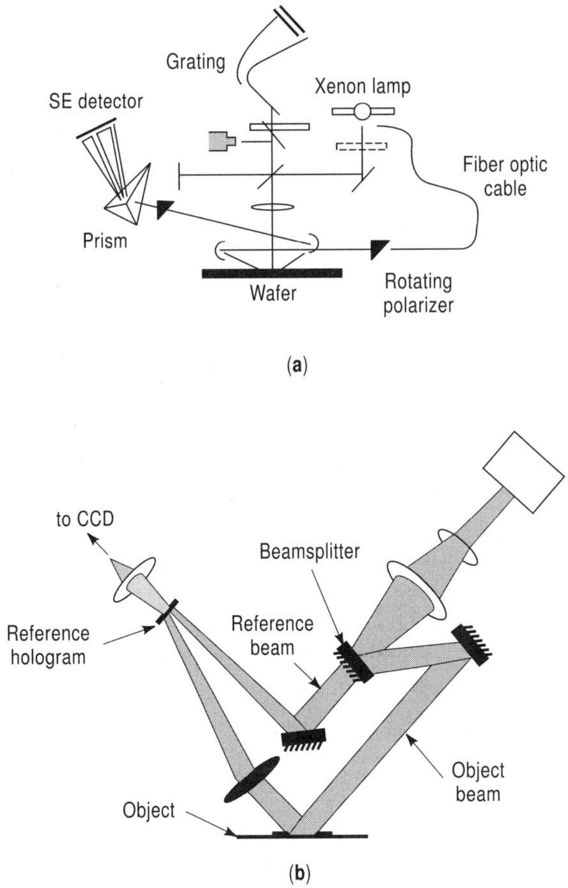

FIGURE 11.2 (A) Spectroscopic ellipsometer for measurement of thin films; (B) CCD-based holographic approach for wafer inspection.

thin, optically transparent films with high accuracy. Full-wafer imaging sensors use one or more independent CCD-based imaging detectors to obtain high spatial resolution during etching and deposition processes. The full images provide visual information about the wafer and the process, whereas the detected signals from thousands of CCD-based detectors quantitatively determine end point, etching, deposition rate, and film uniformity. Optical emission spectroscopic sensors are generally used for end-point detection.

Defect reduction equipment must be designed to capture 95% of all the defects greater than 130 nm on patterned wafers and greater than 43 nm on unpatterned wafers to reduce wafer production costs [3]. Furthermore, defect detection equipment must be capable of scanning 10,000 cm^2 of surface per hour to reduce the inspection cost and improve the profitability of the fabrication process. The spectral range for the next generation of ellipsometric tools must be 190 to 1200 nm. The optical detect detection capability of the 130 nm technology node is strictly dependent on the availability of lower-wavelength laser scattering technology. Studies performed by the author indicate that lasers operating in the UV range from 350 to 550 nm are most cost-effective for this application. Polarization and phase changes of the incident light caused by defects in the wafers must be incorporated into the direct detection tools. The studies further indicate that UV holographic imaging approach is best suited to perform high-aspect ratio defect inspection with minimum time.

11.6 PHOTONIC-BASED ENCRYPTION (PBE) SCHEME

Scientists concerned with security have developed photonic-based techniques to demonstrate the viability of an unbreakable encryption scheme for transmission of secure communications to and from satellites. This particular scheme will be most attractive for transmission of secure communication between various government agencies and between the State Department and U.S. embassies in foreign countries through the LEO or MEO satellite communications systems. This PBE scheme is based on randomly generated characteristics of individual photons and could make major financial transactions, state secret messages, or key military connections not subject to penetration or cracking of coded signals. Researchers have been quite successful in transmitting photonic-based "quantum cryptographics keys" over low-loss optical fibers.

This PBE scheme has been discussed in greater detail a technical article published in *Physical Review Letters* (12 October 1998). This scheme has demonstrated successful transmission of a quantum key through the air over a distance greater than 1 km, thereby revealing its potential for secure satellite communications. It is important to mention that the optical beam will undergo maximum deviation in the lower region of the atmosphere (10 to 20 km). Both the scattering and absorption coefficients for aerosol and molecular distributions decrease with increase in altitude. At altitudes of 100 km or more, optical signals are adversely affected by the turbulence, depending on the severity of the turbulence conditions. One can conclude that if the optical signal remains intact after passing through the turbulence boundary, the rest of the signal travel to a satellite orbiting at an altitude of 300 km or more will experience negligible attenuation, because the overall attenuation coefficient due to scattering and absorbing coefficients at altitudes exceeding 100 km is close to 10^{-6}/km.

The secret of PBE the scheme is that the quantum crytpographic keys are generated as needed between the sender and receiver, thereby generating a random string of numbers, called randomly generated polarization states, of single photons known to no one except the sender and receiver. Since the signals are so low, any attempt to eavesdrop on the transmission will bring the bit-error-rate (BER) to known levels [4].

11.7 IR SENSORS FOR INSECT COUNTING

IR sensors have been designed to count insects in agricultural areas, thereby providing farmers an advance warning to protect their crops. Bugs such as caterpillars can ruin agricultural products including wheat, corn, vegetables, and citrus fruits. Insecticides are effective in removing insects from stored agricultural products such a grain, fruits and vegetables. But the insecticides present a series of health problems, according to health officials. An IR-based device called the electronic grain insect counter (EGIC) takes a common grain-probe trap and exposes it to an IR source to enhance its capability. The sensitive probes are placed inside the traditional tubes located in grain bins and elevators for long periods and are then removed and inspected. The IR device allows the operators to monitor the insect count from a remote or central computer.

The sensor head contains an IR-LED source matched with a phototransistor. If the collector and emitter of the phototransistor are reversed, the transit can be biased in the linear region, thereby making the sensor more sensitive to small insects. The sensitivity of the system can be improved to distinguish between insect types and sizes, which could be of significant importance to farmers. In the stored-grain ecosystem, there can be a whole range of insects; some eat grain and others eat the insects that eat the grain.

11.8 LASER-BASED AERIAL DELIVERY (LBAD) SYSTEM

Improved accuracy of high-altitude aerial delivery is of paramount importance to both military and civilian organizations. Delivery of goods and personnel to a designated spot with minimum cost requires high accuracy. A laser-based (LIDAR) system is capable of providing real-time, three-dimensional maps of wind fields from altitude to ground with range slices of less than 100 m and with velocity accuracy to within ±0.5 m/sec. Precision range slices and high velocity accuracy will facilitate the delivery of goods from the air with great reliability, minimum time, and lowest cost. The first-generation system used a flash-lamp-pumped laser with a peak energy of 50 mJ/pulse at a PRF of 7 Hz/sec and the system was mounted on a scanner. The second-generation system used a diode-pumped transceiver with a peak energy of 3.5 mJ/pulse at a PRF of 200 Hz/sec. The new or the latest version of system uses an eye-safe Tm:YAG 2 μm monostatic diode-pumped laser transceiver with a peak energy level of only 12 nJ/pulse at a PRF of 100 Hz/sec. The entire system is comprised of a transceiver, signal processing unit (SPU), environmental control unit (ECU), optical scanner, inertial navigation system (INS), and power supply. All these subassemblies and components can be mounted into 18 inch racks within the nose cone of a fuel pod. The second-generation airborne LIDAR system weighs slightly over 2000 lb, whereas the latest system weighs around 600 lb and occupies a space of 45 ft^3. The latest LIDAR-based airdrop system offers an improvement of 2:1 over the previous airdrop systems [4].

11.9 LASER-BASED OPTICAL COMMUTATOR (LBOC)

The laser-based optical commutator uses a rotating folded lens comprised of a set of coils arranged on a rotor and stator to generate true-time delays to steer the antenna beam (Figure 11.3). The electronic beam steering is achieved by rotating the lens. The optical commutator can be designed to steer either two-dimensional flat circular or cylindrical arrays over significantly wider bandwidth than any other phased-array antennas. LBOC design offers simplified design with considerable savings in weight, size, cost, and power consumption, compared to other steering techniques. Current true-time delay steering schemes employ a number of switched optical-delay lines for each antenna element. However, this steering suffers from high cost due to the very large number of optical interfaces involving large number of optical switches that must be controlled.

10.9.1 Types of Optical Arrays

Two parabolic distributions of optical fiber lengths provide the time function. Histograms shown in Figure 11.4 indicate the lineup of the rotor and stator fiber elements, which are acting in coil formats. Location of elements in linear and circular arrays are shown in Figure 11.4, along with radiated beam direction. When all three fibers in each channel have the same combined length, the same delay is applied to each of the antenna elements, which directs the transmit beam on boresight. When the combined lengths of the rotor and stator fibers have a linear distribution, it will provide the exact timing required to steer the beam to the left or right, depending on the phase of the optical signals. Two-dimensional optical beam steering can be accomplished with optical commutators connected in series; one provides the steering in the EL plane and the other in AZ plane. The beamwidth of a flat array is dependent on the number of elements in the array. The number of elements in the optical array can be given as

$$N_E = \frac{102}{\theta_B} \tag{11.1}$$

where θ_B is the half-power beamwidth (degree) and N_E is the number of elements. Since each element requires a separate timing control, the number of windings that can be placed on the stator to steer the beam is equal to the elements in the row [5]. The number of usable positions or settings of the optical commutator is expressed as

$$N_p = \frac{(2)(\theta_S)}{\theta_B} \tag{11.2}$$

where θ_S is the equal to half of the scan angle and N_P is the number of positions. The number of windings on the rotor (N_R) can be expressed as

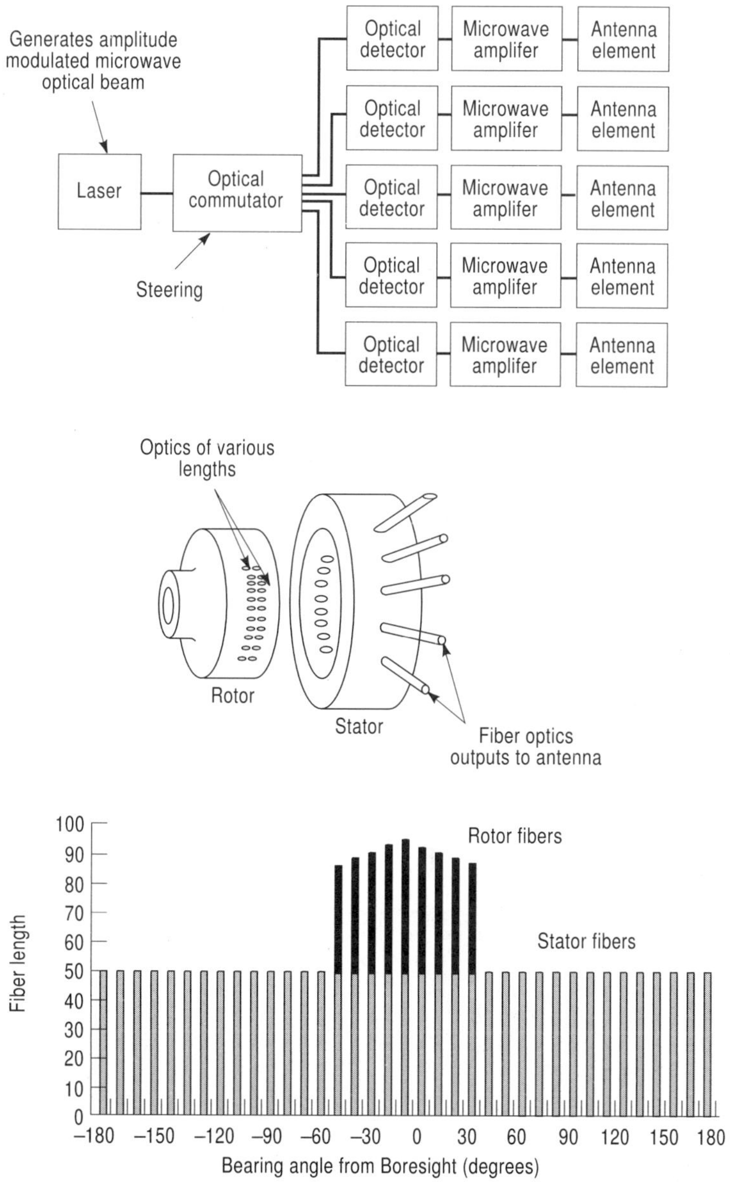

FIGURE 11.3 Optical commutator. (A) Block diagram; (B) major elements; (C) fiber length as a function of steering angle or bearing.

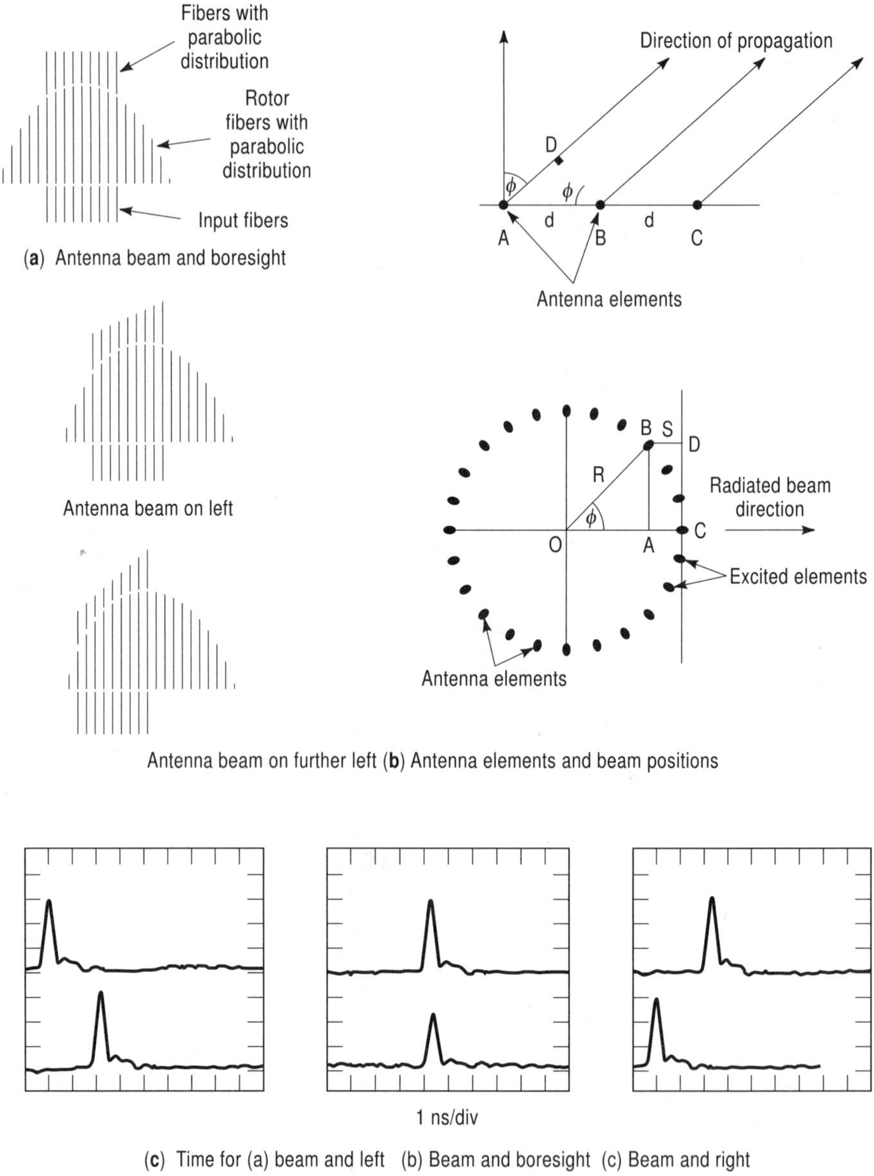

FIGURE 11.4 (A) Histograms of rotor and stator; (B) antenna elements and beam positions; and (C) switching times for beam steering.

$$N_R = N_P + N_E \tag{11.3}$$

Once the system requirements have defined the beamwidth and scan angle, it is easy to compute all the fiber-optic winding lengths for the rotor and stator of the commutator.

From the linear array geometry illustrated in Figure 11.4, it is evident that if the antenna beam is to be steered by an angle ϕ from the boresight, the time required to transmit from point A to D in the direction of propagation is given as

$$T = (d/c)\sin\phi \tag{11.4}$$

where d is the element spacing or separation, c is the speed of light (3×10^{10} cm/sec), and ϕ is the beam-steering angle from the boresight position of the array. The speed of the light is slower in the optical fiber medium by a factor equal to the refractive index (μ) of the fiber material. Typical refractive index values are about 1.47 for a plastic fiber, 1.44 for a fused-silica glass fiber, and 1.53 for a quartz glass fiber. The length of the rotor and stator windings [4] can be expressed as

$$N_R(n) = L_0 - (p)^2(d/\mu)\left(\frac{\sin\theta_S}{2k-1}\right) \tag{11.5}$$

$$N_S(n) = L_S + (p)^2(d/\mu)\left(\frac{\sin\theta_S}{2k-1}\right) \tag{11.6}$$

where L_0 is the length of the fiber in the center slot of the rotor, L_S is the length of the stator winding at the center position of the stator, p is the fiber-slot number, n indicates the nth winding of the rotor or stator, and k is the scan-angle-to-beamwidth ratio ($2\theta_S/\theta_B$).

11.9.2 Critical Design Aspects of Optical Commutators

The rotor and stator windings can be made either with plastic or glass optical fiber. Plastic fibers are less costly and can be easily converted into coils. Glass fibers are costly, but have superior optical qualities that are important for high-performance beam-steering applications. The choice of light sources and optical detectors is dependent on performance requirements, cost, and operating environments. The wavelength of the laser source will influence the choice of the optical fiber lines. If high beam quality and pointing accuracy are desired, coherent laser sources will be required, which can add more cost and complexity. Incoherent laser sources will yield poor performance but with minimum system cost and complexity.

Plastic fibers can be used with laser-diode sources operating at the 800 nm wavelength, but longer-wavelength operation around 1300 nm is not recommended because of high losses. However, most of the high-performance laser sources and detectors are designed to operate around 1330 nm wavelength. This particular wavelength has the minimum absorption in glass fibers (about 0.35 dB/km in a single-mode glass fiber at 1300 nm) and is widely used in the telecommunications industry.

In the case of a 11 element linear array operating at a RF frequency of 1.5 GHz (which means the wavelength λ is 20 cm and the element spacing d is 10 cm) with beam-steering capability of ±45° and rotor-winding length from 21 to 88.6 cm, will have a parabolic distribution shown in Figure 11.4. A multielement distribution system feeds the 11 stator input slots shown in Figure 11.3. The 11 output slots are equipped with fiber optic cables of lengths varying from 25 to 28.5 cm and complimentary parabolic distribution architecture. The plastic fiber cable has a diameter of 0.5 nm surrounded with a 0.25 nm thick protection jacket. The 11 element linear array has an overall length of 100 cm, beam settings of 45° on each side of the broadside, and relative time-shift between the end channels of 2.33 nsec. This means that the 11 element linear array has a 3 dB beamwidth of 9° and takes only 2.33 seconds to steer the beam over ±45° angle from the boresight. Performance capabilities of two linear arrays operating at 1.5 GHz and 9.0 GHz are summarized in Table 11.1.

It is important to mention that such ultrafast beam-steering capability is not possible even with electronic beam-steering schemes using precision array components. Furthermore, cheaper incoherent laser sources, even with 5–6% variations in laser wavelength, can be used without any noticeable degradation in the performance of the array. Architectural studies performed by the author indicate that the total number of optical fibers required to steer a linear array is equal to three times the number of radiating elements plus one extra rotor fiber for each pointing direction or angular setting other than boresight.

The optical commutator provides a practical and inexpensive technique to implement a true-time-delay beam-steering capability with ultrafast speed. Off-the-shelf optical detectors, laser sources, optical filters, and couplers of bandwidths of several gigahertz are readily available with reasonable costs. The optical commutator-based beam-steering concept has potential applications in space surveillance radar, air defense radar, telecommunications equipment, and communications systems, where, speed, reliability, and wide bandwidth are the principal requirements.

TABLE 11.1 Performance comparison of two linear arrays operating at different frequencies

Performance parameter	1.5 GHz	9.0 GHz
Wavelength (cm)	20	3.33
End-to-end scanning time (sec)	2.33	2.36
Scan angle limits (degree)	±45	±45
Element spacing of $\lambda/2$ (cm)	10	1.66
Number of elements	11	68
3 dB beamwidth (degree)	9	1.5
Minimum number of positions	10	60
Number of windings on rotor	21	128

11.10 LASER SYSTEMS FOR REMEDIATION OF NUCLEAR REACTORS

High-power lasers plays a key role in remediation of nuclear reactors when used as cutting tools. Conventional cutting techniques, including plasma arc torch, mechanical saw, etc., are not effective in cutting the thick metals used in nuclear power plants. Nuclear facilities have a wide range of very thick metal structures and components, including complex arrays of steel pipes and stand-alone pipes integrated into reinforced concrete superstructures, which are chemically and radioactively contaminated [5]. After a radioactively contaminated power plant facility ceases operation or is decommissioned, deactivation process starts; this requires immediate and safe removal of nuclear materials. This is followed by decontamination, involving cleanup of the residual radioactively contaminated structures and environments. Because of the large volume of contaminated material from deactivated commercial power reactors and weapon production facilities involved, it is very important to separate various materials based on radioactive level and type of contamination [6] and to store each level with the required safety procedures. This entire process for commercial nuclear power reactors requires the dismantling of large, thick-walled vessels, pipes, supports, and structures. In the case of pressurized-water reactors, the wall thickness varies from 6 to 12 inches of standard carbon steel clad with stainless steel on the inner surface to provide corrosion resistance. The cylindrical vessels are typically 15 ft in diameter and can be as high as 31 ft. Boiling-water reactor vessels have wall thickness of 5 to 9 inches with heights on the order of 65 ft and diameters exceeding 17 ft [6].

The latest technical articles published on the subject [6] indicate that the chemical oxygen iodine laser (COIL) is most suitable for this particular application. This laser is equally attractive for industrial process applications. The COIL system is capable of generating CW power level as high as 50 kW at 1.315 nm wavelength. It is important to point out that very little energy at the COIL wavelength is absorbed by the fused silica fibers, thereby leaving most of the COIL output to be absorbed by the metals. Based on this unique performance capability, it can be concluded that the COIL system will be found most effective in the removal of contaminated debris from decommissioned nuclear power plants.

11.10.1 Capabilities of Laser Systems for Remediation of Nuclear Reactors

Several laser candidates for remediation of nuclear reactors are available, such as the high-power Nd:YAG laser (1.064 μm), CO_2 laser (10.6 μm), and COIL laser (1.315 μm). The first two high-power lasers have been used for cutting and drilling metals of limited thickness, typically not exceeding 4 inches. However, the COIL source has unique performance and has demonstrated cutting and drilling capabilities in steel structures or plates with thicknesses exceeding 10 inches [6]. The COIL laser is capable of producing tens of kW of CW power level with excellent focussed-beam quality at 1.315 μm wavelength that can be transmitted by optical fibers with minimum insertion loss.

In the case of chemical lasers, the reaction times of chemicals dictate that the gas flow velocity be close to the speed of sound (1128 ft/sec), which is equivalent to 1 Mach speed. The latest research activities reveal that the photonic systems have demonstrated chemical flow speeds at supersonic levels, which will accelerate the mixing of chemicals used in a chemical laser system such as the COIL. The main advantage of the COIL laser beam over the CO_2 laser is the wavelength of 1.315 μm, most of which is absorbed by the metals. Furthermore, the fused silica optical fibers exhibit minimum insertion loss at 1.315 μm compared to 10.6 μm wavelength of the CO_2. Published data indicate that the insertion loss varies from 0.55 dB/km for a multimode fused silica fiber to 0.36 dB/km for a single-mode fused silica fiber at the wavelength of 1.315 μm, compared to more than 10 dB/km at CO_2 laser wavelength of 10.6 μm, under similar operating conditions. Laser energy from the Nd:YAG lasers can be transmitted by the fused silica fibers with minimum loss, but these optical fibers suffer from thermal distortion, leading to degradation of both beam quality and output power.

Since the COIL is a low-pressure flowing gas laser, heat from the lasing medium can be removed in minimum time. Furthermore, the COIL laser design can be scaled up or down to achieve desired output power levels. Several 1 kW COIL laser has been in operation for cutting applications. Power handling capability of a COIL laser is dependent on the type of optical fibers. Power handling capability of the fused silica fiber is limited by the core and cladding dimensions. Optical fibers and optical components are susceptible to damage under high-power conditions. Proper selection of core and cladding dimensions can allow transmission of laser power as high as 20 kW, provided an efficient cooling scheme is implemented in the laser design. COIL lasers have demonstrated power levels exceeding 10 kW, approaching 20 kW under controlled laboratory environments. The output power levels can decrease under field operating environments. For comparison purposes, a Nd:YAG laser has delivered power levels up to 2.5 kW through fused silica fibers, which is still significantly lower that that of a COIL.

11.11 PHOTONIC SENSORS FOR BATTLEFIELD SURVEILLANCE

Unmanned aerial vehicles (UAVs) are pilotless aircraft that have potential applications in battlefield surveillance and frontline observations in hostile territory. New generations of UAVs to operate as pilotless fighters are being considered for the 21st century. Top defense planners feel that effective suppression of enemy air defense and other deep-strike missions could be most hazardous for the 21st century manned aircraft. Defense planners are thinking about a new generation of unmanned combat aircraft (UCA) that may be able to outperform a manned aircraft under similar military missions and will not require human pilots. Commercial off-the-shelf (COTS) sensors and components are available for tactical systems to provide real-time battlefield data for the deployment of forces and top-of-the-line (TOTL) strategic systems for the upper command echelons. The COTS systems include powerful onboard processors, navigation subsystems, data links, low-cost global positional systems (GPSs), photonic sensors, optoelectronic components, and satellite and terrestrial line-of-sight (LOS) communication systems. UAV sensors will require narrow bandwidth and data encryption to counter interception or jamming tactics deployed by the enemy.

UAVs can be used as strike aircraft in the forward sector using advanced short-range missiles and munitions including tube-launched, optically tracked, wire-guided (TOW) antitank missiles. For long-range, high-altitude reconnaissance and surveillance missions, IR sensors and synthetic aperture radars (SARs) can be deployed, depending on the mission requirements. Mission requirements will determine the architectural design of the UAV and will define the performance requirements for the analog-to-digital converter (ADC) interface between the sensors and computational assets. Microprocessors with power consumption less than 3 W are available for use in UAVs.

Stabilization with high accuracy is absolutely necessary for the surveillance mission of the UAV. The stabilization package [7] generally includes a high-resolution TV camera with a long-range telephoto zoom lens, a forward-looking IR (FLIR) sensor for night operations, a laser illuminator, stabilization sensor, and laser rangefinder with a resolution capability of about 1 m at 5 km range [8]. All these surveillance sensors can be mounted onto a stabilized platform to achieve high performance missions under the dynamic flight conditions of the UAV. Deployment of a laser-gryo stabilized platform would allow the UAV to fly higher to get a better view of the battlefield with maximum safety. It is important to mention that the UAVs designed for surveillance and reconnaissance missions must be smaller, lighter, and stealthy. Furthermore, to retain the self-defense capability, UAV designs must incorporate IR and RCS reduction techniques needed to operate safely over enemy territory.

11.11.1 Performance Requirements of Sensors for Drones or UAVs

State-of-the art TV cameras, FLIR sensors, and IR line scanners (IRLSs) are available for possible applications in a drone, UAV, or remotely piloted vehicle (RPV). All these vehicles can play key roles as reconnaissance and surveillance platforms for battlefield surveillance. Trade-off studies must be performed in terms of performance level and mission growth for various critical sensors such as IR sensors, laser rangefinders, and high-resolution TV sensors to meet performance levels of a specific mission. Performance capabilities of critical IR sensors for UAVs or RPVs are summarized in Figure 11.5. Endurance capabilities and performance requirements of some EO systems and IR sensors must be defined for specific mission applications [7]. A drone or RPV [8] made from fiberglass material to avoid radar detection can perform vital mission functions, including TV reconnaissance and laser target designation from an altitude ranging from several hundreds of feet to a few thousand feet. The RPV can provide a close support for deep penetration missions in the forward enemy battlefield area (FEBA), because a fully equipped RPV is capable of providing [8] reconnaissance, surveillance, target acquisition, target designation, selected weapon delivery, and bomb damage assessment functions.

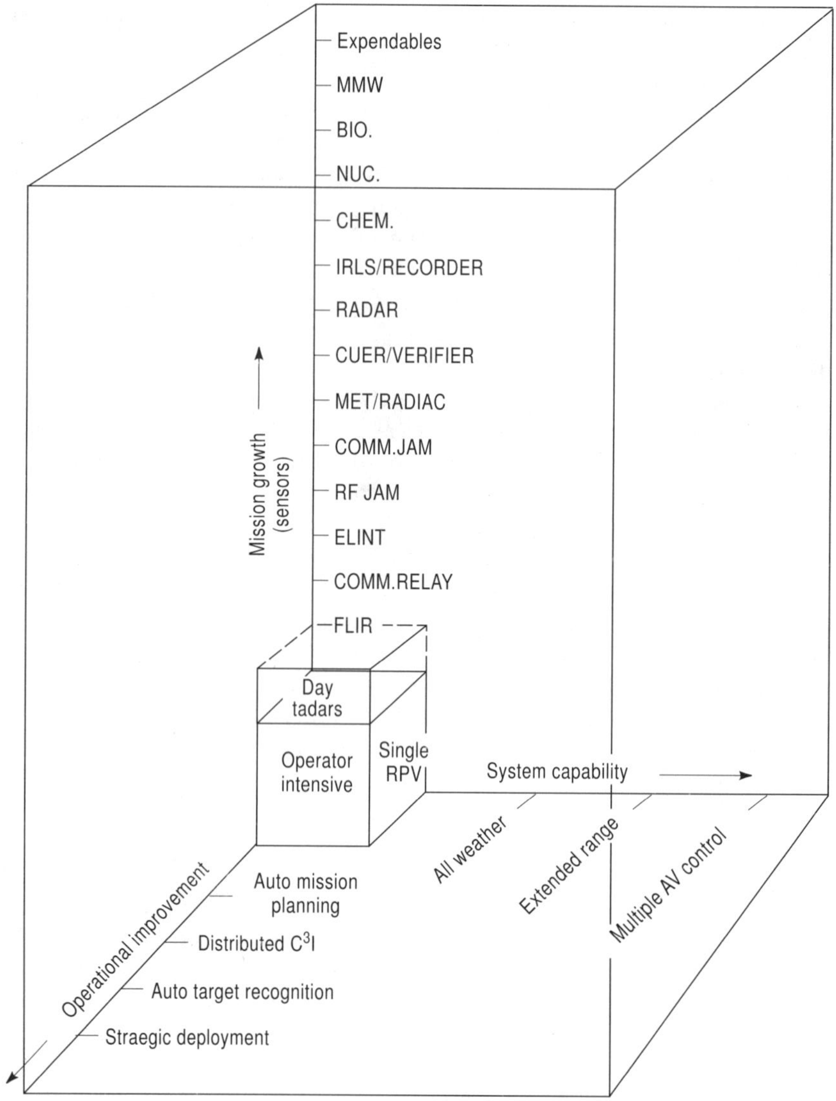

FIGURE 11.5 System performance capability and mission growth for various IR sensors for possible applications in RPV.

11.11.2 Performance Capabilities of a TV Imaging Sensor

This particular sensor provides real-time surveillance of the FEBA, revealing the targets and their locations under day or night conditions. A stabilized TV sensor can be supplemented with a high-resolution 35-mm panoramic camera to provide real-time reconnaissance and surveillance of targets at slant ranges close to 1 mile with a ground resolution better than 3 ft at a slant range of 1 mile. Typical video bandwidth of this sensor is about 5 MHz. A gimbaled TV camera can be used for a reconnaissance role because it provides area search over ±90° in AZ plane and 0° to 45° in the EL plane. A combination of TV camera and laser provides automatic tracking capability at a maximum rate of 10°/sec. The stabilized LOS permits daytime detection of tank-sized targets at slant ranges of more than 3 miles and detection of roads at 2 miles. TV type images are available from a RPV or drone with image quality varying from very low cost CCD-based cameras to highly sophisticated FLIR sensors with day/night imaging capabilties.

11.11.3 Performance Capabilities of a FLIR Sensor

FLIR sensors are highly sophisticated and are very costly, but they provide excellent performance under day or night environments. This sensor provides a wide variation in FOV from narrow FOV to wide FOV needed for various functions including target recognition, fire control, weapon delivery, and navigation with high accuracy. The narrow FOV of the sensor provides target acquisition, tracking, and high-resolution ground mapping necessary for mine detection. A drone or RPV equipped with a FLIR can achieve high-quality images with high S/N ratio most ideal for automatic target tracking and locking functions. The wide FOV of the FLIR is generally used for navigation and passive reconnaissance under day and night conditions. This particular sensor can offer a ground resolution of 2.5 ft and 5 ft at 1 NM range, which is sufficient for identification and recognition, respectively.

11.11.4 Performance Capabilities of Laser Rangefinder and Designator

Laser rangefinders are widely used in weapon delivery applications, where precision weapon delivery under day or night conditions is the principal requirements. A laser designator or illuminator provides target designation in support of terminally guided weapons, including missiles and bombs. A laser designator and TV sensor can be mounted on the same stabilization platform to minimize weight and cost. This sensor, in conjunction with a TV camera, permits target location, artillery adjustment, and boresight calibration functions with high accuracy.

The operating wavelength of a laser rangefinder must be selected such that it offers minimum attenuation at the desired ranges and under heavy haze conditions. Both the Nd:YAG laser operating at 1.06 μm and CO_2 laser operating at 10.6 μm are most suitable for ranging as well as illumination functions. However, the attenuation coefficient under heavy haze conditions is about 1.2 dB/km for a Nd:YAG laser, whereas it is only 0.4 dB/km for a CO_2 laser operating at 10.6 μm, as illustrated in Figure 11.6. A Nd:YAG laser can deliver an energy level of 100 mJ out of the dome with a beam divergence not exceeding 0.5 mrad and pulse repetition rates of 10 or 20 per second.The direct-drive gimbaled platform stabilization technique offers two degrees of freedom, namely, 360° in the AZ plane and 15° in the EL plane.

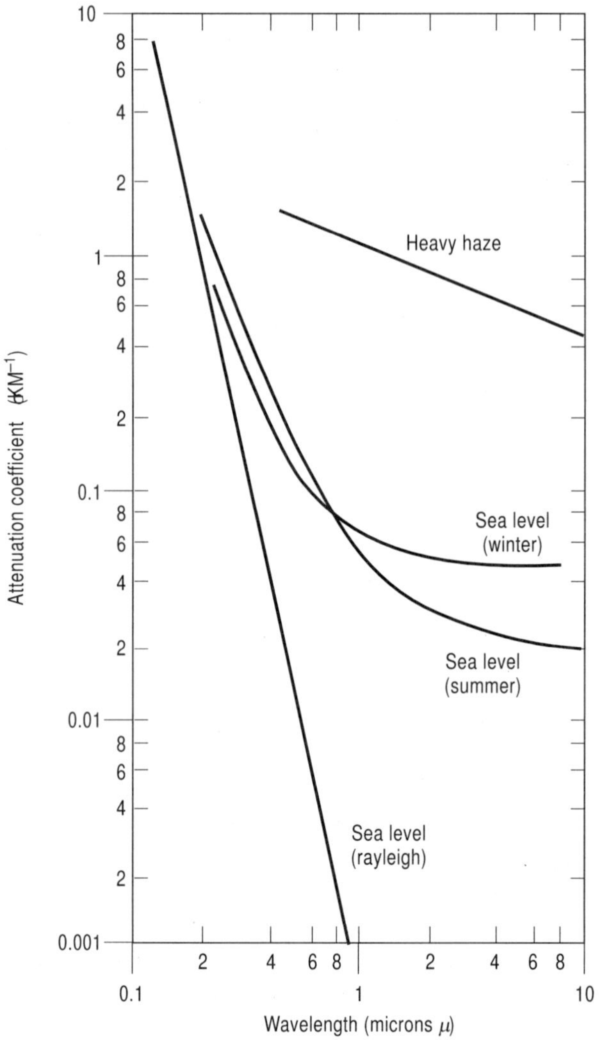

FIGURE 11.6 Attenuation coefficient as a function of wavelength and climatic environments.

11.11.5 Performance Capabilities of IR Line Scanner(IRLS) Sensor

The IRLS sensor is primarily used for wide area reconnaissance missions [7]. This sensor uses cryogenically cooled Hg:Cd:Te detectors to provide high detectivity (better than 10^{10} cm $\sqrt{Hz}$/W). The sensor does not provide real-time dynamic imagery like video cameras or FLIRs, but the IRLS is the most efficient method to obtain long-range, day/night reconnaissance data on a film over wide ranges on both sides of the aircraft. In a typical IRLS swath, a RPV or drone or UAV can cover several square miles of terrain features with sufficient resolution required for target detection and recognition. The IRLS sensor is best suited for long-range reconnaissance beyond the data link range. The wide 10 to 15 km swath available from a high-performance IRLS sensor can eliminate the need for navigation. This sensor provides two operating modes. The IRLS records the reconnaissance data directly onto a cartridge-contained, high-resolution film roll that is retrieved and processed. This sensor capability is most useful in situations where maximum resolution with photographic quality in non-real-time critical situations is required. However, for real-time reconnaissance, the data can be transmitted directly to a film strip recorder for a hard copy or a TV monitor for immediate display, which can provide single-swath coverage in real-time with some degradation in resolution. A video recorder with modified data input format could be used on the air vehicle or to allow "pop-up" playback. This sensor capability is particularly useful for a quick look at the data or information, which can allow the operator to redirect the RPV, if necessary. When an IRLS is integrated with a video recorder, the sensor offers high-resolution capability most ideal for mine detection, wide-area surveillance, target detection, and "pop-up" capability.

11.12 LASER-BASED MICROMACHINING TECHNIQUE

Laser-based micromachining (LBM) tools are beginning to replace traditional computer numerically controlled machines (NCMs). LBM has been recognized as a cutting-edge technology with potential applications in microelectronics, photonics, and optoelectronics devices and microelectrical–mechanical (MEM) systems. Micromachining involves several industrial processes, such as soldering, drilling, and cutting. High-power lasers operating at 10.6 μm have been in use for soldering and drilling for about two decades. The latest research on lasers operating in the lower IR spectrum indicates that a COIL laser operating at 1.315 μm has demonstrated unique cutting capability. A COIL laser has demonstrated the ability to cut steel plates and structures with thicknesses up to 12 inches into small pieces even with a CW power level less than 10 kW.

Many applications require holes to be drilled in wafers, circuits, flex materials, and thin-semiconductor products with high accuracy and without breaking them. Conventional mechanical drilling methods, including lathe machines, milling machines, and drills, are not suitable for operations requiring hole diameters less than 200 μm or 0.0200 cm with stringent tolerances. When the control of finish and the angle of side walls of the holes are critical, laser drilling is the only viable solution.

UV lasers, including excimer lasers operating at 340 nm, the third and fourth harmonics of lamp-pumped Nd:YAG lasers operating at 1060 nm, and Nd:YLF lasers operating at 1047 nm, are best suited for drilling in polyamide materials and packaged multichip modules, where speed, cost, and quality are of paramount importance. Solid-state diode-pumped (SSDP) lasers operating at 1064 nm wavelength are available with power levels ranging from 1–10 W, adequate for most drilling applications. Materials such as ceramics, metals, and woven-glass epoxies may require CO_2, Nd:YAG, Nd:YLF, and COIL lasers operating at 10.6 μm, 1.064

μm, 1.047 μm, and 1.315 μm, respectively for drilling applications, depending on the absorption properties of the materials and material thickness. Multiple laser sources can be incorporated in a drilling machine when multiple materials need drilling on a particular part assembly with minimum time.

Laser drilling has two distinct advantages over conventional drilling techniques involving bits. The bits experience wear and tear, leading to sudden breakage, which does not occur with LBM techniques. Besides high speed and precision drilling operations, the LBM technique provides better safety, improved performance, and good working environments for the operators. Certain dry-cleaning applications such as paint removal from aircraft and semiconductor wafer cleaning are only possible with LBM technology, because of lower cost and high-speed operation compared to other methods.

In summary, both LBM and MEM techniques are critical to rapid advancement and technological progress in manufacturing industry, industrial processes, and medical diagnosis applications. These techniques not only offer state-of-the art performance capabilities, but also significant improvement in unit cost, production rate, product reliability, quality control, and power consumption. Recent surveys on photonic devices and laser technology indicates that the excimer laser is best suited for MEM and LBM applications. Package design studies indicate that an industry-duty excimer laser with built-in power supply will occupy a space not exceeding 18 × 18 × 10 inches based on current packaging technology. An excimer laser can be designed to operate at multiple wavelengths, namely, 157, 193, 248, 307, and 351 nm, thereby allowing a LBM system to undertake multiple industrial and manufacturing operations simultaneously. Such a complete LBM system is comprised of a 19 inch high-resolution color monitor, on-screen multichannel viewing, machine vision, CAD/CAM, laser source, power supply and machining software for specific applications, and the system can be contained in a 64 × 26 × 24 inch package, approximately.

11.13 ACCESSORIES REQUIRED BY IR AND PHOTONIC SENSORS

IR and EO sensors require several accessories, such as power supplies, cryocoolers, control electronics, temperature control circuits, and optical software programs, for their efficient and safe operations. Performance requirements for critical accessories will be described, with emphasis on cost and reliability.

11.13.1 Optical Sources

Several optical sources capable of operating at various wavelengths and output power levels are available and an appropriate source can be selected to meet performance requirements for a specific application with minimum cost and complexity. Tungsten–halogen light sources are available for ratings ranging from 200 W without reflectors to 1000 W with reflectors. Xenon–mercury sources are also available with output power levels ranging from 200 W to 5000 W with spherical and elliptical reflectors. Xenon optical sources are commercially available with output power levels from 150 W to 7000 W with spherical and elliptical reflector. DPSS laser sources are available with moderate power levels, but with higher conversion efficiency and beam quality. Performance characteristics, including power level and reliability, of various DPSS sources and semiconductor sources are described in Chapter 4.

11.13.2 Power Supplies

Power supply requirements are dependent on drive power levels needed to operate specified light sources. Power supplies to drive low-power semiconductor lasers can use air-cooling schemes as long as the power dissipation in the device does not exceed the maximum power dissipation rating. In general, power supplies to drive coherent laser sources have stringent performance requirements, such as maximum ripple amplitude, maximum supply voltage fluctuations, FM noise content, and conversion efficiency. Laser diode drivers or power supplies are designed to meet specific output current requirements ranging from 100 to 400 mA, depending on the laser output power and conversion efficiency. These drivers are designed to operate in constant current or constant power mode operation. Modulation frequencies of 1 MHz in constant power mode operation and 2 MHz in constant current mode operation are generally used.

Generally, the laser diode drivers are equipped with trim pots that allow precision control of output current and laser limit current levels. These drivers feature slow start circuitry, buffered output for laser diode current monitors, and variable supply voltage operation from + 5 V to + 12 V. Such drivers are required to meet stringent performance requirements with modular design configurations. Most of the modular drivers offer output power stability of less than 0.02% over 24 hour duration, current stability less than 50 ppm during 24 hours, RMS current ripple and laser diode forward current noise less than 5 μA, current monitor transfer function of 40 mA/V, power monitor transfer function of 1000 μW/V, and constant power photodiode current range between 5 and 125 μA. These drivers come with built-in internal control circuits for automatic turn-on/off depending on the positive supply voltage amplitude. By bringing the supply voltage up slowly could result in damage to the laser diode. Additional safety and diode protection can be provided through transient detection and filtering, intermittent contact protection techniques, independent current and power level settings, slow turn-on sequence, and shielded cable to protect the laser diode from external radiation levels.

Some precision power supplies have additional key features, including full 16 bit control and characterization of critical parameters of laser diodes, namely, output current and power, via the IEEE-499 or RS-232C interfaces. All critical laser parameters are simultaneously presented on a two-line alphanumeric LCD display on the front panel. The advanced laser diode drivers can be disabled through a pushbutton switch on the front panel, thereby preventing any inadvertent changes in the output by accidental movement of the power control knob. The advanced driver design includes an internal function generator with programmable capability over the 200

Hz to 300 kHz range and has provision for external analog modulation using other modulation waveforms and frequencies in both constant current and constant power modes of operations.

11.13.2.1 Capacitor Charging High-Voltage (CCHV) Power Supplies

CCHV power supplies are widely used for laser-based medical diagnosis equipment, industrial laser systems, high-resolution X-ray equipment, and other pulsed power applications. Modular high-voltage power supplies (HVPSs) are specifically designed for excimer laser systems with average power level around 1500 W and peak power operating voltage as high as 50 kV. Capacitor charged HVPSs are available to meet peak pulsed laser power levels up to 7000 J/sec at operating voltage of 50 kV with single-phase or three-phase input power operations. These power supplies offer regulation of less than 0.1% and repeatability better than 1%. The capacitor-charged HVPSs come with unique features, including high-frequency inverted gate base transit (IGBT) inverters, protection against overly high voltage, current, and temperature operations, fully programmable output voltage, robust air-cooling systems, corrosion resistance chassis, active power factor correction up to 98%, compact size, low starting energy capability, leakage current less than 50 μA, isolation better than 2500 VAC, and efficiency in excess of 85% at full power output.

The capacitor-charged HVPSs use efficient resonant topology involving the latest IGBT power semiconductor technology capable of offering unparalleled reliability and sustained performance over extended periods in a compact protected enclosure. These power supplies are designed for continuous use in harsh ambient and electrical environments. Highly regulated HVPSs are available for YAG laser applications with integral boost and ignition capabilities and power levels ranging from 500 W to 2000 W.

11.14 CONTROL CIRCUITS

Temperature, current, voltage and power control circuits are generally used in the subsystem designs such as power supplies for sensitive IR and photonic systems to maintain high performance levels under the fluctuating operating conditions. Temperature control circuits are widely used by thermoelectric coolers and cryostats to maintain temperature stabilization better than 0.005 °C needed for optimum system performance. A temperature control circuit with a typical power rating of 10 W is adequate for temperature stabilization of laser diodes, compact optics, and high-performance IR detectors. The controller supports thermistor elements with resistance ranging from 1 kΩ to 500 kΩ to provide stability under extreme variations.

An analog input allows remote control of the set point thermistor resistance and the analog output allows remote monitoring of both the actual thermistor resistance and temperature of the device. The thermal control circuits can provide temperature stability over –90 °C to +175 °C. These controls have key features including use of linear low-noise, bipolar current sources, easy adjustment of operating temperature and thermoelectric current limit, control loop with overshoot suppressor, gain adjustment from 1 to 50, integration time adjustment from 1 to 10 seconds, and remote adjustment of set-point thermistor resistance for fine temperature control.

11.14.1 Analog Controller for Coolant Flow Adjustment

Automatic adjustment of coolant flow to a specific component or circuit in a complex IR sensor is desirable to maintain required performance level under fluctuating operating environments. Special cooling procedures are necessary because temperature gradients generate thermoelectron potential differences, resulting in circulating currents. A very slow cooling rate of 4 hours from 10 K to 6 K may be required to minimize spatial variations in temperature, which could adversely affect the IR sensor performance. Gradual temperature reduction is accomplished using a computer integrated with an analog controller. The error signal or the output of the temperature bridge is used to drive the vacuum pump throttle valve or the coolant gas feed valve. The analog controller then continues to adjust the coolant and the pumping speed needed to maintain the set point for flow. When the temperature nearing the desired superconducting transition is reached, the computer issues a set point and waits until the set point is reached. Then it issues a new set point slightly lower in temperature. In this way, any desired cooling profile and IR sensor performance can be monitored through the critical transition region. This type of controller for coolant flow is best suited for IR sensors used by satellites or space stations with operation over extended period.

11.15 CRYOCOOLERS AND MICROCOOLERS FOR IR SENSORS

Cryogenic cooling of optical detectors and optics is necessary if optimum performance of an IR sensor or receiver is the principal requirement. The size, weight, cost and complexity due to cryogenic hardware depend on the cryogenic operating temperature and cooling capacity requirements. Furthermore, regular maintenance and servicing of a cooling system are required to ensure safety and reliable superconducting operation. Commercial, space, and military sensors using cryogenic cooling for optimum performance must address the following design aspects of a cooling system:

- Selection of a cryogenic system configuration must emphasize ease of operation, minimum cost, and less-frequent maintenance service.
- Effective integration of cryogenic scheme in the system is required.
- Proper maintenance in the field with minimum cost must be given serious consideration.

Integration of these design features has been observed in few commercial IR sensors. However, integration of the above design aspects most likely has been materialized in military IR surveillance and target acquisition systems and space reconnaissance sensors.

Closed-cycle and open-cycle cryogenic refrigerators or cryocoolers are available with various cooling capacities. Selection of a cryocooler for a specific application depends on cooling capacity, cost, and maintenance issues. Results of historical review of cryocooler development and survey of commercially available systems and their applications to commercial, space and military systems are summarized in Table 11.2. Performance capabilities of both the high-temperature (77 K) and low-temperature (4.2 K) cryocoolers will be briefly described, with emphasis on cooling capacity and application.

11.15.1 Critical Performance Parameters of Cryocoolers and Minicryocoolers

Cooling capacity, cooling efficiency, input power, weight, size, and maintenance cycle are the key design requirements of a cryocooler. The weight and cost of a cryocooler are strictly dependent on its cooling capacity. For example, the specific power requirements of a 5 W capacity cryocooler are 15, 10, 5, and 0.45 kW at op-

TABLE 11.2 Chronology of systems using closed-cycle refrigeration

Application	Temperature (K)	Refrigerator	User
1960s			
Infrared	80	Stirling	U.S. military
Hydrogen bubble chamber	20	GM	High-energy research
Microwave amplifiers	4.5	GM+JT	NASA
1970s			
High-energy accelerators	4.5	Collins	DOE
DC superconducting motor	4.5	Collins	U.S. Navy
Fusion	4.5	Collins	DOE
1980s			
Fusion magnet test facility	4.2	Collins	DOE
30-MJ SMES	4.5	Collins	BPA
Magnetohydrodynamics	4.5	Collins	Russia
Magnetically levitated train	4.5	Collins	Japan
Magnetic resonance imaging	4.2	GM*	Commercial
Cryovacuum pumping	20	GM	Commercial
1990s (potential)			
SQUID	8.5	Stirling	NBS
SSC	2.0	Supercritical	DOE
MRI	10	GM	Commercial
Magnetically levitated train	—	—	DOT
SMES	—	—	Industry
Infrared	65	Stirling	U.S. military
Infrared	10	Stirling	NASA
High Tc	20-30	—	DOE/military

* Used for shield cooling only.

erating temperatures of 4.2 K, 10 K, 20 K, and 80 K, respectively. The specific weights are 200, 40, 15, and 1.5 kg/W, corresponding to temperatures of 4.2 K, 10 K, 20 K, and 80 K, respectively. These data [8] indicate that significant reduction in weight, input power, and size are possible at higher cryogenic operations. Trade-off studies performed by the author reveal that most benefit can be achieved if the operating temperature is raised from 4.2 K to about 8 K. The trade-off studies further reveal that conductive cooling is most attractive because it is free from most logistic and reliability problems.

11.15.2 Maintenance Requirements for Cryocoolers

Maintenance of a cryocooler is of paramount importance if continuous and optimum performance from a superconducting device or sensor over an extended period is the principal requirement. Maintenance cost and interval must be given serious consideration during the selection process of a cryocooler for a specific application. Limited published data [1] based on the experience of the maintenance personnel of General Electric Company indicate that a cryocooler operating below 5 K requires more elaborate and more frequenct maintenance than a cryocooler operating at 77 K or higher. Higher maintenance costs are due to elaborate helium purification devices required to maintain clean working fluid at all times. Liquid helium cryocoolers run successfully and smoothly under controlled environments in a laboratory, but in commercial applications such as MRI, they failed to perform with high reliability. Cryocoolers operating at higher superconducting temperatures have been more successful in meeting their stated reliability goals and maintenance requirements.

11.15.3 Cryocooler Performance Requirements for Various Applications

Cryocoolers are widely used in commercial, industrial, and military applications and their performance requirements vary from application to application. Operating temperatures for various commercial refrigerators and cryocoolers are shown in Figure 11.7. Three distinct low-temperature cryocoolers operating over the 1–8 K temperature range include Collins helium liquifiers, Gifford-McMahan (GM) cryocoolers, and GM/JT cryocoolers with Joule-Thomson (JT) valves, and they have widely used application. The cooling of high-power lasers can be provided with a Stirling-cycle cryocooler designed with a typical capacity of 250 W at 77 K temperature. This type of cryocooler is best suited for sonar transducer applications at 50 to 55 K temperatures with typical cooldown time less than 4 hours. This cooler is acoustically quiet and has demonstrated a reliability of better than 50,000 hours. The cooler can operate successfully in all directions and does not require periodic replenishment of liquid nitrogen. This cooler is also most suitable for ship propulsion superconducting systems. Closed-cycle cooling schemes have been widely deployed in various commercial, space, and military systems. Typical operating temperatures for commercial refrigerators are shown in Figure 11.7.

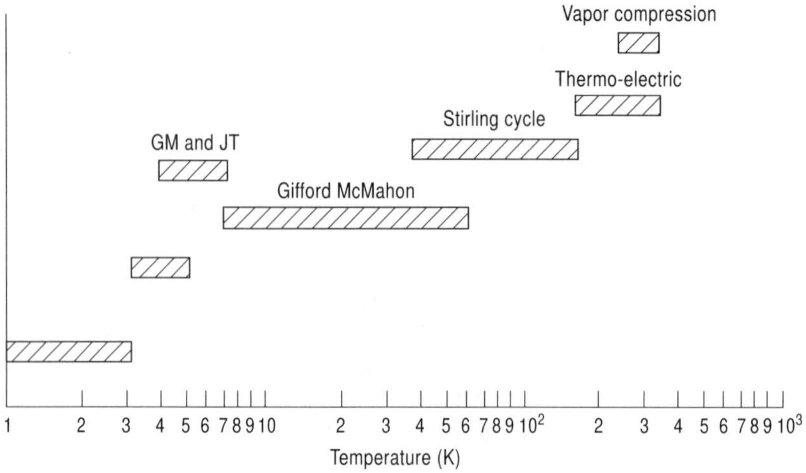

FIGURE 11.7 Temperature ranges for commercial refrigerators.

11.15.4 Performance Capabilities of Microcryocoolers

Miniaturized cryocoolers are known as microcoolers and have potential applications to space sensors and sophisticated military IR systems such as IR line scanners, IR search and track sensors, thermal imaging systems, and forward-looking IR systems. The design of a microcooler uses a Stirling-cycle engine operating principle to generate the liquid nitrogen temperature of 77 K in less than 3 minutes. The microcooler design incorporates a regenerator to create a compression-expansion refrigerator cycle with no valves. The regenerator has a large heat capacity and acts like an efficient heat exchanger. A microcooler design for a NASA thermal imaging application weighs only 15 ounces, consumes electrical power less than 3 W, has a life of 5 years, and demonstrated continuous operation exceeding 8,000 hours. It is important to mention that high reliability requires improved materials with self-lubrication features, very low friction clearance seals, elimination of gaseous contamination, and use of linear drives. A split-Stirling cooler with ratings from 0.25 W to 1 W and using linear drive can provide a continuous operation of better than 7500 hours. A microcooler or miniaturized cryocooler provides a cooling capacity of 150 mW at 77 K temperature with input power less than 3.5 W, heat load of 150 mW, and cooldown time ranging from 1.5 minute to 120 K to 4 minutes to 77 K from a room temperature of 25 °C or 298 K. Input electrical power requirements for a microcooler as a function of heat load and operating temperature are shown in Figure 11.8.

Cryocooler designs using rare earth (RE) regenerator materials have demonstrated significant improvement in the performance level. The first 4.2 K, two-stage GM/RE cooler has used rare-earth materials [8] as regenerator elements. This GM/RE cryocooler can reach the cryogenic temperature of 4.2 K in a very short time because a rare earth material such as erbium nickel has significant heat capac-

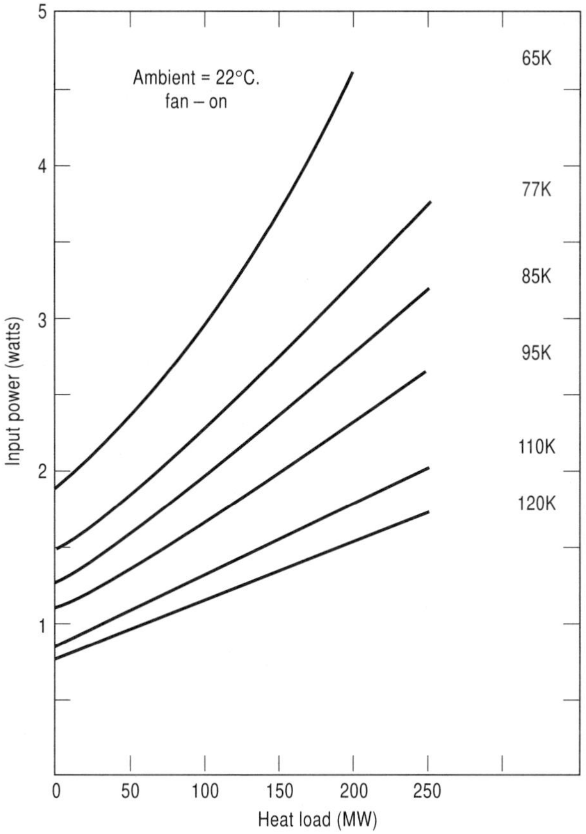

FIGURE 11.8 Input power requirements for cryocoolers as a function of heat load and cooling temperature.

ity below 10 K. This GM/RE microcooler can be of paramount importance in IR missile seekers and other battlefield sensors, where time is the most critical performance parameter. Typical heat load capacity of the GE/RE microcooler is about 100 mW at 4.2 K operation, but a heat capacity close to 1 W has been demonstrated at higher cryogenic operations under controlled environments and using unique combinations of rare-earth materials.

11.15.5 Laser Chillers

Laser chillers are best suited for high-power lasers used in industrial applications such as cutting thick metals and drilling thin plates. Most industrial lasers are either conduction-cooled or air-cooled. If no efficient heat sink or fan is available, a closed-loop chiller should be considered. A typical chiller consists of a water pump, reservoir, compressor, refrigerator, and a temperature-monitoring device. Industrial applications such as cutting, drilling, marking, and welding generally use CO_2 (10.6 μm), excimer (0.350 μm), and Nd:YAG (1.060 μm) lasers with chillers using controlled water flow typically at 10 to 15 °C temperature. According to the laser designers, for every watt of light output, 2 to 10 watts of heat is generated, depending on the conversion efficiency of the laser. In the case of high-power lasers, only 5 to 10% of the total power input is the light output and the rest will be converted into heat. The heat load varies from laser to laser. In the case of a Nd:YAG laser, the heat load varies from 3 to 10 kW.

Most solid-state lasers used for medical applications are cooled by water using air-to-water heat exchangers. This cooling scheme is very effective and there is no need for subambient cooling. A Ho:YAG laser will benefit the most from a chiller, because laser efficiency goes up with subambient cooling of the holmium rod.

Performance specifications of a chiller include laser heat load specifications in kW/hr or BTU/hr, water flow rate, water pressure, coolant temperature, temperature stabilization requirements, and input power requirements. Most chillers for medical application lasers assume an ambient temperature of 20 °C for the air and liquid in the system, but industrial applications can have ambient temperatures from 30 to 43 °C, depending on the heat generated by the laser. Lower ambient operations less than 10 °C may require special controls for the air-cooled condenser.

Different types of lasers require varying degrees of temperature stability. Studies performed by the author on thermal stability indicate that a variation of 4° F could cause significant performance drop in one laser, whereas another laser could tolerate a wide variation in temperature without loss in efficiency. Some lasers such as Ho:YAG are very sensitive to heat and, therefore, large variations are noticed in emission wavelengths and laser efficiencies. The efficiency of a semiconductor laser diode varies with varying temperature. When these laser diode arrays are used for pumping a solid-state laser, the emission wavelength and drive power require tight control to obtain a coherent source.

Thermoelectric (TE) coolers can be used to cool pumping diodes. The TE coolers are compact and are very expensive (about $1500 per unit). However, when the heat load is 200 to 300 W or less, TE coolers must be given serious consideration. Cost-effective studies must be performed to justify the deployment of a TE cooler in a particular device or application. High cost and heat load restriction can limit the application of TE coolers. On the other hand, chillers are simple, less expensive and highly efficient, but they suffer from excessive noise, vibration, and large weight and size.

11.15.6 New Generation of Cryocoolers

Research and development activities are focused to develop a new generation of cryocoolers including pulse tube refrigerators (PTRs) and advanced TE configurations. Research and development studies performed so far seem to indicate that PTR technology has potential application in space and airborne systems, where weight, size and power consumption are the most critical parameters. The PTR cooling scheme (Figure 11.9) is the hottest new generation refrigeration technique and is used in GM and Stirling-cycle refrigerators. PTR represents the latest cryocooler technology and uses no moving parts. The PTR system offers vibration-free operation, which is most ideal for sensitive applications such as IR sensors used for airborne or space missions, high T_c wireless filters, SQUID sensors, YBCO antennas for MRI machines, and cryopumps used for high-density integrated circuits (ICs).

In a PTR system (Figure 11.9), oscillating pressures inside a pulse tube closed at

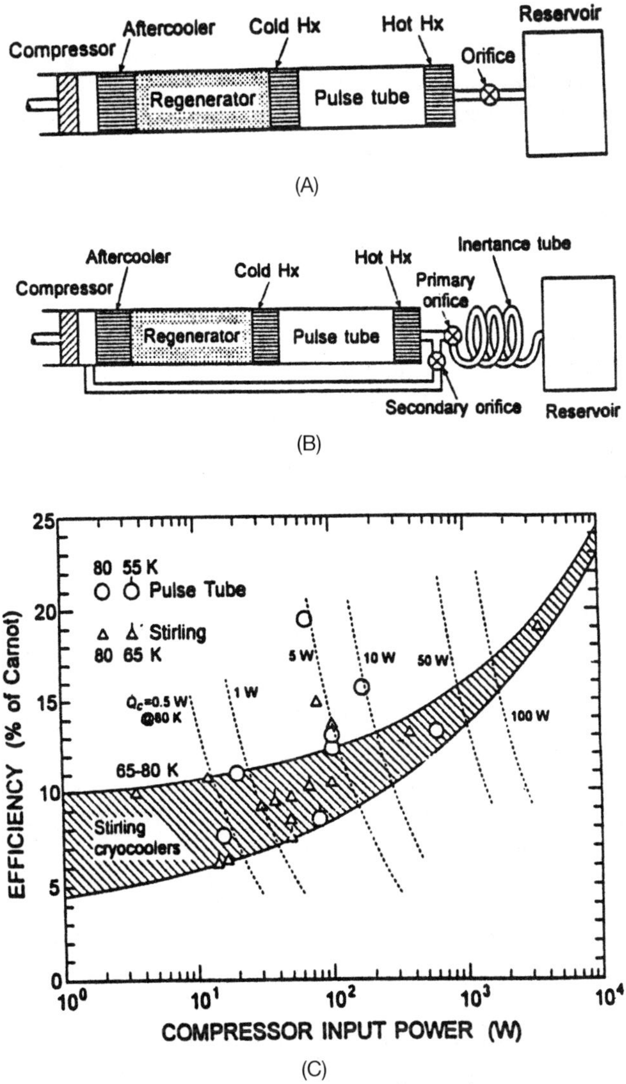

FIGURE 11.9 Block diagram of pulse tube refrigerator (PTR) with (A) single orifice, (B) two orifices and interance tube, and (C) efficiency versus compressor input power requirements for cryocoolers.

one end will cause the gas to be heated at the closed end during compression and cooled at the other end during expansion. The expansion and compression of the gas is responsible for cooling as in other refrigerators. The heat exchange between the gas and pulse tube walls provides significant improvement in the conversion efficiency [9]. It is important to mention that the absence of moving parts in a PTR system offers higher reliability, lower maintenance costs, negligible noise, and improved efficiency not matched by any other system to date. Addition of an orifice at the warm end of the pulse tube brings the temperature down to 105 K, resulting in an increase in the refrigerator power. Introduction of a second orifice (Figure 11.9 B) in the ''double-inlet" design improves the efficiency to very close to that of a Stirling-cycle refrigerator of comparable size. The "double-inlet" PTR design produces a heat lift of 30 W at 80 K with an efficiency of 13% of Carnot's cycle, which is considered the efficiency of an ideal refrigerator. The latest the "double-inlet" PTR incorporating a critical component called the inertance tube, shown in Figure 11.9 B, has demonstrated an efficiency close to 19% of Carnot's cycle, which is significantly higher that the best Stirling refrigerator. Performance comparison data, including efficiency, compressor-input power, and heat capacity for various refrigerators are shown in Figure 11.9 C.

Further performance improvement in the PTR systems is possible by selecting an optimum pulse tube configuration, proper tube orientation with respect to gravity, and helium gas as a working fluid. A thermodynamic model developed for a single-stage PTR system provides optimum values for certain cryocooler performance parameters at a given temperature above 30 K and heat lift. A PTR system using optimum design parameters can achieve better than 30 W of cooling at 60 K with a 1000 W compressor rating. Far lower cryogenic temperatures are possible with multistage PTRs. A three-stage PTR developed by a Japanese professor in 1994 reached a cooling temperature of 4 K. A German scientist designed a two-stage PTR in 1997 that set the world's record for the PTR performance at 2.2 K cryogenic temperature.

Mini-PTRs have been developed recently that are capable of delivering 0.8 W at 80 K and have potential applications for IR sensors used by military, spacecraft, and reconnaissance satellites. The smallest pulse tube [9] developed to date has a rating of 50 mW at 98 K, compressor swept volume of 0.75 cm^3, and mechanical input power less than 10 W. This most compact and efficient PTR system is best suited for future space missions, space communications, and satellite-based reconnaissance. The PTR systems require minimum mechanical power, thereby lowering the battery power requirements in the spacecraft or satellites.

11.16 WAVELENGTH-LOCKING TECHNIQUE FOR FUTURE OPTICAL COMMUNICATIONS SYSTEMS AND IR SENSORS

There are certain commercial and military IR and EO applications that require precision and reliable wavelength stabilization. Wavelength-division-multiplexing (WDM) systems, dense-WDM (DWDM) systems, and certain military IR sensors require technologies and control circuits capable of providing effective wavelength stabilization with high precision and reliability. For example, an IR missile guidance system requires a highly stable IR laser wavelength to provide accurate tracking and guidance capabilities. Similarly, future DWDM telecommunications systems with wavelength spacing of 100 GHz (0.8 nm) or 50 GHz (0.4 nm) will require precision and stable wavelength control to ensure long-term reliability of the system. Currently, distributed feedback lasers (DFLs) provide the accuracy and stability that WDM systems require. As the telecommunications systems put more wavelengths down in each optical fiber, requirements for channel spacing and reliability increase.

A real-time wavelength-locking scheme (WLS) with new heights of sophistication and performance capabilities is required for future DWDM systems. A combination of optics and electronic techniques will be most effective in controlling the laser temperature and creating a feedback loop for precision wavelength management [10]. Conventional DFLs have a typical wavelength drift of about 250 ppm per degree C or about 30 GHz per degree C. Preliminary calculations indicate that to keep the wavelength within 3 GHz per degree C, a dynamic temperature stability of 0.1° is required; to keep within 0.3 GHz per degree C, a dynamic temperature stability of 0.01 °C will be required. Laser-wavelength control accuracy and stability are dependent on certain control parameters, such as sensitivity of the wavelength-selection element, transfer function of the electronic circuits, and control-loop time constant.

Several WLS schemes are available, including interference notch filter, diffraction grating element, and a Fabry–Perot etalon-based (FPEB) locking network. An interference filter approach used two optical filters to form a notch. One filter is placed in front of each IR detector and signals from the amount of drift are determined by comparing the signal from each detector. Combining the signals from these detectors produces an error signal whose amplitude is proportional to the amount of wavelength drift. This error signal drives the laser cooler, which controls the laser temperature and, consequently, the emission wavelength. This approach offers low-cost and compact packaging, but requires some type of thermal compensation device to minimize drift as a function of temperature. The diffraction approach is similar in function to the notch filter concept. It uses a diffraction grating to create two light paths and two photodiode detectors to intercept the diffracted laser beam. When the laser wavelength drifts, one of the photodiodes will see an increase in signal amplitude; the other sees the decrease in the signal strength, depending on the direction of wavelength drift. This approach requires highly precise and fast temperature controlling circuits to avoid large wavelength drifts due to thermal instability of the grating.

The FPEB locking technique involves a pair of partially reflective flat glass plates separated by a parallel spacer. When light enters the etalon, it reflects off both glass plates and forms a wavelength-dependent fringe pattern with an error signal, which is then used to control the temperature of the laser source and, consequently, the emission wavelength. Thermal stability and chromatic dispersion are eliminated using spacers made with a thermally stable material. The finesse of the etalon determines the optical gain of the system. A lower optical gain will exhibit a smaller signal change as a function of wavelength drift, whereas a higher gain will have a much larger signal change for the same wavelength drift. The control range and the accuracy of the wavelength locking circuit are strictly dependent on the design of the electronic control loop and its locking circuit. The EBLS behaves like a comb filter, which generates a continuous range of filters, thereby providing the wavelength lock-on capability over an entire wavelength grid. This lock-on technique will be most effective for a high-density WDM system.

The EBLS lock-on or stabilization technique is ideal for other applications such as real-time wavelength monitoring and multichannel wavelength locking systems. In both these applications, one thermally stable etalon can monitor or lock-on many wavelengths on a 100 GHz grid. In an 8 channel, 50 GHz system configuration, channels 1, 3, 5, and 7 would use a wavelength locker centered on the 100-GHz grid, and channels 2, 4, 6, and 8 would use a wavelength locker with 100-GHz spacing offset by 50 GHz. An EBLS using thermally stable photo detectors and beam splitting will be most beneficial for the next generation of dense-WDM telecommunications systems. In summary, future applications of this technique include photodynamic therapy (PDT), treatment of eye diseases, antisatellite surveillance, and terrestrial communications.

11.17 OPTICAL SOFTWARE PROGRAMS AND OPTOMECHANICAL MODELING

Several optical design software and optomechanical computer programs are available to perform design analysis and to predict performance levels of various optical systems and components. OptiCad modeling software is available from Focus Software Inc. (Tucson, AZ) for prediction of 3-D optical system performance, unconstrained ray tracing, including Monte Carlo ray tracing, surface and volume scattering analysis, and defining and modifying multiple objects. The OptiCad program uses a full 3-D model of every object including edges, facets, holes, mounts, baffles, and many more. This software program is particularly suitable for applications involving reflector/illuminator design, display panel lighting, light pipes, light concentrators, LCD projectors, efficiency computations, baffle analysis, and illumination evaluation.

The Trace Pro software program, developed by Lambda-Research (Littleton, MA) is the first truly visual 3-D illumination software tool with industry-standard modeling engine at its core. This program can share with more than 100 other ACIS-based CAD programs, including Auto CAD, Iron CAD, Bravo, CADKEY-97, VisiCAD solids, solid works-98, PE/Solid-design, and TurboCAD. This program reduces the data entry requirement by 50% over surface-based systems because it uses solids to define the optomechanical geometry. Modifying objects is extremely simple using standard window-based icons, tool bars, and extrude objects onscreen.

This program has potential applications in light pipes, LCD projection systems, LCD background lighting systems for notebook computers, desktop computers, and pagers; telescope analysis, cameras, and optomechanical systems; LED devices for tail lamps; fiber illuminators; brake lamps; switching and stop lamps; scanners; machine vision and target detection systems; arc lamps for airport runways and medical instruments; IR imaging sensors for defense, aerospace, and home security; and copiers.

The Fiber-CAD software program developed by Optiwave Corp. allows an optical design engineer to select optical fiber parameters for given optical systems. Cross-sectional dimensions, material composition, and refractive index profiles that impact the performance level can be selected based on trade-off studies for a given application. This software program is a powerful tool that blends numerical mode solvers for various fiber modes with calculation models for fiber dispersion, insertion loss, and polarization mode dispersion (PMD). This program is for engineers and scientists who are deeply involved in the design of optical fibers, fiber components, and fiber optical communications systems. Other salient features of this program include rapid design of a multilayer fiber using a built-in library of profile functions, models of material losses as a function of wavelength and fiber parameters, composition calculation, calculation of propagation constants and fields of fundamental and higher-order modes in step-index and graded-index fibers, estimation of cutoff wavelengths, calculation of group delay and dispersion as a function of wavelength, calculation of microbending and birefringence effects, and estimation of PMD.

11.18 SUMMARY

Future applications of photonic and infrared technologies are discussed, with emphasis on performance and cost. Performance capabilities and limitations of various photonic devices, optoelectronic components, and electronic control circuits used by various IR sensors and optical communications are summarized. Future applications of IR technology to various systems include free-space intrasatellite communications systems; laser-based surveillance systems; optically steered phased arrays; IRCM techniques; imaging sensors for unmanned aerial vehicles (UAVs); remote pilotled vehicles (RPVs); and Mini-PRV, IR search, and tracking sensors; FLIR systems for navigation and weapon delivery; photonic sensors for process control; laser-based smart weapons; laser-based micromachining tools; photonic-based encryption schemes for secure communication; laser-line scanners for underwater mine detection; and laser-based invasive tools for medical diagnosis. Performance requirements of critical accessories such as power supplies, control circuits, temperature controllers, cryocoolers, thermoelectric (TE) coolers, and monitoring devices used by complex IR systems are summarized. Capabilities of selected software programs for modeling and analysis of IR systems, photonic devices, and optoelectronic components are described.

REFERENCES

1. Staff reporter. "Anthropological information." *Photonics Spectra,* pp. 91–98, September 1998.
2. A. J. Hand. "Photonics analyze food quickly, efficiently." *Photonics Spectra,* pp. 114–119, October 1998.
3. V. Sankaran. "Photonics-based technologies improve semiconductor yield." *Photonics Spectra,* pp. 100–106, December 1998.
4. P. Revely, "Airdrop System Accuracy." *Photonics Spectra,* pp. 38–39.
5. D. J. Page. "An introduction to the optical communications." *IEEE Trans. On Antenna and Propagation, 44,* 5, 652–658.
6. M. Hallader. "Research into COIL systems advances remediation of nuclear reactors." *Society of Photo-Optical Engineering,* pp. 1–6, 23–29 January 1999.
7. T. Vandersteer. "RPV: Battlefield payloads." *Journal of Electronic Defense,* p. 137, October 1985.
8. A. R. Jha. *Superconductor Technology: Applications to Microwave, Electro-Optics, Electrical machines and Propulsion Systems,* pp. 290–291, Wiley, New York, 1997.
9. A. Bitterman et al. "Pulse Tubes foothold in cryocooler markets, *Superconductor and Cryoelectronics,* pp. 12–17, September 1998.
10. E. Miskovic. "Wavelength Locking Schemes." *Photonics Spectra,* pp. 104–110, February 1999.

Index

WILEY SERIES IN MICROWAVE AND OPTICAL ENGINEERING

KAI CHANG, Editor
Texas A&M University

FIBER-OPTIC COMMUNICATION SYSTEMS, Second Edition • *Govind P. Agrawal*

COHERENT OPTICAL COMMUNICATIONS SYSTEMS • *Silvello Betti, Giancarlo De Marchis and Eugenio Iannone*

HIGH-FREQUENCY ELECTROMAGNETIC TECHNIQUES: RECENT ADVANCES AND APPLICATIONS • *Asoke K. Bhattacharyya*

COMPUTATIONAL METHODS FOR ELECTROMAGNETICS AND MICROWAVES • *Richard C. Booton, Jr.*

MICROWAVE RING CIRCUITS AND ANTENNAS • *Kai Chang*

MICROWAVE SOLID-STATE CIRCUITS AND APPLICATIONS • *Kai Chang*

RF AND MICROWAVE WIRELESS SYSTEMS • *Kai Chang*

DIODE LASERS AND PHOTONIC INTEGRATED CIRCUITS • *Larry Coldren and Scott Corzine*

MULTICONDUCTOR TRANSMISSION-LINE STRUCTURES: MODAL ANALYSIS TECHNIQUES • *J. A. Brandão Faria*

PHASED ARRAY-BASED SYSTEMS AND APPLICATIONS • *Nick Fourikis*

FUNDAMENTALS OF MICROWAVE TRANSMISSION LINES • *Jon C. Freeman*

OPTICAL SEMICONDUCTOR DEVICES • *Mitsuo Fukuda*

MICROSTRIP CIRCUITS • *Fred Gardiol*

HIGH-SPEED VLSI INTERCONNECTIONS: MODELING, ANALYSIS, AND SIMULATION • *A. K. Goel*

FUNDAMENTALS OF WAVELETS: THEORY, ALGORITHMS, AND APPLICATIONS • *Jaideva C. Goswami and Andrew K. Chan*

ANALYSIS AND DESIGN OF INTEGRATED CIRCUIT ANTENNA MODULES • *K. C. Gupta and Peter S. Hall*

PHASED ARRAY ANTENNAS • *R. C. Hansen*

HIGH-FREQUENCY ANALOG INTEGRATED CIRCUIT DESIGN • *Ravender Goyal (ed.)*

MICROWAVE APPROACH TO HIGHLY IRREGULAR FIBER OPTICS • *Huang Hung-Chia*

NONLINEAR OPTICAL COMMUNICATION NETWORKS • *Eugenio Iannone, Francesco Matera, Antonio Mecozzi, and Marina Settembre*

FINITE ELEMENT SOFTWARE FOR MICROWAVE ENGINEERING • *Tatsuo Itoh, Giuseppe Pelosi and Peter P. Silvester (eds.)*

INFRARED TECHNOLOGY: APPLICATIONS TO ELECTRO-OPTICS, PHOTONIC DEVICES, AND SENSORS • *A. R. Jha*

SUPERCONDUCTOR TECHNOLOGY: APPLICATIONS TO MICROWAVE, ELECTRO-OPTICS, ELECTRICAL MACHINES, AND PROPULSION SYSTEMS • *A. R. Jha*

OPTICAL COMPUTING: AN INTRODUCTION • *M. A. Karim and A. S. S. Awwal*

INTRODUCTION TO ELECTROMAGNETIC AND MICROWAVE ENGINEERING • *Paul R. Karmel, Gabriel D. Colef, and Raymond L. Camisa*

MILLIMETER WAVE OPTICAL DIELECTRIC INTEGRATED GUIDES AND CIRCUITS • *Shiban K. Koul*

MICROWAVE DEVICES, CIRCUITS AND THEIR INTERACTION • *Charles A. Lee and G. Conrad Dalman*

ADVANCES IN MICROSTRIP AND PRINTED ANTENNAS • *Kai-Fong Lee and Wei Chen (eds.)*

OPTICAL FILTER DESIGN AND ANALYSIS: A SIGNAL PROCESSING APPROACH • *Christi K. Madsen and Jian H. Zhao*

OPTOELECTRONIC PACKAGING • *A. R. Mickelson, N. R. Basavanhally, and Y. C. Lee (eds.)*

OPTICAL CHARACTER RECOGNITION • *Shunji Mori, Hirobumi Nishida, and Hiromitsu Yamada*

ANTENNAS FOR RADAR AND COMMUNICATIONS: A POLARIMETRIC APPROACH • *Harold Mott*

INTEGRATED ACTIVE ANTENNAS AND SPATIAL POWER COMBINING • *Julio A. Navarro and Kai Chang*

ANALYSIS METHODS FOR RF, MICROWAVE, AND MILLIMETER-WAVE PLANAR TRANSMISSION LINE STRUCTURES • *Cam Nguyen*

FREQUENCY CONTROL OF SEMICONDUCTOR LASERS • *Motoichi Ohtsu (ed.)*

SOLAR CELLS AND THEIR APPLICATIONS • *Larry D. Partain (ed.)*

ANALYSIS OF MULTICONDUCTOR TRANSMISSION LINES • *Clayton R. Paul*

INTRODUCTION TO ELECTROMAGNETIC COMPATIBILITY • *Clayton R. Paul*

ELECTROMAGNETIC OPTIMIZATION BY GENETIC ALGORITHMS • *Yahya Rahmat-Samii and Eric Michielssen (eds.)*

INTRODUCTION TO HIGH-SPEED ELECTRONICS AND OPTOELECTRONICS • *Leonard M. Riaziat*

NEW FRONTIERS IN MEDICAL DEVICE TECHNOLOGY • *Arye Rosen and Harel Rosen (eds.)*

ELECTROMAGNETIC PROPAGATION IN MULTI-MODE RANDOM MEDIA • *Harrison E. Rowe*

ELECTROMAGNETIC PROPAGATION IN ONE-DIMENSIONAL RANDOM MEDIA • *Harrison E. Rowe*

NONLINEAR OPTICS • *E. G. Sauter*

ELECTROMAGNETIC FIELDS IN UNCONVENTIONAL MATERIALS AND STRUCTURES • *Onkar N. Singh and Akhlesh Lakhtakia (eds.)*

FUNDAMENTALS OF GLOBAL POSITIONING SYSTEM RECEIVERS: A SOFTWARE APPROACH • *James Bao-yen Tsui*

InP-BASED MATERIALS AND DEVICES: PHYSICS AND TECHNOLOGY • *Osamu Wada and Hideki Hasegawa (eds.)*

DESIGN OF NONPLANAR MICROSTRIP ANTENNAS AND TRANSMISSION LINES • *Kin-Lu Wong*

FREQUENCY SELECTIVE SURFACE AND GRID ARRAY • *T. K. Wu (ed.)*

ACTIVE AND QUASI-OPTICAL ARRAYS FOR SOLID-STATE POWER COMBINING • *Robert A. York and Zoya B. Popović (eds.)*

OPTICAL SIGNAL PROCESSING, COMPUTING AND NEURAL NETWORKS • *Francis T. S. Yu and Suganda Jutamulia*

SiGe, GaAs, AND InP HETEROJUNCTION BIPOLAR TRANSISTORS • *Jiann Yuan*

ELECTRODYNAMICS OF SOLIDS AND MICROWAVE SUPERCONDUCTIVITY • *Shu-Ang Zhou*